W9-BRR-630

# ANNUAL REVIEW OF PHYTOPATHOLOGY

# ANNUAL REVIEW OF PHYTOPATHOLOGY

KENNETH F. BAKER, *Editor*
The University of California, Berkeley

GEORGE A. ZENTMYER, *Associate Editor*
The University of California, Riverside

ELLIS B. COWLING, *Associate Editor*
North Carolina State University

*VOLUME 12*

*1974*

ANNUAL REVIEWS INC. 4139 EL CAMINO WAY PALO ALTO, CALIFORNIA 94306

ANNUAL REVIEWS INC.
Palo Alto, California, USA



*International Standard Book Number: 0-8243-1312-7*
*Library of Congress Catalog Card Number: 63-8847*

| | |
|---|---|
| Assistant Editor | Toni Haskell |
| Indexers | Mary Glass |
| | Susan Tinker |
| Subject Indexer | Donald C. Hildebrand |



PRINTED AND BOUND IN THE UNITED STATES OF AMERICA

# PREFACE

So much has been written about the serious problems facing technical publications (rising costs, shrinking library budgets, copyright, photocopying dilemmas, and so on) that we need not dwell long on them here, except to note that they are of concern to editors, including those for *Annual Review of Phytopathology.* Increasing labor and material costs in publishing, as in other businesses, have necessitated price increases. Libraries find themselves in a particularly difficult situation. All of their operating costs have risen; average journal costs have increased about 20 percent in the last year. In most cases, budgets cannot stretch to cover these increases. Librarians' budgetary problems are, of course, transferred back to publishers, who are then faced with the double squeeze of diminishing sales and increasing costs, at a time when additional increases in book and journal prices could result in further curtailing sales.

In the first *Annual Review of Phytopathology* (1963), J. C. Walker of the University of Wisconsin wrote: "... the time is ripe, if not overripe, to launch an *Annual Review of Phytopathology.* This is really the only way in which we can keep ourselves informed and in which specialists in one area can offer ideas, and inspiration to those in another. This means that there must always be a few of us taking time from busy research and teaching programs to evaluate and summarize the current advances in given areas within plant pathology. I know the reviewers will take their responsibilities seriously. I hope that they will always keep in mind that they are writing for all plant pathologists and not for the specialists in their chosen areas. After all, those specialists should not need such reviews if they are on the job. Let us hope that these reviews will be written not as annotated bibliographies but as sound interpretations in language attractive to those in other areas of plant pathology and to those just outside the field ..."

Since it began publication in 1963 we believe that the *Annual Review of Phytopathology* has been providing an essential reference tool for workers in plant pathology and related fields. Because each volume deals with concepts and critically reviews selected aspects of plant pathology, it perhaps has greater influence in forming and modulating the professional milieu than do individual research journals. Indeed it is commonly said that the profession could not now afford to be without the *Annual Review of Phytopathology.*

In the present (we hope temporary) difficult times, individual plant pathologists have an opportunity to assist their profession. The more individuals and groups that can be encouraged to begin (or continue) subscriptions for themselves or their libraries, the more everyone can help insure the growth and vitality of *Annual Review of Phytopathology* and its continuing utility for the profession as a whole.

We welcome Paul H. Williams as the new member of the Editorial Committee, replacing David Gottlieb, who concludes his 5-year tenure, and we thank A. R. Weinhold and R. G. Grogan, guest Committeemen who helped plan this volume.

THE EDITORIAL COMMITTEE

# CONTENTS

ANNUAL REVIEWS INC. is a nonprofit corporation established to promote the advancement of the sciences. Beginning in 1932 with the *Annual Review of Biochemistry,* the Company has pursued as its principal function the publication of high quality, reasonably priced Annual Review volumes. The volumes are organized by Editors and Editorial Committees who invite qualified authors to contribute critical articles reviewing significant developments within each major discipline.

Annual Reviews Inc. is administered by a Board of Directors whose members serve without compensation.



Annual Reviews are published in the following sciences: Anthropology, Astronomy and Astrophysics, Biochemistry, Biophysics and Bioengineering, Earth and Planetary Sciences, Ecology and Systematics, Entomology, Fluid Mechanics, Genetics, Materials Science, Medicine, Microbiology, Nuclear Science, Pharmacology, Physical Chemistry, Physiology, Phytopathology, Plant Physiology, Psychology, and Sociology (to begin publication in 1975). In addition, two special volumes have been published by Annual Reviews Inc.: *History of Entomology* (1973) and *The Excitement and Fascination of Science* (1965).

# PLANT PATHOLOGY: CHANGING AGRICULTURAL METHODS AND HUMAN SOCIETY

*J. G. ten Houten*

# PLANT PATHOLOGY: CHANGING AGRICULTURAL METHODS AND HUMAN SOCIETY

❖3584

*J. G. ten Houten*
Institute of Phytopathological Research, Wageningen, The Netherlands

> The obligation that the plant pathologist has accepted is to society—world society—not to a crop or a culture . . .
>
> William C. Snyder (21)

Although it is a privilege and an honor to write a prefatory chapter for this distinguished journal it is also a great responsibility, because the readers are critical colleagues who want information that is at least partly new to them.

When I studied biology at the University of Utrecht in the early thirties it was impossible to major in plant pathology anywhere in the Netherlands. Therefore, I majored in plant taxonomy, which in 1935 brought me on a collecting tour to the eastern United States and Canada. Of course I took this opportunity to visit Professor Whetzel and his colleagues at Cornell University and the Boyce Thompson Institute at Yonkers, where Miss Purdy Beale showed me her pioneer serological trials with rabbits.

It was remarkable that one could not major in plant pathology in The Netherlands at that time, since plant pathology had been taught at Amsterdam University in 1895 by Professor Ritzema Bos, who had been appointed director of the Phytopathological Laboratory "Willie Commelin Scholten" the same year. Ritzema Bos may be called the father of plant pathology in The Netherlands, because he was the first professor in this discipline and because he founded the Netherlands Phytopathological Society, the oldest phytopathological society in the world, in 1891.

In 1899, at the request of the Dutch government, Ritzema Bos also founded the Netherlands Phytopathological Service at the little town of Wageningen, where he moved in 1906 after having been appointed Director of the Institute for Phytopathology and teacher at what was then the Agricultural College at Wageningen.

When this College was founded 30 years earlier, two of its teachers, Mayer and Beijerinck, had shown extraordinary interest in plant pathological research. The German-born Adolf Mayer in 1882 gave the name mosaic to a disease of tobacco which he could easily transmit with sap from diseased plants. He implicated the laborers in tobacco fields as the main vectors of the disease and he told us how one of them, who was particularly rough when planting young tobacco plants, was named "motley John" by his colleagues, because of the high percentage of infection with mosaic in his tobacco fields (12).

Beijerinck became interested in this tobacco disease and continued to study its causal agent after he had accepted a professorship in microbiology at the Technical University of Delft.

In 1898 he published a paper in Dutch describing the cause of mosaic as a noncorpuscular *contagium vivum fluidum,* but in the same paper he also used the name virus, and explained that this pathogen was entirely different from all other known pathogens (2).

At about the same time that Burrill in the USA described fire blight as the first infectious bacterial disease of plants (4), the Dutch biologist Wakker found that yellow disease of hyacinth was caused by what he called *Bacterium hyacinthi* (33), now known as *Xanthomonas hyacinthi.*

Ritzema Bos had two collaborators at Amsterdam, Van Hall and Quanjer, and when he moved to Wageningen he was accompanied by Quanjer, a pharmaceutical chemist. In the meantime Johanna Westerdijk had (at the age of 23!), succeeded Ritzema Bos in Amsterdam as director of the Phytopathological Laboratory. In 1917 she was appointed professor in plant pathology at the University of Utrecht. She was the first woman professor in The Netherlands, and later she became a famous president of the International Union of University Women.

Many students, including myself, worked for their doctor's degree in the cheerful stimulating atmosphere of her laboratory. In total she graduated 54 PhDs. Most of the work at the Phytopathological Laboratory "Willie Commelin Scholten," which was moved to Baarn in 1920, had to do with fungal and nutritional diseases. Quanjer, who succeeded Ritzema Bos as professor in plant pathology at Wageningen, concentrated on potato diseases, primarily those caused by viruses.

Like Westerdijk, Quanjer always studied the pathological anatomy of diseased tissues. Westerdijk and her collaborators thus proved that *Ceratocystis ulmi* was the cause of the (Dutch) elm disease that incited vascular tyloses. Furthermore, at Wageningen Fransen proved that the bark beetle *Scolytus scolytus* was one of the vectors. It is rather annoying that as a reward for their careful research this disease was named "Dutch" elm disease as though the researchers had created it. "There is nothing whatever to suggest that it originated in Holland" (5).

Quanjer found phloem necrosis to be a characteristic symptom of potato leaf roll, which he proved to be caused by a virus.

Through the outstanding studies of Mayer, Beijerinck, Quanjer, and others, Wageningen became an international center for plant pathology, attracting scientists from all over the world. They still come to work at the Departments of Phytopathology, Virology, and Entomology of the University, the Plant Protection Service, and the Government Institute of Phytopathological Research (IPO) of which I have the privilege of being director since its establishment in 1949.

The former head of the Virology Department of the IPO, Professor Thung, during his earlier studies of tobacco diseases in Indonesia described a phenomenon which Quanjer, adopting a term used in the medical profession, had called premunity, meaning that further infection of a plant by a virus is impossible if the same plant has already been infected with a milder or related strain of the same virus.

For many years this was only of theoretical significance. However, after careful studies by my collaborator Mr. Rast, premunity has now been introduced to the practical tomato grower, who uses it on a large scale. Thus it is possible to protect the tomato crop under glass against virulent strains of tobacco mosaic virus (TMV) by infecting young plants with a mild strain, resulting in an increase in yield of up to 15%.

Whereas Wageningen originally became well known for its work on virus diseases, and later for its work on nematodes, the Phytopathological Laboratory "Willie Commelin Scholten" soon became famous for its studies on mycological and nutritional diseases, especially for boron deficiency studies with tobacco and sugar beet. Furthermore, in her early days at the laboratory Westerdijk established the Central Bureau for Fungus Cultures at Baarn, which at present houses over 18,000 cultures of species or strains of fungi, actinomycetes, and yeasts.

It is not, however, my intention to give a complete history of plant pathology in The Netherlands. For further details about older work, the reader is referred to Kerling (11), Thung, (29), and ten Houten (26). The work I have mentioned is, of course, only a fragment of the total, and it pays little or no credit to the fine achievements of many of my present-day colleagues. I hope that most of these will be sufficiently known from recent and current publications.

As director of the Institute of Phytopathological Research at Wageningen I have obtained a broad experience—although rather superficial—of most of our discipline, and it always struck me how closely related plant pathology is to agricultural practices.

The tremendous population explosion attributable mainly to better hygiene, and the conquering of epidemics such as typhoid and malaria, has caused serious food deficits unlikely to be solved by traditional agricultural methods.

"Latin America which in the 1930's was still a major world-supplier of corn and wheat had become by 1961 a net deficit region in food grains, even though it had expanded its grain-producing area by nearly one-third over that of the prewar period" (3). I am citing this quotation by the only Nobel prize winner/plant pathologist/plant breeder, Dr. Borlaug, because he and his enthusiastic team of collaborators of Centro Internacional de Mejoramiento de Maiz y Trigo (CIMMYT) were responsible for a tremendous upsurge in food production, not only in Mexico but in many other developing countries. He achieved success through a pragmatic procedure in which a combination of high-yielding dwarf and semidwarf wheat varieties with stem rust resistance (at least for some years), high rates of nitrogen fertilizer, modification of irrigation practices (if necessary), and the use of pesticides were exploited. This "green revolution" had impressive consequences. It changed

Mexico from a wheat-importing to a wheat-exporting country with a production in 1964 6.5 times that in 1945. It brought prosperity to innumerable farmers in the wheat-growing areas of that country. The greatest and most welcome surprise was, however, that the same dwarf varieties behaved very similarly in Africa and Asia, including India and Pakistan with their 600 million inhabitants to feed.

Of course, there were also some disappointments. First, with insufficient soil moisture the Mexican dwarf varieties produced less than did the local drought-resistant varieties. Second, local strains of stripe rust (*Puccinia striiformis*) and certain other pathogens such as *Septoria tritici* attacked the dwarf varieties.

The difficulties depended only partly on the presence of these strains; more important were the enormous areas sown with the high-yielding semidwarf wheat varieties which, at least in the beginning, had a rather narrow genetic base. The area planted with monocultures of these wheats in Afghanistan, India, Nepal, West Pakistan, and Turkey increased from about 9300 hectares in 1966 to more than 12 million hectares in 1971 (23).

To obtain high yields, agronomic practices had to be altered drastically. Plant density and date of planting had to be changed; heavier and more frequent irrigation, higher levels of fertilizers, and a more efficient weed control had to be applied. In some areas only the large land owners could invest the money necessary for these new practices, and only they got the high yields. In the Far East the gap between rich and poor people was thus widened, and this sometimes led to serious riots.

Compared with the local varieties, which are a population consisting of more or less related genetic material, the pure semidwarf varieties suffered much more from attacks of pathogens such as stripe rust, once a virulent strain appeared. We had exactly the same experience in Europe, where new varieties originally resistant to *P. striiformis* became susceptible and suffered tremendous losses. This happened in The Netherlands with the excellent wheat variety Heine VII. When artificially inoculated with the known strains of stripe rust it remained resistant in our field experiments both in 1950 and 1951; when introduced commercially in 1952, 14% of the total wheat area in The Netherlands was sown with it, and no stripe rust occurred. In 1953 the area sown with Heine VII increased to 43% of the total and only one small locus of infection was found in a breeder's farm. In 1955, when 81% of the total wheat area had been sown with Heine VII, it was everywhere heavily infected with stripe rust, and in 1956 70% of the winter wheat area was destroyed (25). In the following years the area was reduced to 15% with the result that infection remained present but was far less severe than before. This is believed to be due to the alternative use of different wheat varieties. Our stripe-rust specialist, Mr. Stubbs is responsible for a worldwide survey on the virulence of *P. striiformis,* for which he uses a number of differential varieties in 250 trials in 48 countries.

As it is impossible to incorporate resistance against all known pathogens of a crop plant it frequently happens that excellent new varieties, even with a good resistance against local pathogens, are severely attacked by other pathogens when exported to other countries. Saari & Wilcoxson (17) treat this subject in greater detail in this volume.

What CIMMYT did in Mexico for the culture of wheat, The International Rice Research Institute (IRRI) did in the Phillippines for the improvement of the productivity of rice. The new IRRI varieties are nonlodging dwarf varieties that respond favorably to high nitrogen fertilization. The green revolution in East Asia followed the same pattern for rice instead of for wheat. Not only the use of new varieties, but also new cultural practices like the application of high quantities of nitrogen fertilizer and the use of pesticides are now widespread in countries such as India, Malaysia, and Indonesia.

Quite soon, however, it became evident that the IRRI varieties and similar Formosan varieties were very susceptible to bacterial blight (*Xanthomonas oryzae*), bacterial leaf streak (*X. oryzicola*), or both. These diseases were either unknown or of no practical importance before the introduction of the new dwarf varieties and the new agricultural methods, including double cropping (20). Srivastava (22) wrote: "Thus, with the introduction of one susceptible variety, a disease which was unknown in all but two (Indian) states earlier, became pandemic within two years. At present, the bacterial blight is a major hurdle in stepping up rice yields." In Indonesia, *Pyricularia oryzae* also suddenly became an important parasite of the new nitrogen-fertilized rice varieties.

When irrigation was expanded in Venezuela it became possible to grow two rice crops instead of one per year. This resulted in a serious outbreak of the virus disease *hoja blanca,* transmitted by the vector *Sogata orizicola.* This insect is much affected by the microclimate in a rice crop; it prefers high humidity. *Sogata orizicola* acquires virus from a weed grass (*Echinochloa colonum*) which frequently is 100% infected. The leafhopper *S. cubana,* another vector of *hoja blanca,* can only transmit the virus from grass to grass, whereas *S. orizicola* has rice as its main host but feeds occasionally on the grass. *Hoja blanca* reduces the rice yield to a fraction of that of healthy crops. Peculiarly enough, the disease is not equally serious in all South American countries, even when the vector is abundantly present. Thus, in Cuba, 8% of the vectors could transmit the virus enough to spread the disease. However, my collaborator Dr. van Hoof (32) found that in Surinam only 0.4% of *S. orizicola* was able to infect rice plants and as a result the *hoja blanca* disease is almost absent.

Unfortunately, there are too many examples of the introduction of very promising new varieties of crops resistant to a well-known major disease but susceptible to another pathogen especially when grown on a large scale. Whereas in 1961 nematodes were of little importance for the banana industry, which then mainly grew the variety Gros Michel, the new Cavendish varieties, introduced for their resistance to Panama disease (*Fusarium oxysporum* f. sp. *cubense*), appear to be very susceptible to the root-rot nematode, *Radopholus similis* (24).

Where it was not possible to breed a resistant variety, sometimes one could graft a susceptible plant on a resistant rootstock. This was normal practice in the cultivation of cucumbers under glass in The Netherlands. They were grafted on the rootstock *Cucurbita ficifolia,* a North African species resistant to *F. oxysporum.* For over 20 years this grafting practice was successful, and until recently was done on a large scale. However, this rootstock is susceptible to *F. javanicum* var. *ensiforme*

(= *F. solani* f. sp. *cucurbitae*) which some years ago became important in greenhouse cucumbers. It originated from rootstock seed taken from diseased fruits, and the disease spread rapidly (28). Thus the long-established practice of grafting had to be abandoned.

The vector-transmitted pear decline was found most frequently in pear trees on oriental rootstocks (*Pyrus ussuriensis* and *P. serotina*) in the western United States. These rootstocks had been introduced and used between 1917 and 1935 to obtain higher resistance to fire blight (*Erwinia amylovora*), a disastrous disease now gradually penetrating into Europe. In this volume Schroth et al (19) will deal in detail with the epidemiology of this disease. When the oriental rootstocks were introduced, no one suspected that their use would lead to a virtually complete loss of the older pear trees between 1952 and 1964 through pear decline, because *Psylla pyricola,* the vector of the virus, was not found in the western United States before 1940.

These two examples show how extremely difficult it is to foresee what results certain changes in agricultural practices may have.

In the foregoing the importance of changing agricultural methods on disease development has been mentioned. Let us consider this subject in more detail. Whereas in the developing countries the changes consist of the importation of new varieties, the use of better irrigation, and more fertilizers and pesticides, the recent changes in the more developed countries in the West are primarily due to mechanization and large-scale specialization. These were the result of the high cost of labor. Industry became financially attractive to the farm worker, and if the farmer wanted to keep him he had to offer higher wages. The result was a rapid reduction in the number of workers on the farm. As the work remained the same or even increased, the farmer had to buy expensive machinery.

This farm mechanization had a number of phytopathological consequences. The introduction of the combine resulted, inter alia, in a tremendous spread of wild oat. This weed is susceptible to *Heterodera avenae* (= *H. major,* the oat-root eelworm); consequently this nematode became very common in all oat fields (16). Furthermore, certain biotypes attacked cereals other than oats.

Combines blow the husks over the field, and probably the growing importance of foot rot and head blight caused by *Fusarium* spp. as well as an increasing occurrence of glume blotch caused by *Septoria nodorum* are partly due to this practice (8). The combine also leaves more and longer straw on the field, and the crop is harvested at a later stage of maturity than before; both must have had an influence on the increase of several diseases.

At the beginning of this century it was common practice in The Netherlands to use the waste of wheat mills as manure on the fields, thus spreading smut on a large scale. Another unfavorable development was that crop rotation decreased sharply because wheat, a major food crop in many countries, was sown much more frequently.

The high cost of labor also resulted in the use of many chemicals on the farm, especially weed killers. The variety of these selective compounds used in agriculture and horticulture is enormous. Their effect on pathogenic fungi, bacteria, and nematodes is only partly known, and one reads many contradictory reports. One of the reasons for this is that most investigators have conducted in vitro trials where the environmental circumstances differ greatly from the situation in the fields. It has been found that effects depend on temperature, soil type, and pH. A number of observations indicate a direct influence of the herbicide on several pathogens or on antagonistic fungi in the soil. Recently Katan & Eshel (10) prepared an extensive literature survey on the interaction of herbicides and plant pathogens.

It is well known that pesticides other than weedkillers may have a special influence on the development of certain plant diseases. When investigating the sudden spread of rubus stunt in raspberry plots in The Netherlands, my collaborators de Fluiter & van der Meer (6) came to the conclusion that the reason for this was that the leafhopper vector, *Macropsis fuscula,* was not destroyed by DDT. Originally the growers had used a tar oil winter wash against certain raspberry fruit insects. This winter wash also killed the eggs of *M. fuscula,* but a DDT spray in the spring did not. It is interesting that growers had said from the beginning of this investigation that they considered DDT to be responsible for rubus stunt. Although they had no idea of the complicated etiology of this disease, this showed again that farmers are good observers.

The use of systemic fungicides not only revolutionized the control of Sigatoka and postharvest disease of the banana (24); it also had unexpected consequences for some other crops, because so far none of these compounds have any activity towards the Phycomycetes (7). *Pythium* species thus become more important as soilborne parasites in greenhouse crops where benomyl is used regularly.

*Botrytis tulipae* could be reasonably well controlled in tulips by storage of planting material under high temperature conditions and by application of fungicidal sprays to prevent spread of the disease in the field. This application of chemicals kept the plants green; as a consequence the lifting date was delayed for 10–20 days. In that period conditions are favorable for infection by *Fusarium oxysporum,* now an important bulb disease, although hardly known before. This is not the only reason for the epidemic occurrence of this pathogen. The revolutionary change in cultural methods of bulb growing (mechanical lifting and handling) has also increased the importance of this disease (18).

Every plant pathologist knows examples of increased disease incidence in certain crops after heavy applications of nitrogen fertilizers. Huber deals with this subject in this volume (9).

These examples may suffice to indicate how great the influence of changing agricultural methods is on disease development. It also shows how complicated the interactions are and how continuous, careful research by plant pathologists will remain important for production of an abundance of high quality agricultural and horticultural crops.

In several countries there is much anxiety about the use of pesticides, partly because people have been told by commentators of the communication media that they are being poisoned by residues of pesticides on their food, and partly because of the far-reaching effects of the persistent pesticides on ecosystems. Public pressure has led to the prohibition of such chemicals in several western countries. On the other hand, during the past decades the consumers have come to expect high quality, spotless fruits (apples, pears, bananas) and vegetables, and would not now accept agricultural products with specks from apple scab, and blemishes or injury caused by insects.

Unless there is a change in the attitude of the consumers, growers will have to go on using pesticides in order to be able to sell their products at reasonable prices. This is especially true for producers for western markets. "There is a real danger of over-emphasizing aesthetic considerations for produce sold under brand names. It could lead to impossible demands being made on pest control techniques, to unacceptably high levels of pesticide usage, to extravagant standards of quality control and hence to a gross wastage of perfectly wholesome foodstuffs" (34). This is a pity, but one should never forget that the consumer really decides the type and quality of the agricultural products he wants.

In western Europe the general trend in modern agriculture is towards bigger farms. The same tendency is seen in horticulture where production units (greenhouses, orchards) become larger and larger. This development is, as a rule, favorable from a sanitation point of view, but this concentration of many plants of the same species carries some special risks which will be considered later.

In agriculture one may say that the larger the average distance between fields the smaller the chance that a disease will spread from one field to another. As van der Plank (30) wrote: "Bringing plants together into fields increases the chance of epidemics; bringing them still further together, by increasing the area of the fields and correspondingly reducing their number, may reduce the chance of a general epidemic." Modern development not only favors larger farms, it also favors specialization on only one crop plant or even on a certain stage of it.

One of the biggest influences in vegetable production nowadays is the special demand for uniform, high quality products for prepacking or processing (quick freezing, canning, dehydrating). The result has been a steady trend from small holdings to large-scale production, mostly on contract, with the factories dictating the details of cultural practices, including pest and disease control operations (34). When vegetables or flowers are grown under glass, as happens on a large scale in The Netherlands (∼5000 ha of vegetables and 2155 ha of flowers in 1972), the trend for specialization goes even further. There are nurseries specializing in the year-round propagation of young planting material, e.g. of lettuce, carnation, chrysanthemum, for commercial growers.

Most vegetables and several flowers under glass are now indeed produced on a year-round basis; this is unfavorable from a sanitary point of view. Chrysanthemum had captured more than 60% of the US flower market by 1964. Only two companies produced the bulk of the American chrysanthemum cuttings (14), one company producing as many as 100,000,000 cuttings in 1958 (1). These cuttings were transported over half the globe. At present, cuttings for the European market are produced in the USA, South Africa, Italy and Malta, the French Riviera, and most of the northwest European countries. They are flown from one country to another, and this incurs the risk that certain plants may carry a disease which, due to quick transportation and a long incubation time, will not show before the plants are grown commercially in the growers' greenhouses. Thus *Ascochyta* ray blight has been spread over the world. In a similar way *Puccinia horiana,* one of the many chrysanthemum rusts endemic in Japan, spread globally from one of the few private cutting producers in South Africa. This white rust disease has a high multiplication rate and a long incubation period and may remain unnoticed for about a month (35). The preparation and distribution of potting soil for greenhouse crops is also done by only a few specialized farms in The Netherlands. This has already resulted in the spread of Arabis mosaic in cucumber, a soil-borne virus transmitted by nematodes (31).

These are only a few out of several examples of the rapid spread of hitherto unknown diseases due to modern agricultural practices and rapid transportation.

I will not consider the production of mycotoxins by phytopathogenic fungi and bacteria, especially those causing postharvest infections, as Patil (15) treats the latter and Mirocha & Christensen (13) the former in this volume. It is clear that this subject is of great importance to human society.

Industrial activities of various kinds, heavy road traffic, and to a minor degree concentrations of larger amounts of farm animals (bio-industries) have led to more or less serious damage to plants. Noxious gases cause leaf injury and reduce yields of many crops by inhibiting vital enzyme systems in plants, and some have also been reported as stimulators or inhibitors of fungal or viral pathogens (27). Much more research is needed before we obtain a better insight into the interaction between certain air pollutants such as $SO_2$, HF, and ozone on the one hand and diseases caused by fungi, bacteria, and viruses on the other.

From the foregoing it will be clear that research by plant pathologists will always be indispensable if human society is to get the optimum yields of agricultural and horticultural products so badly needed for feeding its rapidly increasing millions. We can only succeed by applying crop protection measures, including the use of various chemicals, which must have as little unfavorable ecological side effects as possible. Our final aim should be to protect crops without chemicals, but at the moment this is still impossible.

ACKNOWLEDGMENTS

I am most grateful to my Dutch colleagues Professor Dr. J. Dekker and Mr. G. S. Roosje, deputy director of my institute, for critical reading of the manuscript and to Professor R. K. S. Wood of the Imperial College of Science and Technology, London for his valuable comments. That he moreover took the trouble to polish my English is a precious token of his friendship derived from our close collaboration in the executive committee of the International Society for Plant Pathology. Miss J. M. Krijthe, editor of the publications of my institute, carefully prepared the bibliography.

*Literature Cited*


1. Baker, K. F., Dimock, A. W., Davis, L. H. 1961. Cause and prevention of the rapid spread of the *Asochyta* disease of Chrysanthemum. *Phytopathology* 51:96–101
2. Beijerinck, M. W. 1898. Over een contagium vivum fluidum als oorzaak van de vlekziekte der tabaksbladeren. *Versl. Gewone Vergad. Wis. Natuurk. Afd. Kon. Akad. Wetensch. Amsterdam* 7: 229–35
3. Borlaug, N. E. 1965. Wheat, rust, and people. *Phytopathology* 55:1088–98
4. Burrill, T. J. 1881. Anthrax of fruit trees; or the so-called fire blight of pear and twig blight of apple trees. *Proc. Am. Assoc. Advan. Sci.* 29:583–597
5. Butler, E. J., Jones, S. G. 1949. *Plant Pathology.* London: Macmillan. 979 pp.
6. de Fluiter, H. J., van der Meer, F. A. 1955. De dwergziekte van de framboos, haar verspreiding en bestrijding. *Meded. Landbouwhogesch. Opzoekingsst. Staat. Gent* 20:419–34 (I.P.O.-Meded. 105)
7. Evans, E. 1973. Practical aspects of using systemic fungicides. *Eur. Med. Plant Prot. Org. Bull.* 10:59–67
8. Feekes, W. 1967. Phytopathological consequences of changing agricultural methods. II. Cereals. *Neth. J. Plant Pathol.* 73: Suppl. 1, 97–115
9. Huber, D. M., Watson, R. D. 1974. Nitrogen form and plant disease. *Ann. Rev. Phytopathol.* 12:139–65
10. Katan, J., Eshel, Y. 1974. Interactions between herbicides and plant pathogens. *Residue Rev.* 47: In press
11. Kerling, L. C. P. 1966. Vijfenzeventig jaren fytopathologie in Nederland 1891–1966. *Neth. J. Plant Pathol.* 72: 68–86
12. Mayer, A. 1882. Over de mozaïekziekte van de tabak; voorlopige mededeeling. *Tijdschr. Landbouwk.* 2:359–64
13. Mirocha, C. J., Christensen, C. M. 1974. Fungus metabolites toxic to animals. *Ann. Rev. Phytopathol.* 12:303–30
14. Pag, H. 1962. Stecklingstest bei Chrysanthemen in den U.S.A. *Nachrichtenbl. Deut. Pflanzenschutzdienst Berlin* 14:183–85
15. Patil, S. S. 1974. Toxins produced by phytopathogenic bacteria. *Ann. Rev. Phytopathol.* 12:259–79
16. Richter, H. 1967. Phytopathologischen Konsequenzen der sich ändernden landwirtschaftlichen Anbauverfahren. I. Allgemeine Einleitung. *Neth. J. Plant Pathol.* 73: Suppl. 1, 81–96
17. Saari, E. E., Wilcoxson, R. D. 1974. Plant disease situation of high-yielding dwarf wheats in Asia and Africa. *Ann. Rev. Phytopathol.* 12:49–68
18. Schenk, P. K. 1967. V. Bulb crops. *Neth. J. Plant Pathol.* 73: Suppl. 1, 152–63
19. Schroth, M. N., Moller, W. J., Thomson, S. V., Hildebrand, D. C. 1974. Epidemiology and control of fire blight. *Ann. Rev. Phytopathol.* 12:389–412
20. Singh, K. G. 1969. Bacterial leaf streak in West Malaysia. *FAO Plant Prot. Bull.* 17:64–66
21. Snyder, W. C. 1971. Plant pathology today. *Ann. Rev. Phytopathol.* 9:1–6
22. Srivastava, D. N. 1972. Bacterial blight of rice (*Xanthomonas oryzae*). *Indian Phytopathol.* 25:1–16
23. Stewart D. M., Hafiz, A., Abdel-Hak, T. 1972. Disease epiphytotic threats to high yielding and local wheat in the Near East. *FAO Plant Prot. Bull.* 20:50–57
24. Stover, R. H. 1972. *Banana, plantain and abaca diseases.* Kew (Surrey): Commonw. Mycol. Inst. 316 pp.
25. Stubbs, R. W. 1972. The international survey of virulence of *Puccinia striiformis.* Virulence patterns in the Middle East and Africa and potential sources of resistance. *Regional Wheat Workshop* Vol. I *Diseases* (9):1–15
26. ten Houten, J. G. 1963. Obituary notice Johanna Westerdijk, 1883–1961. *J. Gen. Microbiol.* 32:1–9
27. ten Houten, J. G. 1972. Air pollution and plant health. *Eur. Med. Plant Prot. Org. Bull.* 4:65–77
28. Termohlen, G. P. 1967. III Vegetables, flowers and fruit under glass. *Neth. J. Plant Pathol.* 73: Suppl. 1, 116–29
29. Thung, T. H. 1957. Het virologisch onderzoek aan de landbouwhogeschool, Wageningen. *Tijdschr. Plantenziekten* 63:209–21
30. van der Plank, J. E. 1960. Analysis of epidemics. In *Plant Pathology,* ed. J. G. Horsfall, A. E. Dimond, 3:229–89. New York: Academic 675 pp.
31. van Dorst, H. J. M., van Hoof, H. A. 1965. *Arabis*-mozaïekvirus bij komkommer in Nederland *Neth. J. Plant Pathol.* 71:176–79
32. van Hoof, H. A. 1970. Virusziekten in de tropen. *Landbouwk. Tijdschr.* 82: 12–15
33. Wakker, J. H. 1885. Het geel of nieuwziek der hyacinthen. *Onderzoek der Ziekten van Hyacinthen en andere bol-*

*en knolgewassen gedurende de jaren 1883, 1884 en 1885* (Verslag over 1883). *Alg. Ver. voor Bloembollencultuur:* 4–13
34. Wheatley, G. A. 1971. The role of pest control in modern vegetable production. *World Rev. Pest Contr.* 10:81–93
35. Zadoks, J. C. 1967. International dispersal of fungi. *Neth. J. Plant Pathol.* 73: Suppl. 1, 61–80

# HISTORY OF PLANT PATHOLOGY IN JAPAN

❖3585

*Shigeyasu Akai*
Kyoto University, Kyoto, Japan

## FROM ANCIENT TIMES TO BEFORE THE MEIJI ERA

Scientific development of plant pathology in Japan began with the introduction of European and American civilization in the early years of the Meiji era, although some accounts concerning plant diseases have been found in earlier records. These records show that the ancient people recognized that plant diseases were injuring their crops, but because they had no conception of the nature and cause of these maladies, they superstitiously observed mysteriously deformed diseased plants with surprise and terror. According to Shirai (90) there is an account in the historic book *Zoku-Nihon-Ki* that in January 713 A.D. deformed rice plants were presented to the Emperor. These were thought to be a transformation of grasses into rice (27, 72). According to Shirai (90), these could have been rice ears infected by downy mildew (*Phytophthora (Sclerophthora) macrospora*).

*Ganoderma lucidum* is a comparatively common decay fungus in Japan, attacking stems and roots of coniferous and deciduous trees (21). From very early times development of the fruit body of this fungus was considered to be a sign of happy events. Many descriptions and much information on this are found in the classics; accounts in *Nihon-Shoki* show that in 678 A.D. the fruit body of this fungus was presented to the Emperor and that in 726 A.D. this fungus developed fruit bodies in the Imperial Palace (27).

Many ancient classics indicate that blast disease of rice plants was very serious and, together with unfavorable weather conditions, caused much damage to rice crops. People then suffered severe famine (27, 72). The first record of blast disease of rice in Japan appears in a book, *Kōka Shunju* (45, 101), written by M. Tsuchiya and published in 1707. He emphasized the relation between environmental conditions and occurrence of the disease. Miyanaga (65), Horii (35), Kojima (57), and Konishi (58) also referred to rice blast disease in their books. Konishi (58) emphasized especially the relationship of cultural conditions of rice plants (e.g. nitrogenous fertilizers, deep plowing of soil, density of seeding) to the occurrence of blast disease. According to Hino (27), accounts concerning the following plant diseases or fungi (mushrooms) were often found in agricultural books: honeydew (nectar) containing pycnospores of *Cronartium quercuum;* Goshiki-Mume, chloranthic flowers of *Prunus mume* infected by *Caeoma makinoi* (60, 90); deformed inflorescence of corn caused by *Ustilago zeae* (90); deformed buds of *Camellia* due to *Exobasidium camelliae* (27, 90); snow blight of wheat and barley (65); "Bakanae" disease of rice plants (27, 58); downy mildew of cucumber infected by *Pseudoperonospora cubensis* (27, 101); Sabi-Take (Akagoromo-Byo) caused by *Stereostratum (Puccinia) corticioides* attacking bamboo stems (27, 29); Hyuga-Hantiku, spotted culms of bamboo (*Phyllostachys bambusoides* f. *tanakae*) infected by *Asterinella hiugensis* (28); Tora-fu-Dake, spotted culms of bamboo (*Phyllostachys aurea, P. bambusoides, Semiarundinaria fastuosa*) infected by *Chaetosphaeria fusispora* (26–28, 55); Goma-Take, infected culms of bamboo (*Phyllostachys aurea, P. bambusoides*) by *Munkiella shiraiana* (28); fruit bodies (Mannen-Take) of *Ganoderma lucidum,* Kinugasa-Take (*Dictiophora phalloides*) (27), Matsu-Take (*Tricholoma matsutake*) (27), Shii-Take (*Lentinus edodes*) (27, 99), Kikurage (*Auricularia auricula-judae*) (27, 99); bacterial wilt of eggplant caused by *Pseudomonas solanacearum* (27, 99).

Many diseases of tobacco plants were also observed. Although the nature of these diseases was unknown, one book (5) pointed out the difficulty of tobacco cultivation due to root rot. This may be the oldest record concerning tobacco disease in Japan (71). Nakamura (71) stated that this root rot corresponded to the bacterial wilt caused by *Pseudomonas solanacearum.* Mosaic disease of tobacco also seemed to be prevalent. Tamura (96), a farmer, stated in his book published in 1841 that tobacco leaves were deformed into a bamboo leaf-like appearance (Sasa-Ha Byō) (32) and became useless. This may have been a mosaic or related disease of the tobacco plant (17, 71, 92). Aoe (6) recorded a powdery mildew-like disease of tobacco, and recognized that this disease had been epidemic since about 1850.

The microscope was introduced in Japan about 1700 A.D. Since then, through observation of many biological materials, including filamentous fungi (27), an entirely new field has been developed.

## MEIJI ERA

### *Incipient Stage (1873–1893)*

After the great revolution in 1866, the Japanese government did its utmost to introduce European and American civilization, and the Ministry of Education invited to Japan eminent European and American professors from all branches of science and the arts (91). Plant pathology was introduced into Japan through Tokyo and Sapporo, Hokkaido.

Professor Friedrich M. Hilgendorf (born in 1839 in Brandenburg, Germany) came to Tokyo from Germany in March 1873 and taught botany and zoology in the medical school of the Kaisei Gakko, the predecessor of Tokyo Imperial University. Although Hilgendorf was a zoologist, in Tokyo he lectured on medical botany as well, and often touched on plant diseases (27, 44, 91, 103). After Hilgendorf returned to Germany, Professor Hermann Ahlburg, student of Julius von Sachs, replaced him in May 1876. Ahlburg was a plant physiologist whose speciality was horticultural botany; however, he was very interested in plant pathology and mycology. He delivered special lectures on plant pathology to a private assembly of agriculturists, and was the first to describe Japanese Kōji-fungus, *Aspergillus oryzae,* under the name, *Eurotium oryzae* Ahlburg (27, 44, 91, 103).

In 1875, Shinnosuke Matsubara, a member of the above-mentioned assembly, was appointed as a teacher of the Tokyo Medical School and worked as an interpreter for Hilgendorf and Ahlburg. When the name of the school was changed to College of Medicine, Tokyo Imperial University, Matsubara was appointed associate professor. In 1882 he published a book on botany, containing one chapter on plant pathology. This may have been the first book concerning plant pathology published in Japan.

The Agricultural College of Komaba in Tokyo was established in 1878, and Matsubara was appointed as a teacher of botany and zoology there (91). A special course, Shoku-I-Ka (plant medicine) was started in 1880, but unfortunately was abolished the following year.

Hikotarō Nomura was the first plant pathology specialist in Japan (27). Born in 1859, he entered the new course of plant pathology of Komaba in 1880, but after the course was abolished, he entered the College of Science, Tokyo Imperial University, and majored in botany. From 1906, he studied diseases of mulberry trees and silkworms at Tokyo Kōtō Sanshi Gakko (College of Sericulture). The fungus genus *Nomuraea,* and *Diaporthe nomurai* (the causal fungus of mulberry canker) have been named in his honor.

In 1881, Chujirō Sasaki was appointed as a teacher of zoology and botany in the Komaba Agricultural College, where he studied diseases of silkworms and mulberry trees (91), lectured on plant pathology, and taught botany and zoology (25). At this time (1882) the Tokyo Botanical Society was established, and in February 1887 the first issue of the *Botanical Magazine* was published. This journal, the first botanical periodical in Japan, is still published and has often published papers on plant-disease fungi (91).

Mitsutarō Shirai (born in 1863 in Fukui prefecture) graduated from the College of Science, Tokyo Imperial University in July 1886. In October of the same year, he was appointed a teacher of forest botany and plant pathology in Komaba Agricultural College, renamed the College of Agriculture, Tokyo Imperial University in 1890. He stated (91) that the only reference books on plant pathology then in the college library were the 1874 Sorauer's *Handbuch der Pflanzenkrankheiten* and the first 1882 edition of R. Hartig's *Lehrbuch der Baumkrankheiten.* In 1893–1894 Shirai published a book of plant pathlogy, *Shokubutsu Byōri Gaku,* the most authoritative at that time.

Manabu Miyoshi graduated from Tokyo Imperial University in 1884 and was appointed professor of plant physiology in the College of Science, Tokyo Imperial University in 1895. He was interested in lower fungi. He studied the chemotropism of *Botrytis cinerea* under Professor Pfeffer in Leipzig (66); he discussed penetration by *Botrytis* hyphae of gold leaf or a thin film of mica floated on a nutrient solution (67, 68). This also may be the first report on mechanical penetration of the plant cell wall by fungal hyphae.

During this period many pathogenic fungi found in Japan were identified and described. In the early years of Meiji, cherry trees along the Sumida river in Tokyo were said to be weakened by severe infection of witches' broom (*Taphrina cerasi*). Obuchi (27) recommended cutting off the diseased branches; with this therapy the cherry trees recovered their vigorous growth. This is probably the first written record of surgery on diseased trees in Japan (89).

An epidemic of cucumber downy mildew occurred in 1888 in Tokyo prefecture (34, 64). Professor Kakichi Minosaku, College of Science, Tokyo Imperial University, observing the severe damage to the cucumber crop, recommended Nobujiro Ichikawa (Tanaka) in the Botanical Institute to study the disease. Ichikawa (97) reported it to be a species allied to the late blight fungus of potato. Tamari (95), a professor of Komaba Agricultural College, also studied this disease. He reported a method for its control and identified the causal fungus as *Peronospora cubensis* Berk. & Curtis (7), the same fungus found on wild cucurbitaceae in Cuba. Furthermore, Tamari confirmed that the conidia of this fungus germinated by zoospores (34, 64).

A telegraph service in Japan was established about 1869. A severe decay of wood poles led to their treatment with copper sulfate by the Boucherie process in 1880. Creosote oil injection into railroad sleepers was begun in 1902 (27).

### *Developmental Stage (1894 to the End of the Meiji Era)*

In this period the foundation of plant pathology in Japan was laid. In April 1906 a professorial chair of plant pathology was first provided in the Agricultural College of Tokyo Imperial University,[1] and Mitsutarō Shirai was appointed as professor. This was one of the first professorial chairs for plant pathology in any university.

The Sapporo Agricultural College (Sapporo Nōgakko) was founded in 1876 at Sapporo by the Colonial Government of Hokkaido. William P. Brooks was invited from America as a teacher in 1877, lecturing on plant pathology for about 12 years. Shōtarō Hori, afterwards one of the authorities on plant pathology in Japan, was at that time a student in the college (36).

Kingo Miyabe (born in 1860 in Edo (Tokyo)) was graduated from the Sapporo Nōgakko in 1881. The school later was raised to the status of university. In 1886 Miyabe studied at Harvard University under Professor W. G. Farlow. After his return to Japan in 1889, he was appointed professor at this school, and lectured on botany and plant pathology. In 1907, the year following establishment of a chair for plant pathology in Tokyo Imperial University, a similar chair was provided in the College of Agriculture of Tohoku Imperial University in Sapporo (afterwards renamed Hokkaido Imperial University), and in 1920 another professorial chair was provided. Kingo Miyabe and Seiya Ito occupied the two chairs, and Miyabe there trained many plant pathologists.

In 1899 the section of plant pathology in the Imperial Agricultural Experiment Station at Nishigahara, Tokyo (now the National Institute of Agricultural Sciences, Nishigahara) was organized, and Shōtarō Hori and Yeijirō Uyeda were appointed leaders of the mycological and bacteriological laboratories, respectively. The gov-

[1]The word "Imperial" in the name of the national university was omitted in 1947.

ernment organization for prefectural agricultural experiment stations was announced officially in 1893, and between 1884 to 1911 an agricultural experiment station was established in every prefecture (100).

Noteworthy experiments were carried out in this period, e.g. discovery of flower infection by loose smut of barley and of insect transmission of dwarf disease in rice plants.

### *Discovery of Flower Infection by Loose Smut of Barley*

Formerly, loose smut of barley was thought to develop from seed-borne chlamydospores, and hot water or chemicals were used to disinfect seeds. However, these methods were not effective, and control of loose smut was the subject of extensive discussion. Beginning in 1892, Sato & Yamada (79) carried out experiments in Kyoto prefecture to control loose smut of barley. After several experiments they found that when plants were inoculated at flowering time, a large number of diseased plants resulted from the seed produced. Thus it was recorded for the first time that flowers were infected by loose smut in barley. Independently, Maddox (23, 61–63) proved the occurrence of flower infection by loose smut in wheat, a finding confirmed in 1903 by Brefeld (9, 24, 76). Nakagawa (70) and Hori (33) confirmed flower infection by loose smut in wheat in Japan, thus establishing a scientific basis for Jensen's hot water seed treatment (47, 48) for the control of this disease (2). Furo-Yu Hitashi Hō (hot-water seed treatment) for loose-smut control was devised in 1896 by Kanjirō Shinohara (27, 106), and good results were obtained.

### *Discovery of Insect Transmission of Rice Dwarf Virus*

In the early years of Meiji, a severe outbreak of dwarf disease in rice occurred in the Kansai district, especially in Okayama, Shiga, and Kyoto prefectures. According to Ishikawa (15, 27, 37), dwarf disease of rice plants was first noticed in the Shiga prefecture by Hatsuzo Hashimoto, a conscientious rice grower. He began making detailed observations of this disease in 1883 and suspected that the disease was in some way related to leafhoppers. There was an epidemic of the disease in the Shiga prefecture in 1892. In 1894 Hashimoto experimentally proved the causal relation of leafhoppers to rice dwarf, but did not publish the results, and the leafhoppers he observed were not identified. Takata (94) gave the first report in 1895 of the relationship between the insect and the disease (10, 15), and the leafhopper involved was identified as *Inazuma dorsalis* (Inazuma-Yokobai) (18). However, he did not discover that the disease was caused by a virus of which *Inazuma dorsalis* was a vector.

In 1895 the Shiga Agricultural Experiment Station was established, and Takata was appointed director (27). Research on virus transmission by insects was continued for more than 20 years. In 1900, the Shiga Agricultural Experiment Station published reports (83) that *Nephotettix cincticeps* (Tsumaguro-Yokobai) was the true cause of dwarf disease, while other species, including *Inazuma dorsalis,* had no connection with the disease. Substantiating Takata's early claim, Fukushi (16) later confirmed that *Inazuma dorsalis* was also a vector. During 1902–1908 the Imperial Agricultural Experiment Station and the Shiga Agricultural Experiment Station found both infective and noninfective leafhoppers and also found that noninfective leafhoppers from Tokyo became infective after feeding on diseased plants. From this came the conclusion that rice dwarf was not caused by leafhoppers but by certain unknown agents carried by them (4). This is now considered to be the first plant virus shown to be insect transmitted.

### *Investigation on the Causal Fungus of Rice Blast*

From ancient times, rice blast has been a very serious disease, causing severe damage to rice crops. Shirai (84), observing blasted rice fields in Kyoto (1894) and in Nara prefecture (1895), reported that the causal fungus resembled but was slightly different from *Piricularia oryzae* Briosi & Cavara. As a control measure, he recommended planting resistant cultivars, avoiding planting in shade, spraying with Bordeaux mixture and boric acid solution (88), and not supplying much nitrogenous fertilizer. In regard to the scientific name of blast fungus, Shirai (88) adopted *Dactylaria parasitans* Cavara, but Hori (30) adopted *Piricularia grisea.* Miyabe (27) and Kawakami (53, 54) also used the name *Piricularia grisea,* treating *Piricularia oryzae* as a synonym. Nisikado (73) later asserted that *Piricularia* should be the genus of the blast fungus, and Ito (42) and Hara (19) adopted *Piricularia (Pyricularia) oryzae* Briosi & Cavara.

### *Confirmation of Alternation of Hosts in Cronartium quercuum*

The phenomenon of alternation of hosts in rust fungi was first confirmed respectively by Oersted and DeBary in 1864–1865. In Japan in 1889 Shirai (85, 86) proved experimentally the alternate relationship in pine gall fungus, *Cronartium quercuum.* This fungus was first observed and described by a teacher of the Agricultural College of Tokyo Imperial University, H. Mayr, who came from Germany in 1890. Afterwards, this fungus was reported by von Tubeuf (102) as *Peridermium giganteum* (86). Shirai observed many galls produced on pine trees close to deciduous *Quercus acutissima, Q. serrata,* and *Q. variabilis.* In April 1898 he (85) inoculated aeciospores obtained from pine galls onto leaves of *Quercus.* Uredosori and teliospores developed on the lower surface of the leaves. The teliospores germinated immediately after they matured, forming sporidia. Thus Shirai proved the genetic connection between *Peridermium* stage formed on pine galls and *Cronartium* stage on *Quercus* leaves. He (87) also proved the genetic connection between *Roestelia koraiensis* P. Henn. on Japanese pear leaves and *Gymnosporangium japonicum* Sydow on *Juniperus.* However, subsequent investigations by Ito (41) showed that *G. japonicum,* used by Shirai, was actually *G. haraeanum,* which attacked pear leaves.

### *Epiphytotic of False Smut (Inakōji Byō) of Rice Plant*

The occurrence of false smut (green smut) was long believed to be a sign of a heavy crop year, and severe occurrence was said to be conspicuous in years when weather conditions were favorable for growth of rice plants. During 1894–1907 there were several epidemics of the disease. The fungus was first described by Cooke (11) as a smut fungus and named *Ustilago virens,* based on the specimen from India. The fungus had been described at a very early date in China, but not given a scientific name (75). From materials from Japan, Patouillard (77) named it *Tilletia oryzae,* but Brefeld (8) established a new genus *Ustilaginoidea* and transferred it to *U. oryzae* (Pat.) Brefeld. Tanaka (98) and Hori (29) had previously assumed the fungus to be *Ustilago virens* Cooke, but Takahashi (93) transferred it to *Ustilaginoidea virens* (Cooke) Takahashi, and since then, this name has been generally accepted. Sakurai (78) found sclerotia on diseased grains in 1934; fallen sclerotia on soil produced fruit bodies in which perithecia and ascospores developed (22). Yamashita (104) detected alkaloids in infected grains and reported that the sclerotia of *Ustilaginoidea virens* resembled the ergot formed by *Claviceps.*

### Bakanae Disease of Rice Plants

Bakanae disease (characterized by elongation of seedlings) has been known from ancient times. In 1828 Konishi (58) recorded its occurrence in his book *Nōgyō Yowa,* although he did not know its origin. Hori (31) first reported that it is a parasitic disease, but erroneously identified the causal fungus as *Fusarium heterosporum* Nees, with the comment that he had some doubt about this identification. In 1916 Yosaburo Fujikuro found the perfect stage and it was named *Lisea fujikuroi* by Sawada (81, 82). E. Kurosawa (58a) showed in 1926 that culture media in which the fungus had been growing would induce the disease in rice plants. The active component, gibberellin, was isolated from the fungus and named by T. Yabuta (102a) in 1935. The fungus was later put in the genus *Gibberella* under the name of *G. fujikuroi* (Sawada) Wr. (46, 74, 102b).

## TAISHO ERA (EXPANSION STAGE, 1913–1926)

Discovery of the pathogenic nature of fungi led to an intensive study of these organisms as the cause of plant diseases. Thus plant pathology in the preceding period was considered more or less synonymous with the science of fungus diseases of plants. In the Taisho Era, studies on the physiological nature of pathogenic fungi appeared gradually, and plant pathologists were not necessarily limited to simple etiological works of diseases. Establishment and practice of control measures for the diseases were also needed.

Starting in October 1914 Umenojo Bokura issued a monthly magazine entitled *Byōchu-Gai Zasshi* (Journal of Plant Protection). At that time, Bokura was a pathologist in the Plant Pathology Section of the Imperial Agricultural Experiment Station at Nishigahara. This journal continued until 1943 and contained articles written in Japanese, covering both the fields of plant pathology and economic entomology. It contributed a great deal to the progress and dissemination of knowledge of plant diseases in Japan.

The Phytopathological Society (Nihon Shokubutsu Byōrigaku-Kai) was established in February 1916, and Mitsutarō Shirai was recommended as the first president.

The Sapporo Nogakko (Sapporo Agricultural College) was consolidated with the Tohoku Imperial University in 1907. The Sapporo Nogakko, however, became independent in 1918 and developed into the College of Agriculture, Hokkaido Imperial University. A chair for plant pathology was provided at the Kyushu

Imperial University in Fukuoka in 1922, and Kakugorō Nakata was appointed professor. Nakata was born in 1887, and was graduated from Tokyo Imperial University in 1912. He clarified species of fungi related to sclerotial diseases of rice plants, and the pathogens of hollow stalk (black leg) and bacterial wilt of tobacco plants. A book published in 1934, *Sakumotsu Byōgai Zuhen (Illustrated Monograph of Crop Diseases)* gained public favor. After Nakata's death in 1939, Hazime Yoshii succeeded to the chair of professor. He was born in 1900 and was a graduate of the Agricultural College of Tokyo Imperial University. Yoshii's work was in pathological anatomy and physiology, and his book, *Anatomical Plant Pathology* (*Kaibō Shokubutsu Byōri Gaku*), published in 1947, is still a prominent textbook for graduate study.

A chair for plant pathology was provided in Kyoto Imperial University in 1924, with Takewo Hemmi as professor. Born in Tokyo in 1899, he graduated in 1915 from the College of Agriculture, Tohoku Imperial University in Sapporo. His *Morphological and Physiological Studies on Anthracnose Fungus* was completed under Professor Kingo Miyabe.

During 1902–1924, the Ministry of Education promoted a plan to establish colleges for agriculture and forestry or horticulture in the cities of Morioka (Iwate prefecture), Tottori, Tsu (Mie prefecture), Utsunomiya (Tochigi prefecture), Gifu, Miyazaki, and Matsudo (Chiba prefecture). All these colleges were raised to the status of university after the second world war.

### *Regulation for Control of Imported and Exported Plants* (56, 69)

Since the opening of international commercial activities, many agricultural crops have been imported and were sometimes contaminated with plant pathogens and pests. Thus, new diseases or new races of pathogens were imported, bringing about great damage to crops. Fire blight bacteria of apple, anthracnose fungus of peach, late blight fungus of potato, crown gall bacteria, and powdery mildew and downy mildew of grape were introduced with contaminated imported seedlings. Effective plant quarantine service was thus required, and in 1914 the government announced regulations for plant quarantine, and quarantine stations were established at Yokohama and other areas.

Application of fungicides by motor sprayers was also started in 1914 as a preventive practice in orchards, and disinfection of rice seeds became an important procedure in rice-cultivation practice, especially for preventing seed and seedling rot caused by *Pythium, Achlya,* and other pathogens in the northern part of Japan. Sawada (80) tried dipping rice seeds in 0.5% formalin solution, and Ito (40) treated them in copper sulfate solution.

### *Epidemic of Needle Blight of Cryptomeria japonica*

Needle blight of Sugi (*Cryptomeria japonica*) was first found in 1902 in the Yamaguchi and Hiroshima prefectures. Many pathogenic fungi have been isolated from blighted seedlings. Among them *Cercospora cryptomeriae* is most highly virulent (38), and within a few years after 1910 the disease spread throughout almost all districts in Japan where Sugi was cultivated (39). The fungus, *Cercospora cryp-*

*tomeriae*, is similar to *C. sequoiae*. Ito (39) considered *C. cryptomeriae* synonymous with *C. sequoiae* and suggested that the fungus probably had been introduced from North America with seedlings or stocks of American conifers, especially *Sequoia gigantea.*

## SHOWA ERA (ACTIVE STAGE, 1927 TO THE PRESENT)

In this period plant pathology in Japan has made steady progress as a science of plant disease, i.e. plant medicine, and has not been limited to simple applied mycology.

### *Advances in Experimental Epidemiology*

Since L. R. Jones first pointed out the probable importance of environmental conditions (49, 50) for the occurrence of plant diseases in America, many detailed observations and experimental studies have been made on this relationship. Hemmi (20) studied under Professor Jones and, after returning to Japan, made extensive investigations on the relationship between environmental factors and the occurrence of rice blast. There is no doubt that blast disease of rice plants caused by *Pyricularia oryzae* is the most important and destructive plant disease in Japan, being widely distributed throughout the country and causing great loss in yield. Abe (1) studied effect of soil temperature on development of blast disease. Prior to this work, Hori (30) had pointed out that blast disease was apt to occur more in places with cold running water or cold springs, where it was called "Hie-Imochi" or "Minakuchi-Imochi." Hemmi (20) further elucidated the relationships between occurrence of rice blast disease and environmental factors such as soil water, atmospheric humidity, and temperature. The Ministry of Agriculture and Forestry gave financial support for such epidemiological studies of blast disease, and cooperative experiments were conducted in both the Hokkaido and Kyoto Imperial Universities. This project contributed not only to understanding the basic phenomena associated with blast disease, but also to development of control measures of the disease.

### *Treatments for Rice Blast Control*

The environmental conditions in Hokkaido are unfavorable for cultivation of rice plants, as exemplified by the low soil and air temperatures, and peat-moss soil. Thus, rice crops there suffer from severe blast disease every year. Beginning in 1927, Ito and his co-workers in Hokkaido Imperial University and Hokkaido Agricultural Experiment Station concentrated on the problem of controlling blast disease. They concluded that the most effective protective measures were seed disinfection and treatment of diseased carryover rice straw by scattering them in fields and using a cover spray of Bordeaux mixture. Cultivation of resistant cultivars and soil improvement were also recommended. Ito (43) practiced his theory in the Sorachi district of Hokkaido in 1934 to prevent primary occurrence and spread of the disease. The result was very successful (51, 52) and may be the first case of using a cover spray in a rice field to successfully control blast disease. About 1949 organic mercury fungicide sprays replaced the Bordeaux mixture without noticeable spray injury (3).

### *Forecasting Plant Disease Outbreaks*

For effective control of plant diseases, it is necessary to estimate their occurrence as early as possible. For this reason, the forecasting of rice blast disease has been investigated since 1936 in the Nagano and Toyama prefectures (59). In 1939 the Ministry of Agriculture and Forestry established an organization for forecasting blast disease, in cooperation with both national and prefectural agricultural experiment stations. Many reports have been published on forecasting the disease, and most have discussed weather conditions and number of airborne spores in relation to prediction. Some papers approached this problem from the standpoint of the physiology of rice plants.

### *Advances in Virus Disease Investigations*

During the ten years after Takata (1859) first reported on the relationship between insects and diseases there were few noteworthy investigations on virus disease. However, in the beginning of the Showa Era (1927– ) virus diseases of plants became the subject of research because of their scientific interest and economic importance; dwarf and stripe diseases of rice plants, tobacco-mosaic disease, and others were brought to light by many investigators. Fukushi (14) demonstrated virus transmission through eggs of insect vectors in rice dwarf virus. After the second world war, electron microscopic techniques were introduced into various fields of biological sciences, and in virology these techniques helped elucidate the fine structures of viruses. Serological and immunological approaches to characterization of viruses have also been performed in many laboratories and institutes. However, an especially noteworthy event was the discovery of mycoplasma-like organisms in plants infected by yellows-type diseases, which had been hitherto considered to be virus diseases (12, 13).

Histopathological observations of diseased plants have also progressed. The *Kaibō Shokubutsu Byōrigaku (Anatomical Plant Pathology)* published by Yoshii & Kawamura in 1947 (105) is the first book concerning pathological anatomy of plants in Japan. After the second world war, electron microscopy was introduced in this research area, and extensive anatomical and cytological approaches to host-parasite interaction were made in cooperation with physiological and biochemical analyses. The mechanism of resistance has also been discussed from various viewpoints after the second world war. Studies of the host-parasite interaction also have been promoted from physiological and biochemical standpoints.

## CONCLUDING REMARKS

A long time was required to understand the nature of plant disease and that they were not attributable to God or unfavorable conditions. People always have observed diseased plants with surprise and terror. Scientific knowledge of the nature and cause of plant disease was promoted after the beginning of the Meiji Era (1873–1893). Several universities, colleges, and agricultural experiment stations, both national and prefectural, were established, and many scientific studies have been made.

Investigations on plant pathology recently reached the cellular level, especially in research on host-parasite interactions. However, there has not been much ecological observation in plant pathology, especially regarding the relationship between disease infection and microclimate. This relationship also may be important in understanding principles of disease control.

ACKNOWLEDGMENTS

I wish to express my sincere thanks to Dr. S. Ouchi, Okayama University, and Miss Kyoko Kawamura, Kyoto University, for their valuable suggestions and kind aid in the preparation of this manuscript.

*Literature Cited*

1. Abe, T. 1933. On the influence of soil temperature on the development of the blast disease of rice. *Shokubutsu Byogai Kenkyu* 2:30–54 (In Japanese)
2. Agricultural Experiment Station 1898. Experiments on loose smut control. *Nō-ji Shiken Seiseki* XII. 1:99–110 (In Japanese)
3. Akai, S. 1958. A view of the development of techniques for controlling crop diseases. *Nihon Nōgyo Nenpō* 8:109–20 (In Japanese)
4. Ando, K. 1910. On the dwarf disease of rice plants. *Dai Nihon Nokai Hō* 347:1–3 (In Japanese)
5. Anonymous 1683. *Den Bata Nandai Monogatari.* Cited in Ref. 71
6. Aoe, H. 1881. *Satsu-Gu Tabako Roku.* Kagoshima Prefecture. Cited in Ref. 71
7. Berkeley, M. J. 1878. Contribution to the botany of H.M.S. Challenger, 38. Enumeration of the fungi collected during the expedition of H.M.S. Challenger, 3. *J. Linn. Soc. Bot. London* 16:38–54
8. Brefeld, O. 1895. *Untersuchungen aus dem Gesamtgebiete der Mykologie* 12: 194–205
9. Brefeld, O., Falk, R. 1905. Die Blüteninfektion bei den Brandpilzen und die natürliche Verbreitung der Brand Krankheiten. *Untersuchungen aus dem Gesamtgebiete der Mykologie* 13:1–59
10. Cook, M. T. 1937. Insect transmission of virus diseases of plants. *Sci. Mon.* 44:174–77
11. Cooke, M. C. 1878. Some extra-European fungi. *Grevillea* 7:13–15
12. Doi, Y., Teranaka, M., Yora, K., Asuyama, H. 1967. Mycoplasma- or PLT group-like organisms found in the phloem elements of plants infected with Mulberry Dwarf, Potato Witches' broom, Aster Yellows, Paulownia Witches' broom. *Ann. Phytopathol. Soc. Japan* 33:259–66
13. Doi, Y., Ishiie, T., Yora, K. 1970. Discovery of plant pathogenic Mycoplasma-like organism. *Ann. Phytopathol. Soc. Japan* 36:146–48
14. Fukushi, T. 1933. Transmission of virus through eggs of an insect vector. *Proc. Imp. Acad. Japan* 9:457–60
15. Fukushi, T. 1935. The earliest record on the transmission of plant virus by insects. *Byōchu-Gai Zasshi* 22:38–45 (In Japanese)
16. Fukushi, T. 1937. An insect vector of the dwarf disease of rice plant. *Proc. Imp. Acad. Japan* 13:328–31
17. Fukushi, T. 1950. Advances in the study of plant virus diseases in Japan. *Miyabe Kingo Hakusi 90-Ga Kinen, Shokubutsu Gaku Senshu,* 77–86 (In Japanese)
18. Fukushi, T. 1965. *Relationships Between Leafhoppers and Rice Viruses in Japan.* Presented at Conf. on Relationships between Arthropods and Plant-Pathogenic Viruses, Tokyo, 1–21
19. Hara, K. 1918. *Monograph of Rice Diseases* 1–218. Revised ed. 1959, 1–139. Tokyo: Kōko-Dō Shoten (In Japanese)
20. Hemmi, T. 1933. Experimental studies on the relation of environmental factors to the occurrence and severity of blast disease in rice plants. *Phytopathol. Z.* 6:305–24
21. Hemmi, T., Akai, S. 1945. *Mycology of Wood Decay.* Tokyo: Asakura-Shoten (In Japanese)
22. Hemmi, T., Konishi, S. 1939. Studies on false smut of rice plants. *Byōchu-Gai Zasshi* 26:857–68 (In Japanese)
23. Hino, I. 1927. Early important records on phytopathological science in orient. I, II. *Agr. Hort.* 2:1223–32, 1327–34 (In Japanese)

24. Hino, I. 1928. Frank Maddox Shi no Nōgyō-Kai ni okeru Kōseki. *Agr. Hort.* 3:311–15
25. Hino, I. 1931. Memories in the inchoate stage of plant pathology in Japan. *Kin-Rui (Fungi)* 1:31–34 (In Japanese)
26. Hino, I. 1932. Genus *Miyoshiella* to be included in genus *Chaetosphaeria. Bull. Miyazaki Coll. Agr.* 4:187–92
27. Hino, I. 1949. *History of the Development of Plant Pathology.* Tokyo: Asukura-Shoten (In Japanese)
28. Hino, I. 1961. *Icones Fungorum Bambusicolorum Japonicorum.* Shizuöka: The Fuji Bamboo Garden. 335 pp.
29. Hori, S. 1890. Injurious fungi attacking rice plants. *Bot. Mag.* 4:425–27 (In Japanese)
30. Hori, S. 1898. Blast disease of rice plants. *Imp. Agr. Exp. Sta. Spec. Rept.* 1:1–36 (In Japanese)
31. Hori, S. 1898. Researches on "Bakanae" disease of rice plants. *Nōji Shiken Seiseki* XII. 1:110–19 (In Japanese)
32. Hori, S. 1903. *Nōsakumotsu Byō-Gaku,* 271–74. Tokyo: Seibi-Do
33. Hori, S. 1907. Seed infection by bunt fungus of cereals. *Bull. Imp. Cent. Agr. Exp. Sta.* 1:163–76
34. Hori, S. 1940. Reminiscences in plant medicine 50 years (1, 2) *Ann. Phytopathol. Soc. Japan* 10:72–75, 317–25 (In Japanese)
35. Horii, Y. 1819. *Nōjutsu Kōeki Roku.* Cited in Ref. 45
36. Ideta, A. 1930. A brief history of plant pathology in Japan. *Ann. Phytopathol. Soc. Japan* 2:197–206
37. Ishikawa, R. 1928. Contributions of Hatsuzo Hashimoto, the earliest investigator of dwarf disease of rice plants. *Byōchu-Gai Zasshi* 15:218–22 (In Japanese)
38. Ito, K., Shibukawa, K., Kobayashi, T. 1952. Etiological and pathological studies on the Needle Blight of *Cryptomeria japonica.* I. Morphology and pathogenicity of the fungi inhabiting the blighted needles. *Bull. Gov. Forest. Exp. Sta.* 52:79–152 (In Japanese)
39. Ito, K., Kobayashi, T., Shibukawa, K. 1967. Etiological and pathological studies on the Needle Blight of *Cryptomeria japonica.* III. A comparison between *Cercospora cryptomeriae* Shirai and *Cercospora sequoiae* Ellis et Everhart. *Bull. Gov. Forest. Exp. Sta.* 204:73–90
40. Ito, S. 1913. Effect of copper suphate on rice plants and experiments on the control of rice seed rot. *Sapporo Nōringakkai Hō* 20:563–609 (In Japanese)
41. Ito, S. 1917. Aka-Boshi-Byō Kin ni tsuite (On *Gymnosporangium* spp.) IV. *Byōchu-Gai Zasshi* 4:177–84
42. Ito, S. 1918. On the scientific name of fungi found in Japan, II. *Bot. Mag.* 32:303–8 (In Japanese)
43. Ito, S. 1933. Primary outbreak of the important diseases of rice plants and common treatment for their control. *Rept. Hokkaido Agr. Exp. Sta.* 28:1–204 (In Japanese)
44. Ito, S. 1940. A summary of history of the development of plant pathology in Japan. *Bot. Zool.* 8:153–57 (In Japanese)
45. Ito, S. 1943. *The Blast Disease of Rice Plants and Abstracts of Rice Blast Literature.* Tokyo: Yōkendō (In Japanese)
46. Ito, S., Kimura, J. 1931. Studies on the "Bakanae" disease of the rice plant. *Rept. Hokkaido Agr. Exp. Sta.* 27:1–94 (In Japanese)
47. Jensen, J. L. 1888. The propagation and prevention of smut in oats and barley. *J. Roy. Agr. Soc. Engl.* sec. 2, 24: 397–415
48. Jensen, J. L. 1889. *Le Charbon des Cereales.* Copenhagen
49. Jones, L. R. 1893. The loss from oat smut in Vermont in 1893. *Vermont Agr. Exp. Sta. Ann. Rept.* 7:60–65
50. Jones, L. R. 1895. Some observations regarding oat smut. *Vermont Agr. Exp. Sta. Ann. Rept.* 9:106–12
51. Kaneko, J. 1935. On the result of well-coordinated practice of rice blast control measures. *Ann. Phytopathol. Soc. Japan* 4:224 (In Japanese)
52. Kaneko, J. 1936. On the result of well-coordinated practice of rice blast control measures. *Noji Kairyo Shiryo* 108:1–10 (In Japanese)
53. Kawakami, T. 1901. On the blast disease of rice plants. *Sapporo Nōgakkai Hō* 2:1–47 (In Japanese)
54. Kawakami, T. 1901. On the scientific name of rice blast fungus. *Dai Nihon Nōkai Hō* 237:14–18 (In Japanese)
55. Kawamura, S. 1907. Ueber die Flecken- and Buntbambuse. *J. Coll. Sci. Imp. Univ. Tokyo* 23: Art. 2,1–11
56. Kawamura, T. 1941. A brief history of plant quarantine in ports in the interior of Japan (1, 2). *Byōchu-Gai Zasshi* 18:63–68, 130–39 (In Japanese)
57. Kojima, N. 1814. *Nōka Gyōji.* Cited in Refs. 45 and 27
58. Konishi, A. 1827. *Nōgy Yowa.* Cited in Refs. 45 and 27
58a. Kurosawa, E. 1926. Experimental studies on the secretion of *Fusarium hetero-*

*sporum* on rice plants. *Trans. Nat. His. Soc. Formosa* 16:23–27 (In Japanese)

59. Kuribayashi, K., Ichikawa, H. 1952. Studies on the forecasting of rice blast disease occurrence. *Nagano Agr. Exp. Sta. Bull.* 13:1–229 (In Japanese)
60. Kusano, S. 1906. Notes on the Japanese fungi IV. *Caeoma* on *Prunus. Bot. Mag. Tokyo* 20:47–51
61. Maddox, F. 1895. *Experiments at Eastfield.* Tasmania: Dep. Agr. Cited in Ideta, A. 1929. *Nihon Shokubutsu Byōri Gaku,* 350–51
62. Maddox, F. 1896. Eastfield experiments, smut and bunt. *Agr. Gaz. and J. Counc. Agr. Tasmania* 4:92–95
63. Maddox, F. 1897. Notes and results on agricultural experiments carried on under the auspices of the Council of Agriculture of Tasmania at Eastfield. Newnham, Launceston, Tasmania. pp. 72–84
64. Miyabe, K. 1940. Kai-Kyū-Dan (Reminiscences). *Ann. Phytopathol. Soc. Japan* 10:67–71 (In Japanese)
65. Miyanaga, S. 1788. *Shika Nōgyō Dan.* Cited in Ref. 45
66. Miyoshi, M. 1894. Ueber Chemotropismus der Pilze. *Bot. Z.* 52:1–28
67. Miyoshi, M. 1895. Die Durchbohrung von Membranen durch Pilzfäden. *Jahrb. Wiss. Bot.* 28:269–89
68. Miyoshi, M. 1895. A brief note of a research on the penetration of membranes by hyphae. *Bot. Mag.* 9:243–45 (In Japanese)
69. Mori, N. 1937. Outline of the plant quarantine. *Byōchu Gai Zasshi* 24:182–86, 225–30 (In Japanese)
70. Nakagawa, S. 1898. Studies on loose smut. *Noji Shiken Seiseki* XII, 4:55–58 (In Japanese)
71. Nakamura, H. 1951. Historical review of tobacco disease in Japan. *Shokubutsu Byōgai Kenkyu* 4:23–34 (In Japanese)
72. Nakata, K., Hino, I. 1938. Shokubutsu Byōrigaku Shi. *Shokubutsu Byōrigaku Taikei* I, 63–229. Tokyo: Yōkendō
73. Nisikado, Y. 1917. Studies on the rice blast fungus (1). *Ber. Ōhara Inst. Landw. Forsch. Kurashiki* 1:171–218
74. Nisikado, Y. 1932. Ueber zwei wirtschaftlich wichtige, parasitäre Gramineenpilze: *Lisea fujikuroi* Sawada und *Gibberella moniliformis* Wineland. *Z. Parasitenk.* 4:285–300
75. Ou, S. H. 1972. *Rice Diseases,* 289–95. Kew, Surrey: Commonw. Mycol. Inst. 368 pp.
76. Owens, C. E. 1928. *Principle of Plant Pathology.* New York: Wiley. 629 pp.
77. Patouillard, N. 1887. Contributions à l'études des champignons extra-européens. *Bull. Soc. Mycol. France* 3:119–31
78. Sakurai, M. 1934. On the causal fungus of false smut of rice. *Ann Phytopathol. Soc. Japan* 3:70–71 (In Japanese)
79. Sato, D., Yamada, K. 1896. Studies on loose smut of barley. *Kyoto-Fu Nōkai Hō* 49:1–9 (In Japanese)
80. Sawada, K. 1912. Seed and seedling rot of rice. *Spec. Bull. Agr. Exp. Sta. Formosa* 3:1–84 (In Japanese)
81. Sawada, K. 1917. Contributions to Formosan fungi. No. 14. *Trans. Natur. Hist. Soc. Formosa;* 31:128–35 (In Japanese)
82. Sawada, K. 1919. *Descriptive Catalogue of the Formosan Fungi.* Pt. 1, 251–57. *Spec. Rept. Govt. Res. Inst. Formosa,* No. 19. 695 pp.
83. Shiga Agricultural Experiment Station 1899–1908. *Reports of Experiment on Injurious Insects* II Rept. 1–8 (In Japanese)
84. Shirai, M. 1896. Researches on the control method of diseased rice plants. *Kan-Pō* 3785:333–35 (In Japanese)
85. Shirai, M. 1899. On the genetic connection between *Peridermium giganteum* Tubeuf and *Cronartium quercuum* Miyabe. *Bot. Mag.* 13:74–79
86. Shirai, M. 1899. On a pathogenic fungus causing galls on genus *Pinus* found in Japan. *Bot. Mag.* 13:153–58 (In Japanese)
87. Shirai, M 1900. Ueber den genetischen Zusammenhang zwischen *Roestelia koraiensis* P. Henn. und *Gymnosporangium japonicum* Sydow. *Z. Pflanzenkr.* 10:1–5
88. Shirai, M. 1903. *Saishin Shokubutsu Byōri Gaku* (New Plant Pathology). Revised ed. 1931. Tokyo: Suzanbō
89. Shirai, M. 1914. Unrecognized service in plant protection. *Byōchu-Gai Zasshi* 1:17–20 (In Japanese)
90. Shirai, M. 1914. *Shokubutsu Yōi Kō (Study on Plant Teratology).* Tokyo: Oku-Shoin (In Japanese)
91. Shirai, M. 1918. On the development of plant pathology in Japan. A brief historical sketch. *Ann. Phytopathol. Soc. Japan* 1:1–4
92. Shirai, M. 1924. On mosaic disease of tobacco. *Rigaku Kai* 22:289–91 (In Japanese)
93. Takahashi, Y. 1896. On *Ustilago virens* Cooke and a new spicies of *Tilletia* parasitic on rice plant. *Bot. Mag.* 10:16–20

94. Takata, K. 1895, 1896. Results of experiments on dwarf diseased rice plants. *Dai Nihon Nōkai Hō* 171:1–4, 172: 13–32 (In Japanese)
95. Tamari, Y. 1888. Downy mildew attacking cucumber leaves and its control method. *Dai Nihon Nōkai Hō* 85:11–17 (In Japanese)
96. Tamura, N. 1841, 1857. *Nōgyo Jitoku* I, II. Cited in Ref. 27
97. Tanaka (Ichikawa), N. 1888. A parasitic fungus on cucumber. *Bot. Mag.* 2:143 (In Japanese)
98. Tanaka, N. 1889. Smut fungi parasitic to cereals (continued). *Bot. Mag.* 3:169–70, 197–98 (In Japanese)
99. Terashima, R. 1713. *Wakan Sansai Zue.* Reduced size ed. 1906. Tokyo: Yoshikawa Kōbun-Kan
100. Tomonaga, T. 1966. Chronological history of the present agriculture in Fukui prefecture. *Fukui no Nōgyō* (Agriculture in Fukui) 3:69–75, 4:73–77, 5:90–95, 6:88–94, 7:58–65
101. Tsuchiya, M. 1707. *Kōka Shunju* 4. Cited in Ref. 45
102. von Tubeuf, K. 1895. *Pflanzenkrankheiten durch kryptogamen Parasiten verursacht.* p. 429. Berlin: Springer
102a. Yabuta, T. 1935. Biochemistry of the "bakanae" fungus of rice. *Agr. Hort.* 10:17–22 (In Japanese)
102b. Wollenweber, H. W. 1931. Fusarium-Monographie. Fungi parasiticiet saprophytici. *Z. Parasitenk.* 3:269–516
103. Yamanouchi, T. 1934. On the cradle period of Japanese history of phytopathology. *Ann Phytopathol. Soc. Japan* 3:42–55 (In Japanese)
104. Yamashita, S. 1965. Detection of alkaloids in the unhulled rice infected by the false smut fungus of rice. *Ann. Phytopathol. Soc. Japan* 30:64–66 (In Japanese)
105. Yoshii, H., Kawamura, E. 1947. *Anatomical Plant Pathology* Tokyo: Asakura-Shoten (In Japanese)
106. Yoshii, H., Hamamoto, S. 1957. On the hot bath water treatment of wheat seed for the control of loose smut, *Ustilago tritici* Rostrup. *Ann. Phytopathol. Soc. Japan* 22:169–72

# ASSESSMENT OF PLANT DISEASES AND LOSSES

❖3586

*W. Clive James*[1]
Ottawa Research Station, Agriculture Canada, Ottawa, Ontario

Plant diseases were initially studied because of the losses they cause, yet today it is paradoxical that there are only a few reliable estimates of loss. In 1918 Lyman (67) stated, "How can we expect practical men to be properly impressed with the importance of our work and to vote large sums of money for its support when in place of facts we have only vague guesses to give them and we do not take the trouble to make careful estimates." During the last decade we have witnessed more interest in disease-loss appraisal because of the need to develop more sophisticated disease management schemes that must justify and rationalize the use of fungicides. Funding organizations have also been operating on limited budgets and have required more convincing evidence before supporting research programs. International concern about the inadequacy of present methods for assessing diseases and estimating losses resulted in FAO's sponsoring a Symposium on Crop Losses in 1967 (26) which recommended development of more precise methods (27), so that the limited expertise available could be deployed in the most efficient way. In a world conscious of pollution, loss estimates may achieve new significance by providing evidence that will justify or condemn the use of fungicides to control epidemics. To develop rational and economical control measures, whether by breeding resistant cultivars or using fungicides, it is not sufficient to state that disease causes a loss; the magnitude of the loss must be evaluated so that it can be related to the gain obtained. Only by disease-loss appraisal is it possible to determine the economic loss due to different amounts of disease. Disease-loss appraisal therefore represents an absolutely essential step because until economic loss can be measured, it is not possible to implement disease or pest management schemes aimed at economic control.

The general strategy of disease-loss appraisal has been described by Large (63) and others (15, 26) and involves two phases. Field experiments are conducted in the first phase to characterize the relationship between disease and loss in yield, so that

[1]Contribution No. 370, Ottawa Research Station, Agriculture Canada, Ottawa, Ontario.

a reliable method can be developed to estimate the loss in yield associated with any given amount(s) of disease. It involves studying the disease throughout the season as well as monitoring growth of healthy and diseased plant populations. Growth-stage keys (28, 62) are prepared to identify the different plant growth stages, and disease assessment methods (1, 19, 40, 100) are developed to measure disease. The disease assessment keys illustrate or describe the appearance of a plant organ (40), whole plant (100), or crop (1) when infected by different amounts of disease. Irrespective of which disease assessment method is used, the keys must be designed so that various assessors record similar values for the same diseased crop. A series of field trials is then conducted, and in each trial the disease is allowed to develop in some plots but is controlled in others. Disease is assessed regularly throughout the season, and if necessary, disease progress curves can be generated. The final step is to develop a reliable method for estimating yield loss for any given amount of disease (22, 46, 56, 64, 84, 94) or a given progress curve (10, 51). For practical purposes, a method must be applicable under a range of conditions, and exceptions must be noted for the necessary corrections to be applied. The second phase involves assessing the disease in a survey of a number of fields using the assessment method developed in the first phase. By knowing the quantitative relationship between disease and yield, the loss can be calculated from the disease data recorded in the survey (38).

In 1950 Chester published an extensive supplement (15) on disease-loss appraisal, and a later review by Large in 1967 (63) dealt primarily with the progress made in disease assessment in the United Kingdom since 1950. Several reviews were also presented at the FAO Symposium in Rome in 1967 (26). The first chapter of a recent textbook (79) was devoted to disease assessment. The present paper reviews recent advances in phase one—involving the development of methods for assessing loss—and presents a critical discussion of selected references, which hopefully may serve as a guide for future research. Most of the recently published work on methodology of disease assessment concerns foliage diseases, particularly of cereals and potatoes, and the content of this review reflects this imbalance.

## DEVELOPMENT OF A METHOD FOR ESTIMATING LOSS

The goal in this phase is to measure disease and develop a reliable method for translating this into loss. In this context loss is defined as the measurable reduction in yield and/or quality, but as Ordish & Dufour (76) pointed out, this loss may not always be the same as economic loss. Although field experiments are inherently more difficult to conduct than comparable experiments in growth rooms or glasshouses, they are the only available method for studying the effect of disease on field crops under farming conditions.

### *Location, Design, and Specifications of Field Experiments*

Ideally, identical experiments should be conducted in all geographical areas where the crop is important, over a period of at least 3 years, using the major cultivars under the range of conditions found under normal farming practice. Some experiments have featured paired plots (64, 90), e.g. sprayed and unsprayed or isogenic lines, and disease loss is calculated as the difference in yield between the two treatments expressed as a percentage of the yield on the healthy plot. This design is inferior to the multiple-treatment experiment (51, 84) with several amounts of disease and one healthy treatment, which allows comparison of the effect of different types of epidemics at one location. Standard experimental designs, including randomized block and split plot, are usually used, and the two main variables recorded are disease and yield.

The experimental design, and shape and size of plots, are all factors affecting the precision of yield estimates. The desired precision must be fixed before designing the experiment, and its level should be realistic and based on practical criteria, e.g. the cost of a fungicide application. Plant pathologists assessing losses have not given enough consideration to experimental specifications and have a tendency to use the same plot sizes as their plant-breeding colleagues; the latter often require only a ranking of treatments, whereas the pathologist must know whether a particular difference between treatments is significant. Although the choice of plot size and shape are influenced by many factors including labor and availability of land, these factors should not influence our choice to such an extent that, even before the experiment is begun, it is known that there is a low chance of obtaining the necessary precision. Recently, James & Shih (44) showed that the rod row plot (3 rows, where 16 ft (4.88 m) of the center row is harvested), normally used by cereal pathologists in North America, is too small to allow a reasonable chance of detecting a 10% difference between treatments in experiments for assessing losses due to foliage diseases. Unfortunately, the literature contains very few references on uniformity trials on healthy and diseased plants, and there is little information to guide us in determining the optimum plot size for a given level of precision. Disease considerations can also affect choice of plot size; research on epidemiology of stem rust showed that a minimum plot size of 1000 sq ft (1/40 acre = 93.4 $m^2$ = 1/100 ha) is required to create an epidemic with characteristics similar to that in a large field (Kingsolver 1971, personal communication).

In any design, a treatment with no disease is always necessary to establish the yield of a healthy crop, but in both paired-plot and multiple-treatment experiments interplot interference (16, 50, 101) may result in healthy plots' becoming infected and smaller differences than intended between epidemics in multiple-treatment experiments. Large (63) rightly concluded that interplot interference is not important in disease-loss experiments when the control plots are free from disease (74), but many experimental results (49, 60, 64) show that even with many fungicide sprays, complete control is not possible. Systemic fungicides will obviously suppress interplot interference more than eradicant or protectant fungicides, but even systemics will not give complete control throughout the life of some crops (8). The use of isogenic lines (90, 93) where one line is completely resistant circumvents the interference problem but, in practice, isogenic lines often are not readily available. The cultivars normally used in experiments are susceptible to the disease, and because a healthy plot is essential it is not possible to use van der Plank's solution (101) of minimizing interference by allocating only treatments with similar epidemics to one experiment. A partial solution to this problem may be to use large square plots in a straight line with buffer areas between plots (44, 50). Ideally the buffer zones would contain a crop or cultivar different from the one under study. When dealing with aerial pathogens that can travel many kilometers, single or double guard rows in plots will obviously not hinder the movement of spores; to serve as true guard rows it is very important that they are subject to the same epidemic as the rows designated for yield determinations. Failure to do this may result in competition between guard rows and the rows designated for harvest and will result in invalid treatment comparisons (44).

### *Measuring Yield and Quality*

When suitable experimental specifications to detect a given yield difference between treatments have been chosen, yield and quality should be estimated by the same harvesting techniques and grading systems used by farmers. The adoption by research workers of experimental harvesting techniques, which are more efficient than the commercial harvest methods used by farmers, leads to an underestimate of the actual loss to the farmer (83); this is a common source of error because many diseases affect both yield and quality. Calpouzos et al (13) determined the effect of harvesting technique on recovery of wheat from experimental plots with and without stem rust. They found that the average loss in yield due to machine harvest, as compared with hand harvest, was 13% greater in disease-free plots and 34 and 19% greater for early and late epidemics, respectively. The percentage loss for machine harvest increased with earliness of epidemic, probably because the machine was not as efficient as hands in harvesting the grain, particularly the lighter kernels produced on diseased plants (84). This does not necessarily mean that the hand method should not be used. On the contrary, it provides a better estimate of the actual or "biological" decrease in grain weight, and accordingly it should provide better quantitative data for interpreting the yield decrease relative to the amount of leaf damage. Indeed there is a need for both types of estimates, and both can usually be obtained in the same experiment (83).

### *Disease Assessment*

Diagnosis and assessment of plant disease are equally important functions of plant pathologists. Diagnosis of the more common diseases is based on identification of pathogen and/or symptoms using methods universally known and accepted. By comparison, disease assessment methods have received much less attention, and many published methods are so difficult to apply with uniformity that they are not widely accepted. Many different terms have been used to define disease measurement (15) and endless but fruitless discussions usually result when considering the relative merits of the terms.

In this paper, *disease incidence* is defined as the number of plant units infected, expressed as a percentage of the total number of units assessed, e.g. percentage of diseased plants or leaves, etc.; *disease severity* is defined as the area of plant tissue affected by disease expressed as a percentage of the total area (45). The FAO Crop

Loss Manual (27) uses the term "intensity" to cover either disease incidence or severity.

The only disease-assessment methods that have been applied with adequate uniformity in practice are those that estimate incidence. These methods are usually used for systemic infections, e.g. virus or fungal wilts, or when diseased plants are total losses (14, 29) rather than partial losses (84, 94). The various methods developed to estimate disease severity reflect the diverse interests of specialists studying disease-loss appraisal, epidemiology, or disease resistance.

THE SPECIFICATIONS OF A DISEASE-ASSESSMENT METHOD In the context of loss appraisal, diseases are assessed only as a means to measure and prevent losses. The total leaf area affected by disease, i.e., pustules or lesions, including any accompanying chlorosis, necrosis, or defoliation (1), is likely to be better correlated with losses in yield than with pustule or lesion area alone (40). Samborski & Peturson (88) neatly demonstrated the significance of recording chlorosis or necrosis when they reported substantial losses in yield of wheat cultivars hypersensitive to leaf rust. These cultivars reacted to the disease by forming a necrotic or chlorotic area rather than a sporulating pustule. Accordingly total or percentage leaf area affected, rather than pustule area, is normally used in studies of disease-loss appraisal (40, 46, 64); nevertheless, the standard area diagrams used (40) often show only definite lesions or pustules because it is impractical to illustrate variable chlorosis. This leaf-area approach allows a whole leaf lamina to be recorded as diseased, if a lesion on the petiole or at the junction of lamina and sheath incapacitates the total leaf area.

Many workers (15, 40, 63) have suggested that whenever disease assessments are recorded, the growth stage of the crop should be noted according to a published growth-stage key (62). Similarly if particular organs have been assessed, e.g. a particular leaf or fruit, these should be recorded so that meaningful comparisons can be made at a later date. The simplest disease-assessment method is usually the one least prone to error; an example of such a method is the assessment of disease on individual cereal leaves (40). Each disease present is assessed individually, and because the observer is assessing one disease on one leaf at a time, the error attached to any particular assessment is smaller than when many leaves have to be assessed at the same time, as for example in the Large & Doling cereal-mildew key (64). This field key for cereal mildew describes seven amounts of disease in terms of the percentage lamina affected by mildew on the top four leaves and assumes that the epidemic always develops according to a standard pattern. Such a key is difficult to use if only a fraction of the tillers are infected or if they are infected to varying degrees; further complications arise if the relative infections on the four leaves do not conform with those described in the key. The assessment of one or two leaves is often adequate for cereals; for scald of barley (*Rhynchosporium secalis*), for example, an individual assessment of the top two leaves when the grain is in the "milky ripe" stage of growth provides a reliable estimate of loss because these two leaves produce most of the dry matter in the grain (46). This demonstrates the relevance of securing plant physiological data on the contribution of each plant organ to yield before the specifications of a disease-assessment method are made final. This physiological data also should help solve the problem of developing a rational weighting system to account for the relative importance of symptoms on different plant organs (106). In cereals the individual leaves and growth stages are easy to designate, and assessment of particular leaves therefore presents no problem; but the task is much more difficult for potatoes where growth tends to lack any distinct growth stages except flowering and the meeting of plants in the rows. Epidemics of late blight of potatoes cannot be typified by a standard pattern, and the late-blight assessment key of the British Mycological Society (1) refers only to different amounts of disease as it applies to the whole crop.

Although many arbitrary indices and rating systems have been used (15) to measure disease, their use should be discouraged in favor of percentage scales of disease incidence and disease severity as defined earlier. Only in this way will adequate standardization and consistency of measurements be achieved.

THE PERCENTAGE SCALE The percentage scale should be adopted as a standard because it has many advantages (40): (*a*) the upper and lower limits of a percentage scale are always uniquely defined; (*b*) the scale is flexible in that it can be divided and subdivided conveniently; and (*c*) it is universally known and can be used to record both the number of plants infected (incidence) and the area damaged (severity) by a foliage or root pathogen.

The standard area diagrams developed by Melchers & Parker (71), known as the modified Cobb Scale, labels the maximum possible amount of rust as 100%, but the actual area occupied by the pustules is only 37%. James (40) developed a series of standard area diagrams where the percentage of infection recorded always represents the actual area covered; the keys are also published in a loose-leaf plastic manual (41) that is suitable for rigorous field use. Although 100% infection may never be encountered in practice, when using James' system this is not a disadvantage, because there is no reason to use 100% as a maximum when relating disease incidence or severity to yield loss. Furthermore, in the event that stem rust did occupy more than 37% of the actual area on a particular cultivar, there is no way of recording the observation on the modified Cobb Scale, whereas this presents no problem when the actual area covered is recorded.

All disease assessments are subjective to some extent because they are the results of visual judgments. Some years ago Horsfall & Barratt (35) pointed out that the grades detected by the human eye are approximately equal divisions on a log scale, and generally follow the Webber-Fechner Law which states that visual acuity depends on the logarithm of the intensity of the stimulus; they described twelve grades: 1 = 0%, 2 = 0–3%, 3 = 3–6%, 4 = 6–12%, 5 = 12–25%, 6 = 25–50%, 7 = 50–75%, 8 = 75–87%, 9 = 87–94%, 10 = 94–97%, 11 = 97–100%, 12 = 100% disease. They further noted that when making a percentage disease assessment, the eye actually assesses the diseased area up to 50% and the healthy area above 50% (35); thus the 12–15% grade represents an area of diseased tissue equal to the area of healthy tissue in the 75–87% grade. The method should be used in conjunction with standard area diagrams, which are necessary to define specific levels of disease.

From the standpoint of epidemiology and disease appraisal the use of a logarithmic scale is satisfactory because most pathogens multiply at a logarithmic rate while time advances at a linear rate. For example, the increase from 1 to 10% disease severity may take approximately the same time as the increase from 10 to 100%; the log scale is very appropriate in this case because it too attaches the same importance to the increases from 1 to 10 and from 10 to 100. The log scale is also appropriate for loss appraisal if the relationship between disease and loss in yield can best be characterized by a linear relationship using a log scale for disease and a linear scale for loss, for example, late epidemics of stem rust on wheat (84) and epidemics of cereal mildew except at very low levels of disease (64). The log scale is not suitable when the relationship between disease and yield is arithmetic, i.e. when it can be described by a straight line with linear scales for both disease and time (46). In this particular case, 100% disease gives 10 times the loss of 10% disease; in order to estimate loss with equal and reasonable levels of precision at any level of disease, more subdivisions are needed for the middle categories of Horsfall and Barratt's scale or preferably equal divisions on a linear scale. Chester (15) also noted this deficiency and stated the dilemma to be that the eye could not distinguish the equal divisions on the linear scale that are necessary to achieve the desired precision.

I have observed that, although it may not be possible for assessors to distinguish equal divisions on a linear scale, they can, with the aid of standard area diagrams, often do better than when attempting to record equal divisions on a log scale. By using their ability to judge relative differences in disease, assessors can decide whether a certain leaf is nearer to 50 than 75% or whether it is halfway between the two levels, i.e. approximately 60%. The extent of interpolation should be dictated by the ability of the assessor to detect differences in amount of disease (40), but it is a practice to be recommended because it allows observed differences to be recorded and used. Although the time frame requires assessments to be made before the disease-loss relationship is known, it is always advisable to use a scale with equal divisions on a linear scale, because even if the relationship is shown to be logarithmic, the transformed assessments will still be better than those using equal divisions on a log scale.

REMOTE SENSING TECHNIQUES For the future the whole problem of subjective disease assessments may be aided by the use of microdensitometers (37) and electronic scanners (103) to quantify disease severity as exhibited on infrared (IR) aerial photographs; this technique may even allow previsual disease detection (6). The use of remote sensing to quantify late blight severity illustrates the value and limitations of the technique. The healthy potato crop provides a full chlorophyll canopy, which reflects IR, and when late blight destroys the chlorophyll the decrease in reflectance is measured and can be converted to a measure of disease (34, 37). If the canopy is not full, the exposed soil will also lead to lack of reflectance, and this may be difficult to distinguish from lack of reflectance due to disease. Late blight usually starts on the lower leaves which are covered by the healthy top canopy and therefore not detectable on IR film (37). But this limitation is not likely to be important in surveys designed to monitor the development of the disease, because the disease first appears in foci, which are easily detectable on aerial photographs. The major advantage of this technique is that it allows rapid and reproducible disease assessments to be recorded, which in turn allows whole plots or fields to be scanned, thereby removing the sampling errors normally associated with conventional sampling procedures and disease-assessment methods. Reproducible results are not necessarily the most accurate. When disease cannot be detected on lower leaves, for example, this inherent systematic error in the technique may preclude its use. Compared with visual assessments, remote sensing offers the advantage that disease severity, the independent variable ($X$), is recorded without observer error; this is especially important for regression analysis (45).

INDIRECT METHODS OF MEASURING DISEASE SEVERITY An alternative to disease assessment, based on measurements of symptoms, is to count the spores of the pathogen. Zadoks (106) justifies this approach partly on the logical grounds that one spore may cause one infection and partly on technical grounds that the method is fast, accurate, and nondestructive. For the cereal rusts, several workers (23, 24, 85) have demonstrated that cumulative spore counts are related to disease severity. But more recent work has shown that prediction equations with severity of leaf rust (9) or severity of stem rust (25) as the disease variable were more accurate than those with numbers of spores as the disease variable.

The estimation of low disease severities can often be time consuming because the distribution of disease within the crop is not regular; this often leads to higher standard errors on the mean estimate. Several workers have suggested easier but indirect ways of estimating low severity by relating incidence to severity. James & Shih (45) have used a linear equation to describe the relationship between incidence and severity of leaf rust and mildew on particular leaves of winter wheat. They found that linear regression was adequate for estimating severity for incidence values of up to 65%. A semilog equation could also be used to estimate severity for incidence values of up to 90%, but large errors were possible for incidence values above 65%. Linear regression offers a fast and simple way of assessing severity if the incidence is below 65%; the time saved could be used to examine a larger sample of leaves. Rayner (80) employed a similar principle to estimate the severity of coffee rust from incidence data. He showed that there was an exponential relationship between the percentage of infected leaves (incidence) and the number of rust lesions/100 leaves (severity). Furthermore, because disease incidence also is related to foliage density which in turn is related to crop yield, the estimate of percentage of leaves infected may provide a good estimate of losses due to disease.

Strandberg (98) devised more efficient methods for assessing average infection rates for cabbage black rot by characterizing the spatial and temporal distribution of lesions. Since the patterns of distribution of infected plants and lesions within infected plants were highly aggregated, he devised a modified sequential-sampling method which saved much time and effort.

### *Methods for Generating Epidemics with Different Characteristics*

To study the effect of different amounts of disease on yield, it is necessary to include plots with varying amounts of disease in one field experiment. Epidemics from natural inoculum are preferable to those arising from artificial inoculum, but because the former are difficult to predict, many workers (49, 84) have used artificial inoculation to increase their chances of success in experimentation. Care should be taken to apply the inoculum at a growth stage when natural infection occurs, and preferably to the buffer rows between plots to facilitate "natural spread" to neighboring plots. To obtain different amounts of disease after the epidemic has been initiated it is necessary to delay the progress of the epidemic in some plots. Fungicide and bacteriocide sprays and the differential susceptibility of cultivars to disease are the two main techniques employed. Sprays probably provide a more flexible tool than cultivars for generating a particular disease curve, but both techniques can be used simultaneously (46) to provide a greater range of disease. However, it is important that, irrespective of the technique used, the only effect is to promote or delay the disease and that it does not lead to secondary effects that may invalidate comparisons among treatments.

FUNGICIDES Fungicides have been commonly used to inhibit or delay the progress of certain foliage diseases (49, 64), and Romig & Calpouzos (84) varied both the number and timing of maneb sprays to generate epidemics of wheat stem rust with different characteristics. Economic control of disease is not the aim in experiments in disease-loss appraisal; thus any spray schedule can be used that will allow one treatment to be kept free from disease.

Unfortunately, complete control often is not possible because most of the fungicides used are protectant, rather than eradicant or systemic in action, and the unprotected foliage produced between spray applications often becomes infected. Furthermore, because a plot is always present in which the disease is allowed to develop without hindrance, there is also the problem of interplot interference (healthy plots becoming infected) referred to earlier. If complete control of disease is not achieved, methods must then be found to estimate the yield of the plots had disease not been present. Some workers (64) have assumed that a small amount of disease does not materially affect yield and have used the yield of the sprayed plots to represent that of the healthy crop. Others (46, 94) have characterized the relationship between yield and percentage loss by a linear regression and assumed that the yield of the average healthy crop is the point at which the regression line cuts the yield axis. Increased use of eradicant and systemic fungicides should help to achieve complete control, although even with systemics (8) some infection has been noted at the end of the season.

When the corresponding yields from sprayed and unsprayed plots are compared, there is always the possibility that the decrease or increase in yield in the sprayed plot can be attributed to either a phytotoxicity (86) or a beneficial effect (12) of the fungicide over and above its effect on disease. The possible beneficial effects due to fungicides increasing the persistence of green tissue in cereals has been noted by many workers (46, 53, 72). Increased yield was reported for potato crops sprayed with zinc-containing fungicides (12, 36), and Rosser (86) attributed a yield loss of 3% to copper toxicity. James et al (49) tried to overcome this problem by applying a fungicide that contained no nutritional elements and confirmed that in the absence of disease there was no significant difference in yield between sprayed and unsprayed plots. A problem arising with wide spectrum fungicides, like maneb, is that they may also control several diseases at the same time, and the only solution is to assess each disease separately (40) and correct the results accordingly (46). On the other hand, the healthy green leaves produced with protection from the maneb spray provides more leaf tissue for infection than the unsprayed plots, and consequently diseases not controlled by maneb may become established more rapidly in the sprayed than unsprayed plots. Fungicides that are specific for certain diseases, like ethirimol for barley powdery mildew, do not have the disadvantage that they control other diseases simultaneously, and although they provide a healthy surface for other pathogens to infect, Brooks' data (8) showed that other diseases did not increase. Another problem associated with spraying occurs if a sprayer has to be driven through a plot, where wheel damage can result in additional loss (86). This can be solved either by driving a nonoperative sprayer through all the plots when any plots are sprayed or by designing the experiment so that all the plots can be sprayed from the perimeter (49).

ISOGENIC LINES Isogenic lines with differential susceptibility to disease offer an excellent way to obtain different amounts of disease in loss-appraisal experiments (90, 93). Isogenic lines are two cultivars similar in every respect except that one is susceptible to the disease and the other resistant. They can be produced either by continuous selection of plants heterozygous for resistance to a particular disease in successive generations or, preferably, by backcrossing for seven generations or more (90). The resistant isogenic line is invaluable, because some plots can be sprayed and others left unsprayed to check the phytotoxic or beneficial effect of the fungicide. Furthermore, different spray schedules can be used with the susceptible line to produce different epidemics in a multiple-treatment experiment; in the absence of disease, equal yields for both lines prove their isogenic character (90).

Unfortunately, there has been little cooperative effort between plant breeders and pathologists to produce isogenic lines. Schaller (90) demonstrated the value of the technique when studying losses due to powdery mildew and scald of barley. He did not use fungicides to vary the amount of disease within a trial site but relied on the diversity of the 25 cultivar trials to produce the different amounts of disease. Schaller estimated the severity of scald by assessing the percentage of total leaf area affected by disease and showed that 75% disease severity resulted in a 35% loss in yield. Although he had an excellent opportunity to characterize the relationship between any level of mildew and yield loss, Schaller did not develop a disease-loss model because he used an inadequate disease-assessment method; he merely classified mildew infection as light or heavy which were associated with losses in yield ranging from 3.8 to 17.6%.

CULTIVARS In the absence of isogenic lines, a group of cultivars can be used which vary in susceptibility to the disease under study but whose potential yields under disease-free conditions are fairly similar (46). Characterization of the relationship between amount of disease and yield should not be affected by the small differences in potential yield between cultivars except in the unlikely event that the potential yields under disease-free conditions are highly correlated with disease susceptibility. The advantage that cultivar comparison offers is elimination of possible secondary effects of fungicide. Cultivar trials have been used in conjunction with fungicides (46) to study the relationship between disease and yield loss for scald of barley. The results indicated that the relationship was the same whether cultivars or fungicides were used to vary amount of disease, thus providing circumstantial evidence that both were satisfactory techniques.

If the concept of disease tolerance, as defined by Schafer (89), proves to be real and widespread among cultivars, it will not be practical to develop disease-loss models because each cultivar exhibiting tolerance would have to be characterized separately. It is pertinent, therefore, to consider briefly the evidence for tolerance, defined as "the same epidemic resulting in different percentage yield loss for different cultivars." Tolerance can be expressed only when loss is not a direct function of severity, and most of the evidence is from foliage diseases of grain crops (89). On the one hand there seems to be good evidence for tolerance: e.g. Caldwell et al (11) showed that the oat cultivar, Benton, was more tolerant of crown rust than the cultivar Clinton 95; this was substantiated by Simons (92). On the other hand, when Torres & Browning (99) further investigated some aspects of tolerance to crown rust by improving the disease evaluation techniques, they found that in some cases what had been presumed to be tolerance was in fact slow rusting, i.e. some form of disease resistance. Similarly Brönnimann (7) found that when ten cultivars of wheat were assessed for *Septoria nodorum* at the end of the season the disease levels for cultivars were the same but the losses in yield were different, implying tolerance. But assessments made earlier in the season showed that cultivars differed in disease levels, and these were better correlated with the final losses in yield than with the final disease assessments. It may well be that disease assessments made at one point in time are inadequate for evaluating tolerance and that the characterization of the complete epidemic is necessary. Indeed, the above results of Torres & Browning and Brönnimann could be interpreted as evidence that tolerance simply reflects inadequate methods of disease assessment. It is not possible to conclude whether tolerance constitutes a problem in the development of models for disease-loss appraisal, because tolerance itself is difficult to identify and evaluate.

OTHER MEANS OF GENERATING OR SIMULATING EPIDEMICS For diseases that cause total loss of single plants or any definitive plant unit, e.g. tiller, the effect of disease can often be simulated by removing the plant or plant unit respectively, at a growth stage when natural infection occurs. Hirst et al (32, 33) simulated misses in potato crops by removing plants at emergence and recording the corresponding yield decreases. James et al (47) removed potato plants at random at emergence and at later stages to simulate diseases, like verticillium and rhizoctonia, and showed that losses in yield were not proportional to the percentage of missing plants, indicating that compensation had taken place. For any given percentage of misses many patterns are possible, each associated with different losses in yield; the higher the aggregation of missing plants, the higher the loss. Removal of plants has also been used to simulate virus infection with virus Y and leaf roll in potatoes; Reestman (81) has comprehensively reviewed this literature.

The effect of latent viruses X and S in potatoes was investigated by Kassanis & Varma (55) who produced virus-free clones of X and S by culturing apical meristems that did not include the viruses. These viruses can also be removed by heat treatment and excision of axillary buds (95); Wright (105) has compared tuber yields from virus-free and diseased lots to study the effect of virus X and S on yield. Seed potato stocks free from major virus diseases have been produced for many years, but recently the practice of producing seed from stem cuttings has resulted in seed stocks substantially free from tuber-borne diseases, like skinspot (*Oospora pustulans*) and blackleg (*Erwinia carotovara* var. *atroseptica*). From a disease-appraisal standpoint tubers from virus-free stem cuttings provide the basis for an excellent model where the yield of healthy stock can be compared with that of the same stock when diseased (32). The benefits, in terms of increased yield, accruing from freedom of X virus has been investigated in many countries (2–5, 18, 55, 91, 96, 105) using three methods: comparing two different clones (one virus-free and the other virus-infected); comparing the virus-infected stock with the same stock propagated from virus-free stem cuttings; and comparing the two stocks from the same clone propagated from virus-free stem cuttings but where the virus is reintroduced to one stock. The disadvantage of the first method is that the yielding capacity of the two clones may be different, which will not allow a valid yield comparison. Results from the second method estimate increase in yield due to the absence of both virus X and the tuber-borne fungal and bacterial diseases, which are also eliminated by the stem-cutting technique. It follows that some of the increase in yield attributed to freedom of virus X may be due to freedom from tuber-borne diseases like skin spot which Hide et al (31) has shown to be important. The third method allows the best comparison because the two stocks are treated in exactly the same way, and introduction of the virus is the only difference between the two stocks. I am currently using the third method to estimate the losses from viruses S and X and skinspot when these diseases are present singly or in combination with each other. The technique offers many possibilities because it allows estimation of the following: potential yield under disease-free conditions, yield decrease associated with a single disease, and yield decreases associated with more than one disease; it also provides an opportunity to study interaction effects between diseases, if they are present.

Whereas fungicides and cultivars provide a flexible tool for generating different epidemics of foliage diseases, the same cannot be said for root diseases. Indeed until such tools are developed, we can expect little progress with root diseases because the problems are numerous and difficult. For the foot and root rot of cereals, workers have resorted to different crop rotations to provide different amounts of inoculum (94) or have selected plants or areas with different amounts of disease (65, 68, 87) with the knowledge that cause and effect are difficult to separate.

### *Models for Estimating Yield Losses*

Mathematical models of the relationship between disease and yield loss can be divided into two groups. The critical-point models provide estimates of loss for any given level of disease at a given time (46, 56, 64, 84, 94) or any given time when a particular amount of disease is reached (61, 75). Multiple-point models estimate loss for a disease-progress curve consisting of many (multiple) disease assessments (10, 51).

CRITICAL-POINT MODELS Linear regressions often are used to characterize critical-point models where the independent variable is disease measurement and percentage loss in yield is the dependent variable. Examples of critical-point models for foliage diseases of cereals are as follows. Katsube & Koshimizu (56) estimated loss in yield due to rice blast by the equation $Y = 0.57X$ where $Y$ is percentage loss in yield and $X$ is percentage blasted neck nodes 30 days after heading. For powdery mildew of barley, wheat, and oats (63) the loss in yield was estimated with the formula, mildew severity (percentage) ½ X 2.5, 2.5, and 2.0 respectively, at growth stage 10.5 (62). James et al (46) showed that loss in yield due to barley scald was equivalent to two-thirds and one-half of the percentage disease on the flag and second leaf respectively, at the milky ripe stage of growth. The predicted loss was the average of the two estimates. Romig & Calpouzos (84) reported that the best estimate of loss due to stem rust of wheat was based on the $\log_e$ of disease severity at a growth stage when the developing caryopsis had reached three-quarters its final size; the equation $Y$ (percentage loss) $= -25.33 + 27.17 \log_e X$ (percentage severity) gave the best fit. Critical-point models have also been developed to estimate losses due to late blight of potato by Large (61) and Olofsson (75). In Sweden, Olofsson reported that significant relationships were obtained for yield and length of the blight-free period, i.e. from planting to the date when blight is first noted. For a blight-free period of 65, 80, and 95 days the respective percentage loss in tuber yield was 15, 12, and 9%. Large (61) related loss to the time when 75% of the foliage was killed by blight. Accordingly he reported 25, 15, and 5% loss in yield when 75% of the foliage was destroyed by blight by mid-August, end of August, and mid-September, respectively. The method is based on the assumption that tuber production stops when 75% of the foliage is killed. The principle of the method involves ascertaining the date when 75% of the foliage is affected by late blight and then reading off the percentage of the potential yield normally accumulated after this date from a bulking curve for the tubers; this reading is equivalent to the percentage loss in tuber yield resulting from late blight. Cox & Large (22) maintained that the method was universally applicable, given a mean bulking curve for a cultivar in the region for which loss assessment is made. James et al (49) tested Large's method in Canada and found that the actual losses (42, 33, 52, 17, and 26%) derived by weighing were in poor agreement with the estimated losses (23, 10, 16, 3, and 0%, respectively); various reasons were suggested for the inapplicability of the method.

The work also showed that the method suggested by Olofsson could not be applied in Canada, because epidemics with the same starting date (i.e. the same blight-free period) had completely different characteristics and consequently different losses. Similarly, Large's method gives the same estimates of loss for both early and late epidemics which reach 75% disease on the same date but where loss due to the former is much greater. The above shortcomings demonstrate the major limitation of all critical-point models, all of which assume that neither the infection rate (*r*), which in practice is known to vary for many diseases (101), nor the shape of the progress curve are important variables for determining loss.

MULTIPLE-POINT MODELS The limitation of the critical-point model referred to above can be overcome by the multiple-point model, which measures disease at many points during the epidemic. James et al (51) developed such a model for estimating losses due to late blight of potato. The technique of multiple regression analysis was used to derive an empirical equation using the increase in disease during 9 weekly periods as the independent variable and the loss in yield as the dependent variable. The equation can be used to estimate loss in yield for any given disease-progress curve; the difference between estimated loss, computed from the equation, and actual loss is less than 5% in 9 out of 10 cases (51). The general form of the regression equation used is $Y = b_1X_1 + b_2X_2 + .....$ where $Y$ is percentage loss in yield and $X_1$, $X_2$ are the disease increments for the first and second week respectively; the equation passes through the origin with a multiple correlation coefficient of 0.976. The estimate of loss can be made at any point during the epidemic or for the whole epidemic by multiplying the partial regression coefficient for any given week by the corresponding weekly disease increment to obtain a product. The sum of the products represents the loss at any point in time. Further tests have been carried out at our Research Stations on other major cultivars since the equation was developed in 1970, and these have confirmed the applicability and reliability of the equation. An extension of the method (52), based on further regression analysis, allows estimation of the percentage decrease in marketable tubers, for any given loss in yield (51); this introduces a "quality" aspect to the estimate of loss.

Burleigh et al (10) also developed a mutiple-point model for estimating losses in yield due to wheat leaf rust using a multiple-linear-regression technique. The dependent variable $Y$ was percentage loss in yield and the independent variables ($X_2$, $X_5$, $X_7$) were percentage leaf-rust severity at the boot, early berry, and early dough growth stages, respectively. The equation, $Y = 5.3788 + 5.5260\ X_2 - 0.3308\ X_5 + 0.5019\ X_7$, explained 79% of the variation with a standard error of 9%.

AREA-UNDER-THE-CURVE MODELS Relating the area under the disease-progress curve to loss represents a third type of model which can be described as midway between the critical- and multiple-point models. This third model has been used by Line et al (66); van der Plank (101) has used several examples to demonstrate the usefulness of the model. Compared with the critical-point model it has the advantage of distinguishing between two epidemics with different areas under the curve but which have the same percentage severity at a critical date. But it is not able to apply a different weighting to disease, relative to time, which is essentially a function of the value of the partial regression coefficient in the multiple-point models for late blight (51) and leaf rust (10). Consequently it was not successful for estimating losses due to late blight (51), because it could not distinguish between early light infections and late severe infections that occupied the same area under the disease-progress curve, but where the loss for the former was much greater than for the latter.

COMPARISON OF CRITICAL- AND MULTIPLE-POINT MODELS Having considered examples where the critical- and multiple-point models have been employed, it may be useful to try to analyze why loss for certain diseases can be estimated by a critical model whereas other diseases require a multiple-point model. The rationale is that, if the key factors can be identified, future model builders should benefit by increasing their chances of success.

Duration of the epidemic relative to life of the crop is a primary consideration. For all foliage diseases of cereals where the critical-point model has been successfully applied (20, 46, 64, 84), the epidemics were relatively short and late. Also the critical point occurred at approximately the midpoint of the short period when dry matter is accumulated. Romig & Calpouzos (84) have supported the application of a critical-point model for stem rust of wheat on the basis that major gains in dry weight result from photosynthates temporarily accumulated in the stem during the time of anthesis (97). In terms of yield physiology these late epidemics would be expected to decrease yield by decreasing kernel weight, and this has been confirmed for *Septoria* diseases of winter wheat (7, 21, 54, 72), for late epidemics of mildew (8, 90), and for scald of barley (46).

Unfortunately no models have been constructed for early or long epidemics for cereal diseases, such as those described for barley mildew by Schaller (90) and Brooks (8). But both reported that the early epidemic decreased yield by decreasing number of kernels (by decreasing number of tillers), and the long epidemic decreased the number and weight of kernels, thus proving that cereal yield can be decreased at any growth stage by a long duration epidemic. Potatoes differ from cereals in that yield is accumulated over a much longer period: ~50% of the growing season in the eastern United States and Canada. Late blight can affect the crop anytime from the point where yield is accumulated (49). It follows that any change in disease at any point in the epidemic may affect yield, and because infection rates can also be drastically altered with the application of fungicides, it is not surprising that the multiple-point model was necessary to characterize the loss. The long-duration blight epidemic is analogous to long-duration barley mildew epidemics, and we may speculate that the latter would also require a multiple- rather than a critical-point model. This thesis is supported by the work of Brooks (8) who showed that if mildew attacks spring barley early in the season, it is not possible to correlate loss in yield with severity of mildew at a late growth stage, i.e. critical point.

We may generalize by stating that the following conditions will probably demand a multiple-point model: high variability in infection rates or in shape of disease-progress curves, early and/or long duration epidemics, and a long period for accumulation of yield. In practice the choice between a multiple- and critical-point model usually represents considerable difference in labor, cost, etc, that must be justified, but in the future the use of aerial photography will probably minimize this difference. The increased accuracy of multiple-point models results in part from the fact that even for late epidemics several readings are usually superior to one as a basis for predicting loss. In this connection, Burleigh et al (10) showed that one severity reading of wheat leaf rust at early dough stage explained 64% of the variation in yield, but inclusion of an additional two readings at boot and early berry growth stages explained 79%.

## THE FUTURE

The chain of disease-loss appraisal consists of two links or phases, the first of which represents the strongest link and has been the subject of this review. Disease-yield loss models have been developed almost to the exclusion of disease-quality loss models. The only comprehensive study on quality is that reported by Main et al (69) who used regression analyses to relate the effect of brown spot on tobacco quality, as measured by many parameters. The study is also a valuable model because the many quality factors affected have been incorporated into an assessment equation relating severity to economic impact. Collaborative research with other disciplines, e.g. agronomy and physiology, is likely to be one of the major factors governing progress in developing better models for the future. Cooperative research programs involving mathematicians and statisticians will result in better mathematical models (42), and remote-sensing specialists will probably provide more comprehensive disease data than was ever possible with conventional disease assessment and survey methods.

Plant disease surveys have received little attention, probably because most surveys have been conducted in an ad hoc manner and consequently have been regarded by many plant pathologists as relatively unproductive exercises. Nothing could be further from the truth. Surveys represent the second link in the disease-appraisal chain and are therefore just as important as the development of methods for estimating losses. Well-conducted surveys are the only way of determining the status of disease in crops, and the significance of the results and conclusions may have far-reaching effects.

Comprehensive surveys of barley diseases in England and Wales (38) and southwest England (73) in 1967 demonstrated the relative importance of mildew severity estimated at 11 and 10% respectively (average of percentage leaf area affected on top two leaves at growth stage 11.1). James (38) estimated a yield loss of 13–18% (probably nearer 18) for the 11% severity in 1967, and this compared well with the 16% loss in yield for 10% severity estimated by Melville & Lanham (73). Both are substantially larger than a later estimate (57). Repeat surveys in 1968 and 1969 (57, 73) confirmed the relative importance of mildew relative to other foliage diseases. This stimulated research workers, industry, and farmers to develop, test, and implement chemical control measures for mildew, hitherto considered unnecessary because surveys had not been conducted to establish the severity of mildew and foliage diseases of barley.

An extension of these survey techniques has been employed by Richardson (82) who sampled cereal crops at different growth stages in an attempt to include the effects of other factors affecting crop performance. Accordingly the seed sample was tested for germination and the crop visited after emergence and again at the milky ripe stage in an effort to partition losses occurring at different growth stages. But it is impossible to evaluate the effectiveness of this survey system until the adequacy of the sampling schemes used are evaluated, especially those for detecting final yield of the crop. Although other surveys for fungal diseases on potato tubers (30, 32) and foliage diseases of cereals (70) have also demonstrated the usefulness of monitoring disease in the population, in general, surveys have been neglected.

Recently development of disease-survey systems has been emphasized (17, 58, 59, 77, 104) and this will be even more essential in the future. For crops where the value of the loss does not warrant control by fungicides, the desirability of controlling disease by resistant cultivars or cultural practices can simply be made by comparing the value of the loss, determined by survey, with the cost of a breeding program or cultural control costs (39). For diseases controlled regularly by chemical sprays, like late blight of potato (*Phytophthora infestans*), surveys of fungicide spraying practice (48, 78) must be conducted in conjunction with disease surveys to evaluate the economics of blight control.

For late blight and other diseases controlled by fungicides, future programs for disease-loss appraisal and forecasting must become an integral part of comprehensive disease-management schemes that ensure the rational and economical use of fungicides. At present the decision whether to spray or not to spray is based on projected increase in disease, whereas it should be based on the projected losses in yield and/or quality, as they are the marketable commodities. Ideally this projected loss should be expressed in monetary units. Because of this, economists (Carlson 27, p. 2.3/1–6) have emphasized the important role they can play in the development of modern disease-management programs.

Carlson's work on brown rot of peaches is probably unique because the value of the contracted peach crop is known at the beginning of the season. He developed a model, based on Bayesian decision theory, which utilized disease forecasts and recommended a "spray" or "no spray" decision on the basis of the expected value of the crop. At present the complexity of such a scheme may limit its use in practice, but the increasing role of the economist is to be encouraged.

The use of simulators (102) to project increase in disease and to simulate the effect of a fungicide application, in conjunction with models for estimating loss in yield, should result in a rational disease-control scheme. Such a system is being developed for late blight (43) using a logical disease simulator PHYTOSIM (James and Shih, unpublished information), to forecast a quantitative (rather than qualitative) increase in disease, which is transformed into loss using a disease-loss model (51). If the predicted loss is greater than the cost of a given spray application, the decision to spray is made; conversely, if the predicted loss is less than the cost of the spray, the decision is made not to spray. The system is designed to prescribe a fungicide spray schedule that results in minimum use of chemicals commensurate with rational disease control. This is highly desirable because it results in minimal effects on the ecosystem.

There is no doubt that the need to develop more efficient and economical disease-management schemes has promoted and will continue to promote interest in disease-loss appraisal. At the First International Congress of Plant Pathology in London in 1968, only one session was devoted to disease-loss appraisal compared with five sessions at the Second Congress in Minneapolis in 1973; the well-attended sessions at Minneapolis testified to the marked increase in interest in the last five years. Epidemiologists are becoming increasingly involved because they view disease-loss appraisal as the practical end-result to which they can apply and relate the vast amount of monitored epidemiological data.

For the future, epidemiologists, economists, disease-loss appraisal and plant-protection specialists must combine their efforts to develop modern disease-management strategies which allow both rational and economical disease control. In such schemes, disease-loss appraisal is indispensable, and its status will continue to increase as the need grows for more objective data on economic losses due to plant diseases.

## Acknowledgments

I am grateful to Dr. C. S. Shih for his helpful comments and to Mr. S. I. Wong for checking the manuscript and references.

### *Literature Cited*

1. Anonymous. 1947. The measurement of potato blight. *Trans. Brit. Mycol. Soc.* 31:140–41
2. Bald, J. G. 1943. Potato virus X: Mixtures of strains and the leaf area and yield of infected potatoes. *Counc. Sci. Ind. Res. Bull.* 165
3. Bawden, F. C., Kassanis, B., Roberts, F. M. 1948. Studies on the importance and control of potato virus X. *Ann. Appl. Biol.* 35:250–65
4. Bonde, R. 1954. Potato X Virus causes large loss; better seed is answer. *Maine Farm Res.* 2(2):10–12
5. Borchardt, G., Bode, O., Bartels, R., Holz, W. 1964. Untersuchungen über die Minderung des ertrages von Kartoffelpflanzen durch virusinfektionen. *Nachrichtenbl. Deut. Pflanzenschutzdienst Berlin* 16:150–56
6. Brodrick, H. T., Longshaw, T. G., van Lelyveld, L. J. 1973. Disease detection with spectral analysis. *2nd Int. Congr. Plant Pathol.* 0763 (Abstr.)
7. Brönnimann, A. 1968. Zur Kenntnis von *Septoria nodorum* Berk., dem Erreger der Spetzenbräune and einer Blattdürre des Weizens. *Phytopathol. Z.* 61:101–46
8. Brooks, D. H. 1972. Observations on the effects of mildew, *Erysiphe graminis,* on growth of spring and winter barley. *Ann. Appl. Biol.* 70:149–56
9. Burleigh, J. R., Eversmeyer, M. G., Roelfs, A. P. 1972. Development of linear equations for predicting wheat leaf rust. *Phytopathology* 62:947–53
10. Burleigh, J. R., Roelfs, A. P., Eversmeyer, M. G. 1972. Estimating damage to wheat caused by *Puccinia recondita tritici. Phytopathology* 62:944–46
11. Caldwell, R. M., Schafer, J. F., Compton, L. E., Patterson, F. L. 1958. Tolerance to cereal leaf rusts. *Science* 128:714–15
12. Callbeck, L. C. 1954. A progress report on the effect of zinc as a constituent of potato fungicides. *Am. Potato J.* 31:341–48
13. Calpouzos, L., Madson, M. E., Welsh, J. R. 1971. The effect of harvest technique on recovery of wheat from experimental plots with and without stem rust. *Phytopathology* 61:1022 (Abstr.)
14. Carr, A. J. H., Large, E. C. 1963. Surveys of phyllody in white clover seed crops, 1959–62. *Plant Pathol.* 12:121–27
15. Chester, K. S. 1950. Plant disease losses: their appraisal and interpretation. *Plant Dis. Reptr. Suppl.* 193:189–362
16. Christ, R. A. 1957. Control plots in ex-

periments with fungicides. *Commonw. Phytopathol. News* 3:54, 62
17. Church, B. M. 1971. The place of sample surveys in crop loss estimation. *Crop loss assessment methods; FAO manual on the evaluation and prevention of losses by pests, diseases and weeds.* Rome: FAO. 2.2/1–12 (looseleaf)
18. Clinch, P. E. M., MacKay, R. 1947. Effect of mild strains of virus X in the yield of up-to-date potato. *Sci. Proc. Roy. Dublin Soc. (N.S.)* 24:189–98
19. Cobb, N. A. 1892. Contribution to an economic knowledge of the Australian rusts (uredineae). *Agr. Gaz. N.S. W.* 3:60–68
20. Cooke, B. M., Fozzard, J. T. F. 1973. Development, assessment and seed transmission of *Septoria nodorum. Trans. Brit. Mycol. Soc.* 60(2):211–22
21. Cooke, B. M., Jones, D. G. 1970. The epidemiology of *Septoria tritici* and *S. nodorum.* II: Comparative studies of head infection by *Septoria tritici* and *S. nodorum* on spring wheat. *Trans. Brit. Mycol. Soc.* 54(3):395–404
22. Cox, A. E., Large, E. C. 1960. *Potato Blight Epidemics Throughout the World.* Agr. Handb. No. 174. Washington DC: US Dep. Agr. 230 pp.
23. Dirks, V. A., Romig, R. W. 1970. Linear models applied to variation in numbers of cereal rust urediospores. *Phytopathology* 60:246–51
24. Eversmeyer, M. G., Burleigh, J. R. 1970. A method of predicting epidemic development of wheat leaf rust. *Phytopathology* 60:805–11
25. Eversmeyer, M. G., Burleigh, J. R., Roelfs, A. P. 1973. Equations for predicting wheat stem rust development. *Phytopathology* 63:348–51
26. Food and Agriculture Organization of the United Nations. 1967. Papers presented at symposium of crop losses. Rome: FAO, U.N.
27. Food and Agricultural Organization of the United Nations. 1971. *Crop loss assessment methods; FAO manual on the evaluation and prevention of losses by pests, diseases and weeds.* Rome: FAO. 200 pp. (unnumbered, looseleaf)
28. Hanway, J. J. 1966. How a corn plant develops. In *Special Report 48.* Ames, Iowa: Iowa State Univ.
29. Harris, K. M. 1963. Assessments of the infection of guineacorn *(Sorghum vulgare)* by covered smut (*Sphacelotheca sorghi* (Link) Clint.) in Northern Nigeria in 1957 and 1958. *Ann. Appl. Biol.* 51:367–70
30. Hide, G. A., Hirst, J. M., Salt, G. A. 1968. Methods of measuring the prevalence of pathogenic fungi on potato tubers. *Ann. Appl. Biol.* 62:309–18
31. Hide, G. A., Hirst, J. M., Stedman, O. J. 1973. Effects of skin spot (*Oospora pustulans*) on potatoes. *Ann. Appl. Biol.* 73:151–62
32. Hirst, J. M., Hide, G. A., Griffith, R. L., Stedman, O. J. 1970. Improving the health of seed potatoes. *J. Roy. Agr. Soc. Engl.* 131:87–106
33. Hirst, J. M., Hide, G. A., Stedman, O. J., Griffith, R. L. 1973. Yield compensation in gappy potato crops and methods to measure effects of fungi pathogenic on seed tubers. *Ann. Appl. Biol.* 73:143–50
34. Hodgson, W. A., Tai, G. C. C. 1974. The use of color infrared aerial photographs to estimate the loss in yield caused by potato late blight. *Proc. Fourth Bien. Workshop Aerial Color Photogr.* Orono, Maine: Univ. Maine. In press
35. Horsfall, J. G., Barratt, R. W. 1945. An improved grading system for measuring plant diseases. *Phytopathology* 35:655; Horsfall, J. G. In *Fungicides and Their Action,* 38–41. Waltham, Mass.: Chronica Botanica
36. Hoyman, W. G. 1949. The effect of zinc-containing dusts and sprays on the yield of potatoes. *Am. Potato J.* 26:256–63
37. Jackson, H. R., Hodgson, W. A., Wallen, V. R., Philpotts, L. E., Hunter, J. 1971. Potato late blight intensity levels as determined by microdensitometer studies of false-color aerial photographs. *J. Biol. Photogr. Assoc.* 39: 101–6
38. James, W. C. 1969. A survey of foliar diseases of spring barley in England and Wales in 1967. *Ann. Appl. Biol.* 63:253–63
39. James, W. C. 1971. Importance of foliage diseases of winter wheat in Ontario in 1969 and 1970. *Can. Plant Dis. Surv.* 51:24–31
40. James, W. C. 1971. An illustrated series of assessment keys for plant diseases, their preparation and usage. *Can. Plant Dis. Surv.* 51:39–65
41. James W. C. 1971. A manual of disease assessment keys for plant diseases. *Can Dept. Agr. Publ.* 1458. 80 pp.
42. James, W. C. 1974. Disease appraisal and loss: how do we find answers—a blue print for tomorrow. *Proc. 2nd Int. Congr. Plant Pathol.* 0023

43. James, W. C. 1974. The assessment of economic losses due to late blight of potato. In *Proc. EPPO Integrated Contr. Meet.* Paris: EPPO. In press
44. James, W. C., Shih, C. S. 1973. Size and shape of plots for estimating yield losses from cereal foliage diseases. *Exp. Agr.* 9:63–71
45. James, W. C., Shih, C. S. 1973. Relationship between incidence and severity of powdery mildew and leaf rust on winter wheat. *Phytopathology* 63:183–87
46. James, W. C., Jenkins, J. E. E., Jemmett, J. L. 1968. The relationship between leaf blotch caused by *Rhynchosporium secalis* and losses in grain yield of spring barley. *Ann. Appl. Biol.* 62:273–88
47. James, W. C., Lawrence, C. H., Shih, C. S. 1973. Yield losses due to missing plants in potato crops. *Am. Potato J.* 50:345–52
48. James, W. C., Shih, C. S., Callbeck, L. C. 1973. Survey of fungicide spraying practice for potato late blight in Prince Edward Island, 1972. *Can. Plant Dis. Surv.* 53:161–66
49. James, W. C., Callbeck, L. C., Hodgson, W. A., Shih, C. S. 1971. Evaluation of a method used to estimate loss in yield of potatoes caused by late blight. *Phytopathology* 61:1471–76
50. James, W. C., Shih, C. S., Callbeck, L. C., Hodgson, W. A. 1973. Interplot interference in field experiments with late blight of potato (*Phytophthora infestans*). *Phytopathology* 63:1269–75
51. James, W. C., Shih, C. S., Hodgson, W. A., Callbeck, L. C. 1972. The quantitative relationship between late blight of potato and loss in tuber yield. *Phytopathology* 62:92–96
52. James, W. C., Shih, C. S., Hodgson, W. A., Callbeck, L. C. 1973. A method for estimating the decrease in marketable tubers caused by potato late blight. *Am. Potato J.* 50:19–23
53. Jenkins, J. E. E., Melville, S. C., Jemmett, J. L. 1972. The effect of fungicides on leaf diseases and on yield in spring barley in South-west England. *Plant Pathol.* 21:49–58
54. Jenkins, J. E. E., Morgan, W. 1969. The effect of *Septoria* diseases on the yield of winter wheat. *Plant Pathol.* 18:152–56
55. Kassanis, B., Varma, A. 1967. The production of virus-free clones of some British potato varieties. *Ann. Appl. Biol.* 59:447–50
56. Katsube, T., Koshimizu, Y. 1970. Influence of blast disease on harvests in rice plant. I: Effect of panicle infection on yield components and quality. *Bull. Tohoku Nat. Agr. Exp. Sta.* 39:55–96 (in Japanese, with English summary)
57. King, J. E. 1972. Surveys of foliar diseases of spring barley in England and Wales 1967–70. *Plant Pathol.* 21:23–35
58. Kranz, J. 1972. Zur Ermittlung von Befalls/Verlust-Relationen in Feldversuchen. Überarbeitete Fassung eines auf der Tagung "Biometrie in der Phytomedizim" am 9.3.72 in Fulda gehaltenen Vortrages
59. Kranz, J. 1973. Sampling and data processing in survey systems. *2nd Int. Congr. Plant Pathol.* 0760 (Abstr.)
60. Large, E. C. 1945. Field trials of copper fungicides for the control of potato blight. I. Forage protection and yield. *Ann. Appl. Biol.* 32:319–29
61. Large, E. C. 1952. The interpretation of progress curves for potato blight and other plant diseases. *Plant Pathol.* 1:109–17
62. Large, E. C. 1954. Growth stages in cereals: illustration of the Feekes scale. *Plant Pathol.* 3:128–29
63. Large, E. C. 1966. Measuring plant disease. *Ann. Rev. Phytopathol.* 4:9–28
64. Large, E. C., Doling, D. A. 1962. The measurement of cereal mildew and its effect on yield. *Plant Pathol.* 11:47–57
65. Ledingham, R. J. et al 1973. Wheat losses due to common root rot in the Prairie provinces of Canada, 1969–71. *Can. Plant Dis. Surv.* 53:113–22
66. Line, R. F., Peet, C. E., Kingsolver, C. H. 1967. The effect of stem rust on yield of wheat at Stillwater, Oklahoma. *Phytopathology* 57:819 (Abstr.)
67. Lyman, G. R. 1918. The relation of phytopathologists to plant disease survey work. *Phytopathology* 8:219–28
68. Machacek, J. E. 1943. An estimate of loss in Manitoba from common root rot in wheat. *Sci. Agr.* 24:70–77
69. Main, E. C., Nusbaum, C. J., Lucas, G. B., Chaplin, J. F. 1973. Tobacco quality related to severity of foliar disease. *2nd Int. Congr. Plant Pathol.* 0590 (Abstr.)
70. McDonald, W. C. et al 1969. Losses from cereal diseases and value of disease resistance in Manitoba in 1969. *Can. Plant Dis. Surv.* 49(1):114–21
71. Melchers, L. E., Parker, J. H. 1922. Rust resistance in winter-wheat varieties. *Bull. US Dep. Agr.*, No. 1046
72. Melville, S. C., Jemmett, J. L. 1971. The effect of glume blotch on the yield of winter wheat. *Plant Pathol.* 20:14–17
73. Melville, S. C., Lanham, C. A. 1972. A survey of leaf diseases of spring barley

in South-west England. *Plant Pathol.* 21:59–66

74. Murphy, P. A. 1939. A study of the seasonal development of the potato in relation to blight attack and spraying. *Sci. Proc. Roy. Dublin Soc.* 22:69–82
75. Olofsson, B. 1968. Determination of the critical injury threshold for potato blight (*Phytophthora infestans) Medd. Wäxtskyddsanst,* Stockholm. 14:81–93
76. Ordish, G., Dufour, D. 1969. Economic bases for protection against plant diseases. *Ann. Rev. Phytopathol.* 7:31–50
77. Populer, C. 1973. Developing survey systems. *2nd Int. Congr. Plant Pathol.* 0759 (Abstr.)
78. Potato Marketing Board, Rothamsted Experimental Station and National Institute of Agricultural Engineering. Survey of main crop potatoes, 1968. *Potato Marketing Board Rept., London,* 1970
79. Preece, T. F., 1971. Disease assessment. In *Diseases of Crop Plants,* ed. J. H. Western 8–20. London: Macmillan. 404 pp.
80. Rayner, R. W. 1961. Measurement of fungicidal effects in field trials. *Nature* 190:328–30
81. Reestman, A. J. 1970. Importance of the degree of virus infection for the production of ware potatoes. *Potato Res.* 13:248–68
82. Richardson, M. J. 1971. Yield losses in wheat and barley—1970. *Scot. Agr.* 50:72–77
83. Richardson, M. J., Rennie, W. J. 1970. An estimate of the loss of yield caused by *Cephalosporium gramineum* in wheat. *Plant Pathol.* 19:138–40
84. Romig, R. W., Calpouzos, L. 1970. The relationship between stem rust and loss in yield of spring wheat. *Phytopathology* 60:1801–5
85. Romig, R. W., Dirks, V. A. 1966. Evaluation of generalized curves for number of cereal rust uredospores trapped on slides. *Phytopathology* 56:1376–80
86. Rosser, W. R. 1957. Potato blight control trials in the West Midland province, 1950–1954. *Plant Pathol.* 6:77–84
87. Rosser, W. R., Chadburn, B. L. 1968. Cereal diseases and their effects on intensive wheat cropping in the East Midland Region, 1963–65. *Plant Pathol.* 17:51–60
88. Samborski, D. J., Peturson, B. 1960. Effect of leaf rust on the yield of resistant wheats. *Can. J. Plant Sci.* 40:620–22
89. Schafer, J. F. 1971. Tolerance to plant disease. *Ann. Rev. Phytopathol.* 9:235–52
90. Schaller, C. W. 1963. The effect of mildew and scald infection on yield and quality of barley. *Agron. J.* 43:183–88
91. Schultz, E. S., Bonde, R. 1944. The effect of latent mosaic (virus X) on yield of potatoes in Maine. *Am. Potato J.* 21:278–83
92. Simons, M. D. 1966. Relative tolerance of oat varieties to the crown rust fungus. *Phytopathology* 56:36–40
93. Slinkard, A. E., Elliot, F. C. 1954. The effect of bunt incidence on the yield of wheat in eastern Washington. *Agron. J.* 46:439–41
94. Slope, D. B., Etheridge, J. 1971. Grain yield and incidence of take-all (*Ophiobolus graminis* Sacc.) in wheat grown in different crop sequences. *Ann. Appl. Biol.* 67:13–22
95. Stace-Smith, R., Mellor, F. C. 1968. Eradication of potato viruses X and S by thermotherapy and axillary bud culture. *Phytopathology* 58:199–203
96. Stapp, C. 1942. Uber serologische virusforschung und den diagnostischen wert serologischer methoden Zum nachweis der pflanztechen, insbesondere der am Kartoffelabbau beteriligten Viren. *Sonderabdruck aus Jour. fur Landwirtschoft Bd.* 89, H. 3
97. Stoy, V. 1965. Photosynthesis, respiration, and carbohydrate accumulation in spring wheat in relation to yield. *Physiol. Plant. Suppl.* 4:125 pp.
98. Strandberg, J. 1973. Spatial distribution of cabbage black rot and the estimation of diseased plant populations. *Phytopathology* 63:990–1003
99. Torres, E., Browning, J. A. 1968. The yield of uredospores per unit of sporulating area as a possible measure of tolerance of oats to crown rust. *Phytopathology* 58:1070 (Abstr.)
100. Ullstrup, A. J., Elliott, C., Hoppe, P. E. 1945. Report of the committee on methods for reporting corn disease ratings. *Wash. Div. Cereal Crops Dis.* Wash. DC: US Dep. Agr. 5 pp. (unnumbered, mimeographed)
101. van der Plank, J. E. 1963. *Plant Diseases: Epidemics and Control.* New York/London: Academic. 349 pp.
102. Waggoner, P. E. 1968. Weather and the rise and fall of fungi. In *Biometerol. Proc. Ann. Biol. Coll., 28th, 1967,* ed. W. P. Lowry, 45 66. Corvallis, Oreg.: Oregon State Univ. Press
103. Wallen, V. R., Jackson, H. R. 1971. Aerial photography as a survey technique for the assessment of bacterial

blight of field beans. *Can. Plant Dis. Surv.* 51:163–69
104. Wallen, V. R., Jackson, H. R. 1973. Quantification of remote sensing systems. *2nd Int. Congr. Plant Pathol.* 0762 (Abstr.)
105. Wright, N. S. 1970. Combined effects of potato viruses X and S on yield of Netted Gem and White Rose potatoes. *Am. Potato J.* 47:475–78
106. Zadoks, J. C. 1972. Methodology of epidemiological research. *Ann. Rev. Phytopathol.* 10:253–76

# PLANT DISEASE SITUATION OF HIGH-YIELDING DWARF WHEATS IN ASIA AND AFRICA

❖3587

*E. E. Saari and Roy D. Wilcoxson*
CIMMYT/ALAD, The Ford Foundation, Beirut, Lebanon, and Department of Plant Pathology, University of Minnesota, St. Paul, Minnesota 55101

During the past eight years wheat production in many developing nations of Asia and Africa has undergone a historic revolution (9–12, 20, 120, 126). The entire production system has changed from a traditional agriculture using local varieties planted haphazardly, often in a mixture with other crops and with little use of fertilizers or irrigation, to a sophisticated system utilizing modern dwarf varieties sown at specified rates and depths on recommended dates. The timely application of fertilizer and irrigation is now routine unless supplies are limited by external factors (16). In this region, wheat is grown in environments ranging from humid to arid, from subtropical to temperate, and from sea level to altitudes of 3000 meters. Two species are grown commercially, *Triticum aestivum* L. and *T. durum* Desf. The centers of origin for wheat are found here (65), and natural stands of various species and subspecies can be found adjacent to intensively cultivated commercial plantings. There are both winter and spring varieties depending upon the elevation and latitudes. Furthermore, in India and Pakistan wheat was introduced more than 1000 years ago, while in East Africa the crop was introduced about 100 years ago (83). All these circumstances provide an ideal setting for dramatic changes in diseases (50). This review describes diseases of wheat that occur in this vast region and indicates changes that have occurred during the past eight years.

The countries concerned in this review are grouped by zone—South Asia: Afghanistan, Bangladesh, India, Nepal, Pakistan; West Asia: Iran, Iraq, Israel, Jordan, Lebanon, Saudi Arabia, Syria, Turkey; North Africa: Algeria, Egypt, Lybia, Morocco, Tunisia; and East Africa: Ethiopia, Kenya, Sudan, Tanzania. We have not included Russia or China and other far eastern countries because of lack of information. Other reviews (20, 23, 32, 76, 120) should be consulted for documentation of social and economic changes brought about by the "wheat revolution" or the "Green Revolution" as it is sometimes called.

The information to be presented was gleaned from literature that is generally available; from conference and progress reports and local publications; from oral and written communications with colleagues; and from first hand observation of nurseries and fields in many nations during the past six years. We hope we have correctly presented the views of those who have so generously given us information.

## *Analysis of Agronomic Practices in Relation to Wheat Diseases*

IRRIGATION The introduction of dwarf wheats has resulted in greatly increased use of irrigation in some areas (46). If properly done, irrigation does not significantly increase diseases in wheat. In arid areas, however, irrigation may provide the moisture for dew formation which in turn may favor some diseases such as rust (22). In general, the greater the need for irrigation, the less the disease hazard (69, 94, 112). An exception to this may be in areas like Egypt where irrigation is required for wheat production and the rusts are an important threat. Aside from rusts, other diseases are a minor threat to wheat production in Egypt. Downy mildew of wheat has been reported with increasing frequency, but it is restricted to fields that are excessively irrigated or where the soil is waterlogged, and provides little or no threat to wheat production (123).

FERTILIZER Fertilization favors development of obligate parasites and rust fungi in particular (21). We have observed increases in rust severities at higher nitrogen levels with all three rusts only if the variety was susceptible to the races of the pathogen. If the variety is resistant, fertilizer does not alter the response (113). Some diseases such as Alternaria leaf blight are less of a problem under higher nutritional levels when there is good plant development.

TILLAGE PRACTICES Yarwood (127) suggested that when production is taken from a natural to a cultivated state, diseases may increase in importance. Once crops are in production, further changes in cultural practices may have virtually no effect on disease incidence (79) or there may be a notable increase, as reported for Cercosporella foot rot of wheat (21, 26). The tillage requirements for dwarf wheats have substantially changed a number of practices (15, 108), but we have seen no indications that this has altered the severity or prevalence of wheat diseases.

CROP ROTATION The availability of early maturing dwarf varieties has permitted later sowing dates and has therefore affected the possible crop rotations. It is now common practice on the Indian subcontinent to harvest two or more crops per year in irrigated areas, making it necessary to sow wheat later than is recommended. Early maturity allows wheat to escape damage by leaf and stem rusts (26). Late sowing increases the potential hazard from leaf and stem rust but decreases the significance of leaf blights caused by *Septoria* and *Alternaria* species (101). In rotations following planting of rice, poor soil structure and excess moisture at wheat seeding time has increased the frequency of seedling blights (100) and some poor stands may result from improper seed bed preparation.

In general, multiple cropping decreases soil-borne diseases if the alternate crops do not sustain the pathogen. Crop debris decomposes rapidly in the subtropics and soil-borne diseases tend to be less of a problem on wheat. In a few cases the overlap of one crop with wheat has resulted in the appearance of disease that does not normally occur. Thus a new virus disease of wheat was recently reported, although it is primarily a disease of millet and sorghum (104). The disease is insignificant in wheat because the environment becomes unfavorable to the vector soon after it attacks the wheat, preventing the spread of the virus.

SEED INDUSTRY During the last 10 years the seed industry in a number of the developing countries has greatly improved. Seed originating from these industries is often treated with fungicides and foreign matter is removed. As a result, diseases like bunt, ear cockle, and yellow ear rot have decreased in importance when this seed is used.

VARIETY CHANGES The dwarf wheats that replaced local wheat varieties were successful because of their wide adaptability, yield potential, and disease resistance. At the time of their release they were resistant to most of the predominant diseases of the region or were no more susceptible than the local varieties. Only in the high rainfall areas of the Mediterranean basin did *Septoria tritici* generally prove to be a more serious problem on dwarf wheats than on local varieties.

Because a common dwarfing source was used in the development of the dwarf wheats used throughout the region of Asia and Africa, concern has been raised about the possible widespread development of diseases (26, 113). We are not aware of any linkage between susceptibility to disease and dwarfing. The dwarf varieties have not increased the incidence of diseases and no new diseases have arisen because of them. With the possible exception of *Septoria tritici* in the Mediterranean region, old diseases have not become more severe. In our surveys it was usually possible to find more foliage diseases in the farm plantings of local varieties than in the dwarf varieties.

AWARENESS OF DISEASE Holton (50) suggested in 1967 that additional scientists were needed to study wheat diseases in India. We estimate there are now about 125 full- and part-time scientists and technicians working on wheat diseases throughout the entire region, mostly in India and Pakistan. Because these people are now reporting their results, the impression is created that wheat diseases have increased in prevalence and importance. Most of the diseases being reported from various parts of the region were previously known (29, 30, 77) and are merely being recognized more frequently.

## *Analysis of Individual Disease Situations*

Many diseases and their pathogens have been reported on wheat in North Africa, East Africa, West Asia, and South Asia (29, 30). The importance of these diseases and their impact on wheat production during the last six years varies greatly and the information available is summarized by disease.

RUST DISEASES Stem, leaf, and yellow rusts (*Puccinia graminis* Pers. f. sp. *tritici* Eriks. & E. Henn., *P. recondita* Rob. ex Desm., and *P. striiformis* West.) are the most important diseases of wheat in the region. Leaf and stem rust are found wherever wheat is grown, and yellow rust occurs in the cooler areas (2–7, 27, 43–45, 61, 73, 75, 77, 81, 90, 115). Today rusts are the diseases with the greatest potential for epidemic and pandemic development (49, 53, 112).

The dwarf wheats were derived from a broad spectrum of parents (9, 10, 67, 86) that had wide adaptability, yield potential, and disease resistance. Mexican Cross No. 8156, one of the important sources of dwarf wheat varieties distributed by the International Maize and Wheat Improvement Center (CIMMYT), yielded several varieties, for instance Kalyansona, Mexipak, Siete Cerros, Super X, and Indus 66. These varieties were resistant to stem rust, yellow rust, loose smut, bunt, and powdery mildew at the time of release. They were less resistant to leaf rust but reselection provided moderately resistant lines. These varieties still provide outstanding resistance to the smuts and powdery mildew, but the situation with the rusts has changed drastically because there are now races or biotypes of all three rusts in the region that can attack them (47, 48, 55, 56, 105, 113).

On the Indo-Pak subcontinent epidemics of leaf rust occurred in 1971–1972 and 1972–1973 that were most severe in the northern areas on Kalyansna and Mexipak, dwarf varieties that occupied 75% of the wheat acreage. Throughout the subcontinent all of the local varieties were severely attacked.

In 1971–1972 the losses due to leaf rust in India's Punjab state were estimated at 5 to 7% of the potential production for dwarf wheats and 10 to 15% for indigenous wheats. In the remainder of the country, losses were less severe and the All-India loss was about 3% (Joshi, unpublished data). The situation in 1972–1973 has not yet been fully analyzed, but the rust was more widespread and severe than in 1971–1972 and yield losses must be expected to be as great.

Leaf rust epidemics similar to those just described in India also occurred in Pakistan during 1971–1972 and 1972–1973. Loss estimates are not available but they would probably be similar to those for northern India, because most of Pakistan's wheat is grown adjacent to India's major production area. Losses estimated for Afghanistan and Nepal are not available.

In the epidemic of leaf rust on the subcontinent, a number of dwarfs such as Sonalika, Choti Lerma, UP301, and others were resistant. A number of advanced lines in India and Pakistan were also resistant, and show promise as future varieties.

Stem rust races and biotypes that can attack many dwarf wheats have been identified in Asia and Africa, but have not increased in prevalence (4, 48, 55). These new races and biotypes should not be discounted as threats to the dwarf wheats because stem rust has a well-known potential for suddenly becoming epidemic during favorable weather (112). For the immediate future, it appears that the stem rust resistance of the dwarf wheats is adequate in the areas where they are being grown.

Stem rust in East Africa is somewhat unique and different from that in the remainder of the region because of the occurrence of many different races, rapid shifts in the occurrence of races, the presence of inoculum throughout the year, and continuous growing of wheat (43, 45, 83). Historically wheat improvement has been hampered because of stem rust (8, 83), and dwarf varieties have not been successfully established because of it (8, 83).

Races and biotypes of yellow rust have evolved in West and South Asia that can attack the 8156 varieties and may present a serious threat (62, 88). No serious epidemics have been reported on the dwarf wheats, but yellow rust has been recorded with increasing frequency and has been severe in some individual fields (6, 48). While the loss in individual fields has been substantial, national losses have been minor. In North Africa, yellow rust has not been a serious disease on bread wheats except in Egypt (5). The virulence spectrum is relatively narrow and yellow rust does not present a problem to the dwarf wheats at this time (88, 116, 117). Egypt recorded serious yellow rust epidemics in 1967 and 1968 (4). Yellow rust must reestablish itself each year in Egypt but the epidemiological patterns for yellow rust in the Middle East have not been determined (5, 104, 116, 117).

The rust fungi are capable of evolving new races that can attack previously resistant varieties (53, 112) and the effective resistance of wheat varieties has been estimated to last about 5 years (17, 26). The dwarf wheats introduced into Asia and Africa have been grown for more than five years and new varieties are being developed as rapidly as possible to replace them.

Because plant pathogens, especially the rust fungi, may quickly develop new races to attack hitherto resistant varieties (125), a means for early detection of new races and their distribution and potential importance is needed. To meet this need, systematic surveys and trap disease nurseries have been established (4, 54, 87, 98). The racial identity of rust isolates obtained from the nurseries and surveys is being made in South Asia (47, 48, 55, 56), West Asia (18, 62, 116), and Africa (4, 45), and inoculum of important races is supplied to plant breeders.

International nurseries are used for assessing wheat diseases on a global basis (99). In the region the Regional Trap Nursery (RTN) containing about 40 commercial wheat varieties with resistant and susceptible check varieties was planted in 1972 in 20 countries (87). Another nursery, the Regional Disease and Insect Screening Nursery (RDISN), is planted in "hot spots" where disease epidemics occur most years (98). The RDISN in 1971–1972 contained 2400 entries and was planted in 30 locations throughout the region. Many agencies cooperate with making these and other nurseries successful, including all of the national wheat improvement programs in the area; the Regional Office of the FAO of the United Nations, Cairo, Egypt; the Arid Lands Agricultural Program of the Ford Foundation, Beirut, Lebanon; The Rockefeller Foundation and the Wheat Research and Training Program, Ankara, Turkey; and CIMMYT. Data and materials from these nurseries are available to the various national breeding programs.

Plant pathologists hope to use information from these nurseries to develop a system to predict the relative importance of diseases and the possible danger to varieties. The possibilities may be seen from a preliminary analysis of some data on rust occurrence collected in 1971–1972 (Tables 1 and 2).

The data are summarized by zones which are assumed to be epidemiologically different (112, 117), and the average coefficient of infection was calculated to provide

**Table 1** Average coefficients of infection for three rust fungi found on local, improved, and dwarf wheat varieties in four geographical zones of Asia and Africa. Data from the Regional Trap Nursery of 1971–1972 (88)

| Rust Fungus | Wheat Variety Group[a] | Average Coefficient[b] of Rust Infection by Zones[c] | | | |
|---|---|---|---|---|---|
| | | South Asia | West Asia | North Africa | East Africa |
| *P. graminis* | local | 24 | 32 | 48 | 56 |
| f. sp. | improved | 2 | 15 | 13 | 46 |
| *tritici* | dwarf | < 1 | 5 | 1 | 39 |
| *P. striiformis* | local | 11 | 20 | 7 | 24 |
| f. sp. | improved | 5 | 14 | 1 | 14 |
| *tritici* | dwarf | 2 | 8 | 1 | 11 |
| *P. recondita* | local | 50 | 33 | 50 | 9 |
| f. sp. | improved | 34 | 21 | 32 | 3 |
| *tritici* | dwarf | 18 | 10 | 6 | < 1 |

[a] Includes 7 local, 12 improved, and 12 dwarf varieties.
[b] Coefficient of infection = resistance rating × severity.
[c] Data based on rust readings from 6 locations in South Asia, 14 locations in West Asia, 5 locations in North Africa, and 4 locations in East Africa.

a standard measure of the rust problems (67). A coefficient of infection greater than 10 indicates that rust is out of control on the varieties tested, a value between 5 and 10 indicates that new varieties are needed, and a value below 3 indicates that resistance is still adequate. Coefficients of infection averaged for all varieties in the RTN nurseries (Table 1) indicate that the local indigenous wheats are all in danger of attack by all three rusts in each zone. The improved nondwarf wheats are also in danger of attack by the three rusts in most places. The dwarf wheats are in danger of attack by stem rust in East Africa, yellow rust in West Asia and East Africa, and leaf rust in South and West Asia.

The potential for rust attacking the principal dwarf varieties varies with the zone and the variety (Table 2). The stem rust resistance of the six dwarf wheat varieties is inadequate in East Africa, the resistance of the 8156 varieties in West Asia and North Africa is endangered, and Penjamo 62 and Sonalika in West Asia are threatened. The leaf rust resistance of the 8156 varieties and Chenab 70 is no longer effective in most of the zones, while the reactions of the other dwarfs vary greatly depending upon the zone. A similar comparison can be made for yellow rust. In East Africa the dwarf wheats, with the exception of Tobari 66, do not have adequate resistance to yellow rust; in West Asia 8156 varieties and Penjamo 62 are subject to attack.

The only variety with adequate resistance to the three rusts in Asia and North Africa is Tobari 66. The resistance of Sonalike is also good in these three zones, but

**Table 2** Average coefficients of infection for stem, leaf, and yellow rust on important dwarf wheats grown in Asia and Africa in 1971–1972. Data from the Regional Trap Nursery (88)

| Variety | Rust[e] | Average Coefficient of Infection by Zones[f] | | | |
|---|---|---|---|---|---|
| | | South Asia | West Asia | North Africa | East Africa |
| Cross No. 8156[a] | SR | <1 | 5 | 8 | 27 |
| | YR | 0[g] | 15 | 3 | 17 |
| | LR | 16 | 28 | 16 | 4 |
| Penjamo 62[b] | SR | <1 | 7 | 0 | 38 |
| | YR | 0 | 19 | 0 | 5 |
| | LR | 17 | 12 | <1 | 0 |
| Inia 66[b] | SR | 0 | 1 | 0 | 40 |
| | YR | 0 | 2 | 0 | 13 |
| | LR | 13 | 4 | <1 | 1 |
| Tobari 66[b] | SR | 0 | <1 | 0 | 38 |
| | YR | 0 | 1 | 0 | 3 |
| | LR | <1 | <1 | 3 | 0 |
| Sonalika[c] | SR | 0 | 6 | 0 | 38 |
| | YR | 0 | 2 | 0 | 5 |
| | LR | 1 | 4 | <1 | 0 |
| Chenab 70[d] | SR | 0 | 1 | 0 | 55 |
| | YR | 0[g] | 9 | <1 | 10 |
| | LR | 52 | 11 | 7 | <1 |

[a] Released by several countries under different names. (See text.) Results represent pooled data for Mexipak 65, Mexipak 69, and Indus 66.

[b] Semidwarf wheats released by Mexico.

[c] Semidwarf released by India.

[d] Semidwarf released by Pakistan.

[e] SR = stem rust; YR = yellow rust; LR = leaf rust.

[f] Based on data from 6 locations in South Asia, 14 locations in West Asia, 5 locations in North Africa, and 4 locations in East Africa.

[g] This value was calculated from nursery data, but the rust was seen on these varieties in other nurseries.

indications are that the resistance to stem rust in West Asia will be effective only for a limited period.

LEAF BLOTCH AND GLUME BLOTCH These diseases are found throughout the region but their importance varies greatly (57, 73, 82, 101, 103, 106). *Septoria tritici* Rob. ex. Desm. and *S. nodorum* (Berk) Berk. have been identified from most of the countries. In India *S. nodorum* is important at higher elevations where it may be severe in individual fields (57). Early reports from East Africa indicate that *S. nodorum* was important in Kenya and Tanzania (106), but the behavior of varieties in the RDISN nurseries suggests that *S. tritici* is probably more important now. *S. avenae* Frank f. sp. *triticeae* T. Johnson has also been reported (83), and studies should be made to establish the relative importance of the *Septoria* species in East Africa.

*Septoria tritici* is the species most important economically. In South Asia it is found at higher elevations and on the plains adjacent to the mountains where higher precipitation occurs (122). In the vast plains of India and Pakistan and in the irrigated areas of Afghanistan, *S. tritici* does not occur because temperature and amount of moisture are not favorable. Also in this area, crop debris decomposes readily, thereby destroying inoculum. In a local epidemic in north India next to the Himalayas in 1967–1968 (122), local wheats were susceptible and the dwarf wheats were resistant. Severe attacks occasionally occur in mountainous areas, but there is little information available from such areas.

*S. tritici* is found at higher elevations and in the coastal areas of the Caspian Sea and the Persian Gulf in West Asia. It occurs in irrigated areas of Saudi Arabia, Iraq, Egypt, and Sudan in most of the rain-fed areas of Turkey, Syria, Jordan, and Iran. Its importance is not well understood, but it is probably unimportant except locally.

Septoria blight is one of the major disease problems of wheat on the coasts of the Mediterranean Sea and in the highlands of East Africa. In Morocco, Algeria, Tunisia, Turkey, and Israel, Septoria blight has epidemic potential in areas receiving 700 mm or more of rainfall annually. As the annual rainfall decreases, the disease becomes progressively less of a problem. *S. tritici* is the major pathogen in this area; *S. nodorum* has rarely been identified (101).

In 1968–1969 a severe *S. tritici* epidemic occurred in North Africa when the growing season was unusually wet and cool. In Morocco the epidemic was especially severe and early planted dwarf wheats were badly damaged. Fields of Siete Cerros in Northern Morocco which had a potential yield of 4000 kg per hectare produced 500 to 800 kg per hectare (68). Substantial yield losses also occurred in Tunisia and Turkey (113).

The dwarf wheats as a group have been susceptible to *S. tritici* because they were selected in the absence of Septoria blight. There are, however, a number of dwarf wheat varieties whose resistance probably came from South American wheats in the breeding program; the varieties Tobari 66, Soltane, Zaafrane, and Utique are good examples (9).

The suggestion has been made that the Septoria problem in the Mediterranean zone resulted from the introduction of dwarf wheats (113), but *S. tritici* was a problem 30–40 years ago (39, 81) and epidemics were observed during the decade of the 1950s in Tunisia (Auriau, personal communications). While the indigenous varieties in the Mediterranean area appear adequately resistant when grown under traditional conditions, when grown with the use of additional fertilizer (40) and denser plant populations they also suffer severe attacks by *S. tritici* (Saari, unpublished data). Thus we conclude that the dwarf wheats have not caused the *Septoria* problem but rather have called attention to it.

When the importance of *S. tritici* became apparent, an intensive effort to identify sources of resistance and to develop resistant dwarf varieties was undertaken by the countries concerned and by CIMMYT (9, 97). A number of sources of resistance and potential varieties have been identified in *T. aestivum* (96, 97). According to S'Jacob (109) some of the sources resistant to *S. tritici* may also be resistant to *S. nodorum;* the varieties Nova Prata and Veranopolis are examples. Other sources of resistance are found in the winter wheats and in spring wheats from southern Europe and South America.

*S. tritici* is also recognized as a major disease problem in the Ethiopian Highlands (82) and is probably a problem in Kenya (66) and Tanzania (91) as well, but current information is limited. Some of the best sources of resistance to *S. tritici* in the 1971–1972 nurseries were Kenya wheats selected in Ethiopia (82) and submitted for regional testing in the RDISN (96).

In North Africa where durum wheats are preferred and extensively grown, *S. tritici* resistance is badly needed if semidwarf varieties are to be grown under intensive cultivation. Resistance in the durum wheats is not common but some older varieties, Jaafari, Mouri, Lobeiro, Amarlelejo are promising sources (96).

Physiological races of *S. tritici* have been demonstrated in the laboratory (3[illegible]) but there is no clear evidence that races are important in the field (106). Usually materials resistant at one location are resistant elsewhere. Differences in varietal reactions occur in different areas but are not distinct, except as indicated from North India (122) and Australia (93). The dependence on weather favorable to *S. tritici* probably accounts for many of the observed differences in varietal reaction (122).

In our experience the symptoms produced by *S. tritici, S. nodorum,* other fungi, and physiological disorders are often confused, leading to reports that seem to be in conflict. For example, *S. nodorum* and *S. tritici* can be distinguished at early stages of infection, but after the lesions coalesce, microscopic examination is required to distinguish them. *S. avenae* f. sp *triticea* has been reported from Kenya (83) and its presence may add to the problem of identifying the pathogen. A number of different leaf spots have been attributed to *Septoria,* according to Hosford (52). The authors have received leaf samples reported to be infected with *S. tritici,* but in fact the lesions were caused by other factors and were covered by sporodochia of *Epicoccum* spp. which had been mistaken for pycnidia of *S. tritici.* Other common errors are that infection of the spike is caused only by *S. nodorum* and that all leaf lesions are caused by *S. tritici.* Both organisms attack all above ground parts of the plant.

Losses caused by *Septoria* in the West-Asian region are difficult to determine. Limited data (33, 38) suggest a great variation in loss even though Eyal (38) recently reported that 50% infection of tissue produced about 20% loss. Local epidemics probably occur frequently but seldom does the severe epidemic situation of 1968–1969 occur.

BUNT Bunt is the most serious smut disease of wheat (27, 44, 58, 73, 81, 121). *Tilletia foetida* Wallr. and *T. caries* (D.C.) Tul. occur throughout the region (64), but *T. controversa* Kühn. has been reported only from Turkey and Iran (42, 44, 121). Since *T. controversa* can be confused with *T. caries,* the reports of its absence may be incorrect. For example, we have observed severe stunting of bunt-infected wheat in several countries in South and West Asia but we attributed the dwarfing to common bunt.

Because bunt requires relatively cool weather it is commonly found at higher elevations or in the more northern latitudes of the region. Fields with 40 to 60% infection have been observed in Nepal and Afghanistan.

The pathogen survives as seed- or soil-borne inoculum, depending upon the environment. In south Asia the source of inoculum is principally seed-borne in stocks maintained by the farmers. In West Asia and North Africa, soil-borne inoculum may be a major source of infection. The etiology of bunt in East Africa is not well documented, but seed-borne inoculum is probably common.

The dwarf wheats, Penjamo 62, and the 8156 family of varieties all possess outstanding resistance to the bunts. Wherever these varieties have been grown, the incidence of bunt has been reduced (54). In the higher elevations of South Asia these varieties are increasing in popularity because of their high yield performance and resistance to powdery mildew and bunt. The bunt fungi are capable of genetic change, and new physiologic races may evolve that will overcome this resistance (51).

Fungicidal seed treatment to control bunt is not widely used in the developing nations but it should be encouraged (51). Turkey has a comprehensive program of wheat seed treatment and attempts to replace all seed stocks every 4 to 5 years. This program, according to Turkish authorities, has reduced the incidence of bunt to a manageable level. It has been estimated that without the seed treatment program, bunt would probably be the major disease problem in Turkey.

Losses caused by bunt are difficult to determine because it occurs mostly in hilly and remote areas; for this reason its economic importance is limited to special situations.

PARTIAL BUNT This disease (*Neovossia indica* [Mitra] Mundkur) occurs in Northwestern India and adjacent Pakistan (14, 27, 49, 54, 58) and is a flower-infecting smut. Most varieties of *T. durum* and *T. aestivum* are susceptible when inoculated, but in the field only a few florets per spike become infected. The prevalence of this disease is usually quite low and it appears to offer no threat to wheat production even though most dwarf wheats have been rated susceptible.

LOOSE SMUT Loose smut (*Ustilago tritici* [Pers.] Rostr.) is present in all of the countries (27, 44, 58, 73, 81, 121) and is one of the best known wheat diseases in the developing nations. The degree of infection varies greatly with the wheat varieties grown and with the environment. Prior to the extensive cultivation of dwarf wheats in northern India and Pakistan, loose smut infection on the local varieties averaged 1 to 5%, depending on the year (47, 48, 58). Individual fields with 30–50% smutted heads have been observed (58). Severely infected fields are usually found where seed stocks of a susceptible variety are maintained by the farmer for extended periods of time. The dwarf wheats now being grown are either highly resistant or the levels of infection have been at a very low level with the exception of Inia 66. In the Caspian Sea area of Iran, Inia 66 is often severely infected but it is also grown in several countries without a serious loose smut problem (62). In general, loose smut does not appear to be a serious threat to dwarf wheat production, but highly susceptible varieties should be avoided.

FLAG SMUT *Urocystis agropyri* (Preuss.) Schroet. is found in each of the zones at higher elevations, in the cooler areas adjacent to mountains, and in northern latitudes (27, 37, 47, 48, 58, 81, 90, 121). It is considered a minor disease, although infected plants display severe symptoms and individual fields may contain 10% or more infected plants (48). The inoculum is soil-borne and the continuous culture of wheat could lead to an increase in the importance of flag smut in some areas. The dwarf wheats are susceptible to flag smut (58).

POWDERY MILDEW *Erysiphe graminis* DC. f. sp. *tritici* E. Marchal is widely distributed on wheat in all countries in Asia and Africa (19, 47, 48, 57, 64, 73, 90, 101, 103), but it is apparently important only in certain localities such as the higher elevations of Asia and Africa and the coastal areas of the Mediterranean and Caspian Seas. In such areas powdery mildew can be severe on both bread and durum wheats. Losses probably occur locally, but figures are unavailable.

Since 1968–1969 powdery mildew has been noted with increasing frequency on susceptible lines and varieties at experiment stations and on occasional farm fields in the plains immediately adjacent to the Himalayan Mountains of India and Pakistan. This situation may be the result of longer cool periods near the mountains, the use of more water, fertilizer, and denser plant populations. This situation is not of economic importance at present, but the planting of highly susceptible varieties could present a future disease problem (54). Many of the dwarf wheats are susceptible to powdery mildew (57).

Physiologic races have been identified (54, 85), but their importance and distribution are unknown. Cleistothecia and ascospores are produced readily in most areas where powdery mildew develops and may serve as the mechanism for surviving unfavorable seasons.

LEAF SPOT OR BLIGHT In this category are leaf diseases other than the rusts, smuts, and Septoria diseases. Many different pathogens are reported (13, 54, 57, 72, 74, 81, 91, 103, 114), including 12 species of Helminthosporium, *Alternaria tenuis* Nees. ex Cda., *Ascochyta graminicola* Sacc., *Phoma insidiosa* Tassi, *Cercospora secalis* Chupp, *Cochliobolus sativus* (Izo & Kuribayashi) Dresch. ex Dastur, *C. tritici* Dastur, *Cladosporium herbarum* (Pers.) Link, *C. graminum* Cda. *Pyricularia oryzae* Sacc., *Nigrospora sphaerica* (Sacc.) Mason, *Dilophosphera alopecuri* Fr., and *Stagnospora hemmebergii* (Kuehn) Petrak & Syd., but *Helminthosporium sativum* Pam. King & Bakke, and *Alternaria triticina* Prasada & Prabhu are probably the most important (54, 70, 77). Most pathogens have been reported only a few times, usually only once, and their importance is slight except perhaps occasionally in local areas. These diseases have been studied primarily in India though similar foliage diseases occur in all the other countries of the region.

*Alternaria triticina* is reported only from India and is most severe in eastern and central India (84). It attacks susceptible varieties at any stage of growth, but usually appears on older plants. It is common on senescing leaves or plants weakened because of poor agronomic practices (57). The dwarf wheats are susceptible in greenhouse trials but in field tests they are moderately resistant (54). When dwarf wheats are grown under good agronomic conditions, the incidence of Alternaria blight is minor. Resistant sources have been identified, but the disease is unlikely to be of major importance if good agronomy is practiced and if highly susceptible varieties are avoided.

At least twelve Helminthosporium species (57) are involved in spotting leaves of wheat but the most commonly reported species is *H. savitum,* which occurs throughout the region (7, 41, 73, 81, 90–92, 119, 121). Severe attacks occur in the hot humid areas of eastern and southern India outside the normal range of wheat cultivation (54, personal observations). The Helminthosporium leaf spots also occur in Northern India and Pakistan but are not economically important (78). In West Asia, East Africa, and North Africa Helminthosporium leaf spot is reported but again the economic importance is unknown (80, 90, 91, 119, 121).

The most extensive identification of leaf spotting pathogens has been done in India (70), where many plant pathologists trained in mycology have been working. The pathogens of most of the leaf spots in the region are assigned to *H. sativum* (also *A. triticina* in India), probably because these pathogens are fairly common and because many workers do not have facilities or expertise for mycological examinations. Thus some diseases and pathogens may be overlooked. We have repeatedly observed a leaf spot on wheat in experimental plots in northern Afghanistan and in the Caspian Sea area of Iran that resembles the symptoms described for *H. tritici-repentis* Died., but the pathogen has not been positively identified. Hosford (52) suggested that *H. tritici-repentis* may be important on durum in North Africa. Recently, Peregrine & Siddiqi (80) reported that spots caused by *Cercospora secalis* Chupp resembled those caused by *H. savitum.*

ROOT ROT, FOOT ROT, AND SEEDLING BLIGHT Many pathogens have been associated with these diseases including species of *Fusarium, Gibberella, Helminthosporium, Rhizoctonia, Ophiobolus graminis* Sacc., *Sclerotinia sclerotiorum* (Lib) de Bary, *Sclerotium rolfsii* Sacc., *Pythium graminicolum* Subram., and *Cercosporella herpotrichoides* Fron. (54, 81, 91–92, 95, 101, 113, 119). Most of these pathogens have been described in all of the countries of the region, but there is little information on the seriousness and extent of the diseases.

There are some indications that root and seedling diseases may be increasing in importance, expecially in India and North Africa. In India, Saksena (100) has noted an increased incidence of Rhizoctonia root rot and seedling blight in localities where wheat is sown after rice in a system of double cropping. In the Mediterranean region, root rots and seedling blights frequently occur (81, 90). Perhaps the climate tends to favor survival of the pathogens and predisposes the wheat to attack. *Fusarium* and *Helminthosporium* species are usually encountered.

In Eastern Africa where root and foot rots have been a concern for years (92), the usual pathogens are species of *Fusarium* and *Helminthosporium.* Perhaps these diseases will be important on the dwarf wheats when their use expands in the area.

The concern with seedling blights and root rots in wheat has been brought about indirectly through the use of dwarf wheat varieties. They are not more susceptible than the older local wheats, but the modern agronomic methods developed for their use have greatly improved the uniformity, vigor, and dependability of stands, improving the ability to recognize the problems. In addition, wheat cultivation has been extended into warmer areas and seasons; the higher temperatures favor the development of wheat seedling blights and root rots. Fungicidal seed treatment by farmers is not common practice, but this may become necessary if these diseases continue to increase in prevalence and importance.

DOWNY MILDEW This disease, caused by *Sclerophthora macrospora* Thir. Shaw and Narasimham, has been observed in all countries (27, 59, 81, 123). It occurs only sporadically on a few plants, usually in fields where the soil has been waterlogged by excessive irrigation. No sources of resistance have been identified, but proper irrigation should eliminate the disease. It is doubtful that any yield loss occurs at present.

ERGOT *Claviceps purpurea* (Fr.) Tul. is commonly found throughout the region on grasses and rye (27). It seldom occurs on wheat but has been reported in Turkey (111), Morocco (90), and Iran (103). While ergot is not an economic problem of wheat in the region, the recent experience in the United States with ergot on Waldron wheat indicates that a highly susceptible variety can be seriously affected in certain seasons.

SCAB Scab, caused by several *Fusarium* spp., has been observed in most countries (35, 37, 81, 91, 111, 114, 119) and is prevalent in the coastal areas of North Africa. Moisture at heading and flowering time favors the disease; it is seldom encountered in arid or irrigated areas. The significance of scab also appears to increase in areas where maize and wheat are rotated or exist as companion crops (21). The potential increase of scab appears to be slight but the dwarf wheats as a group are susceptible; only a few of the local or improved varieties appear to have some resistance (35).

BLACK POINT This disease, also known as kernel smudge, can be caused by several fungi including *Alternaria* spp., *Helminthosporium* spp., and *Fusarium* spp. (34, 57, 59, 90, 91, 103, 119). It occurs in all the countries in the various zones included in this summary. Rainfall late in the season just before harvest favors the development of black point. Most varieties are susceptible but the disease is of little economic importance because of prevalent dry weather during harvest time (34, 54).

BACTERIAL DISEASES No unusual or unique bacterial diseases have been reported or observed on dwarf wheats (27, 28, 37, 54, 91, 102, 103, 114, 119, 124) probably because of the warm dry weather that often prevails during the latter part of the growing season. In North Africa late rains occur and in areas of East Africa where rainfall is more abundant, bacterial black chaff (*Xanthomonas translucens* f. sp. *cerealis* Hagborg) and basal glume rot (*Pseudomonas atrofaciens* [MuCulloch] Stevens) are encountered.

Yellow ear rot or spike blight is caused by *Corynebacterium tritici* (Hutchinson) Burkholder and related species (102), and it is associated with the nematode, *Anguina tritici* (28, 118). The disease is most important in individual fields where farmers maintain their own seed, especially in India and Pakistan. Similar situations undoubtedly occur in Iran, Afghanistan, and the rest of West Asia, but reports are limited or the disease has been overlooked. Yellow ear rot has been reported from Egypt but not from the rest of North Africa, although it may also occur there, especially in wet seasons. Yellow ear rot is readily controlled by removal of the nematode galls from the seed (28, 124).

VIRUS DISEASES Several virus and virus-like diseases (barley yellow dwarf, mosaic, striate, pearl millet streak, African cereal streak, and streak) occur on wheat in the region (1, 8, 27, 54, 57, 104, 110, 121). However, their economic impact is probably slight. Little work has been done other than to establish their presence and their mode of transmission, except in the case of mosaic streak of wheat, which has been studied in some detail in India (89).

In Iran and Turkey a severe yellows-type disease has been observed, and from the appearance of the symptoms we believe it to be caused by a virus or mycoplasma, but identification of the causal agent or vector has not to our knowledge been attempted.

No major changes in virus or virus-like diseases have been identified that can be associated with the introduction and increased use of dwarf wheats. A greater number of viruses have been identified recently, but this probably reflects a growing awareness of and study of the wheat crop. It is likely that additional virus and virus-like diseases will be reported from wheat in the future because of the increased use of fertilizer and irrigation which results in more normal appearing plants and permits more ready detection of diseased plants. Furthermore, these agronomic inputs may allow larger populations of insects to develop in wheat as a result of the physiological and microclimatic changes produced (21, 118). Since many of the aphids and leafhoppers on wheat are potential vectors of viruses, an increase in their numbers may increase the hazard from these diseases.

NEMATODES Ear cockle [*Anguina tritici* (Steinbach) Filipiev] and the cereal cyst nematode (*Heterodera avenae* Wollenweber) are the most easily recognized and probably the most important nematode diseases of wheat. Ear cockle is important in the Indo-Pak subcontinent (124), though it is occasionally found in other parts of Asia and Africa (28). Invariably the disease is associated with farmers who maintain their own seed. The nematode is reported to survive in the soil for periods up to 7 months (25). It can be effectively controlled by using clean seed or by the removal of the nematode galls from the seed (118, 124). Healthy seed should be sown on fields that have been free of wheat or related cereals for a year (25). Resistant wheats have been identified, but it is questionable whether the effort to develop resistant varieties is merited when effective management can virtually eliminate the disease.

The cereal cyst nematode is important in certain limited areas of India (63) and Tunisia (24, personal observation). In heavily infested soils, losses can be substantial (71, 107).

As more intensive nematode studies develop, the cereal cyst nematode will probably be identified in most countries, and in those areas where the nematode is an economic problem, resistant or tolerant varieties will have to be developed if wheat cultivation is to be intensified. Coupled with the development of such varieties, rotations and other agronomic practices may be also required. A number of resistant sources have been identified (63, 71, 107), but the possibility of races of the nematode has not received adequate attention.

Several other nematodes have been reported as pathogens on wheat, but these reports require confirmation (37, 54). There is no information available on the significance of these nematodes to production.

MINERAL DEFICIENCIES Nitrogen and phosphorous are deficient in most of the wheat-growing areas of the region. In addition, zinc deficiency is becoming increasingly evident in North India and adjacent Pakistan, and copper-deficient areas occur in Kenya and probably other countries of the Rift Valley. As wheat cultivation continues to intensify throughout the region, we expect an increased awareness of nutrient deficiencies, especially as diagnostic capabilities improve (36, 60).

### *Concluding Remarks*

The dwarf wheats introduced into Asia and Africa stimulated production, particularly in areas where the required inputs were available. Despite the changes that have occurred, diseases of recognized importance have not been altered significantly. The case of *Septoria tritici* in the Mediterranean area illustrates the growing awareness of epidemic disease potentials and emphasizes the need to adequately evaluate all varieties to the endemic pathogen populations. This task is great and is one of the major challenges confronting plant pathology in developing nations.

The wheat diseases now found in this region of the world occur on both indigenous and dwarf wheats. These reports reflect both the interest stimulated by increased production and the effort of a greater number of professionals working on wheat.

The significance of some diseases has decreased with the cultivation of dwarf varieties and no disease has increased in significance. However, the magnitude of losses has not been well documented in the region (31) and this aspect of plant pathology requires immediate attention.

The age-old problem of the wheat rusts still exists, and presents the major concern to intensified wheat cultivation. New races have appeared that are capable of attacking many dwarf wheats, and the leaf rust situation on the Indo-Pak subcontinent threatens the continued use of currently grown varieties.

In Asia and Africa resistant varieties are the only practical means of controlling wheat diseases. Some of the dwarf wheats retain adequate rust resistance, but a number of extensively grown varieties are now subject to rust attack. To provide resistant varieties, comprehensive breeding programs are being developed in most of the countries; India, Pakistan, Iran, Turkey, and Egypt are examples. CIMMYT assists with materials, training, and mutual assistant programs such as the monitoring of diseases and variety testing on a regional basis. Furthermore, multilineal resistant varieties will soon be available from these cooperative endeavors. It is imperative that all such efforts to guard against wheat disease epidemics be expanded in countries dependent on cereal grains as the major food source because of increasing population pressures and the lack of food supplies.

*Literature Cited*


1. Abdel-Hak, T. 1966. *Diseases of Field Crops,* Vol. I. Cairo, Egypt: Anglo Book Shop. 455 pp. (In Arabic)
2. Abdel-Hák, T. 1970. Importance of leaf rust in the Near East, *Proc. 3rd FAO/ Rockefeller Found. Wheat Seminar,* Ankara, Turkey, 239–46. 386 pp.
3. Abdel-Hak, T. 1970. Importance of wheat stem rust in the Near East. See Ref. 2, pp. 226–31
4. Abdel-Hak, T., Kamel, A. H. 1972. Present status of wheat stem rust in the Near East Region. *Proc. Reg. Wheat Workshop. Vol. I. Diseases,* Chap. 15. Beirut, Lebanon: Ford Found. (Mimeo.)
5. Abdel-Hak, T. A., Stewart, D. M., Kamel, A. H. 1972. The current stripe rust situation in the Near East region. See Ref. 4, Chap. 10
6. Abdel-Hak, T., Stewart, D. M., Kamel, A. H. 1973. Wheat diseases and their relevance to the improvement and production programmes in the Near East. *Proc. 4th FAO/Rockefeller Found. Wheat Seminar, Tehran, Iran.* (Mimeo.) In press
7. Ahmadi, G. S. 1973. Wheat Disease in Afghanistan. See Ref. 6
8. *Annual Report* for 1972. Plant Breeding Station, Njorio, Kenya. 11 pp.
9. *Annual Report* of the International Maize & Wheat Improvement Center (CIMMYT), Mexico, 1970–1971. 114 pp.
10. *Annual Report* of the International Maize & Wheat Improvement Center (CIMMYT), Mexico, 1969–1970, 136 pp.
11. *Annual Report* of the International Maize & Wheat Improvement Center (CIMMYT), Mexico, 1968–1969, 122 pp.
12. Athwal, D. S. 1971. Semidwarf rice and wheat in global food needs. *Quart. Rev. Biol.* 46:1–34
13. Bamdadian, A. 1973. *Check list of wheat diseases in Iran* (Compiled from personal experience and reports written in several languages). Evin, Tehran, Iran: Plant Pests and Dis. Res. Inst.
14. Bedi, K. S., Sikka, M. R., Mundkur, B. B. 1949. Transmission of wheat bunt due to *Neovossia indica* (Mitra) Mundkur. *Indian Phytopathol.* 2:20–26
15. Bhardwaj, R. B. L., Gautam, R. C., Singh, A. 1971. Depth and method of sowing dwarf and semi-dwarf wheats. *Indian J. Agron.* 16:28–32
16. Borlaug, N. E. 1968. Wheat breeding and its impact on world food supply. *Proc. 3rd Int. Wheat Genet. Symp.,* ed. K. W. Finlay, K. W. Shephard. Sydney, Australia: Butterworth. 479 pp.
17. Borlaug, N. E. 1965. Wheat, rust and people. *Phytopathology.* 55:1088–98
18. Boskovic, M. 1972. Short communication of the results in the International wheat leaf rust project. See Ref. 4, Chap. 16
19. Boughey, A. S. 1946. A preliminary list of plant diseases in the Anglo-Egyptian Sudan. *Mycological Papers No. 14.* Kew, Surrey, Engl.: Imp. Mycol. Inst. 16 pp.
20. Brown, L. R. 1970. *Seeds of Change.* New York: Praeger. 205 pp.
21. Bruehl, G. W. 1967. Diseases other than rust, smut, and virus. *Wheat and Wheat Improvement,* ed. K. S. Quisenberry, L. P. Reitz, Monogr. No. 13: 375–410. Madison, Wisc.: Am. Soc. Agron. 560 pp.
22. Butler, E. J., Hayman, J. M. 1906. Indian wheat rusts. *Mem. Dep. Agr. India, Bot. Ser.* 1:57
23. Byres, T. J. 1972. The dialectic of India's Green Revolution. *South Asian Rev.* 5:99–116
24. Cheikh, M. 1968. *Biologie des nematodes parasites des graminees,* No. 40. Tunisie: Inst. Rech. Agron. Tunisie. 30 pp.
25. Christie, J. R. 1959. *Plant Nematodes: Their Bionomics and Control.* Jacksonville, Florida: Drew. 256 pp.
26. Committee Report. 1972. *Genetic Vulnerability of Major Crops.* Washington DC: Nat. Acad. Sci., 307 pp.
27. Commonwealth Mycological Institute. 1971. *Distribution maps of plant diseases.* Pathogen index (wheat, p. 24) and maps issued through map no. 80, 1973. Kew, Surrey, Engl.: Commonw. Mycol. Inst.
28. Commonwealth Mycological Institute. 1973. *Corynebacterium tritici* (Hutchinson) Burkholder. *Descriptions of Pathogenic fungi and bacteria,* No. 377. Kew, Surrey, Engl.: Commonw. Mycol. Inst.
29. Commonwealth Mycological Institute. 1968. A bibliography of lists of plant

diseases and fungi. I. Africa. *Rev. Appl. Mycol.* 47:553–58

30. Commonwealth Mycological Institute. 1970. A bibliography of lists of plant diseases and fungi. II. Asia. *Rev. Plant Pathol.* 49:103–8
31. Cramer, H. H. 1967. Plant protection and world crop production. *Pflanzenschutz Nachr.* 1967/1. Leverkusen: Bayer. 524 pp.
32. Dalrymple, D. G. 1972. Imports and plantings of high yielding varieties in less developed nations, *Foreign Econ. Develop. Rept.-14. Foreign Econ. Devel. Ser., US Agency Int. Develop.* 56 pp.
33. Devecioglu, B. 1972. Prospects and methods of breeding varieties resistant to *Septoria.* See Ref. 4, Chap. 5
34. Dharam Vir, Adlakha, K. L., Joshi, L. M., Pathak, K. D. 1968. Preliminary note on the occurrence of black point disease of wheat in India. *Indian Phytopathol.* 21:234–35
35. Djerbi, M. 1971. Ecological study of the parasitical actions on wheat of *Fusarium* species. *Bull l'Inst. Nat. Agron. Tunise* 28–29:155–56
36. Einkerton, A., Barrett, M. W., Guthrie, E. J. 1965. A note on copper deficiency in the Njoro area, Kenya. *East Afr. Agr. Forest. J.* 30:257–58
37. El-Helaly, A. F. et al 1966. General survey of plant diseases and pathogenic organisms in the U.A.R. (Egypt) until 1965. *Alexandria J. Agr. Res. Fac. Agr. Egypt. Res. Bull. No. 15.* 134 pp.
38. Eyal, Z. 1972. Effect of *Septoria* leafblotch on the yield of spring wheat in Israel. *Plant Dis. Reptr.* 56:983–86
39. Eyal, Z., Amiri, Z., Wahl, I. 1973. Physiologic specialization of *Septoria tritici. Phytopathology* 63:1087–91
40. Fellows, H. 1962. Effects of light, temperature and fertilizer on infection of wheat leaves by *Septoria tritici. Plant Dis. Reptr.* 46:846–48
41. Gattani, M. L. 1964. Plant diseases of economic importance in Afghanistan. *FAO Plant Prot. Bull.* 10:30–35
42. Gobelez, M. 1963. La mycoflore de Turquie. *Mycopathol. Mycol. Appl.* 19: 296–314
43. Green, G. J., Martens, J. W., Ribeiro, O. 1970. Epidemiology and specialization of wheat and oat stem rusts in Kenya in 1968. *Phytopathology.* 60: 309–14
44. Hafiz, A. 1970. Economic significance of wheat diseases and insect pests and methods of testing varieties against various diseases. See Ref. 2, pp. 207–25
45. Harder, D. E., Mathenge, G. R., Mwaura, K. 1972. Physiologic specialization and epidemiology of wheat stem rust in East Africa. *Phytopathology* 62:166–71
46. Harrington, R. E. et al October 1972. *Agricultural Mechanization in India.* New Delhi, India: Ford Found. Rept.
47. Hassan, S. F. 1970. Cereal disease situation in Pakistan. See Ref. 2, pp. 255–58
48. Hassan, S. F. 1973. Wheat diseases and their relevance to improvement and production in Pakistan. See Ref. 6
49. Hogg, W. H., Hounam, C. E., Mallik, A. K., Zadoks, J. C. 1969. Meterological factors affecting the epidemiology of wheat rusts. *World Meteorol. Org. Tech. Note,* No. 99. 143 pp.
50. Holton, C. S. 1967. Plant disease pitfalls for high yielding wheats in India. *Indian Phytopathol.* 20:183–88
51. Holton, C. S. 1967. Smuts. See Ref. 21, pp. 337–53
52. Hosford, R. M. Jr. 1972. Propagules of *Pyrenophora trichostoma. Phytopathology* 62:627–29
53. Johnson, T., Green, G. J., Samborski, D. J. 1967. The world situation of cereal rusts. *Ann. Rev. Phytopathol.* 5:183–200
54. Joshi, L. M. 1974. Wheat Diseases. *Wealth of India: A Dictionary of Indian Raw Materials and Industrial Products.* Delhi, India: Counc. Sci. Ind. Res. In press
55. Joshi, L. M., Ahmed, S. T., Sinha, V. C., Mohan, M. 1971. Note on the threat to Kalyan Sona and other improved wheat varieties from races 42, 42-B and 122 of *Puccinia graminis tritici* (Pers.) Erikss. & Henn. *Indian J. Agr. Sci.* 41:640–42
56. Joshi, L. M., Goel, L. B., Sharma, R. C., Mohan, M. 1970. Prevalence and distribution of physiologic races of wheat rusts on dwarf wheats during 1967–68 crop. *Indian Phytopathol.* 23:637–43
57. Joshi, L. M., Renfro, B. L., Saari, E. E., Wilcoxson, R. D., Raychaudhuri, S, P. 1970. Diseases of wheat in India other than rusts and smuts. *Plant Dis. Reptr.* 54:594–97
58. Joshi, L. M., Renfro, B. L., Saari, E. E., Wilcoxson, R. D., Raychaudhuri, S. P. 1970. Rust and smut diseases of wheat in India. *Plant Dis. Reptr.* 54:391–94
59. Kamel, M., Moghal, S. M. 1968. *Studies on Plant Diseases of Southwest Pakistan.* Karachi: W. Pakistan Govt. Press. 207 pp.
60. Kanwar, J. S. 1972. Twenty-five years of research in soil, fertilizer and water

management in India. *Indian Farming* 22:16–25

61. Khadka, B. B., Shah, S. M. 1967. Preliminary list of plant diseases recorded in Nepal. *Nepalese J. Agr.* 2:47–76
62. Khazra, H., Bamdadian, A. 1973. Wheat disease situation in Iran. See Ref. 6
63. Koshy, P. K., Swarup, G. 1971. Distribution of *Heterodera avenae, H. zeae, H. cajani* and *Anguina tritici* in India. *Indian J. Nematol.* 1:106–11
64. Kranz, J. 1965. A list of plant pathogenic and other fungi of Cyrenaica (Libya) *Phytopathological Papers* No. 6. Kew, Surrey, Engl.: Commonw. Mycol. Inst. 24 pp.
65. Leppik, E. E. 1970. Gene centers of plants as sources of plant disease resistance. *Ann. Rev. Phytopathol.* 8:323–44
66. Little, R. 1969. Blotch (*Septoria* spp.) parental collection. *FAO Inform. Bull. Near East Wheat Barley Impr. Prod. Proj.* 6(3):31–32
67. MacKenzie, D. R., Mexas, A. G., Borlaug, N. E., Finlay, K. W. 1971. Results of the fifth international spring wheat yield nursery, 1968–1969. *Int. Maize Wheat Impr. Center Mexico City Mexico Res. Bull.* No. 19. 33 pp.
68. McCuistion, W. L. 1972. *Septoria:* Experience in North Africa. See Ref. 4, Chap. 4
69. Menzies, J. D. 1967. Plant diseases related to irrigation. *Irrigation of Agricultural Lands,* ed. R. M. Hagan, H. K. Haise, T. W. Edminister. Madison, Wis.: Am. Soc. Agron. Monogr. No. 11. 1180 pp.
70. Misra, A. P. 1973. *Studies on Helminthosporium species occurring on cereals and other Gramineae with special reference to species occurring on corn and sorghum in India.* Final Report 1966–71. US Public Law 480 Res. Proj. Dholi, Mazuffarpur, Bihar, India: 386 pp.
71. Mukhopadhyaya, M. C., Dalal, M. R., Saran, S., Kharub, S. S. 1972. Studies on the 'Molya' disease of wheat and barley. *Indian J. Nematol.* 2:11–20
72. Munjal, R. L., Kaul, T. N. 1961. *Dilophosphora* leafspot of wheat in India. *Indian Phytopathol.* 14:13–15
73. Nattrass, R. M. 1961. Host lists of Kenya fungi and bacteria. *Mycological Papers* No. 81. Kew, Surrey, Engl.: Commonw. Mycol. Inst. 46 pp.
74. Nema, K. G., Dave, G. S., Khosla, H. K. 1971. A new glume blotch of wheat. *Plant Dist. Reptr.* 55:95
75. Ozkan, M., Prescott, J. M. 1972. Cereal rusts in Turkey. *Proc. Eur. Mediter. Cereal Rusts Conf. Prague Czech,* 2: 183–84. 311 pp.
76. Paddock, W. C. 1970. How green is the green revolution. *Bio. Sci.* 20:897–902
77. Pal, B. P. 1966. *Wheat.* New Delhi: Indian Counc. Agr. Res. 370 pp.
78. Palmer, L. D. 1970. Physiological yellowing of wheat. *1970 All-India Wheat Res. Workers' Workshop New Delhi India* 5:383–88
79. Pendleton, J. W. December 1969. Trends in practices and production of major corn belt crops. Disease Consequences of Intensive and Extensive Culture of Field Crops. *Spec. Rept. No. 64 of Iowa State Univ. Sci. Technol.,* 6–11. Ames, Iowa. 55 pp.
80. Peregrine, W. T. H., Siddiqi, M. A. 1972. A revised and annotated list of plant diseases in Malawi. *Phytopathological Papers* No. 16. Kew, Surrey, Engl.: Commonw. Mycol. Inst. 51 pp.
81. Petit, A. 1935. Les maladies cryptogamiques du ble. *Ann. Serv. Bot. Agro.,* Tunisie 11:195–234
82. Pinto, F. F. 1972. Development of *Septoria tritici* in wheat and sources of resistance in Ethiopia. See Ref. 4, Chap. 6
83. Pinto, F. F., Hurd, E. A. 1970. Seventy years with wheat in Kenya. *East Afr. Agr. Forest.* J. V. 36. (Special Issue) 24 pp.
84. Prabhu, A. S., Prasada, R. 1966. Pathological and epidemiological studies on leaf blight of wheat caused by *Alternaria triticina. Indian Phytopathol.* 19:95–112
85. Prabhu, A. S., Prasada, R. 1963. Physiologic races of wheat powdery mildew (*Erysiphe graminis tritici*) in Simla and Nilgiri hills. *Indian Phytopathol.* 16: 201–4
86. Poulsen, K. K. 1969. Parents used in the development of Mexican wheat varieties. FAO *Inform. Bull. Near East Wheat Barley Impr. Proj.* 6:33–37
87. Prescott, J. M. 1972. The Regional Trap Nursery 1971–72. See Ref. 4, Chap. 18
88. Prescott, J. M. 1973. Results of Regional Trap Nursery. 1971–72. *Mimeogr. Rept. Wheat Res. Training Center Ankara, Turkey.* 26 pp.
89. Raychaudhuri, S. P., Ganguly, B. 1968. A mosaic streak of wheat. *Phytopathol. Z.* 62:61–65
90. Rieuf, P. 1971. Parasites et saprofytes des plantes au Maroc. *Cahiers Rech. Agron.* 30:469–570
91. Riley, E. A. 1960. A revised list of plant diseases in Tanganyika territory. *Mycol.*

*Papers* No. 75. Kew, Surrey, Engl.: Commonw. Mycol. Inst. 42 pp.

92. Rosella, A. 1948. Les pietins des cereales au Moroc. Les organisms que l'on rencontre le plus frequemment. Les conditions qui facilitent ou qui provoquent l'attaque. *C. R. Seances Acad. Agr. Fr.* 34:681–83
93. Rosielle, A. A. 1972. Sources of resistance in wheat to speckled leaf blotch caused by *Septoria tritici. Euphytica* 21:152–61
94. Rotem, J., Palti, J. 1969. Irrigation and plant diseases. *Ann. Rev. Phytopathol.* 7:267–88
95. Saad, A. T., Nienhaus, F. 1969. Plant Diseases in Lebanon. *Z. Pflanzenkr. Pflanzenschutz* 76:539–51
96. Saari, E. E. 1974. Results of studies on *Septoria* in the Near East and Africa. *FAO Plant Prot. Bull.* In press
97. Saari, E. E. 1972. Septoria: Review of the current situation in the region, methods for evaluating resistance, sources of resistance identified in the first regional disease and insect screening nursery. See Ref. 4, Chap. 3
98. Saari, E. E. 1972. The Regional Disease and Insect Screening Nursery. Objectives and Mode of Operation. See Ref. 75, pp. 213–17
99. Saari, E. E., Kingma, G., Prescott, J. M. 1972. Identifying sources of resistance through international testing. See Ref. 75, pp. 219–23
100. Saksena, H. K., Kumar, K. 1970. Rhizoctonias on wheat causing root rot and seedling blight and sharp eyespot. *All-India Wheat Research Workers' Workshop New Delhi India* 5:377–78
101. Santiago, J. C. 1970. *Resultado das observacoes effectuadas em Marroco e na Tunisia respeifantes as doencas e pragas does cereals, principalmente em relacao aos trigos Mexicanos cultivados nestes paises: Missoes de estudo relizadas em 1969 a expensas das Missoes Americanas de auxilio tecnico a Marracos e Tunisia.* Elvas, Portugal, Estacao de Melharamento de Plantas. 22 pp. (Mimeo.)
102. Scharif, G. 1961. *Corynebacterium iranicum* sp. nov. on wheat (*Triticum vulgare* L.) in Iran and a comparative study of it with *C. tritici* and *C. rathayi. Entomol. Phytopathol. Appl.* 19:1–24
103. Scharif, G., Ershad, D. 1966. *A list of fungi on cultivated plants, shrubs and trees of Iran.* Plant Pests and Diseases. Research Institute Publ. Evin Tehran, Iran. 89 pp.
104. Seth, M. L., Raychaudhuri, S. P., Singh, D. V. 1972. Bajra (Pearl millet) streak: A leafhopper-borne cereal virus in India. *Plant Dis. Reptr.* 56:424–28
105. Sharma, S. K., Joshi, L. M., Singh, S. D., Nagarajan, S. 1972. New virulence of yellow rust on Kalyansona variety of wheat. *Proc. Eur. Mediter. Cereal Rusts Conf. Prague Czech.* 1:263–65. 308 pp.
106. Shipton, W. A., Boyd, W. R. J., Rosielle, A. A., Shearer, B. I. 1971. The common septoria diseases of wheat. *Bot. Rev.* 37:231–62
107. Sikora, R. A., Koshy, P. K., Malek, R. B. 1972. Evaluation of wheat selections for resistance to the cereal cyst nematode. *Indian J. Nematol.* 2:81–82
108. Singh, A., Bhardwaj, R. B. L., Gautam, R. C., Singh, M. 1971. Optimum seed rate and nitrogen requirement of dwarf wheat varieties under different dates of sowing. *Indian J. Agron.* 16:23–27
109. S'Jacob, J. O. 1967. Verslagen van literatuuronderzoek op het gebied der graanziekten ten dienste van de veredeling. IV. Septoriosen van granen: A kafjesbruin (Glume blotch. Spelzenbraune) veroorzaakt door *Leptosphaeria nodorum* Muller (*Septoria nodorum* Berk.) Stichting Nederlands Graan-Centrum, Wageningen. 48 pp.
110. Slykhuis, J. T. 1962. An international survey for virus diseases of grasses. *FAO Plant Prot. Bull.* 10:1–16
111. Sprague, R. 1950. *Diseases of Cereals and Grasses in North America.* New York: Ronald. 538 pp.
112. Stakman, E. C., Harrar, J. G. 1957. *Principles of Plant Pathology.* New York: Ronald. 581 pp.
113. Stewart, D. M., Hafiz, A., Abdel-Hak, T. 1972. Disease epiphytotic threats to high-yielding and local wheats in the Near East. *FAO Plant Prot. Bull.* 20:50–57
114. Stewart, R. B., Yirgou, D. 1967. Index of Plant Diseases in Ethiopia. *Exp. Sta. Bull.* No. 30. Ethiopia: Coll. Agr. Haile Sallasie Univ. 95 pp.
115. Stubbs, R. W. 1973. Significance of stripe rust in wheat production and the role of the European-Mediterranean cooperative programme. See Ref. 6
116. Stubbs, R. W. 1972. The international survey of virulence of *Puccinia striiformis* virulence patterns in the middle east and Africa and potential sources of resistance. See Ref. 4, Chap. 9
117. Stubbs, R. W. 1972. The international survey of factors of virulence of *Puccinia striiformis* Westend. See Ref. 105, pp. 283–88

118. Swarup, G., Singh, N. J. 1962. A note on the nematode bacterial complex in tundu disease of wheat. *Indian Phytopathol.* 15:294–95
119. Thorold, M. A. 1935. Diseases of cereal crops in Kenya colony. *Dep. Agr. Bull.* No. 2. Colony and Protectorate of Kenya. 66 pp.
120. Tsu, S. K. 1971. High-yielding varieties of wheat in developing countries. *USDA Econ. Res. Serv. Foreign.* 322 pp.
121. *Turkrye Kultur Bitkilerinde Farar Yapan Haslatik, Zararli ve Yabanci Otlar.* 1969. Tarim Bakanligi zirai mucadele ve zirai Karantina zenel Mudurlugu Arastirma Subesi Sayi: 2. Anakar, Turkey. 122 pp.
122. Tyagi, P. D., Joshi, L. M., Renfro, B. L. 1969. Reactions of wheat varieties to *Septoria tritici* and report of an epidemic in North Western Punjab. *Indian Phytopathol.* 22:175–78
123. Tyagi, P. D., Anand, S. C. 1968. Downy mildew of wheat in India. *Plant Dis. Reptr.* 52:569
124. Vasudeva, R. S., Hingorani, M. K. 1952. Bacterial disease of wheat caused by *Corynebactrium tritici* (Hutchinson) Bergey et al. *Phytopathology* 42:291–93
125. Watson, I. A. 1970. Changes in virulence and population shifts in plant pathogens. *Ann. Rev. Phytopathol.* 8:209–30
126. *World agricultural production and trade statistical report.* 1973. USDA Foreign Agr. Serv. 33 pp.
127. Yarwood, C. E. 1968. Tillage and plant diseases. *Bio Science* 18:27–30

# PLANT RESPONSE TO ROOT-KNOT NEMATODE

❖3588

*Alan F. Bird*
Division of Horticultural Research, CSIRO, Adelaide, South Australia

## INTRODUCTION

This review is restricted to a discussion of the responses of susceptible plants to root-knot nematodes and is based largely on the recent research relevant to the hypotheses discussed here. This review considers only very briefly the responses of the nematode to the plant and does not cover plant resistance to nematodes because these topics have been dealt with in recent reviews (12, 71). Also, the early literature on the responses of plants to nematode infections is not discussed, as it has been well documented in a series of reviews in the ten year period from 1961 (27, 33, 50, 55, 75, 89). Some of the statements made in this review are speculative and in time will probably be shown to be incorrect. Nevertheless, they are included here as a possible stimulus to further research.

## RESPONSES OF WHOLE INTACT SUSCEPTIBLE PLANTS

The response of plants to parasitism by the genus *Meloidogyne* can perhaps best be considered under two major headings: First, the response of the whole plant, and second, that of the plant cells in the immediate vicinity of the parasite. The first type is the more obvious because it may involve the stem and leaves, as well as the roots.

### *Morphological Responses in Whole Plants*

Species of *Meloidogyne* have a very wide host range, including both crop plants and weeds. Their economic effects on crop production have well documented in a recent text (90). In addition to a decline in fruit production, heavily infected plants are stunted and can exhibit symptoms of nutritional deficiency. These responses are related to the numbers of second stage infective larvae ($L_2$s) entering and becoming established within the plant. At high inoculum levels the growth of tomato plants is significantly reduced when compared with controls and with plants infected with fewer $L_2$s of *M. javanica* (11). Experiments on the effect of different inoculum levels of eggs of *Heterodera rostochiensis* on the growth and yield of potato plants (77) have given somewhat similar results in that plant weight is reduced at medium densities by necrosis of root tips caused by penetration of the $L_2$s. Growth stopped entirely in the first weeks after planting, at high inoculum densities.

Wallace (86) has proposed the hypothesis that differences in plants' responses to parasitism by *M. javanica* are the result of an interaction between stimulatory and inhibitory processes in the plant. There is no information on the precise nature of these two opposing physiological processes in host plants. The level of plant response must depend on factors such as the ratio between the number of nematodes and the food resources supplied by the plant. Clearly, a plant growing rapidly in a suitable environmental situation will do better than one struggling in a poor environment and exposed to the same number of nematodes. The rate of increase of nematode populations is density dependent up to a level known as the equilibrium density (76), the population being regulated by competition for food.

The most obvious morphological response to infection by *Meloidogyne* is the characteristic galling or knotting of the host plant's roots, hence its common name. Normally the $L_2$ can reach and infect only the roots; however, they have been reported on stems and leaves of plants when they have been able to reach these under natural conditions (51, 54, 80, 81). *M. javanica* can be induced to grow in these unusual situations either by injection using a hypodermic syringe (3, 69) or by applying the nematodes in a mist to leaves and covering the leaves with a plastic bag (54). The amount of galling in roots varies depending on the species of both plant and nematode involved in the host-parasite relationship. For instance, in some plants such as dwarf beans, galling is not very pronounced and the turgid females protrude from the roots in much the same way as members of the genus *Heterodera* normally do (3). At the other extreme, enormous galls are formed by multiple infections in the genera *Rheum, Begonia,* and *Thunbergia.* In the latter case, galls have been known to reach a diameter of 2 ft (81). These very large galls are often produced at the base of the stem just above ground level. Since the root galls on these species are somewhat smaller, there may be some relationship between gall size and their location on their host's anatomy.

Galling starts relatively rapidly, i.e. sometimes several hours after invasion, and the response for different species of host and parasite varies. Galling is not essential for nematode growth and development and can take place in plants in which the nematode does not become established or reproduce (35, 52). The cellular changes associated with galling are dealt with later in this review.

### *Physiological Responses in Whole Plants*

Most physiological and biochemical measurements made on host plant responses to *Meloidogyne* have been made on dissected, extracted, or homogenized infected material simply because this has been the most practical way to obtain material for these measurements. Much information has been obtained using these methods, and I consider this later, particularly with regard to syncytia. Several types of physiological measurements have been made on whole, infected plants. These experiments are divided rather loosely into two categories: first, those in which various types of specific activators or inhibitors of physiological processes are applied to infected plants in which the growth rates or reproductive rates of the nematodes in these plants are compared with similar rates in untreated infected plants, and second, measurements of normal physiological functions in infected plants and uninfected controls.

Dropkin and his co-workers (30) first showed that exogenously supplied cytokinins made resistant tomato plants more susceptible to *M. incognita* and increased gall formation, while various exogenously supplied substances including indoleacetic acid and gibberellic acid neither increased the number of larvae that grew nor decreased the extent of host cell necrosis. Subsequently it has been shown (48) that *M. javanica* will grow and reproduce in resistant cultivars of peach if these plants are wick-fed with kinetin or 1-naphthylacetic acid. Kochba & Samish (49) subsequently found that cytokinin and auxin levels were significantly lower in these resistant root stocks than in susceptible ones. Furthermore, root-knot nematode can apparently cause a decrease in cytokinins and gibberellins in the root tissues and xylem exudates of its host tomato plants (19). These results suggest that the nematode may use these growth hormones for its own purposes and may require a certain level of them in its host plant before it can grow and reproduce properly. The manner in which this balance between the nematode and its host plant's hormones is achieved probably varies in different host plants and for different species of nematodes.

Breakdown of resistance to *Meloidogyne* has also been shown to occur in some plants if the temperature rises above 28°C (28). Presumably, changes of this type may occur under natural conditions in these plants' normal environments if they are exposed to prolonged hot weather. Physiological measurements on the responses of whole intact susceptible plants to *Meloidogyne* have been made only fairly recently (53). This is a reflection of the apparatus available for these types of measurements and the degree of cooperation that has existed between plant nematologists and plant physiologists.

Plants can respond to parasitism by the root-knot nematode by decreasing their photosynthetic rate. This response can take place quite rapidly, at least with heavy infestations, and highly significant differences between infected and control plants have been recorded within 1–2 days (53) (Figures 1 and 2). These measurements were done on intact control and test plants whose photosynthetic rates were similar prior to infection. The relatively rapid responses are thought to be physiological and not directly related to any change in the plant's morphology. For instance, the mean fresh weight of the control plants in the short-term experiment was 1.10 g ± 0.12 (SE) compared with 1.14 g ± 0.06 for the infected plants, whereas at the conclusion of the longer term experiment (Figure 1), the mean fresh weights were 2.29 g ± 0.10 for the control and 1.77 g ± 0.04 for the infected plants. Clearly these infected plants were showing signs of growth inhibition after several weeks of infection. Further evidence that root-knot nematode can inhibit photosynthesis in its host has come from another laboratory (87). Here the effect of different population levels of *Meloidogyne* on plants that had been infected from 2 weeks onward and had well-developed syncytia was examined. At this age, photosynthesis was found to be inhibited by as little as 4000 nematodes per inoculum per plant.

How does the root-knot nematode bring about a reduction in the photosynthetic rate of its host plant? At the moment it seems that there may be two different ways in which this could operate. First, root damage caused by the nematodes could lead to water stress in the plant causing partial closure of the stomates; this in turn would result in a decreased entry of $CO_2$ into the leaf, which could lead to a reduction in the rate of $CO_2$ fixation. At the moment this mechanism of nematode inhibition of photosynthesis cannot be dismissed, although neither infected nor control plants showed any signs of wilting in our experiments. An alternative hypothesis is that the nematode interferes with the synthesis and/or translocation of growth hormones produced in the roots. This second hypothesis seems more likely, as it has been shown not only that are hormones such as cytokinins and gibberellins produced in roots and translocated to the aerial parts of the plant (47, 68) but also that their concentration decreases in plants infected with root-knot nematodes as compared with uninfected controls (19). Furthermore, as can be seen in the longer-term experiment (Figure 1), after pruning there is a marked increase in photosynthetic

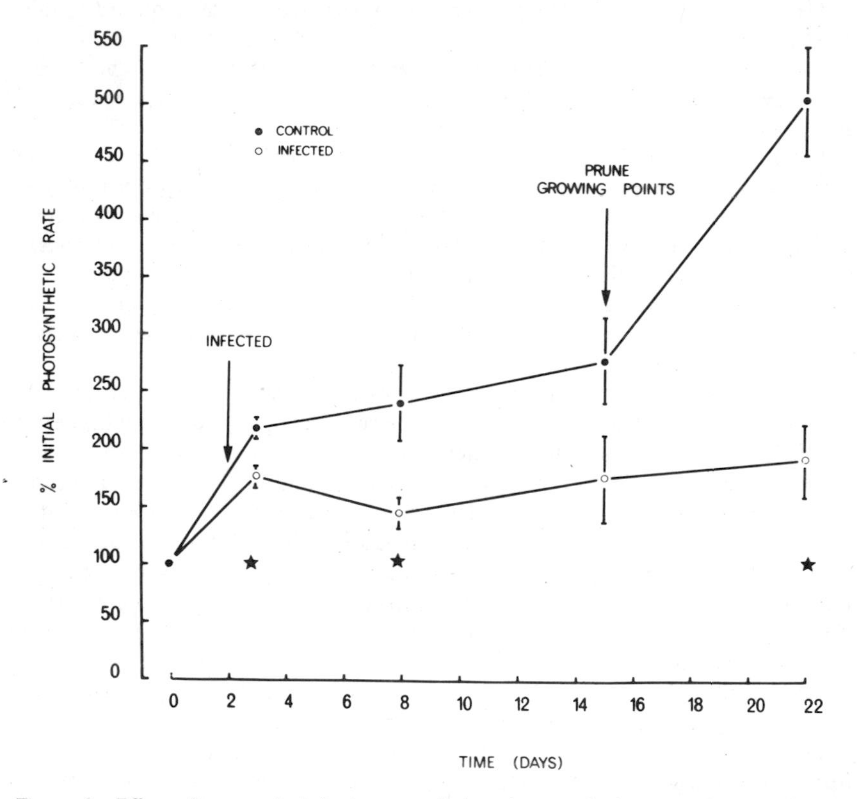


*Figure 1* Effect of nematode infection on relative photosynthetic rates of whole tomato seedlings over a period of 22 days. Bars represent 2X standard error of the means and * indicates significant differences ($P < 0.05$) (53).

rates of the remaining leaves of the control seedlings, an effect well documented in other plants and thought to be due to an enhanced supply of root-derived factors to the remaining leaves (92). This effect was not observed in the nematode-infected plants, which suggests that there may be an interference with production and/or translocation of these factors.

## RESPONSES OF CELLS IN SUSCEPTIBLE PLANTS

In susceptible plants pronounced morphological and physiological changes take place in response to stimuli from the nematode. As already mentioned, the swelling or hyperplasia varies from host to host and, although it is a normal and obvious symptom, the nematode may grow and reproduce perfectly satisfactorily in its absence. The most important criterion for successful development of the nematode is the production of syncytia, also known as giant cells or multinucleate transfer cells (2, 44, 64). These are highly specialized cellular adaptions induced and maintained by the nematode. Their structure, particularly at the light microscope level, has been examined by numerous authors and has been the subject of recent reviews (27, 33, 89). It seems that if *Meloidogyne* species are to reproduce normally, they must be able to initiate and maintain these syncytia.

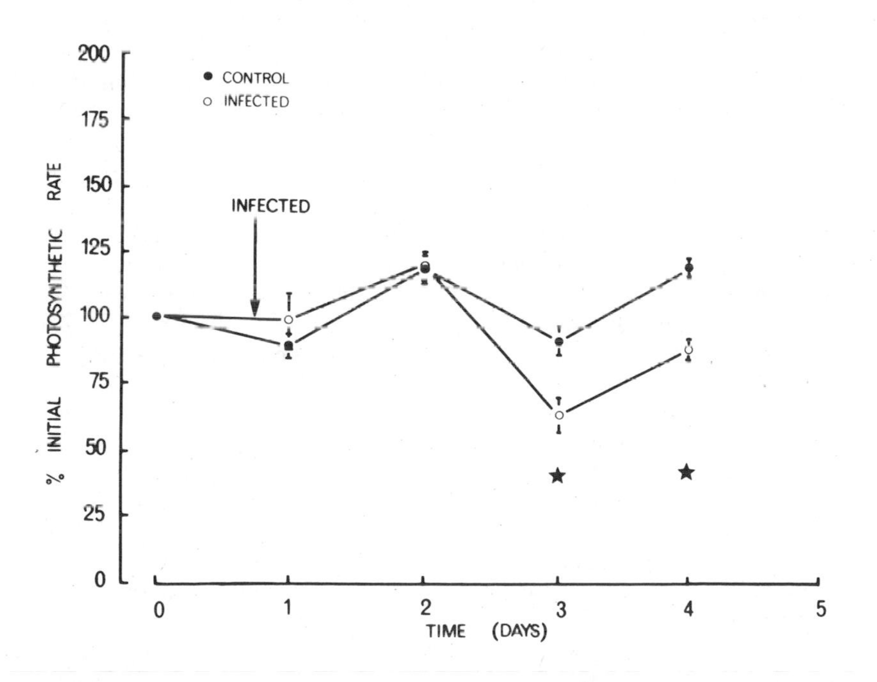


*Figure 2* Effect of nematode infection on relative photosynthetic rates of whole tomato seedlings over a period of 4 days. Bars represent 2X standard error of the means and * indicates significant differences ($P < 0.01$) (53).

This genus apparently can feed and maintain itself on plant cells without bringing about any marked changes in the morphology (22, 23). For instance, when rooted cuttings of *Pelargonium graveolens* were grown in root-knot–infested soil, small woody galls were formed in which no parasites could be found (22). However, at the bases of some of these cuttings were small thimble-shaped succulent outgrowths made up of thin-walled, more or less undifferentiated cells. These outgrowths contained normal egg-laying females. Christie (22) stated that he was not prepared to say what these outgrowths should be called; however, it seems to me that they could have been callus tissue. Although root cultures from resistant plants are known to retain this resistance after numerous passages through culture solutions (66), callus from resistant plants may be susceptible to *Meloidogyne.* No mention was made in Christie's papers of microscopic examination of the cells surrounding the nematode head. Such an examination might have revealed small syncytia. Where growth has taken place in undifferentiated callus tissue, it is associated with the establishment of syncytia (73). Most authorities agree that syncytial formation is necessary for the establishment of a successful host-parasite relationship (89).

How is this structure (synctium) formed? And what is known about the chemical forces associated with its formation and maintenance? How are these forces involved in syncytial formation and maintenance? The first of these questions can probably be answered quite satisfactorily if one restricts oneself to a rather narrow morphological point of view. The second question is probably easier to answer, but in attempting to answer the third question, I am forced to take the reader of this review into the fields of conjecture and speculation. I can only hope that this may serve as a stimulant to further research and not prove too misleading.

## *Manner of Syncytial Formation*

Since the first published description by Treub in 1886 (82) these structures have been examined by numerous workers, most of whom are listed in the reviews mentioned above as well as in two relevant papers (2, 21). The view of these earlier workers could, perhaps, be best summarized as follows: Infective second stage larvae ($L_2$s) migrate through the cortex of the root tip and usually come to lie with their heads in the plerome or potential vascular tissue. Usually only cells from the stele are transformed, but syncytia can also develop from cortical cells. The $L_2$s tend to pass between cells, minimizing destruction of the latter (21). Cell multiplication and cell wall breakdown give rise to large multinucleate syncytia with thick walls, dense cytoplasm, and large irregularly shaped nuclei containing large nucleoli.

Christie (21) cut serial sections, 10 $\mu$m thick, through young galls containing developing syncytia (giant cells) and stated, as a result of his observations, that "at the beginning of giant-cell formation it is usually several adjacent members of a row of undifferentiated cells in the central cylinder that first coalesce through the dissolution of the separating cell walls." This work received further support and was extended at the level of resolution of the electron microscope and further studies with the light microscope (2) and in a comprehensive review (27) in 1969. Later in 1969, two papers published by Huang & Maggenti (39, 40) caused a reappraisal of the method of syncytial formation and led to the statement in the most recent review of this subject (33) that "some clarification is still needed, however, on the rather fundamental question of cell wall dissolution."

Briefly, this alternative hypothesis to syncytial formation is that their nuclei "are derived from repeated mitoses of the original diploid cells without subsequent cytokinesis," i.e. all the nuclei are derived from a single nucleus, and the cell increases its size by swelling and not by incorporation of adjacent cells. This situation gives rise to a ploidy sequence in which chromosome numbers of 4, 8, 16, 32, and 64 *n* are found when syncytial nuclei are formed in this manner. Chromosome counts by Huang & Maggenti (39) revealed that these numbers always fit this ploidy sequence, although considerable difficulty must have been experienced in accurately counting those chromosomes of nuclei with very high ploidy. They also provided further support for this hypothesis in being unable to demonstrate cell wall breakdown during syncytial formation in material observed with the aid of the electron microscope. Some breakdown of cell walls was observed, but this was attributed to mechanical breakdown caused by the penetration and migration of $L_2$s. As mentioned earlier, $L_2$s of *Meloidogyne* tend to move between cells, rather than through them, in their initial migration in the root. It is difficult to determine precisely whether cell wall breakdown results from examination of $L_2$ migration through the cell or whether syncytial initiation has commenced.

A subsequent study of syncytial formation using both the light and electron microscopes (65) did not resolve this problem. Although Paulson & Webster (65) noted that in some cases walls between developing syncytia were partially broken down or irregularly thickened, they concluded that syncytia possibly "originate from expansion of a single cell." Hesse (38) on the other hand suggests that both cell wall breakdown and swelling of the giant cell occur. However, in a recent paper on the induction of giant cells by *Meloidogyne arenaria* in *Coleus blumei,* Jones & Northcote (44a) supply more positive support for Huang & Maggenti's hypothesis. As a result of their inability to detect cell wall dissolution, these workers conclude that giant cells "form solely by the expansion of single cells." Cell wall breakdown at the periphery of giant cells, an occasional probability, was thought to be due to rapid expansion of giant cells and was not thought to contribute to their formation.

Support for the original hypothesis has come recently from a series of papers (13–15) reporting studies at the light microscope level but using quantitative histochemical techniques and thinner serial sectioning (2 μm) of material embedded in epoxy resin than was previously possible with material embedded in wax. They show that syncytia are connected to adjacent cells so that what appears to be part of a syncytium in one section appears to be an adjacent cell in one or two sections further on.

Quantitative measurements of DNA in individual nuclei of syncytia in both beans and tomato plants revealed that the amount of DNA was extremely variable at all harvests. This variability within a single syncytium does not support the regular ploidy sequence hypothesis. How can this irregularity in DNA content be explained other than by assuming that additional cells are incorporated into the syncytium as it grows, or that some nuclei fuse? It seems that both may be the case, as nuclear fusion can readily be observed in syncytia (13).

Recently (15) both cell fusion and cell wall breakdown have been observed occurring simultaneously in whole developing syncytia that were dissected out of the host plant after fixation and treatment with pectinase. It could be argued that the pectinase was responsible for the observed cell wall breakdown in syncytia. However, this seems most unlikely in view of the observed state of the cell walls of adjacent cells and the degree of cell fusion taking place in parts of syncytia where cell wall breakdown had occurred. The material was fixed for 24 hr prior to treatment with the enzyme so that cell fusion must have taken place before enzyme treatment and could not be equated with the spontaneous fusion described when plant cells were treated with cell wall-degrading enzymes (70, 91). Furthermore, synchronous mitoses were observed in these syncytia, and some of the chromosome numbers did not fit into the hypothetical ploidy sequence. It seems reasonable to postulate that syncytia are formed and grow by a combination of cell fusion and synchronous mitosis without cytokinesis. The manner in which this takes place is best illustrated by a scale diagram (Figure 3). This drawing is based on a series of photographs showing early stages of syncytial formation in the root of *Vicia faba.*

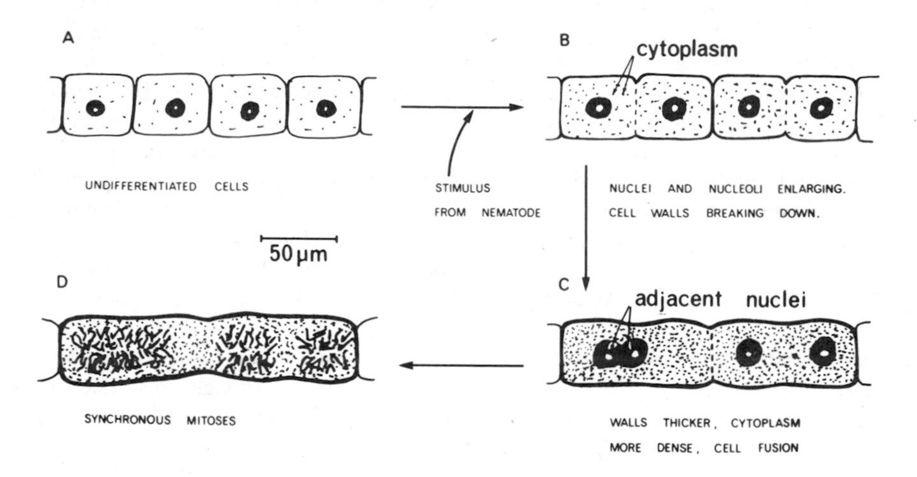


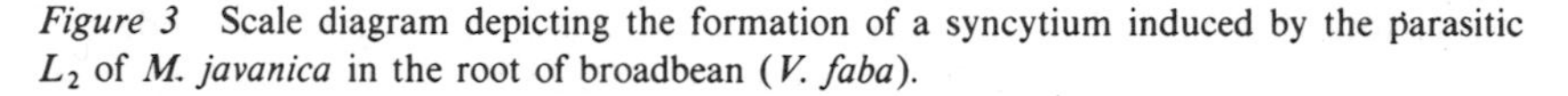

*Figure 3* Scale diagram depicting the formation of a syncytium induced by the parasitic $L_2$ of *M. javanica* in the root of broadbean (*V. faba*).

In recent years biologists have shown considerable interest in cell fusion, and a number of books and reviews have appeared on this topic (36, 70). It has been shown that animal cells from widely divergent species can be fused together (36). This has also been shown to be true for plant cells (70), although the techniques are somewhat different. Briefly, plant cells of different species may be induced to fuse by separating the cells using pectinase and then degrading their walls using cellulase. The cells are then treated with chemical agents such as sodium nitrate which activate the plasmalemma of the cell and induce fusion, provided there is sufficient contact and depth of cytoplasm between these cells. Hence there is more likelihood of fusion between meristematic cells than between differentiating cells that contain less cytoplasm per unit area and are more vacuolated.

Syncytial formation usually starts in the vicinity of cells of the meristematic type. The $L_2$s of *Meloidogyne* contain the enzymes required for cell wall dissolution (26), although it has not yet been possible to demonstrate their exudation from the buccal stylets of these larvae (6).

Fusion in plant cells may be induced by the method briefly outlined above, but it may also occur spontaneously during treatment of plant tissues by the cell wall-degrading enzymes, giving rise to large multinucleate protoplasts (91). Spontaneous fusion results from removal of the cell wall constrictions on plasmodesmata, which in turn permits their expansion so that these cytoplasmic connections become quite broad, containing endoplasmic reticulum elements and mitochondria. This leads to a mixing of the cytoplasms of the fusing protoplasts. Spontaneous fusion of cells in plants is always intraspecific and it seems probable that a similar mechanism operates during the formation of nematode-induced syncytia.

## *Potential Chemical Inducers Exuded from the Nematode*

In attempting to answer the question posed above on the nature of the chemical forces responsible for syncytial formation and maintenance, it must be remembered that the amounts involved are minute and will probably be resolved only by the use of sensitive histochemical and bioassay techniques, which have yet to be developed. Thus, information on the types of enzymes identified from exudates in plant parasitic nematodes is sparse, and a recent review (25) lists only a few such as amylase, pectinase, and a protease. However, the methods used to determine these enzymes are open to the criticism that bacterial contaminates could have contributed to their release.

A cellulase has been reported (56) that is exuded from infective $L_2$s of *Meloidogyne incognita acrita.* This cellulase did not appear to be of bacterial origin, because the response was much greater than that achieved using only the supernatant fluid containing the nematodes. However, subsequent tests carried out elsewhere (6) at much the same time did not support this contention. Nevertheless, work in progress in this laboratory using more sensitive methods than were used previously (6) lends support to Myer's findings (56). The exact nature of this stimulus from the plant is unknown, but it leads to dramatic changes in the internal morphology of the $L_2$ (7, 8, 12, 18). These changes, which take place within two days of entry into the plant, make it possible to distinguish between the infective $L_2$ and the parasitic $L_2$ before the latter starts to grow. The principal changes occur in the morphology and chemical composition of the subventral esophageal glands, which appear to be implicated in the production of protein and glycoprotein which may be enzymatic in nature and involved in the initiation of syncytia.

As this nematode grows and molts, the dorsal esophageal gland becomes more prominent than the two subventrals, and this structure is thought to be responsible for the buccal exudations of the adult female, which contain basic protein, possibly histone (9), 10). As stated previously (9) it seems likely "that nematodes do produce enzymes in their esophageal gland exudates which could influence the cells of their hosts and have an important bearing on the host-parasite relationship," so it is not surprising that a peroxidase has been identified in the stylet exudate of the adult female *Meloidogyne* (41). It seems quite reasonable to suspect that other enzymes may be detected in these exudations as histochemical methods are developed. The peroxidases are a versatile group, being involved in numerous physiological reactions. They are thought to be involved in the formation of lignin by plants, in catalyzing oxidations of protein and amino acids, and in the aerobic oxidation of indoleacetic acid (IAA) (74), to name a few of the compounds associated with syncytial formation. Unraveling the pathways and roles of these ubiquitous enzymes in nematode host-parasite studies will undoubtedly prove a complex and difficult task.

Hormones associated with the nematodes' normal growth mechanism may have an influence on plant tissues (6) in much the same way as ecdysone-L and gibberellic acid have been reported to have (20, 32). There is some disagreement on this point, as recent workers (37) have stated that $\beta$-ecdysone did not mimic the effects of $GA_3$ in their bioassays. The point is an interesting one and will be well worth looking into when more information on hormones in nematodes becomes available.

Certainly growth hormones have important roles in these host-parasite relationships. Whether they are exuded from the nematode or are synthesized in syncytia as a result of an exuded stimulus remains to be determined. This section has dealt only with substances known to have exuded from the nematode. *Meloidogyne* of course contains many enzymes including some that could have obvious functions during syncytial development, such as cellulase (26). However, such enzymes have yet to be observed exuding from the nematode.

### *Physiological Responses of Plant Cells to Meloidogyne*

Before attempting to answer the question posed on the mechanism of syncytial induction, I shall briefly summarize some of the work on the physiology and biochemistry of cell response. The series of papers by Owens and his co-workers (59–62, 72) have provided a great deal of information on this topic. They reaffirmed that syncytia were multinucleate and underwent nuclear and nucleolar hypertrophy, cytoplasmic densification, cell wall dissolution, and synchronous mitosis. They showed by means of radiotracers that the developing syncytium was a region of intense ribo- and deoxyribonucleic acid biosynthesis. This was reflected in increased amounts of DNA, RNA, and total phosphorus, a result subsequently supported by the work of Ishibashi & Shimizu (42).

Respiration in galls of *Meloidogyne*-infected tomato roots is similar to that in adjacent host root tissue and to that of uninfected roots of similar age (17, 61). The respiration of the syncytia themselves has not yet been measured. Undoubtedly it could be argued that removal of synctia by microdissection from the root could lead to mechanical injuries which could mask their true physiological state. There seems to be no doubt, however, that the rate of intermediary metabolism is accelerated in the galls, especially in pathways leading to nucleic acid and protein synthesis. For instance (61), carbon from glucose-U-$^{14}$C was found to be incorporated more rapidly into gall tissue, particularly in the case of amino acids, proteins, and nucleic acids where the increase in labeling ranged from 54 to almost 300%. Subsequently other workers (34, 83) have demonstrated by histochemical techniques that a variety of enzymes function more actively within the limits of the syncytia than in the surrounding tissue. It now seems fairly well established that more growth-promoting substances are located in galls than in adjacent or uninfected roots (1, 3, 78, 84) and that these are also found in the parasitizing nematode (85, 93).

The use of various metabolic inhibitors has given further insight into the complexity of this host-parasite relationship. Thus, responses to the growth-regulating agent ethrel (57) suggest the involvement of ethylene. The systemic growth inhibitor maleic hydrazide (24, 67) inhibits nematode development and influences sexual differentiation. The antimitotic morphactin (58) suppresses syncytial formation, and the antimetabolites 6-azauridine and 5-bromo-2-deoxycytidine (16) inhibit nematode growth. Unfortunately all these compounds are either too toxic to the plant or too costly to be used successfully as nematicides. However, they offer hope for the future when it may become possible, as a result of further knowledge of syncytial induction and growth, to block a specific metabolic pathway in a much more precise and rewarding manner.

### *Suggested Mechanism for Syncytial Induction*

Information on the manner in which syncytia are formed and on the chemical forces responsible for initiating and maintaining them, and an understanding of the physiological responses of these syncytia, are necessary for proper understanding of these host-parasite relationships. However, the really interesting problems of the future will be resolved by concerning ourselves more with the biology of cell differentiation. Views similar to these have been expressed previously: "In essence, the problem of understanding the host-parasite interaction between nematodes of the family *Heteroderidae* and their hosts is the problem of control of differentiation, in which the nematode holds the reins" (29).

During syncytial formation the nematode induces specialized cells to be formed from unspecialized cells. These unspecialized cells would normally form specialized cells of different types with specific functions in response to various repressors and inducers operating within the plant. The nematode also represses part of the cells' genetic coding and activates other parts so that a special type of cell is induced.

The synctium is a highly specialized cell resembling in many respects transfer cells which function as intensive and selective transporters over short distances (44, 64). This highly specialized cell is induced, maintained, and completely dependent on a continuous stimulus from the nematode; the syncytium is not self replicating, and removal of the stimulus leads to its atrophy (3).

The nematode is responsible for influencing normal coding so that reactions take place in the developing root to give rise to cells similar to those formed naturally in a wide variety of anatomical situations in plants (64). In this respect the situation is quite unlike that which takes place during the development of cancer, which has been broadly defined as uncontrolled cell growth and is thought to be caused by accumulation of either somatic mutations, viruses, or irreversible differentiation (88). Only the last of these hypotheses would seem to have a bearing on syncytial formation, although the fact that virus particles have not been detected in syncytia does not preclude their possible involvement.

Of the chemicals listed (6, 9) that are exuded from, or can be found in, the anterior region of the nematode, two seem to be of particular interest. The first of these are the esterases which have been detected in the amphid pouches of all stages of *Meloidogyne* (15), and the second is a histone-like protein exuded from the buccal stylet (9, 10). With regard to the first group, pharmacological and chromatographic techniques have shown that acetylcholine is not only found in plants but also occurs at its highest concentrations in actively growing areas such as bud and root tips. Acetylcholine is thought to influence membrane permeability in these areas and to reduce the formation of secondary roots. Acetylcholine esterase inhibits these responses. It seems reasonable to postulate that the esterase produced in the amphids of *Meloidogyne* may have similar effects on developing syncytia. This could help to explain the common occurrence of secondary roots in the vicinity of syncytia and leads one to suggest that measurement of bioelectric potentials in developing and establishing syncytia could provide some interesting information. In fact, some experiments of this type have already been done on syncytia in *Impatiens balsamina* induced by *M. incognita* (45, 46). It appears that these syncytia possess an electrogenic ion pump at the plasmalemma and some of them exhibit trains of action potentials. Apparently they respond to changes in the external solution in a manner similar to that of normal cells of higher plants. Because of their size, however, it is much easier to obtain these measurements in syncytia, so that they may prove useful models for biophysicists interested in membrane potentials.

It would be interesting to know if the action potentials that can be measured in syncytia undergo changes in pattern during their formation, development and decline. The rate at which these processes take place in syncytia varies depending on the host plant, but in hosts such as the broad bean (*V. faba*) it can all occur within a few weeks. (13).

The second chemical mentioned above is the histone-like protein exuded from the buccal stylet of adult female *Meloidogyne* (9). In view of the interest shown in histones as regulators of gene action, this might be a profitable area for future research. Histones (31) are small basic proteins found in association with DNA in all eukaryotes. They are classified according to their lysine/arginine ratios into four classes containing a total of about a dozen different types whose molecular weights vary from 11,300 to 21,000. It is thought (31, 79) that they may have numerous roles, particularly that of gene repression, Thus, if the buccal exudate of *Meloidogyne* functions as a histone by keeping some of the genes in undifferentiated plant cells incompetent (94) this could explain the first steps of syncytial formation, namely a cessation of normal differentiation of root cells. However, it has been established only that basic proteins are exuded from the nematode, and more precise chemical methods are needed to establish clearly that these are histones.

As more specific histochemical techniques are developed, more types of enzymes will be shown to be present in addition to the peroxidase mentioned earlier (41), and a more concise picture of the chemical composition of the stylet exudate will be gradually pieced together. There is little or no information available on the nature of the stylet exudation from infective and parasitic $L_2$s beyond knowing that these substances have antigenic properties and appear to be proteins resembling those produced by adult females (4). The exudations from this particular stage should receive a thorough study, as these chemicals control initiation and growth of syncytia. The problem is compounded by the minute amount of material available for testing and the need for development of suitable and sensitive bioassays.

In conclusion, I wish to consider some of the facts and hypotheses advanced for cell differentiation (63) and to attempt to relate these to syncytial formation and growth. It is generally accepted that cells of living organisms have totipotency: the potential of expressing all genes in the embryonic cell. What prevents the expression of these genes in many instances? Their expression appears to be determined by the nature of the surrounding cytoplasm, and the selective nature of this implies the presence of substances that can mask the remainder of the genes. These are thought to be either histones or acidic nucleoproteins called hertones (31). In fact, the isoelectric points of the hertones range from acidic through neutral to very slightly basic. They have a much wider range of molecular weight than the histones, varying from 12,000 to several hundred thousand. It has been anticipated that hertones may be involved in gene activation, but so far there has been no direct evidence of this (31). Furthermore, no satisfactory histochemical methods have been developed for detecting hertones. The present hypothetical role for histones exuded from *Meloidogyne* may best be established by demonstrating their inhibitory effect on DNA transcription and replication in plant cells. Again the minute quantities of material available for testing may prove to be a stumbling block in attempts to develop suitable techniques.

## SUMMARY

The responses of plants to root-knot nematodes are considered under two major headings. First, the responses of whole intact plants are considered and, second, the responses of their cells. Whole plants respond to infection by a reduction in their photosynthetic rate, growth, and yield. It is suggested that the nematode influences the physiology of the plant by interfering with the synthesis and translocation of growth hormones produced in the roots. Cells of susceptible plants are changed from normal undifferentiated cells to highly specialized syncytia, also known as giant cells or multinucleate transfer cells. This process involves a chain of events starting with nuclear and nucleolar enlargement followed by cell wall breakdown, synchronous mitoses, and incorporation of adjacent cells. This highly specialized syncytium is induced, maintained, and completely dependent on a continuous stimulus from the nematode. Exudations from *M. javanica* are responsible for gene regulation leading to the above-mentioned changes in cell form and function. The nature of these exudations is discussed in relation to their hypothetical functions.

*Literature Cited*


1. Balasubramanian, M., Rangaswami, G. 1962. Presence of indole compound in nematode galls. *Nature* 194:774–75
2. Bird, A. F. 1961. The ultrastructure and histochemistry of a nematode-induced giant cell. *J. Biophys. Biochem. Cytol.* 2:701–15
3. Bird, A. F. 1962. The inducement of giant cells by *Meloidogyne javanica. Nematologica* 8:1–10
4. Bird, A. F. 1964. Serological studies on the plant parasitic nematode, *Meloidogyne javanica. Exp. Parasitol.* 15: 350–60
5. Bird, A. F. 1966. Esterases in the genus *Meloidogyne. Nematologica* 12:359–61
6. Bird, A. F. 1966. Some observations on exudates from *Meloidogyne* larvae. *Nematolotica* 12:471–82
7. Bird, A. F. 1967. Changes associated with parasitism in nematodes. I: Morphology and physiology of preparasitic and parasitic larvae of *Meloidogyne javanica. J. Parasitol.* 53:768–76
8. Bird, A. F. 1968. Changes associated with parasitism in nematodes. III: Ultrastructure of the egg shell, larval cuticle and contents of the subventral esophageal glands in *Meloidogyne javanica,* with some observations on hatching. *J. Parasitol.* 54:475–89
9. Bird, A. F. 1968. Changes associated with parasitism in nematodes. IV: Cytochemical studies on the ampulla of the dorsal esophageal gland of *Meloidogyne javanica* and on exudations from the buccal stylet. *J. Parasitol.* 54:879–90
10. Bird, A. F. 1969. Changes associated with parasitism in nematodes. V: Ultrastructure of the stylet exudation and dorsal esophageal gland contents of female *Meloidogyne javanica. J. Parasitol.* 55:337–45
11. Bird, A. F. 1970. The effect of nitrogen deficiency on the growth of *Meloidogyne javanica* at different population levels. *Nematologica* 16:13–21
12. Bird, A. F. 1971. Specialized adaptions of nematodes to parasitism. *Plant Parasitic Nematodes.* ed. B. M. Zuckerman, W. F. Mai, R. A. Rohde, 2:35–49. New York: Academic. 347 pp.
13. Bird, A. F. 1972. Quantitative studies on the growth of syncytia induced in plants by root-knot nematodes. *Int. J. Parasitol.* 2:157–70
14. Bird, A. F. 1972. Cell wall breakdown during the formation of syncytia induced in plants by root-knot nematodes. *Int. J. Parasitol.* 2:431–32
15. Bird, A. F. 1973. Observations on chromosomes and nucleoli in syncytia induced by *Meloidogyne javanica. Physiol. Plant Pathol.* 3:387–91
16. Bird, A. F., McGuire, R. J. 1966. The effect of antimetabolites on the growth of *Meloidogyne javanica. Nematologica* 12:637–40
17. Bird, A. F., Millerd, A. 1962. Respiration studies on *Meloidogyne*-induced galls in tomato roots. *Nematologica* 8:261–66
18. Bird, A. F., Saurer, W. 1967. Changes associated with parasitism in nematodes. II: Histochemical and microspectrophotometric analyses of preparasitic and parasitic larvae of *Meloidogyne javanica. J. Parasitol.* 53:1262–69
19. Brueske, C. H., Bergeson, G. B. 1972. Investigation of growth hormones in xylem exudate and root tissue of tomato infected with root-knot nematode. *J. Exp. Bot.* 23:14–22
20. Carlisle, D. B., Osborne, D. J., Ellis, P. E., Moorhouse, J. E. 1963. Reciprocal effects of insect and plant-growth substances. *Nature* 200:1230
21. Christie, J. R. 1936. Development of root-knot nematode galls. *Phytopathology* 26:1–22
22. Christie, J. R. 1949. Host-parasite relationships of the root-knot nematodes, *Meloidogyne* spp. III: The nature of resistance in plants to root-knot. *Proc. Helminthol. Soc. Wash.* 16:104–8
23. Christie, J. R. 1950. *Meloidogyne* life history, biology, races, pathogenicity and other topics dealing with root-knot nematode research. *Proc. S. 19. Workshop in Phytonematology,* ed. E. J. Cairns, Auburn, Alabama
24. Davide, R. G., Triantaphyllou, A. C. 1968. Influence of the environment on development and sex differentiation of root-knot nematodes. *Nematologica* 14:37–46
25. Deubert, K. H., Rohde, R. A. 1971. Nematode enzymes. See Ref. 12, pp. 73–90
26. Dropkin, V. H. 1963. Cellulase in phytoparasitic nematodes. *Nematologica* 9:444–54
27. Dropkin, V. H. 1969. Cellular responses of plants to nematode infections. *Ann. Rev. Phytopathol.* 7:101–22
28. Dropkin, V. H. 1969. The necrotic reaction of tomatoes and other hosts resistant to *Meloidogyne:* Reversal by temperature. *Phytopathology* 59: 1632–37

29. Dropkin, V. H., Boone, W. R. 1966. Analysis of host-parasite relationships of root-knot nematodes by single-larva inoculations of excised tomato roots. *Nematologica* 12:225–36
30. Dropkin, V. H. Helgeson, J. P., Upper, C. D. 1969. The hypersensitivity reaction of tomatoes resistant to *Meloidogyne incognita:* Reversal by cytokinins. *J. Nematol.* 1:55–61
31. Elgin, S. C. R., Bonner, J. 1973. *The Biochemistry of Gene Expression in Higher Organisms,* ed. J. K. Pollak, J. W. Lee, 142–63. Sydney: Australia & New Zealand Book. 656 pp.
32. Ellis, P. E., Carlisle, D. B., Osborne, D. J. 1965. Desert Locusts: Sexual maturation delayed by feeding on senescent vegetation. *Science* 149:546–47
33. Endo, B. Y. 1971. Nematode-induced syncytia (giant cells). Host-parasite relationships of *Heteroderidae.* See Ref. 12, pp. 91–117
34. Endo, B. Y., Veech, J. A. 1969. The histochemical localization of oxidoreductive enzymes of soybeans infected with the root-knot nematode *Meloidogyne incognita acrita. Phytopathology* 59:418–25
35. Gaskin, T. A. 1959. Abnormalities of grass roots and their relationship to root-knot nematodes. *Plant Dis. Reptr.* 43:25–26
36. Harris, H. 1970. *Cell Fusion.* Oxford: Clarendon. 108 pp.
37. Hendrix, S. D., Jones, R. L. 1972. The activity of $\beta$-ecdysone in four gibberellin bioassays. *Plant Physiol.* 50:199–200
38. Hesse, M. 1970. Cytological studies of nematode galls. *Österr. Bot. Z.* 118:517–41
39. Huang, C. S., Maggenti, A. R. 1969. Mitotic aberrations and nuclear changes of developing giant cells in *Vicia faba* caused by root-knot nematode, *Meloidogyne javanica. Phytopathology* 59:447–55
40. Huang, C. S., Maggenti, A. R. 1969. Wall modifications in developing giant cells of *Vicia faba* and *Cucumis sativus* induced by root-knot nematode, *Meloidogyne javanica. Phytopathology* 59:931–37
41. Hussey, R. S. 1973. Peroxidase from *Meloidogyne incognita. Physiol. Plant Pathol.* 3:223–29
42. Ishibashi, N., Shimizu, K. 1970. Gall formation by root-knot nematode, *Meloidogyne incognita* (Kofoid and White, 1919) Chitwood, 1949, in the grafted tomato plants, and accumulation of phosphates on the gall tissues (Nematoda:Tylenchida). *Appl. Entomol. Zool.* 5:105–11
43. Jaffe, M. J. 1970. Evidence for the regulation of phytochrome-mediated processes in bean roots by the neurohumor, acetylcholine. *Plant Physiol.* 46:768–77
44. Jones, M. G. K., Northcote, D. H. 1972. Nematode-induced syncytium—a nultinucleate transfer cell. *J. Cell Sci.* 10:789–809
44a. Jones, M. G. K., Northcote, D. H. 1972. Multinucleate transfer cells induced in Coleus roots by the root-knot nematode, *Meloidogyne arenaria. Protoplasma* 75:381–95
45. Jones, M. G. K., Novacky, A., Dropkin, V. H. 1973. Membrane potentials of transfer cells. Presented at Can. Soc. Plant Pathol., Calgary, Alberta
46. Jones, M. G. K., Novacky, A., Dropkin, V. H. 1973. Nematode induced giant cells—a model system for membrane potential measurements. Presented at 2nd Int. Congr. Plant Pathol., St. Paul, Minnesota
47. Kende, H. 1965. Kinetin-like factors in the root exudate of sunflowers. *Proc. Nat. Acad. Sci. USA* 53:1302–7
48. Kochba, J., Samish, R. M. 1971. Effect of kinetin and 1-naphthylacetic acid on root-knot nematodes in resistant and susceptible peach rootstocks. *J. Am. Soc. Hort. Sci.* 96:458–61
49. Kochba, J., Samish, R. M. 1972. Level of endogenous cytokinins and auxin in roots of nematode resistant and susceptible peach rootstocks. *J. Am. Soc. Hort. Sci.* 97:115–19
50. Krusberg, L. R. 1963. Host response to nematode infection. *Ann. Rev. Phytopathol.* 1:219–39
51. Linford, M. B. 1941. Parasitism of the root-knot nematode in leaves and stems. *Phytopathology* 31:634–48
52. Loewenberg, J. R., Sullivan, T., Schuster, M. L. 1960. Gall induction by *Meloidogyne incognita incognita* by surface feeding and factors affecting the behavior pattern of the second stage larvae. *Phytopathology* 50:322
53. Loveys, B. R., Bird, A. F. 1973. The influence of nematodes on photosynthesis in tomato plants. *Physiol. Plant Pathol.* 3:525–29
54. Miller, H. N., DiEdwardo, A. A. 1962. Leaf galls on *Siderasis fucata* caused by the root-knot nematode *Meloidogyne incognita incognita. Phytopathology* 52:1070–73
55. Mountain, W. B. 1965. Pathogenesis by soil nematodes. *Ecology of Soil-borne Plant Pathogens.* ed. K. F. Baker, W. C.

Snyder, 285–301. Berkeley, Calif.: Univ. Calif. Press. 571 pp.
56. Myers, R. F. 1964. *Organic substances discharged by plant-parasitic nematodes.* PhD thesis. Univ. Maryland, College Park. 64 pp.
57. Orion, D., Minz, G. 1969. The effect of ethrel (2-chloroethane phosphonic acid) on the pathogenicity of the root-knot nematode *Meloidogyne javanica. Nematologica* 15:608–14
58. Orion, D., Minz, G. 1971. The influence of morphactin on the root-knot nematode, *Meloidogyne javanica,* and its galls. *Nematologica* 17:107–12
59. Owens, R. G., Specht, H. N. 1964. Root-knot histogenesis. *Contrib. Boyce Thompson Inst.* 22:471–90
60. Owens, R. G., Bottino, R. F. 1966. Changes in host cell wall composition induced by root-knot nematodes. *Contrib. Boyce Thompson Inst.* 23:171–80
61. Owens, R. G., Rubinstein, J. H. 1966. Metabolic changes induced by root-knot nematodes in host tissues. *Contrib. Boyce Thompson Inst.* 23:199–214
62. Owens, R. G., Specht, H. N. 1966. Biochemical alterations induced in host tissues by root-knot nematodes. *Contrib. Boyce Thompson Inst.* 23:181–98
63. Pasternack, C. A. 1970. *Biochemistry of Differentiation.* London: Wiley-Interscience. 189 pp.
64. Pate, J. S., Gunning, B. E. S. 1972. Transfer cells. *Ann. Rev. Plant Physiol.* 23:173–96
65. Paulson, R. E., Webster, J. M. 1970. Giant cell formation in tomato roots caused by *Meloidogyne incognita* and *Meloidogyne hapla* (Nematoda) infection. A light and electron microscope study. *Can. J. Bot.* 48:271–76
66. Peacock, F. C. 1959. The development of a technique for studying the host-parasite relationship of the root-knot nematode *Meloidogyne incognita* under controlled conditions. *Nematologica* 4:43–55
67. Peacock, F. C. 1963. Systematic inhibition of root-knot eelworm (*Meloidogyne incognita*) on tomato. *Nematologica* 9:581–83
68. Phillips, I. D. J., Jones, R. L. 1964. Gibberellin-like activity in bleeding-sap of root systems of *Helianthus annuus* detected by a new dwarf pea epicotyl assay and other methods. *Planta* 63:269–78
69. Powell, N. T., Moore, E. L. 1961. A technique for inoculating leaves with root-knot nematodes. *Phytopathology* 51:201–2
70. Power, J. B., Cocking, E. C. 1971. Fusion of plant protoplasts. *Sci. Progr. Oxford* 59:181–98
71. Rohde, R. A. 1972. Expression of resistance in plants to nematodes. *Ann. Rev. Phytopathol.* 10:233–52
72. Rubinstein, J. H., Owens, R. G. 1964. Thymidine and uridine incorporation in relation to the ontogeny of root-knot syncytia. *Contrib. Boyce Thompson Inst.* 22:491–502
73. Sandstedt, R., Schuster, M. L. 1963. Nematode-induced callus on carrot discs grown in vitro. *Phytopathology* 53:1309–12
74. Saunders, B. C., Holmes-Siedle, A. G., Stark, B. P. 1964. *Peroxidase.* London: Butterworth. 271 pp.
75. Seinhorst, J. W. 1961. Plant-nematode inter-relationships. *Ann. Rev. Microbiol.* 15:177–96
76. Seinhorst, J. W. 1970. Dynamics of populations of plant parasitic nematodes. *Ann. Rev. Phytopathol.* 8:131–56
77. Seinhorst, J. W., Den Ouden, H. 1971. The relation between density of *Heterodera rostochiensis* and growth and yield of two potato varieties. *Nematologica* 17:347–69
78. Setty, K. G. H., Wheeler, A. W. 1968. Growth substances in roots of tomato (*Lycopersicon esculentum* Mill.) infected with root-knot nematodes (*Meloidogyne* spp). *Ann. Appl. Biol.* 61: 495–501
79. Shepherd, G. R., Noland, B. J., Hardin, J. M. 1973. See Ref. 31, pp. 164–76
80. Steiner, G. 1940. The root-knot nematode attacking stems and leaves of plants. *Phytopathology* 30:710
81. Steiner, G., Buhrer, E. M., Rhoads, A. S. 1934. Giant galls caused by the root-knot nematode. *Phytopathology* 24: 161–63
82. Treub, M. 1886. Some notes on the effects of parasitism of *Heterodera javanica* in the roots of the sugar cane. *Ann. Jard. Bot. Buitenz* 6:93–96
83. Veech, J. A., Endo, B. Y. 1969. The histochemical localization of several enzymes of soybeans infected with the root-knot nematode *Meloidogyne incognita acrita. J. Nematol.* 1:265–76
84. Viglierchio, D. R., Yu, P. K. 1965. Plant parasitic nematodes: A new mechanism for injury of hosts. *Science* 147:1301–3
85. Viglierchio, D. R., Yu, P. K. 1968. Plant growth substances and plant parasitic nematodes. II: Host influence on auxin content. *Exp. Parasitol.* 23: 88–95

86. Wallace, H. R. 1971. The influence of the density of nematode populations on plants. *Nematologica* 17:154–66
87. Wallace, H. R. Personal communication
88. Watson, J. D. 1970. *Molecular Biology of the Gene.* New York: Benjamin. 2nd ed. 662 pp.
89. Webster, J. M. 1969. The host-parasite relationships of plant-parasitic nematodes. *Advan. Parasitol.* 7:1–40
90. Webster, J. M., Ed. 1972. *Economic Nematology.* New York: Academic. 563 pp.
91. Withers, L. A., Cocking, E. C. 1972. Fine-structural studies on spontaneous and induced fusion of higher plant protoplasts. *J. Cell Sci.* 11:59–75
92. Woolhouse, H. W. 1967. The nature of senescence in plants. *Symp. Soc. Exp. Biol.* 21:179–213
93. Yu, P. K., Viglierchio, D. R. 1964. Plant growth substances and parasitic nematodes. I: Root-knot nematodes and tomato. *Exp. Parasitol.* 15:242–48
94. Zalokar, M. 1964. The relationship of gene action to embryonic induction and competence. In *The Nucleohistones,* ed. J. Bonner, P. Ts'O. San Francisco: Holden-Day. 398 pp.

# LATENT INFECTIONS BY BACTERIA

❖3589

*A. C. Hayward*
Department of Microbiology, University of Queensland, St. Lucia 4067, Brisbane, Queensland, Australia

## INTRODUCTION

Bacteria may be present in plant tissues without effecting any gross change in form or function. According to Gaumann (25) an infection of this type is termed inapparent or symptomless, and the individuals affected are known as carriers. However, whether this condition of the plant is recognized as latent or diseased depends on the sensitivity of the physiological or microscopic techniques by which the state of the host plant is assessed (86). In any disease there is a latent period during which symptoms are not evident; in some latent infections there may be a greatly prolonged period of incubation due to resistance in the host or presence of some environmental factor, such as temperature, which is unfavorable for the pathogen. When conditions change, the disease is expressed. A change in the virulence of the pathogen while resident in healthy tissues may also lead to an expression of disease symptoms in a previously latent condition. Latent infection may imply low numbers of the pathogen. Internal tissues of some plants usually considered to be sterile may contain relatively large numbers of bacteria that may become parasitic under certain conditions. In this review the concept of latency is restricted neither to temporarily symptomless infections that subsequently break out nor to latent pathogenic bacteria.

Latent bacteria occur within, rather than upon the surface of the plant, but this distinction is not clear; there are, for example, bacteria in the depressions at the juncture of anticlinal walls of epidermal cells, sometimes in clusters and often in lines (45, 47). Pathogenic and saprophytic bacteria may occur within the buds of healthy plants (17, 48–51). Survival of pathogenic bacteria in buds is also a factor in the overwintering of certain diseases of perennial hosts, such as blight and canker of stone and pome fruits caused by *Pseudomonas syringae* (20, 23, 48).

Distribution of bacteria on root surfaces is three-dimensional rather than planar. Examination of sections of epidermal cells of the root system of plants has shown the presence of bacteria embedded in a discontinuous matrix up to 5 μm thick and adjacent to the epidermal cells (13, 14). This outer boundary of the root has been termed the mucigel layer (36). Microorganisms can become embedded in the surface of the root and may not be displaced even by vigorous shaking.

There are many reports of healthy tissues that contain a mixed internally borne bacterial flora, whose significance is largely unknown (2, 11, 19, 24, 34, 59, 62, 68, 71, 73, 74, 78, 80, 81). In most cases the bacteria isolated from stems, roots, fruits, and vegetables resemble the common saprophytic bacteria found in the rhizosphere or phylloplane of hosts from which the bacteria were isolated; in other cases bacteria demonstrably pathogenic either on the host of origin or on another host have been isolated.

## LATENT BACTERIA IN VEGETABLE CROPS

Bacteria are commonly present within healthy tomato fruits (24, 78) and to an equal extent in green- and red-ripe fruits (73). Larger numbers of bacteria have been found in the connective tissues at the stem end and at the center of the stylar column than at the apex or the periphery of the fruit. It is most probable that the bacteria enter through microlesions on the surface where the stalk is attached and from there penetrate to the core of the fruit. The saprophytic bacteria present within fruits are similar to those found on leaves of the tomato plant, on the sepals, and at times also upon the surface of the fruit. The bacterial content of tomatoes obtained from different fields shows considerable fluctuation; these differences could be due to anatomical differences between cultivars and to the favorable microclimate produced through use of overhead irrigation (24, 72).

Cucumbers similarly contain a diverse bacterial flora (59, 71); however, in this vegetable the bacterial population has been shown to be lower near the stem and higher at the center and blossom end of the fruit. It is possible that the bacteria find their way into the developing fruit through the blossom (71). No correlation between bacterial numbers and either fruit size or season of sampling has been found.

A disease may be overt in one part of a plant and latent in another part. For example, *Xanthomonas vesicatoria,* the cause of bacterial spot of tomato and pepper, has been isolated from the interior of symptomless fruits on plants that showed evidence of infection elsewhere (12, 24). Ercolani & Casolari (24) sprayed tomato blossoms with a suspension of *X. vesicatoria* containing $10^5$ organisms/ml. One month later, after fruit set, typical symptoms of the disease were observed on the leaves whereas the fruit were outwardly healthy. The pathogen was isolated from the center of these symptomless fruits.

The stems of healthy pinto bean plants have been shown to harbor six different species of bacteria including three normally pathogenic for bean (81). A cool season preceded the isolations in this test, possibly preventing expression of disease symptoms, because isolates of *Xanthomonas phaseoli* and *Corynebacterium flaccumfaciens* produced typical symptoms of bacterial blight or wilt on the same varieties grown under greenhouse conditions. Similarly, in the case of soybean, Kennedy (39) has shown that *Pseudomonas glycinea,* the cause of a leafspot, is often present within host stems during the growing season, but without profuse proliferation and without evidence of systemic infection. Other bacteria may be endemic in soybean (22a).

There are many examples of seed-borne bacterial pathogens in which the bacteria are within the seed, in some cases without the appearance of any obvious symptom. In the case of halo blight of bean caused by *Pseudomonas phaseolicola* it is known that the hilum and cracks in seed coats from threshing injuries or from wetting the seed provide sites at which the bacteria may enter the seed and be protected from surface sterilization procedures (29). The subject of seed pathology has been extensively reviewed recently and is not further dealt with here (3).

Fruits and vegetables may also serve as a vehicle for the transport of organisms of significance in human disease. In a study of the bacterial flora of homogenates prepared from various fresh vegetables in a hospital kitchen, tomatoes yielded the highest counts and greatest frequency of isolations of *Pseudomonas aeruginosa.* It was estimated that a patient consuming an average portion of tomato salad might ingest as many as $5 \times 10^3$ colony-forming units of *P. aeruginosa* (41). This work did not differentiate between surface flora and bacteria of endogenous origin. The possible occurrence of *P. aeruginosa* or of other bacteria of medical significance in fruits or vegetables should not be ignored. It is known that irrigation or fertilization with inadequately treated municipal, animal, or food processing wastes can lead to a reduction in the bacteriological quality of farm produce. When fruits and vegetables raised or processed under these conditions are eaten raw, transmission of disease is a matter for concern (26). Furthermore, ornamental plants have recently been implicated as a means by which *P. aeruginosa* may be introduced into the hospital environment (84), although this conclusion has been questioned (75a).

### *Saprophytic and Soft Rot Bacteria in Potato Tubers*

Low numbers of several kinds of saprophytic bacteria have been shown to occur in subterranean storage organs. In mature healthy potato tubers and aerial stems, bacteria occurred in the xylem of the vascular tissue and with greatest frequency at the stem end (74); bacteria were less frequent in the cortex and central pith (34). This suggests that the bacteria may enter through natural wounds caused by emergence of secondary roots and that they may then be translocated to the upper parts of the plant. The prevalence of bacteria in seed pieces and tubers is proportional to the degree of vascularization of the tuber; bacteria are more numerous in seed pieces than in tubers of the same potato lot (34). Hollis (34) found that *Bacillus megaterium* and *Aerobacter (Enterobacter) cloacae* were the bacteria most frequently isolated from the vascular tissue of tubers and stems. Montuelle & Beerens (62) found that *Bacillus cereus* and *B. licheniformis* were always present within healthy tubers, whereas Gram-negative rods representing several different genera were isolated inconsistently.

Pérombelon (66) has shown that stored stocks of apparently healthy potato tubers of all seed grades and cultivars in the east of Scotland are extensively contaminated by *Erwinia carotovora* var. *carotovora* and *E. carotovora* var. *atroseptica,* the cause of blackleg. Most tubers from blackleg-infected or blackleg-free crops were contaminated with *E. carotovora;* about 80% of the isolates were identified as var. *atroseptica,* the rest being var. *carotovora.* These organisms were shown to survive in and on tubers for 6–7 months of bulk storage over winter and up to time of planting the following spring. Determination of the numbers of soft-rotting bacteria in surface-sterilized tubers showed that most contamination was located at the tuber surface and in the lenticels, whereas bacteria were rare in the vascular ring and absent from the cortex. Winter storage resulted in substantial reduction of surface contamination whereas the numbers of bacteria in lenticels were only slightly affected (67). In half of the stocks examined each lenticel contained more than $10^2$ soft-rotting bacteria.

These results have had an impact on understanding the epidemiology of potato blackleg and soft rot. Because most tubers going into store are already contaminated by soft-rotting *Erwinia* spp., breakdown in storage need not be attributed to release of bacteria from old rotting tubers. The incidence of contamination with the blackleg pathogen is far higher than could be attributable to release from the low numbers of blackleg-infected plants present in the crops from which the tubers were harvested. Lenticels are the most probable point of entry for soft-rotting *Erwinia* spp., and the site of survival overwinter. Crops apparently free from blackleg were as extensively contaminated by var. *atroseptica* as those evidently infected (67). Recent evidence has confirmed latent infection of potato tubers by blackleg bacteria (84a).

*Erwinia* spp. are not the only soft-rotting bacteria associated with vegetables in storage. Pectolytic representatives of *Pseudomonas, Bacillus,* and *Xanthomonas* have been detected in healthy cucumbers, and they were activated by incubation at a temperature of 37°C (59). The pectolytic microflora of potato tubers is frequently diverse (66) and may include anaerobic bacteria. *Clostridium* spp. may be a significant cause of soft rot in stored potato tubers (55, 56). Pectolytic clostridia have been found in numbers of $8 \times 10^5 - 1 \times 10^8$/g of rotting potato tissue in the presence of $1\text{–}4 \times 10^8$ *Erwinia carotovora*/g (55).

### *Ring Rot of Potato Caused by Corynebacterium sepedonicum*

In this disease the pathogen may persist in tubers without causing obvious damage. Experiments carried out by Bonde & Covell (8) showed that the pathogen may be present in the susceptible Katahdin variety without the appearance of symptoms in either tubers or foliage for at least one year. It is possible for a tuber to carry the bacteria internally and produce symptomless plants and tubers, although all three growth stages are infected. Under conditions where both spring and summer are cool, masking of symptoms is more likely to occur, and top symptoms may not appear before harvest (83). No top symptoms appeared in plants from infected tubers grown at an air temperature of 16°C, whereas stunting was severe in plants grown at 24°C (42).

## LATENCY IN BACTERIAL DISEASES OF FRUIT TREES

Stone, pome, and citrus fruit trees may be infected with certain bacterial pathogens without the occurrence of visible symptoms (4, 11, 16, 21, 24a, 28, 38, 82). Cameron (11) examined the systemic distribution and frequency of *Pseudomonas* spp. including *P. syringae* within diseased and apparently healthy trees of sweet cherry (*Prunus avium*). Pathogenic bacteria were isolated from apparently healthy trees. Highest populations occurred in the early spring, and there was a moderate increase after the first autumn rains; populations were lowest in midsummer and during the coldest weeks of winter. These differences appeared to be correlated with variation in temperature and available moisture. *Pseudomonas* spp. were found throughout many of the trees. Bacteria were not only isolated from near cankers, but from as far as 20 feet from obviously diseased tissue. The highest bacterial counts were obtained from the trunk, roots, and lower scaffold limbs. The significance of these results lies in the fact that protectant sprays designed to reduce surface inoculum will be effective only where the pathogen is not systemic. If the bacterium is already within the host, then the effectiveness of the protectant will be reduced in proportion to the extent of systemic infection (11).

Latent infection of apple and pear trees by *Erwinia amylovora* has recently been demonstrated (38, 82). Highly virulent isolates of the fireblight pathogen were obtained from internal tissues of symptomless side shoots on artificially inoculated apple and pear trees in the greenhouse. Virulent bacteria were also recovered from branches of pear trees in orchards without any evidence of fireblight. After the fireblight pathogen enters a tree artificially or naturally, the bacterium may remain dormant for periods of 1–6 months and be translocated without producing blight. These observations may explain the frequent unexpected appearance of the disease in nurseries, young plantings, and well-managed orchards. A form of masked infection has been reported (53) in which pockets of infection up to 2 cm in diameter were found in the center of outwardly healthy fruit on apparently unaffected trees in an affected orchard.

Goto (28) has shown in the lemon variety *Citrus unshu* var. *sugiyama* that *Xanthomonas citri* can survive for long periods in dead bark tissues and in apparently healthy periderm or epidermis of young twigs. Bacteria surviving in the bark may be responsible for the sporadic nature of citrus canker in Japan and be the cause of severe outbreaks even in citrus groves where the disease had not been detected for several years.

Without citing any published evidence, Cameron (10) states that pathogenic isolates of *Agrobacterium tumefaciens* have been isolated from symptomless trees. However, *A. tumefaciens* is a wound pathogen and could remain latent within a host until internal wounding occurred. In this regard the observations of Lehoczky (52) are relevant. He showed in chronically infected grapevine and raspberry plants that secondary tumors on aerial parts of the plant could result from twisting injuries to the vascular system that release the pathogen into the internode tissue from the damaged xylem vessels.

## LATENT BACTERIA IN RELATION TO THE PHYSIOLOGY OF THE HOST PLANT

Little is known about the influence of endophytic bacteria on the growth of healthy plants. Montuelle & Beerens (62) have proposed that saprophytic bacteria may play a role in the germination of potato tubers; they were led to this conclusion by their observation that bacteria were most numerous in the vascular tissue near the point of bud emergence. During the dormant period the number of endogenous bacteria was low; during germination of the tuber the number of bacteria per gram of bud fresh weight showed a marked increase. Many of the bacteria in culture produced substances such as indoleacetic acid and gibberellins that acted as plant growth hormones (61). The hypothesis that potato tubers are dependent on bacteria for provision of certain plant growth hormones requires verification using sterile (axenic) plant tissues. There is evidence that epiphytic bacteria on leaf surfaces produce indoleacetic acid from tryptophan and contribute to the hormone content of plant tissues (54). In some plants it is supposed that the majority of the indoleacetic acid synthesized is produced by contaminating bacteria. However, bacteria may not be involved in auxin synthesis within plant roots because supposedly sterile root tips have been shown to synthesize indoleacetic acid from tryptophan, though in extremely small amounts (60).

There is evidence that the bacterial symbiosis in *Ardisia* species is based on the fact that the plant is obligately dependent on the associated bacterium for provision of cytokinins that it is unable to synthesize itself (69).

Plant physiologists and biochemists working with whole plants, tissue slices, and subcellular fractions have sometimes not taken into account the effect of bacteria on their experiments (6, 7, 31, 33, 35, 57, 74a). In most instances aberrant results have been attributed to growth of external contaminants rather than to the presence of bacteria already present in plant tissues. The $Qo_2$ (dry weight) of bacteria is 10–100 times greater than that of plant tissues; accordingly, even low levels of bacterial infection could contribute significantly to the oxygen uptake of tissue slices or of subcellular fractions. Similarly, in studies of macromolecular synthesis bacterial contamination may have a disproportionate effect because bacterial cells are generally some 20–100-fold more active than plant cells in nucleic acid and protein synthesis. Bendich (6, 7) has shown that when roots of seedlings are exposed to labeled precursors of nucleic acids, such as $^{32}P$ phosphate, large quantities of labeled bacterial DNA can be isolated from the shoot tissue. There are several possible explanations of this phenomenon. It may be that bacteria on leaf surfaces have access to the labeled phosphate in the vascular system or that bacteria present in the vascular system of the plant incorporate the precursor.

Uptake of bacterial DNA (63) and of bacteria (15) by plant protoplasts has been demonstrated. As the transfer and expression of certain bacterial genes into plant cells has been shown recently, it is reasonable to hypothesize that, during the course of evolution, gene transfer from endogenous bacteria to their plant hosts may have occurred (22, 79).

## DISCUSSION

Many bacteria that cause foliar diseases seem to function as saprophytes on plant surfaces for protracted periods; they are residents, as defined by Leben (43, 46). Leben (46) has stressed the difficulty of determining precisely whether assumed epiphytes are "in" as well as "on" their hosts. In general, studies on the epiphytic phase of bacterial pathogens have not excluded the possibility that a given species may also be within the plant and reach the surface by means of natural openings. On certain plants a pathogen may be harbored exclusively as an epiphyte; such plants may serve as a source of inoculum for a neighboring susceptible host (30). It is also possible that the occurrence of pathogens in low numbers in the leaves of symptomless resistant hosts is a factor of greater significance in the epidemiology of foliar plant pathogens than has been realized hitherto. For example, Laub & Stall (44) have shown that *Physalis minima* and *Solanum nigrum* are both resistant to *Xanthomonas vesicatoria.* At high levels of inoculum, isolates of *X. vesicatoria* induced a hypersensitive reaction in these hosts; at low levels of inoculum no macroscopic symptoms were obtained on infiltration of leaves, although it was possible to isolate the pathogen 35 days after inoculation. The bacterial spot pathogen may therefore survive as an internal resident in these resistant plants, possibly survive the intercrop period under Florida conditions, and serve as a primary source of inoculum on tomato and pepper plants. Other workers have drawn attention to the need for a certain threshold level of inoculum concentration in order to elicit a macroscopic response, and to the survival and multiplication of bacteria in resistant or immune plants (1, 18, 37, 75).

## *Leaf Scald of Sugar Cane*

A bacterial disease of sugar cane, leaf scald caused by *Xanthomonas albilineans* is a notorious example of the problems that arise in international exchange of plant material when a particular disease is prone to remain dormant for long periods, particularly in resistant or tolerant varieties (58). In recent years there have been reports of the spread of this disease in the Caribbean area, to North America, and among certain African countries where previously the disease was not known to occur (40). The disease has now been found in Barbados, a major breeding center. This situation is a reflection of the difficulty experienced in detecting symptoms in plants undergoing quarantine and of the fact that resistant varieties can harbor the pathogen for several months at least (83). Often the disease has been in a territory for a long period before being recognized; it is apparent that further spread may have occurred prior to recognition and that some countries have inadvertently exported the leaf scald pathogen (40). Interception of leaf scald in quarantine is more likely to be successful where the practice of ratooning all introductions is adopted (76).

## *Techniques of Isolation*

Several recent studies have illustrated the importance of technique in determining the presence of relatively low numbers of plant pathogenic bacteria in soil or in host plants. Goto (27) has concluded that because of the low efficiency of the applied techniques, results from many earlier investigations on survival of plant pathogenic bacteria in association with vegetation generally referred to populations containing more than $10^4$–$10^5$ cells per gram. Hildebrand & Schroth (32) were able to isolate *Pseudomonas phaseolicola* consistently from bean leaves showing systemic chlorosis using an infiltration-centrifugation technique, and much less consistently and in lower numbers using a comminution technique. Comminution techniques, in common with most others in use, do not differentiate between the population within as well as on the surface of leaves. Ultraviolet irradiation of leaf surfaces has been used successfully for this purpose (5). Woolhouse (87) used an elaborate system of surface sterilization to free the leaves of *Perilla frutescens* of all superficial bacteria. He concluded that he was either dealing with very resistant bacteria on the surface or with bacteria below the surface of the leaf, because he could never obtain total bacterial counts below $10^3$ organisms per $cm^2$ of leaf area.

A recent study of soft-rotting bacteria associated with stored potato tubers was made possible by the use of a new method for detection of *Erwinia* spp. (65) and counting of soft-rotting bacteria in plant material. Populations of *E. carotovora* in the soil that could be estimated by a quantal method were 100–1000th of those detectable by a spot plate, viable count method. Numbers as low as ~25 organisms/g of soil were detected by the former method (64).

The cultural conditions and nutrient composition of culture media used to detect resident endophytic bacteria will condition the nature of the bacteria isolated. Few studies have taken into account the occurrence of anaerobic bacteria in plant tissue (55, 81) that may occur in subterranean storage organs. Lactic acid bacteria require enriched culture media, and may frequently have been overlooked in healthy or diseased fruits and vegetables (77) because the conditions used in isolation would generally favor the less nutritionally exacting, aerobic, Gram-negative bacteria. Similarly, acetic acid bacteria associated with the latent condition of pineapple fruit known as pink disease (9, 70), and which gain entry into the nectary ducts before or at the time of flowering, may occur in fruits with an acid pH.

*Literature Cited*


1. Allington, W. B., Chamberlain, D. W. 1949. Trends in the population of pathogenic bacteria within leaf tissues of susceptible and immune plant species. *Phytopathology* 39:656–60
2. Bacon, M., Mead, C. E. 1971. Bacteria in the wood of living aspen, pine, and alder. *Northwest Sci.* 45:270–75
3. Baker, K. F. 1972. Seed pathology. In *Seed Biology,* ed. T. T. Kozlowski, 2:317–416. New York: Academic. 425 pp.
4. Baldwin, C. H., Goodman, R. N. 1963. Prevalence of *Erwinia amylovora* in apple buds as detected by phage typing. *Phytopathology* 53:1299–1303
5. Barnes, E. H. 1965. Bacteria on leaf surfaces and intercellular leaf spaces. *Science* 147:1151–52
6. Bendich, A. J. 1972. Effect of contaminating bacteria on the radiolabeling of nucleic acids from seedlings: false DNA "satellites." *Biochim. Biophys. Acta* 272:494–503
7. Bendich, A. J. 1972. Assessing contribution of bacteria to satellite DNA in Cucurbitaceae. *Plant Physiol.* 49:12
8. Bonde, R., Covell, M. 1950. Effect of host variety and other factors on pathogenicity of potato ring-rot bacteria. *Phytopathology* 40:161–72
9. Buddenhagen, I. W., Dull, G. G. 1967. Pink disease of pineapple fruit caused by strains of acetic acid bacteria. *Phytopathology* 57:806
10. Cameron, H. R. 1972. Relationship of host metabolism to bacterial infection. *3rd Int. Conf. Plant Pathog. Bacteria Proc., Wageningen, 1971,* ed. H. P. Maas Geesteranus, 59–61. Wageningen: Centre Agr. Publ. Document. 365 pp.
11. Cameron, H. R. 1970. *Pseudomonas* content of cherry trees. *Phytopathology* 60:1343–46
12. Crossan, D. F., Morehart, A. L. 1964. Isolation of *Xanthomonas vesicatoria* from tissues of *Capsicum annuum. Phytopathology* 54:358–59
13. Darbyshire, J. F., Greaves, M. P. 1970. An improved method for the study of the interrelationships of soil microorganisms and plant roots. *Soil Biol. Biochem.* 2:63–71

14. Darbyshire, J. F., Greaves, M. P. 1971. The invasion of pea roots, *Pisum sativum* L., by soil microorganisms, *Acanthamoeba palestinensis* (Reich) and *Pseudomonas* sp. *Soil Biol. Biochem.* 3:151–55
15. Davey, M. R., Cocking, E. C. 1972. Uptake of bacteria by isolated higher plant protoplasts. *Nature* 239:455–56
16. Davis, J. R., English, H. 1969. Factors related to the development of bacterial canker in peach. *Phytopathology* 59:588–95
17. de Lange, A., Leben, C. 1970. Colonization of cucumber buds by *Pseudomonas lachrymans* in relation to leaf symptoms. *Phytopathology* 60:1865–66.
18. Diachun, S., Troutman, J. 1954. Multiplication of *Pseudomonas tabaci* in leaves of burley tobacco, *Nicotiana longiflora* and hybrids. *Phytopathology* 44:186–87
19. Dickey, R. S., Nelson, P. E. 1970. *Pseudomonas caryophylli* in carnation. IV. Unidentified bacteria isolated from carnation. *Phytopathology* 60:647–53
20. Dowler, W. M., Petersen, D. H. 1967. Transmission of *Pseudomonas syringae* in peach trees by bud propagation. *Plant Dis. Reptr.* 51:666–68
21. Dowler, W. M., Weaver, D. J. 1972. Characterization of green fluorescent pseudomonads isolated from apparently healthy dormant peach trees. *Phytopathology* 62:754
22. Doy, C. H., Gresshoff, P. M., Rolfe, B. G. 1973. Biological and molecular evidence for the transgenosis of genes from bacteria to plant cells. *Proc. Nat. Acad. Sci. USA* 70:723–26
22a. Dunleavy, J. M., Urs, N. V. R. 1973. Isolation and characterization of bacteriophage SBX-1 and its bacterial host, both endemic in soybeans. *J. Virol.* 12:188–93
23. Ercolani, G. L. 1969. Sopravvivenza di *Pseudomonas syringae* van Hall sul Pero in rapporto all'epoca della contaminazione, in Emilia. *Phytopathol. Mediter.* 8:207–16
24. Ercolani, G. L., Casolari, A. 1966. Ricerche di microflora in pomodori sani. *Ind. Conserve Parma* 41:15–22
24a. Gardner, J. M., Kado, C. I. 1973. Evidence for systemic movement of *Erwinia rubrifaciens* in Persian walnuts by the use of double-antibiotic markers. *Phytopathology* 63:1085–86
25. Gäumann, E. 1950. *Principles of Plant Infection.* London:Crosby Lockwood. 543 pp.
26. Geldreich, E. E., Bordner, R. H. 1971. Faecal contamination of fruits and vegetables during cultivation and processing for market. A review. *J. Milk Food Technol.* 34:184–95
27. Goto, M. 1972. The significance of the vegetation for the survival of plant pathogenic bacteria. See Ref. 10, pp. 39–53
28. Goto, M. 1972. Survival of *Xanthomonas citri* in the bark tissues of citrus trees. *Can. J. Bot.* 50:2629–35
29. Grogan, R. G., Kimble, K. A. 1967. The role of seed contamination in the transmission of *Pseudomonas phaseolicola* in *Phaseolus vulgaris. Phytopathology* 57:28–31
30. Hagedorn, D. J., Rand, R. E., Ercolani, G. L. 1972. Survival of *Pseudomonas syringae* on hairy vetch in relation to epidemiology of bacterial brown spot of bean. *Phytopathology* 62:762
31. Hallaway, M. 1968. The enzymology of cell organelles. *Plant Cell Organelles,* ed. J. B. Pridham, 1–15. New York/London: Academic. 261 pp.
32. Hildebrand, D. C., Schroth, M. N. 1971. Isolation of *Pseudomonas phaseolicola* from bean leaves exhibiting systemic symptoms. *Phytopathology* 61:580–81
33. Hill, H. M., Rogers, L. J. 1972. Bacterial origin of alkaline L-serine dehydratase in French beans. *Phytochemistry* 11:9–18
34. Hollis, J. P. 1951. Bacteria in healthy potato tissue. *Phytopathology* 41: 350–67
35. Ingram, M., Riches, J. P. P. 1951. The preparation of sterile carrot discs for prolonged physiological experiments. *New Phytol.* 50:76–83
36. Jenny, H., Grossenbacher, K. 1963. Root-soil boundary zones as seen in the electron microscope. *Soil Sci. Soc. Am. Proc.* 27:273–77
37. Kawamoto, S. O., Lorbeer, J. W. 1972. Multiplication of *Pseudomonas cepacia* in onion leaves. *Phytopathology* 62: 1263–65
38. Keil, H. L., Van der Zwet, T. 1972. Recovery of *Erwinia amylovora* from symptomless stems and shoots of Jonathan apple and Bartlett pear trees. *Phytopathology* 62:39–42
39. Kennedy, B. W. 1969. Detection and distribution of *Pseudomonas glycinea* on soybean. *Phytopathology* 59:1618–19
40. Koike, H. 1968. Leaf scald of sugarcane in continental United States—A first report. *Plant Dis. Reptr.* 52:646–49

41. Kominos, S. D., Copeland, C. E., Grosiak, B., Postic, B. 1972. Introduction of *Pseudomonas aeruginosa* into a hospital via vegetables. *Appl. Microbiol.* 24: 567–70
42. Larson, R. H., Walker, J. C. 1941. Temperatures affect development of ring rot. *Wis. Agr. Exp. Sta. Res. Bull.* 451: 62–63
43. Last, F. T., Warren, R. C. 1972. Non-parasitic microbes colonizing green leaves: their form and functions. *Endeavour* 31:143–50
44. Laub, C. A., Stall, R. E. 1967. An evaluation of *Solanum nigrum* and *Physalis minima* as suscepts of *Xanthomonas vesicatoria*. *Plant Dis. Reptr.* 51:659–61
45. Leben, C. 1965. Influence of humidity on the migration of bacteria on cucumber seedlings. *Can. J. Microbiol.* 11:671–76
46. Leben, C. 1965. Epiphytic microorganisms in relation to plant disease. *Ann. Rev. Phytopathol.* 3:209–30
47. Leben, C. 1968. Colonization of soybean buds by bacteria: observations with the scanning electron microscope. *Can. J. Microbiol.* 15:319–20
48. Leben, C. 1971. The bud in relation to the epiphytic microflora. In *Ecology of Leaf Surface Micro-organisms,* ed. T. F. Preece, C. H. Dickinson, 117–27. New York/London: Academic. 640 pp.
49. Leben, C. 1972. Microorganisms associated with plant buds. *J. Gen. Microbiol.* 71:327–31
50. Leben, C., Rusch, V., Schmitthenner, A. F. 1968. The colonization of soybean buds by *Pseudomonas glycinea* and other bacteria. *Phytopathology* 58: 1677–78
51. Leben, C., Schroth, M. N., Hildebrand, D. C. 1970. Colonization and movement of *Pseudomonas syringae* on healthy bean seedlings. *Phytopathology* 60:677–80
52. Lehoczky, J. 1968. Spread of *Agrobacterium tumefaciens* in the vessels of the grapevine, after natural infection. *Phytopathol. Z.* 63:239–46
53. Lelliott, R. A. 1968. Fireblight in England: its nature and its attempted eradication. *Eur. Med. Plant Prot. Org. Publ. Ser. A,* No. 45-E:10–14
54. Libbert, E., Kaiser, W., Kunert, R. 1969. Interactions between plants and epiphytic bacteria regarding their auxin metabolism. VI: The influence of the epiphytic bacteria on the content of extractable auxin in the plant. *Physiol. Plant.* 22:432–39
55. Lund, B. M. 1973. Isolation of pectolytic clostridia from potatoes. *J. Appl. Bacteriol.* 35:609–14
56. Lund, B. M., Nicholls, J. C. 1970. Factors influencing the soft-rotting of potato tubers by bacteria. *Potato Res.* 13:210–14
57. Margulies, M. M., Parenti, F. 1968. In vitro protein synthesis by plastids of *Phaseolus vulgaris.* III: Formation of lamellar and soluble chloroplast protein. *Plant Physiol.* 43:504–14
58. Martin, J. P., Robinson, P. E. 1961. Leaf Scald. In *Sugar-Cane Diseases of the World,* ed. J. P. Martin, E. V. Abbott, C. G. Hughes, 1:79–101. New York: Elsevier. 542 pp.
59. Meneley, J. C., Stanghellini, M. E. 1972. Occurrence and significance of soft-rotting bacteria in healthy vegetables. *Phytopathology* 62:778
60. Mitchell, E. K., Davies, P. J. 1972. Indoleacetic acid synthesis in sterile roots of *Phaseolus coccineus. Plant Cell Physiol.* 13:1135–38
61. Montuelle, B. 1966. Synthèse bactérienne de substances de croissance intervenant dans le métabolisme des plantes. *Ann. Inst. Pasteur Paris* 111:136–46
62. Montuelle, B., Beerens, H. 1964. Les bactéries des organes tuberises de divers végétaux. Étude et évolution. *Ann. Inst. Pasteur Lille* 15:131–36
63. Ohyama, K., Gamborg, O. L., Miller, R. A. 1972. Uptake of exogenous DNA by plant protoplasts. *Can. J. Bot.* 50:2077–80
64. Pérombelon, M. C. M. 1971. A quantal method for determining numbers of *Erwinia carotovora* var. *carotovora* and *E. carotovora* var. *atroseptica* in soils and plant material. *J. Appl. Bacteriol.* 34:793–98
65. Pérombelon, M. C. M. 1972. A reliable and rapid method for detecting contamination of potato tubers by *Erwinia carotovora. Plant Dis. Reptr.* 56:552–54
66. Pérombelon, M. C. M. 1972. The extent and survival of contamination of potato stocks in Scotland by *Erwinia carotovora* var. *carotovora* and *E. carotovora* var. *atrospetica. Ann. Appl. Biol.* 71:111–17
67. Pérombelon, M. C. M. 1973. Sites of contamination and numbers of *Erwinia carotovora* present in stored seed potato stocks in Scotland. *Ann. Appl. Biol.* 74:59–65
68. Philipson, M. N., Blair, I. D. 1957. Bacteria in clover root tissue. *Can. J. Microbiol.* 3:125–29

69. Rodrigues Pereira, A. S., Houwen, P. J. W., Deurenberg-Vos, H. W. J., Pey, E. B. F. 1972. Cytokinins and the bacterial symbiosis of *Ardisia* species. *Z. Pflanzenphysiol.* 68:170–77
70. Rohrbach, K. G., Pfeiffer, J. 1971. Temperature and moisture requirements for infection of detached pineapple inflorescences by pink disease bacteria. *Phytopathology* 61:908
71. Samish, Z., Dimant, D. 1959. Bacterial population in fresh, healthy cucumbers. *Food Mf.* 34:17–20
72. Samish, Z., Etinger-Tulczynska, R. 1963. Distribution of bacteria within the tissue of healthy tomatoes. *Appl. Microbiol.* 11:7–10
73. Samish, Z., Etinger-Tulczynska, R., Bick, M. 1961. Microflora within healthy tomatoes. *Appl. Microbiol.* 9:20–25
74. Sanford, G. B. 1948. The occurrence of bacteria in normal potato plants and legumes. *Sci. Agr.* 28:31–35
74a. Sarrouy-Balat, H., Delseny, M., Julien, R. 1973. Plant hormones and DNA synthesis: evidence for a bacterial origin of rapidly labelled heavy satellite DNA. *Plant Sci. Lett.* 1:287–92
75. Scharen, A. L. 1959. Comparative population trends of *Xanthomonas phaseoli* in susceptible, field tolerant, and resistant hosts. *Phytopathology* 49:425–28
75a. Schroth, M. N., Cho, J. J., Kominos, S.D. 1973. No evidence that *Pseudomonas* on chrysanthemums harms patients. *Lancet* 1973:906–7
76. Sheffield, F. M. L. 1969. Leaf scald again. *Sugarcane Pathol. Newslett.* 3:10
77. Smith, M. A., Niven, C. F. 1957. The occurrence of *Leuconostoc mesenteroides* in potato tubers and garlic cloves. *Appl. Microbiol.* 5:154–55
78. Stall, R. E., Hall, C. B. 1969. Association of bacteria with graywall of tomato. *Phytopathology* 59:1650–53
79. Stroun, M., Anker, P. 1971. Bacterial nucleic acid synthesis in plants following bacterial contact. *Mol. Gen. Genet.* 113:92–98
80. Tervet, I. W., Hollis, J. P. 1948. Bacteria in the storage organs of healthy plants. *Phytopathology* 38:960–67
81. Thomas, W. D., Graham, R. W. 1952. Bacteria in apparently healthy pinto beans. *Phytopathology* 42:214
82. Van der Zwet, T., Keil, H. L. 1972. Importance of pear-tissue injury to infection by *Erwinia amylovora* and control with streptomycin. *Can. J. Microbiol.* 18:893–900
83. Walker, J. C. 1969. *Plant Pathology.* New York: McGraw-Hill. 3rd ed. 819 pp.
84. Watson, A. G., Koons, C. E. 1973. *Pseudomonas* on the chrysanthemums. *Lancet* 1973:91
84a. Webb, L. E., Wood, R. K. S. 1974. Infection of potato tubers with soft rot bacteria. *Ann. Appl. Biol.* 76:91–98
85. Wismer, C. A. 1968. Resistant varieties can harbor bacteria that cause leaf scald disease. *Hawaii Sugar Plant. Assoc. Exp. Sta. Ann. Rept.* 1968:69–70
86. Wood, R. K. S. 1967. *Physiological Plant Pathology.* Oxford/Edinburgh: Blackwell. 570 pp.
87. Woolhouse, H. W. 1971. Discussion. See Ref. 48, pp. 221–22

# LATENT INFECTIONS BY FUNGI

❖3590

*K. Verhoeff*

Phytopathological Laboratory Willie Commelin Scholten, Baarn (Departments of Plant Pathology of the Universities of Amsterdam and Utrecht) The Netherlands

## INTRODUCTION

Before a pathogenic fungus establishes a parasitic relationship with its host plant, a number of stages in the life cycle of the fungus will have been passed. After spores land on the surface of the host, germination takes place if circumstances are favorable. A spore can germinate immediately after landing, or it can take some time before germination occurs. During this time, the spores must remain viable; the fungus can be isolated by washing the spores from the host and plating out this suspension or, when stuck to the surface, by plating out pieces of the host tissue. This period between landing and germination is sometimes referred to as a latent infection, although no infection has yet taken place. After germination has occurred, the germ tube can penetrate directly into the host tissue, with or without the development of superficial mycelium preceding penetration. In both cases, infection hyphae can be produced from appressoria. Sometimes these appressoria remain viable on the surface of the host for some time, but infection has not yet taken place.

Penetration into the host occurs in various ways. In many cases, an infection hypha penetrates the host cuticle and then the outer epidermal cell wall. This can be the start of a parasitic relationship. However, some time may pass between penetration and the start of such a relationship. This stage is correctly referred to as a latent, dormant, or quiescent infection. The actual infection has taken place, though macroscopically not yet visible, but further growth of the infection hypha is delayed. Nothing is known about the relationship between host and parasite at this stage; there might be an equilibrium between these two. This implies that the term *latency* should be described as a quiescent or dormant parasitic relationship which, after a time, can change into an active one.

The definition of a latent infection given above, first stated by Gäumann (32), is used in this paper, although I do not extend the definition, as Gäumann did, to those stages in the life cycle of a number of *Ustilago* spp. during which no external symptoms can be seen. Similarly, I do not regard the symptomless growth of *Botrytis allii* in young leaves of onion plants as a latent infection, because the mycelium continues to grow towards the bulb (65). The same holds true for *B. narcissicola,* where mycelium spreads through the flower stalk towards the bulb, without producing symptoms (W. R. Jarvis, unpublished data). Neither is this the case with some other plant-fungus combinations where no symptoms are seen, although the fungus can be isolated after surface sterilization, e.g. *Fusarium oxysporum* f. sp. *cepae* on *Oxalis corniculata* (1) and *F. oxysporum* of cotton on a number of other plants (6). These relationships with nonhosts might play a role in survival of some fungi (37), but these are not true latent infections, as no parasitic relationship eventually occurs.

Generally speaking, most of the work on latent infection has been done with tropical fruits, apples, and recently, some fruit rots caused by *Botrytis cinerea.* In banana, papaw, mango, and citrus fruits, *Colletotrichum* spp. (*Gloeosporium* spp.) produce latent infections, the transition into active growth taking place during ripening. *Gloeosporium* spp. (*Pezicula* spp.) infect young apples and establish latent infections. Here again, renewed growth takes place during ripening. In more recent years, *Botrytis cinerea* was found to produce latent infections in fruits of strawberry, raspberry, grape, and tomato plants. This paper deals mainly with these host-fungus combinations.

## LATENT INFECTIONS IN BANANA AND OTHER SUBTROPICAL FRUITS

After surface sterilization, *Colletotrichum gloeosporioides* (stat. conid. of *Glomerella cingulata*) and *Gloeosporium musarum* (*Colletotrichum musae*), fungi known to produce storage fruit rots, can be isolated from the skin of a number of apparently healthy fruits. Dastur (23) in India in 1916 was the first to find *G. musarum* as a dormant infection in the skin of fruit of the plantain (*Musa paradisiaca*). Later *C. gloeosporioides* was isolated from the skin of healthy-looking grapefruits and oranges (*Citrus* spp.). *G. musarum* was also found in the skin of unripe bananas (*Musa* spp.) as latent infections (73), while *C. gloeosporioides* was isolated from the skin of mango (*Mangifera indica*) and papaw (*Asimina triloba*) (74) and from the skin of avocado (*Persea* spp.) and citrus fruits (7). Both fungi were also isolated from the skin of grapefruit, cacao (*Theobroma cacao*), and tomato (*Lycopersicon esculentum*) (7, 8).

Conidia of *G. musarum* germinate in a water film on the fruit surface of the banana. After the production of a germ tube, an appressorium is formed which, according to some authors, is developed just above the junction of two epidermal cells. The appressorium is closely pressed against the cuticle and surrounded by a mucilaginous sheath. From the appressorium a thin infection hypha penetrates the cuticle but not the outer epidermal cell wall. Some small subcuticular hyphae can be seen, sometimes accompanied by a brown discoloration of a few epidermal cells. Occasionally a thickening of the cellulose layer under the appressorium and the subcuticular hyphae can be seen (20, 49, 60).

A similar infection process was found in papaw and mango inoculated with *G. musarum* (60), but it is sometimes difficult to find infection hyphae penetrating the cuticle (9, 62).

A more or less similar pattern occurs in other tropical or subtropical fruits, leading to a latent infection. According to Adam et al (2), infection hyphae are produced from a peg-like thickening of the appressorium in the cuticle of grapefruits and oranges. The infection hypha grows intercellularly two or three cells deep in the skin, where it stops and remains dormant. Usually some cells around the infection hyphae appear to stain differently from the other cells.

On grapefruit, mango, and avocado, *C. gloeosporioides* follows the same pattern as *G. musarum* on banana (7). However, it was found in Israel (15) that appressoria sunken in the cuticle form latent infections on avocado fruits, infection hyphae being produced only when the fruits are ripening.

*Sclerotinia fructicola* produces on young immature fruits of apricot (*Prunus armeniaca*), a dormant infection from germ tubes of conidia within the cells surrounding the stomatal cavity (70).

In all cases examined, wound inoculation does not lead to latent infections; usually an inter- and intracellular mycelium develops in both unripe and ripe fruits, although the riper the fruit at inoculation, the faster the mycelium spreads (61, 73). Sometimes wound inoculation leads to infection only when the fruits are ripening (15).

## LATENT INFECTIONS IN APPLES BY *GLOEOSPORIUM* SPECIES

Latent infections by *Gloeosporium perennans* (stat. conid. of *Pezicula malicorticis*) and *G. album* (stat. conid. of *P. alba*) are known to occur in apples, although they have been demonstrated histologically for *G. perennans* only (25). *Gloeosporium perennans* seems to be more pathogenic than *G. album*, and fruit rots caused by the former appear earlier (26).

Conidia of *G. perennans* germinate on young unripe fruits when conditions are favorable. A germ tube is formed which ends in a rather thick-walled appressorium that is firmly attached to the fruit surface, and sometimes even sunken into lenticels. The appressorium produces a thin infection hypha which grows into the parenchyma cells of lenticels, but it never passes through the meristematic layer under the cork of the lenticel. A few hyphae are produced in the lenticels and here the fungus remains dormant until the fruit reaches a certain stage of ripeness.

Bompeix (16) found a positive correlation between the thickness of the cork layer lining a lenticel and susceptibility to fruit rot; the continuity of the cork layer with the epidermis seems to be especially important in this respect. When lenticels are completely enclosed by cork layers, and no gas exchange with the underlying tissue is possible, there is no change from dormancy to active mycelial growth leading to fruit rot (50). Furthermore, such enclosed lenticels do not become infected unless nutrients are added to the inoculum (24). When fruits are put under vacuum for a short time, the cork layer is damaged, probably together with some of the underlying parenchyma tissue, and mycelium is then able to penetrate into deeper layers of parenchyma cells of the unripe fruits without a period of latency (50). Direct penetration of hyphae through the cuticle is possible only when the cuticle is damaged (45).

## LATENT INFECTION CAUSED BY *BOTRYTIS CINEREA*

*Botrytis cinerea* has long been known as a fungus infecting plants primarily through dead or dying tissues, while in some cases a leaf spotting occurs. Due to difficulties in controlling *B. cinerea* in strawberry, Powelson (53) studied the infection process on flowers and fruits. He demonstrated a latent infection of the fungus in the stem end of the developing fruit, where the fungus remained dormant until the fruit ripened. A high proportion of necrotic flower parts contained mycelium of *B. cinerea,* and via these parts the fungus apparently grew into the immature fruit. Consequently, when flower parts were removed after pollination, the incidence of fruit rot was considerably reduced. Jarvis (41) confirmed these findings in Scotland, where he found a similar latent infection in strawberries and raspberries. Here again, rotting of the fruit is not seen until ripening, sometimes even after picking. A positive correlation was established between the incidence of blossom blight and latent infections in strawberry, as determined on the number of fruit rots during ripening (42). Schönbeck (57) also demonstrated the latent infection in strawberries. According to him, infection by *B. cinerea* of the anthers where the tops are open at the release of the pollen, and germination and germ tube growth are stimulated by pollen (21), a rapid growth of mycelium through the anthers occurs. Schönbeck (57) did not find a direct penetration of mycelium through the style into the developing fruit. More recently, a dormant infection was found in grapes, where *B. cinerea* remains latent in the necrotic stigmatic tissue, after infection at bloom time (48).

In tomato plants, *B. cinerea* produces a latent infection in young fruits, where germ tubes penetrate into epidermal cells, but unlike the examples described above, no further growth takes place when the fruits are ripening, although the fungus is still alive (68).

## POSSIBLE EXPLANATIONS FOR LATENCY

The transition of a dormant parasitic relationship into an active one usually takes place only when fruits are ripening, a phenomenon difficult to explain. As pointed out by Wardlaw et al (74), the cause must be physiological. Simmonds (61), reviewing the work done with bananas in this respect, discussed four possible explanations for the latency of an infection: 1. A toxin is present in unripe but not in ripe fruits. 2. The nutritional requirements of the pathogen are not met by the composition of the green unripe fruit. 3. The energy requirements of the fungus are met only when the metabolism of the host has passed from the unripe into the ripening phase. 4. The enzyme potential of the fungus is not great enough to permit invasion of immature fruits, but is sufficient to permit invasion of ripe fruits.

It is difficult to make clear separations between these four possibilities, especially between 2 and 3. I will therefore distinguish between and discuss, using the host-parasite combinations described above, the following three points: (*a*) toxic compounds are present in unripe, but not in ripening fruits; (*b*) there are nutritional differences between unripe and ripening fruits, with regard to fungal development; and (*c*) the enzyme potential of the fungus is insufficient to invade unripe fruits.

## THE SIGNIFICANCE OF TOXIC COMPOUNDS

Concerning the presence of toxic substances, the banana fruit contains physiologically active tannins in latex-containing vessels and, especially in the skin, in a number of parenchyma cells. During the ripening process, the extractable active tannins undergo a tenfold decrease in concentration in the skin (13). It was demonstrated earlier that tannins have a toxic effect upon germination and mycelial growth of various fungi, including *G. musarum,* when moderate concentrations of this compound were present in the medium (22). Low concentrations of tannins, however, stimulate mycelial growth; when the agar medium contained 0.25% tannins, germination and mycelial growth were not inhibited. Even on agar with 0.6% tannins, some germination and germ tube growth occurred. The toxic effect of higher concentrations of tannins could be reduced to some extent by adding nutrients to the medium: the higher the concentration of tannins, the more nutrients had to be added to overcome the effects of the toxin. Thus there could be a correlation between the reduction in tannin concentration during ripening and the transition of the latent infection into active mycelial growth, although Toro (66) could not explain latency by assuming tannins to be responsible. On the other hand, tannins appear to inactivate extracellular enzymes produced by *G. musarum,* in this way inhibiting the establishment of an active parasitic relationship (35).

Simmonds (60) could find no difference in germination or appressorium formation in *G. musarum* when extracts of green or ripe banana fruits were used as nutrient media. Neither was any evidence obtained that oxidized polyphenols are responsible for the resistance of unripe banana fruit. Use of the diffusion drop technique revealed no differences in growth of *G. musarum* in diffusates from inoculated or uninoculated skins, nor was there any inhibition of germination when thin slices of the skin of unripe fruits were added to a spore suspension (61). However, when slices of skin of an unripe banana fruit were put into water at 80–95°C and subsequently cooled, there was nearly complete inhibition of germ tube growth when conidia of *G. musarum* were suspended in this extract. A catechol-like compound appeared to be present in this extract, together with an oxidase system. In extracts of skin slices of ripe fruits, the same compounds appeared to be present, though not at toxic levels. Apparently, no oxidizing system was present then, or else it could not operate (61). There is no conclusive evidence as to whether toxic compounds are involved in establishing latent infections in the banana, but they might play a role.

With regard to toxic compounds in immature apples, ethanol extracts of unripe apples decrease the activity of polygalacturonases and of macerating enzymes produced by *G. perennans* (26). Similar extracts of ripe fruits have no such effect, unless extracts of ripe cider apples are used. Since cider apples contain rather higher concentrations of phenolic compounds, inactivation of the enzyme system of the fungus might be the cause of latency in unripe fruits (26). Moreover, the decrease in total phenolic compounds in fruits during the process of ripening coincides with increased susceptibility to attack by *G. perennans.*

The decrease in phenols in ripening apples is mainly caused by a decrease in concentration of chlorogenic acid in the peel of the fruits (39), so the acid could be an important factor in latency (18). However, when chlorogenic acid is added to a nutrient medium, no inhibition of pectolytic enzyme activity of *G. perennans* occurs (25), but the germination of conidia is inhibited. It is not quite clear, therefore, what role phenolic compounds play in relation to latency. Borecka & Pieniazek (18) suggest that the plant hormone abscissic acid might play a role in breaking the latency, but their experiments do not give conclusive results. During ripening, abscissic acid remains in the pulp but disappears from the skin and, in vitro, it stimulates germination of conidia of *G. perennans.*

With regard to latent infections by *B. cinerea,* no information is available on whether toxic compounds are present in unripe strawberry or raspberry fruits.

In young fruits as well as leaves of tomato plants, the fungitoxic compound tomatine is present (30). The concentration of tomatine is much higher in the peel than in other parts of the fruits, e.g. 0.95% and 0.002% of the dry weight, respectively (K. Verhoeff, unpublished data). In vitro experiments showed that concentrations up to 1000 ppm had no effect upon germination of conidia, while a reduction, but not a complete inhibition of mycelial growth occurred at concentrations of 200 ppm and higher. However, sap expressed from the skin of young fruit gave only a slight inhibition of germination and germ tube growth. It is, of course, questionable whether all the tomatine is present in the sap, and the effect of the pH of the tomatine medium upon toxicity to *B. cinerea* is not yet known. In ripe fruits, no tomatine can be found in the skin or the peel (56). Sap expressed from ripe fruits strongly stimulates germination of conidia and germ tube growth (K. Verhoeff, unpublished data). As yet, no conclusions can be drawn from these experiments as to whether tomatine plays a role in latency.

*Colletotrichum phomoides* also produces a quiescent infection in tomato fruits. Conidia of this fungus germinate on the fruit surface and, after formation of an appressorium, an infection hypha penetrates into an epidermal cell of the young fruit. When the fruit ripens, transition into active growth takes place (31). Allison (4) suggested a correlation with the fungitoxic compound solanine, which reaches higher concentrations in the skin than in the pulp but disappears during the ripening process. However, the data presented are far from conclusive.

In contrast to the examples mentioned above, polyphenol concentration increases during maturation and ripening in tomato fruits. The concentration of these compounds in green full-sized fruits is about 30 mg per 100 g fresh weight, increasing to about 50 mg per 100 g fresh weight when the fruits are red (71). This increase in concentration might explain the fact that infection of *B. cinerea* in tomato fruits remains latent during ripening. *B. cinerea* might be more sensitive to these compounds than *C. phomoides,* which does establish an active parasitic relationship at ripening (31).

## DIFFERENCES IN NUTRIENTS IN UNRIPE AND RIPENING FRUITS

During ripening a number of other changes occur in the chemical constitution of fruits. These changes were studied in banana by Barnell (10–12). During development of the banana fruit on the plant, sucrose, glucose, and fructose are absent from both pulp and skin, and the main carbohydrate is starch. At harvest, the starch content of the pulp is much higher than that of the skin, 16 and 2.5% of the fresh weight respectively. A slight increase in titratable acids occurs in both pulp and skin during growth of the fruit.

The harvested fruits are usually stored for some time at about 11.5°C (i.e. during transport) and are then ripened at about 20°C. During these treatments, a rapid decrease in starch content occurs in both pulp and skin, with a corresponding increase in total sugars. This increase in sugar concentration of the pulp is followed by a decrease, but the sucrose concentration in the skin continues to increase during the final stages of the ripening process from about 0.1 to 2.5% of the fresh weight.

Sacher (54, 55) reported that the onset of senescence in banana fruit is marked by changes in permeability of cell membranes. At the respiratory peak, the free space of the tissue was almost entirely occupied by various carbohydrates, such as sucrose and fructose. A similar increase in leakage of carbohydrates was found by Burg et al (19), but according to them it was the result of an overall increase in endogenous levels of sugars and not of an increased permeability of the membranes.

*Gloeosporium musarum,* when cultured in vitro, utilizes starch as a carbon source; further, when inoculated onto pulp of unripe fruits, conidia germinate and produce a dense mycelial mat in a relatively short time (61). It thus seems that nutrients are not the limiting factor in unripe banana fruits responsible for latent infections.

During the ripening process of apples the amount of soluble sugars increases (5, 63). When the fruits are subsequently stored at low temperatures (2 or 12°C), the concentration of sucrose decreases, but that of glucose and fructose increases still further (50). Also the dry matter content of the fruits gradually decreases with increasing storage temperature, as does the total amount of acids, although pH increases slightly. *Gloeosporium perennans* grows poorly on starch as the only carbon source in an agar medium, whereas on simple sugars good growth is obtained (17). From these results, it seems that differences in chemical composition between unripe and ripe fruits might be responsible for breaking latency. However, when fruits of various varieties differing in susceptibility to fruit rot were analyzed chemically, there appeared to be no nutritional basis for susceptibility (16, 17).

A number of changes occur during the ripening process in strawberries. One of them is the increase in soluble sugars (47). When ripe, the fruits contain about 5% w/w sugars (76), mostly reducing sugars (about 4.1% w/w) (34). As *B. cinerea* develops very well on sugars as a sole carbon source, it has been suggested that this increase in soluble sugars might explain the transition from quiescence into active mycelial growth.

From the examples cited above, no conclusive evidence can be drawn concerning the significance of nutrients available to the fungus in latency. Nutrients might play a role in the transition of a quiescent parasitic relationship into an active one, but they do not seem to be of primary importance.

## THE ENZYME POTENTIAL OF THE FUNGUS IN RELATION TO LATENCY

After conducting a number of experiments on the production of pectolytic enzymes by *G. musarum,* Simmonds (61) states that the results "do not invalidate the hypothesis that latency is due to insufficient enzyme production, but rather shifts the emphasis towards the general metabolic state of the fungus in relation to its substrate and energy sources." I regard the latter part of this statement as very important, when this is seen in relation to another aspect of fruit ripening, e.g. the changes that occur in the cell wall.

Simmonds mentioned earlier that possibly the cell wall material in unripe banana fruits, especially that of the middle lamellae, cannot be dissolved by enzymes produced by the fungus. During ripening, the various compounds become more loosely bound and might only then become degraded by pectolytic enzymes of the fungus, leading to a breaking of the dormancy (60).

Such changes in the middle lamellae of fruit tissue during ripening are well known. Wallace et al (72) found a negative correlation between the ripeness of apples and the amount of water-insoluble material of the pulp. The amount of water-insoluble protein decreases sharply at the time when apples usually start to rot. Their results suggest that the protopectin in the middle lamella is an unsuitable substrate for fungal enzymes. Later, softening of the fruit is caused by degradation of the protopectin by apple enzymes (28, 40), and the middle lamella becomes more susceptible to degradation by fungal enzymes.

In strawberry and tomato, too, the ripening process of the fruit is accompanied by increased solubility of the protopectin (38), leading to decrease in firmness of the fruit (52).

The term *protopectin* is used for the pectic material in cell walls which is insoluble in cold water (75). The insolubility of the protopectin could result from linkages between COOH groups of pectic substances and basic groups of proteins. Another reason for this insolubility might be cross-linkages between adjacent polygalacturonic acid chains and hemicelluloses (14, 44) or through linking COOH groups via polyvalent ions, particularly calcium and magnesium. Such a stabilizing effect of cations, leading to a chelate formation, was also demonstrated (33). In this insoluble state, middle lamellae are not hydrolyzed by pectolytic enzymes produced by fungi. The pectin-cation-protein complex or the pectin-protein-hemicellulose complex has to be changed first, and this is normally accomplished during ripening by fruit enzymes. Only if a fungus produces a wall-modifying enzyme (43) can this barrier be reached before ripening.

It is known that in many fungi, pectolytic enzyme production is greatly stimulated by soluble pectic materials in the medium. Thus it can be expected that when the middle lamella is changed during ripening, and there is an increase in water-soluble compounds, pectolytic enzyme production will be stimulated. However, such a stimulation does not take place in some incompatible host-parasite combinations, where cell wall proteins inhibit pectolytic enzyme activity (3, 29).

*Gloeosporium perennans* and *G. album* produce pectolytic enzymes especially when grown on pectin-containing media (58), so that when such material becomes available during ripening, the fungus can spread rapidly in the tissue of the fruit. *Colletotrichum gloeosporioides* even produces an endopeptidase on apple fruit tissue, by which the pectin-cation-protein complex is hydrolyzed (46). However, such an hydrolysis will be possible only when some of the complex material has become soluble. On the other hand, pectolytic enzymes may play a role in removing the calcium from the complex middle lamellae (51). Like *Gloeosporium* and *Colletotrichum, B. cinerea* produces pectolytic enzymes, the activity being related to the type of carbon source in the medium (36, 69). This explanation for the transition of a quiescent infection into an active parasitic relationship has been suggested earlier (31, 60, 72).

That the chemical constitution of the middle lamella plays an important role in keeping infections quiescent is supported by the fact that apples low in Ca rot more quickly (27), and Ca treatments of apples decrease fruit rot by *G. perennans* (59). Similarly, grey mold in tomatoes, caused by *B. cinerea,* can be reduced by spraying the plants with calcium (64). Also the reduction of infections by *B. cinerea* in glasshouse-grown tomato plants by higher nitrogen levels in the soil supports this, as senescence of the tissues and hence cell wall changes is delayed by this treatment (67).

The breakdown of the protopectin in tomato fruits during ripening (38) does not lead to an active parasitic relationship with *B. cinerea.* Apparently, the phenolic compounds present in the ripe fruits remain sufficient to inhibit fungal enzyme activity.

The question remains why penetration occurs so easily in these cases. *Botrytis cinerea* conidia before germination contain endopolygalacturonase (K. Verhoeff, unpublished data) and, together with available nutrients in fruit exudates, this compound ensures that penetration will occur. When in the host, the inhibiting factors such as the chemical constitution of the middle lamella, possibly together with fungitoxic compounds, come into effect and stop further spread.

Although it is possible to recognize that latency and its transition are broadly connected with physiological changes in maturing host tissues, especially with changes in the middle lamella of cell walls, no single hypothesis fits all cases in detail. Each host-parasite combination has a unique and delicately balanced set of factors in the latent phase, which changes in a special way to permit the transition from passive to active parasitic relationship. Thus, each combination has to be investigated separately to obtain details of the mechanism involved in latency.

Acknowledgments

The author is indebted to his colleagues of the laboratory at Baarn for their interest and suggestions, and especially to W. R. Jarvis (Scottish Horticultural Research Institute, Invergowrie) for his criticism, advice, and correction of the English text.

*Literature Cited*

1. Abawi, G. S., Lorbeer, J. W. 1972. Several aspects of the ecology and pathology of *Fusarium oxysporum* f. sp. *cepae*. *Phytopathology* 62:870–76
2. Adam, D. B., McNeil, J., Hanson-Merz, B. M., McCarthy, D. F., Stokes, J. 1949. The estimation of latent infection in oranges. *Austr. J. Sci. Res., Ser. B* 2:1–18
3. Albersheim, P., Anderson, A. 1971. Proteins from plant cell walls inhibit polygalacturonases secreted by plant pathogens. *Proc. Nat. Acad. Sci. USA* 68:1815–19
4. Allison, P. V. 1952. Relation of solanine content of tomato fruits to colonization by *Colletotrichum phomoides*. *Phytopathology* 42:1
5. Archibald, H. K. 1932. Chemical studies in the physiology of apples. 12. Ripening processes in the apple and the relation of time of gathering to the chemical changes in cold storage. *Ann. Bot.* 46:407–59
6. Armstrong, G. M., Armstrong, J. K. 1948. Nonsusceptible hosts as carriers of wilt fusaria. *Phytopathology* 38: 808–26
7. Baker, R. E. D. 1938. Studies in the pathogenecity of tropical fungi. 2. The occurrence of latent infections in developing fruits. *Ann. Bot. n.s.* 2:919–31
8. Baker, R. E. D., Wardlaw, C. W. 1937. Studies in the pathogenicity of tropical fruits. 1. On the types of infection encountered in the storage of certain fruits. *Ann. Bot. n.s.* 1:59–65
9. Baker, R. E. D., Crowdy, S. H., McKee, R. K. 1940. A review of latent infections caused by *Colletotrichum gloeosporioides* and allied fungi. *Trop. Agr.* 17:128–32
10. Barnell, H. R. 1940. Studies in tropical fruits. 8. Carbohydrate metabolism of the banana fruit during development. *Ann. Bot. n.s.* 4:39–71
11. Barnell, H. R. 1941. Studies in tropical fruits. 11. Carbohydrate metabolism of the banana fruit during ripening under tropical conditions. *Ann. Bot. n.s.* 5: 217–47
12. Barnell, H. R. 1941. Studies in tropical fruits. 13. Carbohydrate metabolism of the banana fruit during storage at 53° and 68° F. *Ann. Bot. n.s.* 5:608–46
13. Barnell, H. R., Barnell, E. 1945. Studies in tropical fruits. 16. The distribution of tannins within the banana and the changes in their condition and amount during ripening. *Ann. Bot. n.s.* 9:77–99
14. Bauer, W. D., Talmadge, K. W., Keegstra, K., Albersheim, P. 1973. The structure of plant cell walls. 2. The hemicellulose of the walls of suspension-cultured sycamore cells. *Plant Physiol.* 51:174–87
15. Binyamini, N., Schiffermann-Nadel, M. 1972. Latent infection in avocado fruit due to *Colletotrichum gloeosporioides*. *Phytopathology* 62:592–94
16. Bompeix, G. 1966. Contribution a l'étude de la maladie des taches lenticellaires des pommes 'Golden Delicious' en France. *Mém. Fac. Sci. Rennes. Univ., Brest* 121 pp.
17. Borecka, H. 1962. *Pezicula malicorticis* as a pathogen on apple during the period of storage. *Acta Agrobot.* 12:13–66
18. Borecka, H., Pieniazek, J. 1968. Stimulatory effect of abscissic acid on spore germination of *Gloeosporium album* Osterw. and *Botrytis cinerea* Pers. *Bull. Acad. Polonaise Sci.* 16:657–61
19. Burg, S. P., Burg, E. A., Marks, R. 1964. Relationship of solute leakage to solution tonicity in fruits and other plant tissues. *Plant Physiol.* 39:185–95
20. Chakravarty, T. 1957. Anthracnose of banana (*Gloeosporium musarum* Cke. et Massee), with special reference to latent infection in storage. *Trans. Brit. Mycol. Soc.* 40:337–45
21. Chou, M. C., Preece, T. F. 1968. The effect of pollen grains on infections caused by *Botrytis cinerea* Fr. *Ann. Appl. Biol.* 62:11–22
22. Cook, M. T., Taubenhaus, J. J. 1911. The relation of parasitic fungi to the contents of the host plants. 1. The toxicity of tannin. *Delaware Agr. Exp. Sta. Bull.* 91. 77 pp.
23. Dastur, J. F. 1916. Spraying for ripe rot of the plantain fruit. *Agr. J. India* 11:142
24. Edney, K. L. 1956. The rotting of apples by *Gloeosporium perennans* Zeller & Childs. *Ann. Appl. Biol.* 44:113–28
25. Edney, K. L. 1958. Observations on the infection of Cox's Orange Pippin apples by *Gloeosporium perennans* Zeller & Childs. *Ann. Appl. Biol.* 46:622–29
26. Edney, K. L. 1964. A comparison of the production of extra cellular enzymes and rotting of apples by *Pezicula alba* and *P. malicorticis*. *Trans. Brit. Mycol. Soc.* 47:215–25
27. Edney, K. L., Perring, M. A. 1973. Study of farms with high incidence of *Gloeosporium* infection. *E. Malling Res. Sta. Rep.* 1972, p. 157

28. Fisher, D. V. 1943. Mealiness and quality of Delicious apples as affected by growing conditions, maturity and storage techniques. *Sci. Agri.* 23:569–88
29. Fisher, M. L., Anderson, A. J., Albersheim, P. 1973. Host-pathogen interactions. 6. A single plant protein efficiently inhibits endopolygalacturonases secreted by *Colletotrichum lindemuthianum* and *Aspergillus niger*. *Plant Physiol.* 51:489–91
30. Fontaine, T. D., Irving, G. W., Ma, R. M., Poole, J. B., Doolittle, S. P. 1948. Isolation and partial characterization of crystalline tomatine, an antibiotic agent from tomato plants. *Arch. Biochem. Biophys.* 18:467–75
31. Fulton, J. P. 1948. Infection of tomato fruits by *Colletotrichum phomoides*. *Phytopathology* 38:235–46
32. Gäumann, E. 1951. *Pflanzliche Infektionslehre.* Aufl., Basel: Verlag Birkhauser 2nd ed. 681 pp.
33. Ginzburg, B. Z. 1961. Evidence for a protein gel structure cross-linked by metal cations in the intercellular cement of plant tissue. *J. Exp. Bot.* 12:85–107
34. Green, A. 1971. Soft fruits. In *The Biochemistry of Fruits and their Products,* ed. A. G. Hulme, 2:375–410. London: Academic
35. Greene, L., Morales, C. 1967. Tannins as the cause of latency in anthracnose infections of tropical fruits. *Turrialba* 17:447–49
36. Hancock, J. G., Millar, R. L., Lorbeer, J. W. 1964. Role of pectolytic and cellulolytic enzymes in *Botrytis* leaf blight of onion. *Phytopathology* 54:932–35
37. Hendrix, F. F., Nielsen, L. W. 1958. Invasion and infection of crops other than the forma suscept by *Fusarium oxysporum* f. *batatas* and other formae. *Phytopathology* 48:224–28
38. Hobson, G. E., Davies, J. N. 1971. The tomato. See Ref. 34, pp. 437–82
39. Hulme, A. C., Edney, K. L. 1960. Phenolic substances in the peel of Cox's Orange Pippin apples with reference to infection by *G. perennams.* In *Phenolics in Plants in Health and Disease,* ed. J. B. Pridham, 87–94. London: Pergamon
40. Hulme, A. C., Rhodes, M. J. C. 1971. Pome fruits. See Ref. 34, pp. 333–73
41. Jarvis, W. R. 1962. The infection of strawberry and raspberry fruits by *Botrytis cinerea* Fr. *Ann. Appl. Biol.* 50:569–75
42. Jarvis, W. R., Borecka, H. 1968. The susceptibility of strawberry flowers to infection by *Botrytis cinerea* Pers. ex. Fr. *Hort. Res.* 8:147–54
43. Karr, A. L., Albersheim, P. J. 1970. Polysacharide-degrading enzymes are unable to attack plant cell walls without prior action by a wall-modifying enzyme. *Plant Physiol.* 46:69–80
44. Keegstra, K., Talmadge, K. W., Bauer, W. D., Albersheim, P. 1973. The structure of plant cell walls. 3. A model of the walls of suspension-cultured sycamore cells based on the interconnections of the macromolecular components. *Plant Physiol.* 51:188–96
45. Kidd, M. N., Beaumont, U. 1925. An experimental study of the fungal invasion of apples in storage, with particular reference to invasion through lenticels. *Ann. Appl. Biol.* 12:14–33
46. Kuç, J., Williams, E. B. 1962. Production of proteolytic enzymes by four pathogens of apple fruit. *Phytopathology* 52:739
47. Letzig, E., Handschack, W. 1962. Die Veränderungen von Fruchtfleischfestigkeit und Inhaltsstoffen von sechs Erdbeersorten während der Reife. *Arch. Gartenb.* 10:419–33
48. McClellan, W. D., Hewitt, W. B., La Vine, P., Kissler, J. 1973. Early *Botrytis* rot of grapes and its control. *Am. J. Enol. Viticult.* 24:27–30
49. Meredith, D. S. 1964. Appressoria of *Gloeosporium musarum* Ckc. & Massee on banana fruits. *Nature* 201:214–15
50. Moreau, C., Moreau, M., Chollet, M. M., Bompeix, G. 1964. Recherches sur la maladie des taches lenticellaires de la pomme Golden. *Fruits d'Outre Mer* 19:507–19
51. Morre, D. J. 1968. Cell wall dissolution and enzyme secretion during leaf abscission. *Plant Physiol.* 43:1545–59
52. Neal, G. E. 1965. Changes occurring in the cell walls of strawberries during ripening. *J. Sci. Food Agr.* 16:604–11
53. Powelson, R. L. 1960. Initiation of strawberry fruit rot caused by *Botrytis cinerea. Phytopathology* 50:491–94
54. Sacher, J. A. 1962. Relations between changes in membrane permeability and the climacteric in banana and avocado. *Nature* 195:577–78
55. Sacher, J. A. 1966. Permeability characteristics and amino acid incorporation during senescence (ripening) of banana tissue. *Plant Physiol.* 41:701–8
56. Sander, H. 1956. Studien über Bildung und Abbau von Tomatin in der Tomatenpflanze. *Planta* 47:374–400
57. Schönbeck, F. 1967. Untersuchungen über Blüteninfektionen. Fruchtfäulen der Erdbeere. *Z. Pflanzenkrankh. Pflanzenschutz* 74:72–75

58. Schulz, F. A. 1972. Vergleichende Darstellung der Bildung von Polymethylgalakturonase (PMG) durch drei *Gloeosporium* spp. *Mitt. Biol. Bundesanst. Land-Forstw.* 146:193–94
59. Sharples, R. O., Little, R. C. 1970. Experiments on the use of calcium sprays for bitter pit control in apples. *J. Hort. Sci.* 45:49–56
60. Simmonds, J. H. 1939. Latent infection in tropical fruits discussed in relation to the part played by species of *Gloeosporium* and *Colletotrichum*. *Proc. R. Soc. Queensl.* 52:92–120
61. Simmonds, J. H. 1963. Studies in the latent phase of *Colletotrichum* species, concerning ripe rots of tropical fruits. *Queensl. J. Agr. Sci.* 20:373–424
62. Simmonds, J. H., Mitchell, R. S. 1940. Black end and anthracnose of the banana, with special reference to *Gloeosporium musarum* Cke. and Mass. *Counc. Sci. Ind. Res. Aust. Bull.* 131: 1–63
63. Sitterly, W. R., Shay, J. R. 1960. Physiological factors affecting the onset of susceptibility of apple fruit to rotting by fungus pathogens. *Phytopathology* 50: 91–93
64. Stall, R. E., Hortenstine, C. C., Hey, J. R. 1965. Incidence of *Botrytis* grey mold of tomato in relation to a calcium-phosphorus balance. *Phytopathology* 55:447–49
65. Tichelaar, G. M. 1967. Studies on the biology of *Botrytis allii* on *Allium cepa*. *Neth. J. Plant Pathol.* 73:157–60
66. Toro, R. A. 1922. Studies on banana anthracnose. *J. Dep. Agr. Porto Rico* 6:1–23
67. Verhoeff, K. 1968. Studies on *Botrytis cinerea* in tomatoes. Effect of soil nitrogen level and of methods of deleafing upon the occurrence of *B. cinerea* under commercial conditions. *Neth. J. Plant Pathol.* 74:184–92
68. Verhoeff, K. 1970. Spotting of tomato fruits caused by *Botrytis cinerea. Neth J. Plant Pathol.* 76:219–26
69. Verhoeff, K., Warren, J. M. 1972. In vitro and in vivo production of cell wall degrading enzymes by *Botrytis cinerea* from tomato. *Neth. J. Plant Pathol.* 78:179–85
70. Wade, G. C. 1956. Investigations on brown rot of apricot caused by *Sclerotinia fructicola* (Wint.) Rehm. *Aust. J. Agr. Res.* 7:504–14
71. Walker, J. R. L. 1962. Phenolic acids in 'cloud' and normal tomato fruit wall tissue. *J. Sci. Food Agr.* 13:363–67
72. Wallace, J., Kuç, J., Draudt, H. M. 1962. Biochemical changes in the water-insoluble material of maturing apple fruit and their possible relationship to disease resistance. *Phytopathology* 52: 1023–27
73. Wardlaw, C. W. 1931. Banana diseases. 3. Notes on the parasitism of *Gloeosporium musarum* (Cooke & Massee). *Trop. Agr.* 8:327–31
74. Wardlaw, C. W., Baker, R. E. D., Crowdy, S. H. 1939. Latent infections in tropical fruits. *Trop. Agr.* 16:275–76
75. Wood, R. K. S. 1967. *Physiological Plant Pathology.* Oxford, Engl.: Blackwell 570 pp.
76. Woodward, J. R. 1972. Physical and chemical changes in developing strawberry fruits. *J. Sci. Food Agr.* 23:465–73

# TEMPORAL PATTERNS OF VIRUS SPREAD

❖3591

*J. M. Thresh*
East Malling Research Station, Maidstone, England

There have been outstanding advances recently in work on the structure and replication of plant viruses, but studies on their spread and control have been relatively neglected, and progress has been limited. Nevertheless, there is an increasing awareness of the prevalence of virus diseases in diverse crops of many countries.

Cacao swollen shoot, cotton leaf curl, hop nettlehead, and many other virus diseases have long caused serious losses, and these continue. Other viruses including plum pox, citrus tristeza, and sugar cane mosaic have increased in importance because of spread into new areas or crops. As a result of detailed surveys or the introduction of new tests, newly detected and in some instances widespread and prevalent viruses have been discovered in maize, rice, apple, and many other hosts.

The well characterized plant viruses are of several distinct morphological types and there is great diversity in their mode of spread. Recent reviews are available on transmission and spread of viruses by pollen and seed (64) and by vectors (20, 52, 56), especially aphids (73, 90), leafhoppers (7, 92), white flies (27), thrips (62), beetles (87), mites (55), nematodes (17, 75), and fungi (76). This paper emphasizes general features in the epidemiology of plant viruses that lead to characteristic curves of disease progress with time.[1] Spatial patterns of infection into and within crops will be considered in a subsequent review.

## RATE OF SPREAD

The rate at which a virus spreads between plants varies widely according to the type of virus, crop, environment, and mode of transmission. In extreme instances large plantings become almost totally infected within a few weeks. By contrast, the spread of many viruses among woody plants is relatively slow, or even imperceptible. Such

[1]It is uncertain whether blackcurrant reversion and the diseases transmitted by whiteflies are caused by viruses. Otherwise the review is restricted mainly to economically important virus diseases and excludes those now associated with mycoplasmas or other organisms.

infection may have originated as a result of rare instances of spread from other hosts. Once established in commercial clones, viruses do not have to spread rapidly to survive, because they are perpetuated inadvertently by vegetative propagation (58). Cucumber mosaic virus, for example, occurs throughout certain clones of blackcurrant, despite an apparent inability to spread between bushes and despite the low rate of infection by aphids entering the crop from elsewhere (79).

Some viruses spread solely from sources outside the crop, there being no plant-to-plant spread within the crop, at least during the first year. Other viruses spread both into and within crops, and newly infected plants soon become foci for secondary spread. It is convenient to distinguish between these two main types of spread and refer to "simple-" and "compound-interest" diseases (83). These terms come from the analogy between the increase of disease with time and the increase of capital by simple or compound interest. This concept facilitates the analysis of data, although a particular virus does not always spread in the same way in all crops, at all sites, or at all stages of the growing season.

Six features are particularly important in distinguishing complex biological systems from simple mathematical models:

1. Spread does not occur at a uniform rate and depends upon such factors as seasonal or other changes in the size and mobility of vector populations.
2. There are finite limits to the amount of disease that can develop at a particular site.
3. An increasing number of plants receive more than one infective dose as spread proceeds. The extent of this "multiple infection" can be calculated, and appropriate values have been tabulated for transforming percentage of infection into infection units (39).
4. Newly infected plants do not immediately contribute to further spread, and the plants infected first become increasingly remote from the remaining healthy ones.
5. Disease is seldom randomly distributed and tends to occur in localized areas of high intensity (often referred to as foci).
6. There are seasonal changes in the size and susceptibility of plants and in the virus content of infected tissue.

## *Simple-Interest Diseases*

Few virus diseases are of the simple-interest type that spread into crops solely from outside sources. There are few data on the progress of such diseases; this subject is discussed by Van der Plank (83).

Capital increases at a fixed rate of simple interest to give a linear increase on plotting the accumulated sum invested against time, the slope of the gradient depending on rate of interest ($r$). No such simple relationships have been observed between the proportion or percentage of virus-infected plants ($x$) and time ($t$). Values of $x$ increase in curvilinear or sigmoid fashion; the influx of infection tends to be low at the outset and to increase with time as seasonal or other conditions affecting spread become more favorable. The absolute rate at which new infections appear ($x_{t_2}-x_{t_1}/t_2-t_1$) later declines as the influx decreases or conditions deteriorate. An important distorting factor is that progressively fewer healthy plants remain to be infected and multiple infection becomes increasingly important (39). Hence, the correction factor (1–$x$) is used when obtaining values of $r$ by plotting $\log_e [1/(1-x)]$ against $t$. Van der Plank has presented the detailed mathematics and tabulated appropriate values for each unit of $x$ (83).

Several diseases caused by tomato spotted wilt virus are of the simple-interest type. Infection is carried into tobacco, tomato, and pineapple plantings by adult thrips that acquire virus as nymphs while feeding on infected weeds (6, 51, 84).

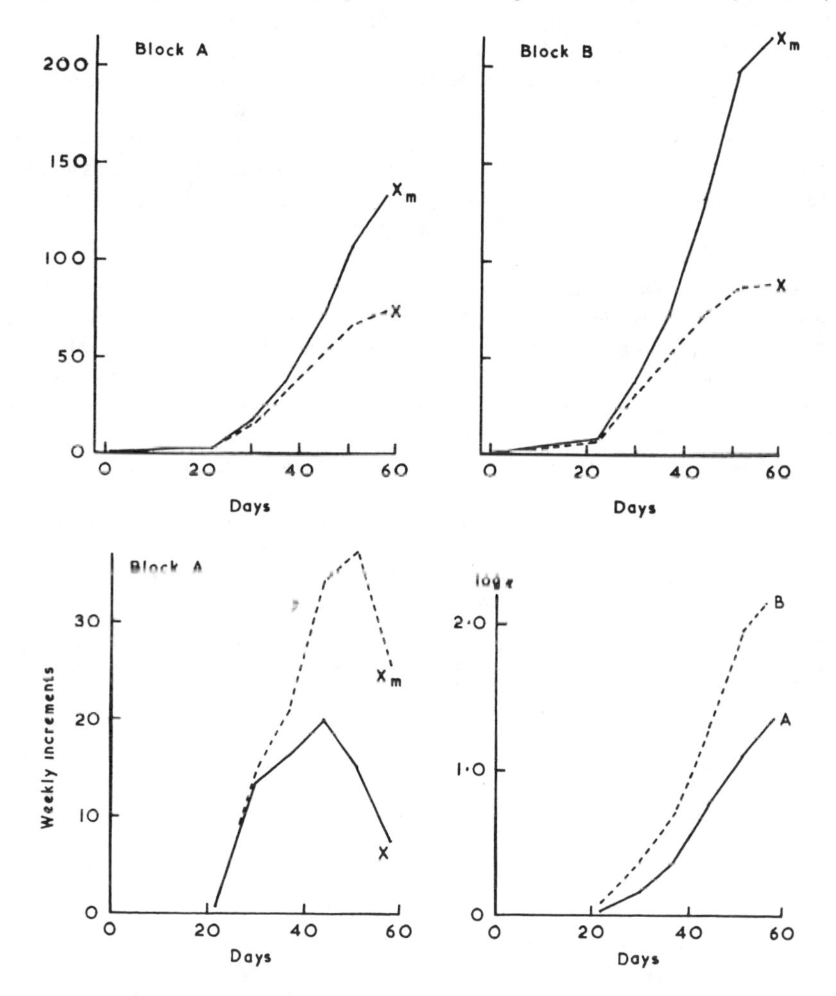


*Figure 1* The spread of spotted wilt virus into two blocks (*A* and *B*) of tomato in Australia (6). (*top*) Weekly totals of infected plants in each block as a percentage ($x$) and after transformation ($x_m$) to allow for the progressively increasing amount of multiple infection (39). (*bottom*) Successive values of $\log_e [1/(1-x)]$ for each block (*right*) and weekly increments of $x$ and $x_m$ in block *A* (*left*).

Spread is influenced by weather conditions, principally temperature, which affects the development and influx of winged adults. Once spread begins it occurs in surges, and periods of high temperature are followed within days by the appearance of many new infections. These infections developed at random at sites in Australia, and statistical analyses revealed no evidence of the grouping that would have occurred as a result of spread between adjacent plants (6). A later method developed for the analysis of similar data from South Africa (81) involved a comparison between the observed number of pairs of infected plants and the number that would have been expected from a totally random distribution. This "doublet" test has been widely quoted and used to investigate the spread of other diseases, despite the limitations discussed by Freeman (36), who proposed an alternative method.

Apart from short-term variations in flight activity there are overall trends in thrip populations as the number of adults increases to a peak when the weed hosts mature, flower, and begin to senesce (Figure 2). Hence it is advantageous to increase the initial plant population and to delay thinning until the main influx of thrips is over (84).

Necrotic yellows is an aphid-borne virus of lettuce that is spread exclusively from sowthistle (*Sonchus oleraceus*), the primary host of virus and vector (60, 69, 70). Similarly, maize rough dwarf virus is carried into maize by planthoppers, which do not breed or acquire virus within the crop (41). Common mosaic of cotton in Brazil is an example of a virus transmitted by white flies and spread exclusively into crops from nearby malvaceous weeds (26).

In certain western states of the USA, sugar beet curly top virus is carried into bean, tomato, flax, melon, and other crops that are not breeding hosts of the leafhopper vector (*Circulifer tenellus*). The main influx is by hoppers from indigenous desert plants and introduced weeds that are the principal winter and spring hosts of virus and vector (4, 8).

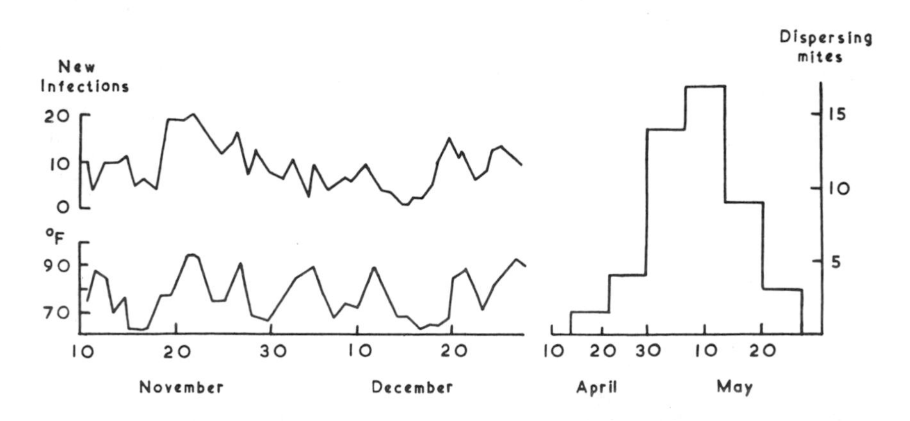


*Figure 2* (*left*) Multiple peaks in the spread of spotted wilt virus in tomato (*top*) and maximum screen temperatures 12 days previously (*bottom*). (*right*) Single peak in the spread of the gall mite vector of blackcurrant reversion virus (79).

Curly top behaves like a compound-interest disease (21) when *C. tennelus* breeds and spreads virus within sugar beet crops. The spread of blackcurrant reversion virus by its mite vector is also complex and not wholly within a single category. In young plantations, spread is mainly from outside sources, and the newly infected bushes tend to be widely scattered and not obviously grouped around those present originally (72). Secondary spread occurs eventually, unless the early infections are removed before they have been invaded systematically and become particularly vulnerable to mite infestation (78).

Compound-interest diseases behave initially in the simple-interest manner when they first appear in crops and before secondary spread occurs. Thus cantaloup plantings were almost totally infected with watermelon mosaic virus 2 by an early and heavy influx of aphids from a nearby source (54). By contrast, there was little secondary spread of leaf roll of potatoes when aphid infestations occurred late in the season (18, 30).

### *Compound-Interest Diseases*

Most viruses spread into and within crops and cause diseases of the compound-interest type. However, such diseases seldom spread for long in a manner closely analogous with the logarithmic increase of capital at compound rates of interest. With virus diseases the total amount of infection ($x$) usually increases in a sigmoid manner with time ($t$). Initially the increase of disease is limited by the few sources of infection present and/or by the lack of sufficient active vectors. There is a similarly low rate of increase when spread has continued until few uninfected plants remain. At intermediate values of $x$ around 0.5 (50% infection) the absolute rate

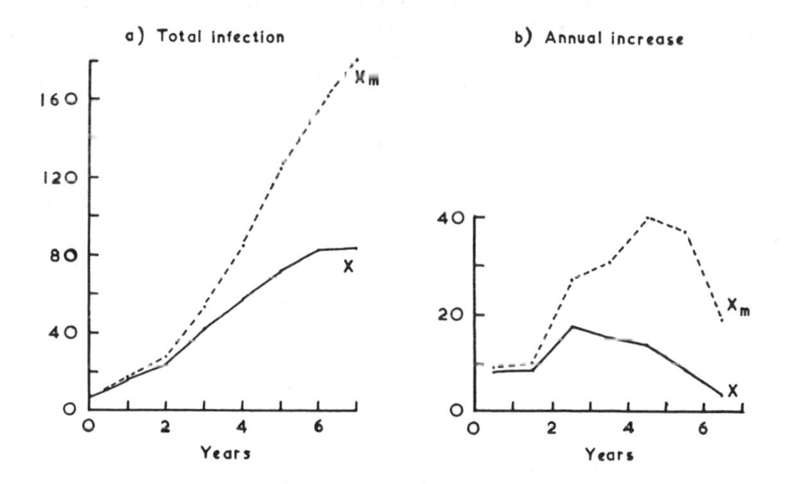


*Figure 3* The spread of cacao virus in a Trinidad plantation (28). (*left*) The total number of infected trees at the end of each year as a percentage ($x$) of the total stand and after transformation ($x_m$) to allow for the progressively increasing amount of multiple infection. (*right*) The annual increments in $x$ and $x_m$.

of increase is relatively fast because spread is not restricted by a lack of inoculum, vectors, or healthy plants.

The initial phase is truly logarithmic when the number of new infections that appear is directly proportional to the total infection already present. Hence, there is a linear relationship between $\log_{10} x$ and $t$, with a slope indicating infection rate ($r$). Any deviation from the line indicates that $r$ has varied and the slope tends to flatten with time. (Such behavior distinguishes disease spread from the growth of fixed-interest investments and from autocatalytic chemical reactions proceeding at a fixed rate.) In calculating $r$ it is again appropriate to use the correction factor $1-x$ to allow for the diminishing proportion of uninfected plants and plot $\log_{10} [x/(1-x)]$ against $t$ (Figure 4). Appropriate values have been tabulated for each unit of $x$ (83).

Values of $r$ are usually greatest during early spread when there is a progressively increasing number of infected plants from which further spread can occur and a corresponding decrease in the importance of outside sources. Spread is facilitated, especially in annuals, by an increase in plant size. This may be accompanied by an increase in the number and activity of the vectors. There is also decreased separation and sometimes increased contact between individual plants, together with a great increase in the amount of both infected and vulnerable tissue accessible to vectors.

As spread proceeds, the oldest infections become increasingly remote from the remaining healthy plants. Consequently there is an increase in the amount of infected tissue and in the number of vectors that do not contribute to spread. This effect is particularly pronounced for nematode and other vectors of limited mobility and for those that do not thrive on virus-infected plants. For example, cacao trees infected with virulent strains of swollen shoot virus become unfavorable hosts of the vector and eventually die (68). Spread is mainly by mealybugs that move onto the branches of adjacent trees from new infections on the periphery of outbreaks (25, 68, 77). Thus the annual spread in a plantation is directly proportional to the

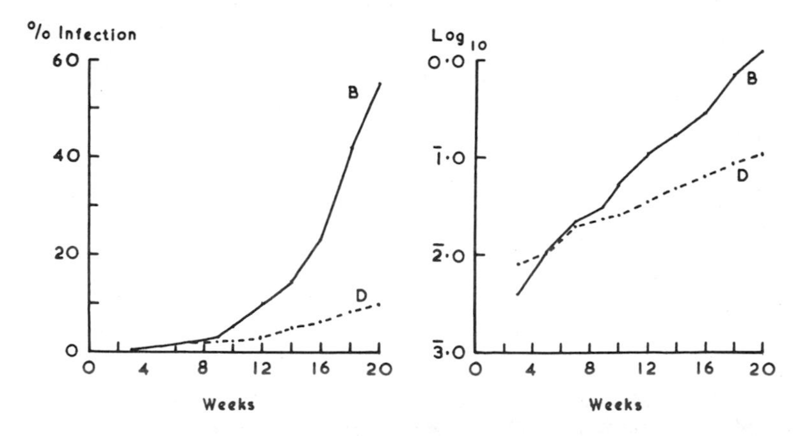


*Figure 4* The spread of leaf curl virus in two varieties of cotton (*B* and *D*) in the Sudan (38). Infection plotted as percentage ($x$) of the total stand (*left*) and as $\log_{10} [x/(1-x)]$ (*right*).

number of healthy trees in contact with infected ones at the beginning of the year and not to the total value of $x$ (77).

With all diseases, the amount of multiple infection increases as progressively fewer healthy plants remain (Figure 3); but spread may be checked or even halted before this becomes important, especially in annual crops. These mature or otherwise become resistant to infection, as with leaf roll of potato (18, 74). There is often a tendency for the virus content of infected plants to decrease with age, and this decrease may be accompanied by a seasonal decline in vector populations or a decrease in their ability to acquire or transmit the virus (23). For example, the spread of maize rough dwarf is checked by high summer temperatures that decrease the virus content of the planthopper vector (41). Such factors contribute to the generally similar sigmoid shape of disease-progress curves, whether or not much spread occurs (Figure 5).

Vector populations may decline because of chemical control measures or seasonal trends. In some areas aphid numbers are decreased by high summer temperatures and, with irrigation, hot arid areas can be used to grow virus-free seed crops or planting material (71). Elsewhere the main check on the population comes with the onset of cool autumn conditions or as parasites and predators become numerous (10). Occasionally spread is halted by sudden storms (63) or by prolonged periods of unfavorable weather (33, 91).

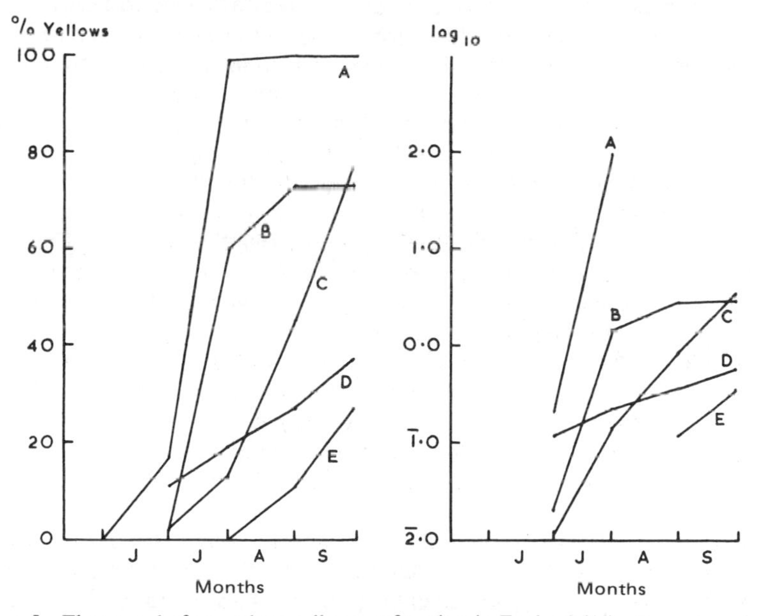


*Figure 5* The spread of sugar beet yellows at five sites in England (45). Percentage ($x$) of plants infected at monthly intervals (*left*) and successive values of $\log_{10} [x/(1-x)]$ (*right*). Infection appeared early and spread rapidly at *A*, whereas spread was slow early or late in the season at *E* and *B* and throughout the season at *D*.

The ultimate proportion of infected plants and the rate at which new infections appear vary widely among different viruses and for different crops. There are also major differences between sites and seasons for particular diseases such as sugar beet yellows (31, 45, 65, 89) and plum pox (49).

Viruses that infect annual crops spread much more rapidly than those of trees and shrubs (82). This explains why eradication measures that control the spread of virus diseases of some tree crops are largely ineffective against those of herbaceous annuals. In a typical orchard in California, tristeza virus spread to an average of two citrus trees a year for each infected one already present (29). By contrast cauliflower mosaic virus spread from a single infected plant to as many as 131 others in one season (48). Often there is almost total infection of tobacco crops with mosaic (93) and of cotton with leaf curl virus transmitted by white flies (38). Similarly with the aphid-borne viruses of cucurbits (Figure 6), lettuce (15), groundnut (1, 10), pepper (50, 66), carrot (91), brassicas (13, 48), and other vegetables and cereals (12, 57, 67). The ability to spread rapidly and become established in new areas has obvious survival value for viruses and vectors that depend on exploiting ephemeral hosts and habitats. This is particularly important for viruses that are not seed-borne and for vectors that have no special drought-resistant or overwintering form.

There are relatively few papers on the spread of virus diseases among woody perennials. Compared with herbaceous plants they provide a stable long-lived substrate for virus and vector. Invariably such viruses as citrus tristeza (9), cacao swollen shoot (77), plum pox (49), peach mosaic (47), and blackcurrant reversion (3) take several years to spread throughout plantations (Figure 7). Nevertheless, they cause serious losses because individual trees are far larger and take longer to produce a crop and consequently are much more valuable than herbaceous plants. They are also slow growing and sometimes difficult or impossible to replace.

The actual rate of spread between trees depends on such factors as the number and proximity of the main sources of infection and the mode of spread (49). Tree size and age are also important. Prunus necrotic ringspot virus, which is pollen-borne, cannot spread until flowering commences (64). Similarly, young cacao trees support few mealybugs, and there is little spread of virus until the branches form a continuous interlocking canopy (25, 68).

Woody perennials tend to be difficult to infect, and the generally wide spacing between them hampers the movement of vectors and impedes virus spread. Moreover, virus is slow to become systemic and there is generally a long interval that may extend for many months before the virus becomes available to vectors. This delays spread compared with that in herbaceous hosts in which viruses may become systemic in a few days.

The date infection first appears within a crop is of crucial significance in epidemiology. The existence of viruses within crops from the outset poses a particular problem. These viruses may be due to the presence of weed hosts (32), to regeneration from the residues of previous crops, or to the use of infected stocks of seed (64) or other planting material. As a basic control measure such foci of infection should be eliminated; this is a major objective of the various seed and stock certification schemes being developed in many countries (44). These aim to defer the onset of disease and delay or decrease the ultimate amount of infection, as with the use of mosaic-free lettuce seed (94). Such measures decrease crop losses, especially because the plants that are infected longest are usually the worst affected.

## *Soil-Borne Viruses*

Diseases caused by viruses with nematode or fungal vectors behave unlike typical simple- or compound-interest diseases. Usually the roots of susceptible plants soon become infected when grown in soil containing infective populations of the vector, but there is some delay before virus reaches the aerial parts and causes symptoms (17, 75). The delay may extend to several months or more with fanleaf of grapevine or nettlehead disease of hop (80). These and other soil-borne diseases occur suddenly in patches that coincide with the distribution of infective vectors. The patches may extend over whole fields in the case of the wheat mosaic transmitted by the fungus *Polymyxa graminis* (Plasmodiophorales).

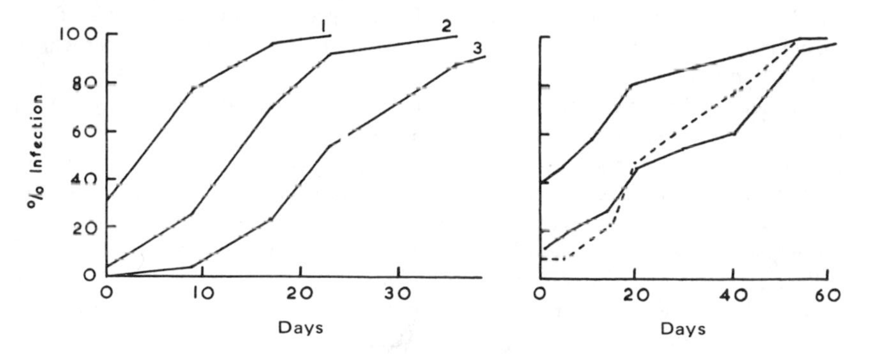


*Figure 6* The spread of curly top virus into three sugar beet plantings in Idaho (*right*) and of watermelon mosaic virus 2 into a cantaloup field in Arizona (*left*). Sites 1–3 were at increasing distances from the original source of infection (54).

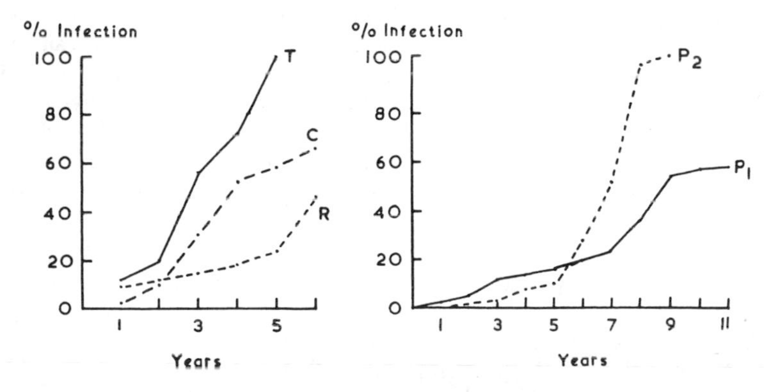


*Figure 7* The spread of viruses of perennial crops. (*left*) C = cacao swollen shoot (77), R = blackcurrant reversion (3), T = citrus tristeza (9). (*right*) $P_1$ and $P_2$ = plum pox at different sites (49).

Soil-borne virus diseases tend to spread slowly after their initial appearance, due to the limited mobility of their vectors. There is no detailed information on fungal vectors, but it has been estimated that the nematode vector of arabis mosaic and strawberry latent ringspot viruses *(Xiphinema diversicaudatum)* moves through the soil only 30 cm per year (40). Viruses move more rapidly than this through the root systems of such perennials as grapevine, hop, and cherry. This explains the unexpectedly rapid spread that occurs when growers unintentionally plant infected stocks in soils containing noninfective nematode vectors of the virus. Increasing numbers of vectors have access to infected roots, from which they spread the virus to nearby healthy plants.

## SEASON OF SPREAD

There is great diversity in the periodicity of virus spread, as may be expected from the different types of vector involved and the wide range of conditions under which crops are grown.

Weather conditions are most uniform in the humid tropics. Perennial crops grow almost continuously throughout the year in the forest areas of West Africa except during the short dry season and during the relatively cool conditions at the height of the rains. Hence there is at least some spread of viruses of sugar cane, cassava, pineapple, banana, abaca, cacao, and other crops at all times of year. Seasonal trends in the rate and pattern of spread of cacao swollen shoot virus are due to differences in the size and activity of populations of the mealybug vectors (25, 68, 77). These trends are not readily apparent because most new infections show during peak flushes of new growth after a long and variable latent period.

Elsewhere in the tropics and subtropics there are clearly defined seasons of growth separated by dry periods that may be intense and in some areas prolonged. The main rain-fed crops are annuals including many cereals and legumes. Many of these are of short duration, and yet they may become severely affected by virus diseases. Groundnut rosette virus has received particular attention in many African countries, because it is soon carried into and spread rapidly within plantings by aphids that originate from distant crops or from nearby perennials that survive the dry season (2). At Nigerian sites, the incidence of rosette was closely correlated with the proportion of plants that become infested with *Aphis craccivora.* Predators were numerous, and aphid populations fluctuated rapidly, with few colonies persisting more than a week (10).

There is a similar pattern of invasion of tobacco, cotton, pepper, cucurbit, maize, vegetable, and other crops with viruses transmitted by aphids, whiteflies, leafhoppers, or beetles.

Spread between crops is facilitated in regions with a bimodal distribution of rain and two overlapping growing seasons per year. This contributes to the prevalence of sugar cane mosaic and other viruses of maize in parts of Kenya. Similarly, the increasing use of irrigation to obtain crops during the dry season is likely to increase the overall incidence of disease, as noted already in rice and maize (5, 61, 85).

Drought provides the main check on growth in the tropics and subtropics, whereas cropping patterns in the temperate regions are mainly determined by winter temperatures. The principal crops are deciduous perennials or annuals with a restricted season of growth. The activity of virus vectors is similarly restricted, especially in areas with severe winters where the topsoil is frozen for long periods and where insects survive only as eggs or in diapause. In less extreme conditions there is spread of some viruses during the winter months, although the movement of vectors between plants is less than at higher temperatures and many crops are not planted until the spring or early summer. The increased survival of crop and weed hosts and of vectors in mild winters and at protected sites is important in the epidemiology of many viruses and leads to much spread early in the main growing season (13, 15, 46, 88, 91).

The spread of blackcurrant reversion virus is unusual in that it is restricted to a few weeks in late spring (Figure 2). At this time, the gall mite vector emerges from overwintered galls and disperses to young buds of the new season's growth (78, 79). Other mite vectors are less confined to buds, some being free living with a prolonged dispersal period (55).

Many insect vectors and especially aphids (73) have a complex life cycle that determines the main period of virus spread. This often coincides with the appearance of particularly active and in some instances specialized forms of the vector that are adapted for dispersing to new habitats. For example, the peak periods of spread of aphid-borne viruses of strawberry in southern England coincide with the appearance of alate forms, and apterae predominate at other times (59). Two main periods of spread were detected by exposing successive batches of potted plants at intervals throughout the growing season (Figure 8). Some plants were protected by sticky

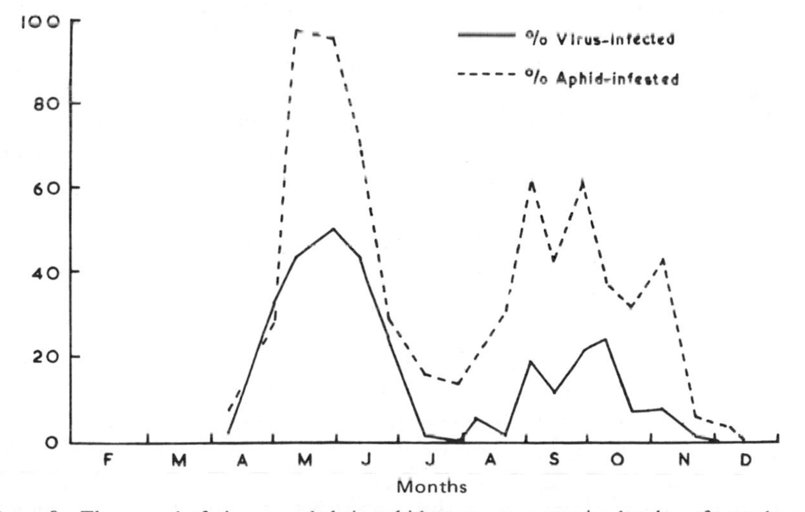


*Figure 8* The spread of viruses and their aphid vector to successive batches of strawberry plants exposed at fortnightly intervals throughout the growing season in southern England (59).

surfaces, as in previous work with potato (14), in an attempt to distinguish spread due to alates and apterae from that due to alates alone.

A limitation of the exposure technique is that it is not always possible to ensure that the experimental plants are exactly comparable in size and susceptibility to those of the crop in which they are placed. Nevertheless, the method is more generally applicable than others used to investigate the spread of potato viruses. These involve either the sequential sampling of tubers (30), or the use of insecticides in attempts to halt spread at different stages of the season (16). Knowledge of the main period of spread is important in developing control measures and methods of forecasting the progress and prevalence of a disease.

## FORECASTING

Many environmental and other factors influence the complex interactions between virus, host, and vector; hence forecasting the appearance and development of a disease is difficult. Attempts at forecasting are justified because success indicates an understanding of the main factors influencing epidemiology. There are also great potential benefits to growers, processors, and advisory officers who can anticipate losses and apply appropriate chemical or other control measures. Moreover it may be possible to operate an early warning system or to select growing seasons or areas for special crops where infection is unlikely.

There has been little progress in forecasting virus disease with many crops. A major problem has been the difficulty in getting the necessary sequence of data on disease incidence from representative sites over a sufficient number of years. Standard methods are required for assessing virus infection and vector abundance throughout the season, and it is desirable to have continuity of suitably trained personnel.

Few crops have received such detailed attention as sugar beet. In Idaho a correlation was soon established between severe winters and relative freedom from damaging outbreaks of the leafhopper vector of beet curly top (19). Winter and early spring temperatures affect the rate at which the vectors mature on their winter hosts and the main dispersal period was predicted from data on accumulated degree-days above 45°F (24).

Elsewhere, trap catches of the leafhopper vectors of maize streak virus in Rhodesia were correlated with the amount of rain at the end of the preceding wet summer season (61). More complex relationships have been used to predict the incidence of rice dwarf virus and its leafhopper vector in Japan. In this case, the amount of infection the previous season and counts of overwintering individuals, as amended by later population trends (42, 53), were used as the basis of prediction.

In Britain, aphids carry yellows viruses into the sugar-beet crop from outside sources. Further spread is mainly within or between crops until there is almost total infection or until growing conditions become unfavorable. Early infection increases in the compound-interest manner, and it is possible to predict the ultimate amount of infection from earlier counts of yellows and sticky trap catches of winged *Myzus persicae* (89) (Figure 9).

Aphid counts were weighted according to collection date, because infection and loss of crop are greatest when virus appears early. This occurs after mild winters, when many aphid colonies survive on herbaceous plants including some that are hosts of virus. Alates develop later and are least numerous after cold winters, when survival is mainly as eggs on primary woody hosts immune to virus. For this reason there is a good correlation between winter temperatures (46) or number of freezing days (88) and the later incidence of virus (Figure 10). Weather conditions are partially related to cycles of sunspot activity, which have also been associated with the incidence of yellows (37).

In concurrent work, yields of carrots were correlated with sticky-trap catches of the aphid vector of carrot motley dwarf virus (91). Similar factors affecting the overwintering of this aphid and *Myzus persicae* explain the association between the prevalence of motley dwarf and beet yellowing viruses (91).

Sticky-trap catches have also been used to predict the incidence of barley yellow dwarf virus in winter wheat in New Zealand. Growers are advised to spray when many cereal aphids are caught in the autumn. Insecticides decrease overwintering populations and check the secondary spread that otherwise leads to heavy crop losses (22).

Quite different conditions affect the appearance of barley yellow dwarf virus in Minnesota and adjacent states where the main vectors do not overwinter. Severe losses occur only in seasons permitting rapid increase and spread following an early and heavy influx of aphids from southern states (12). The particular conditions favoring such long distance dispersal have been established, and heavy infestations can be anticipated if suitable weather occurs in the critical April–May period (34, 43, 86).

Wheat streak mosaic virus has also received much attention in North America. The eventual prevalence of infection and loss of yield in autumn-sown crops was predicted with fair accuracy by assessing early infection. This assessment was done in winter when cold weather had virtually halted spread by the mite vector until

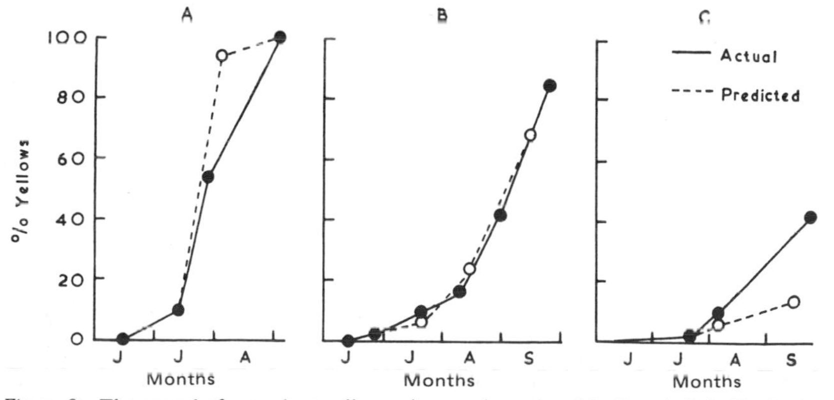


*Figure 9* The spread of sugar beet yellows virus at three sites (*A*, *B*, and *C*) in England as observed and as predicted from previous observations and sticky trap catches of winged *Myzus persicae* (89).

the following spring. Few, small samples were adequate because the disease only became serious when there was much early infection (35).

The incidence of cotton leaf curl in the Sudan has been associated with the severity of the preceding dry season. Severity determines the amount of regeneration from the infected stumps of previous crops that act as primary foci. Regeneration was greater after additional dry-season irrigations than when water was withheld prematurely (11). Similarly, rainfall influences the survival of overwintering hosts of curly top virus in Washington state. This explains the correlation between winter rain and the later incidence of infection in sugar beet. Infection is also related to the number of summer days with above average temperatures that increase the activity of vectors (21).

These examples show that specific features early in the season may be of crucial importance in the epidiomology of a disease, despite the apparent complexity of the overall situation. Ultimately it may be possible to use remote scanning techniques to provide data on a comprehensive or even global scale for detailed computer assessment, analysis, and prediction of disease incidence. There is much interest in the possibilities of such methods, and high-flying aircraft and satellites are being used with sophisticated photographic and electronic equipment. These developments have yet to make a major impact in virus epidemiology, and much remains to be done in reconciling aerial photographs with ground observations.

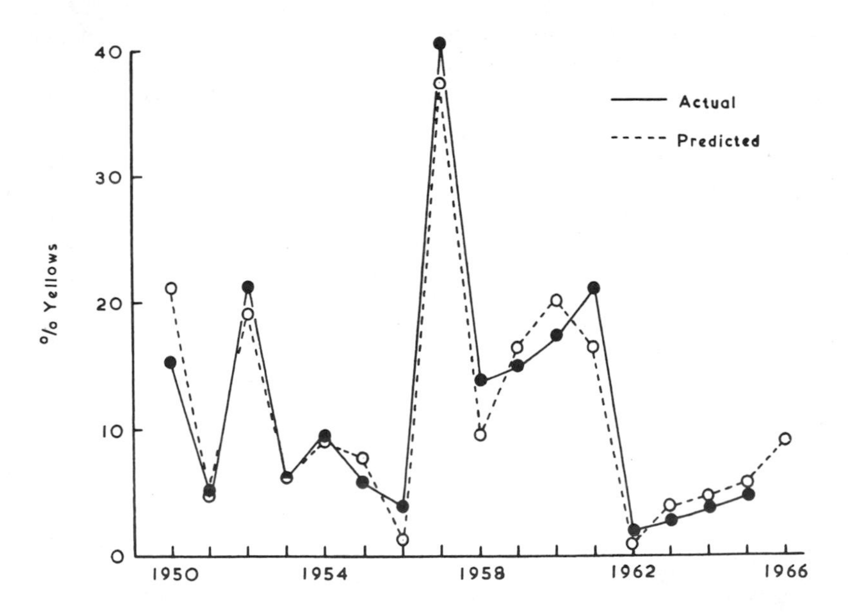


*Figure 10* Overall incidence of yellowing viruses in English sugar beet crops for the years 1950–1966. Observed levels at the end of August compared with those calculated from the multiple regression involving numbers of freezing days between January and March and deviations from average April temperatures (88).

*Literature Cited*


1. A'Brook, J. 1964. The effect of planting date and spacing on the incidence of groundnut rosette disease and of the vector, *Aphis craccivora* Koch. at Mokwa, Northern Nigeria. *Ann. Appl. Biol.* 54:199–208
2. Adams, A. N. 1967. The vectors and alternative hosts of groundnut rosette virus in central province, Malawi. *Rhodesia Zambia Malawi J. Agr. Res.* 5:145–52
3. Amos, J., Hatton, R. G. 1928. Reversion in blackcurrants. II. Its incidence and spread in the field in relation to control measures. *J. Pomol. Hort. Sci.* 6:282–95
4. Annand, P. N., Chamberlain, J. C., Henderson, C. F., Waters, H. A. 1932. Movement of the beet leafhopper in 1930 in Southern Idaho. *US Dep. Agr. Circ.* 244:1–24
5. Bakker, W. 1970. Rice yellow mottle a mechanically transmissible virus disease of rice in Kenya. *Neth. J. Plant Pathol.* 76:53–63
6. Bald, J. G. 1937. Investigations on "spotted wilt" of tomatoes. III. Infection in field plots. *CSIRO Commonw. Aust. Bull.* 106:1–32
7. Bennett, C. W. 1967. Epidemiology of leaf-hopper-transmitted viruses. *Ann. Rev. Phytopathol.* 5:87–108
8. Bennett, C. W. 1971. The curly top disease of sugar-beet and other plants. *Am. Phytopathol. Soc. Monogr.* 7:1–81
9. Bennett, C. W., Costa, A. S. 1949. Tristeza disease of citrus. *J. Agr. Res.* 78:207–37
10. Booker, R. H. 1963. The effect of sowing date and spacing on rosette disease of groundnut in Northern Nigeria with observations on the vector, *Aphis craccivora. Ann. Appl. Biol.* 52:125–31
11. Boughey, A. S. 1947. The causes of variations in the incidence of cotton leaf curl in the Sudan Gezira. *Mycol. Paper Imp. Mycol. Inst.* 22:9 pp
12. Bruehl, G. W. 1961. Barley yellow dwarf. *Am. Phytopathol. Soc. Monogr.* 1:1–52
13. Broadbent, L. 1957. *Investigations of Virus Diseases of Brassica Crops.* London and New York: Cambridge Univ. Press. 94 pp.
14. Broadbent, L., Tinsley, T. W. 1951. Experiments on the colonization of potato plants by apterous and by alate aphids in relation to the spread of virus disease. *Ann. Appl. Biol.* 38:411–24
15. Broadbent, L., Tinsley, T. W., Buddin, W., Roberts, E. T. 1951. The spread of lettuce mosaic in the field. *Ann. Appl. Biol.* 38:689–706
16. Burt, P. E., Heathcote, G. D., Broadbent, L. 1964. The use of insecticides to find when leafroll and Y viruses spread within potato crops. *Ann. Appl. Biol.* 54:13–22
17. Cadman, C. H. 1963. Biology of soil-borne viruses. *Ann. Rev. Phytopathol.* 1:143–72
18. Cadman, C. H., Chambers, J. 1960. Factors affecting the spread of aphid-borne viruses in potato in eastern Scotland. III. Effect of planting, roguing and age of crop on the spread of potato leafroll and Y viruses. *Ann. Appl. Biol.* 48:729–38
19. Carter, W. 1930. Ecological studies of the beet leafhopper. *US Dep. Agr. Tech. Bull.* 206:1–115
20. Carter, W. 1973. *Insects in Relation to Plant Disease.* New York, London, Sydney, Toronto: Wiley. 759 pp.
21. Clark, R. L. 1968. Epidemiology of tomato curly top in the Yokima valley. *Phytopathology* 58:811–13
22. Close, R., Smith, H. C., Lowe, A. D. 1970. Cereal virus warning system. *Commonw. Phytopathol. News* 10:7–9
23. Cohen, S., Nitzany, E. E. 1963. Identity of viruses affecting curcurbits in Israel. *Phytopathology* 53:193–96
24. Cook, W. C. 1945. The relation of spring movements of the beet leafhopper (*Eutettix tenellus* Baker) in central California to temperature accumulations. *Ann. Entomol. Soc. Am.* 38: 149–62
25. Cornwell, P. B. 1958. Movements of the mealybug vectors of virus diseases of cacao in Ghana. I. Canopy movement in and between trees. *Bull. Entomol. Res.* 49:613–30
26. Costa, A. S. 1965. Studies on abutilon mosaic in Brazil. *Phytopathol. Z.* 24: 97–112
27. Costa, A. S. 1969. White flies as virus vectors. In *Viruses, Vectors and Vegetation,* ed. K. Maramorosch, 95–119. New York and London: Interscience. 666 pp.
28. Dale, W. T. 1953. Further notes on the spread of virus in a field of clonal cacao in Trinidad. *Cocoa Research 1945–1951,* pp. 130–31. Trinidad: Imperial College of Tropical Agriculture
29. Dickson, R. C., Johnson, M. McD., Flock, R. A., Laird, E. F. 1956. Flying

aphid populations in southern California citrus groves and their relation to the transmission of the tristeza virus. *Phytopathology* 46:204–10
30. Doncaster, J. P., Gregory, P. H. 1948. The spread of virus diseases in the potato crop. *ARC Rept. Ser. No. 7.* London: HMSO. 189 pp.
31. Duffus, J. E. 1963. Incidence of beet virus diseases in relation to overwintering beet fields. *Plant Dis. Rept.* 47: 428–31
32. Duffus, J. E. 1971. Role of weeds in the incidence of virus diseases. *Ann. Rev. Phytopathol.* 9:319–40
33. Dunn, J. A. 1965. Studies on the aphid *Cavariella aegopodii* Scop. I. On willows and carrot. *Ann. Appl. Biol.* 56: 429–38
34. Evans, D. A., Medler, J. T. 1967. Flight activity of the corn leaf aphid in Wisconsin as determined by yellow pan trap collections. *J. Econ. Entomol.* 60: 1088–91
35. Fellows, H., Sill, W. H. 1955. Predicting wheat streak mosaic epiphytotics in winter wheat. *Plant. Dis. Reptr.* 39: 291–95
36. Freeman, G. H. 1953. Spread of diseases in a rectangular plantation with vacancies. *Biometrika* 40:287–96
37. Gibbs, A. J. 1966. A possible correlation between sugar beet yellows incidence and sunspot activity. *Plant Pathol.* 13:150–52
38. Giha, O. H., Nour, M. A. 1969. Epidemiology of cotton leafcurl virus in the Sudan. *Cotton Grow. Rev.* 46: 105–18
39. Gregory, P. H. 1948. The multiple infection transformation. *Ann. Appl. Biol.* 35:412–17
40. Harrison, B. D., Winslow, R. D. 1961. Laboratory and field studies on the relation of arabis mosaic virus to its nematode vector *Xiphinema diversicaudatum* (Micoletzky). *Ann. Appl. Biol.* 49: 621–33
41. Harpaz, I. 1972. *Maize Rough Dwarf.* Jerusalem: Israel Univ. Press. 251 pp.
42. Hashizuma, B. 1968. Studies on forecasting and control of the rice green leafhopper *Nephotettix cincticeps* Uhler with special reference to eradication of the rice dwarf disease. *Rev. Plant Prot. Res. Japan* 1:78–82
43. Hodson, A. C., Cook, E. F. 1960. Long-range aerial transport of the harlequin bug and the greenbug into Minnesota. *J. Econ. Entomol.* 53:604–8
44. Hollings, M. 1965. Disease control through virus-free stock. *Ann. Rev. Phytopathol.* 3:367–96
45. Hull, R. 1953. Assessment of losses in sugar beet due to virus yellows in Great Britain, 1942–52. *Plant Pathol.* 2:39–43
46. Hurst, G. W. 1965. Forecasting the severity of sugar beet yellows. *Plant Pathol.* 14:47–53
47. Hutchins, L. M., Bodine, E. W., Thornberry, H. H. 1937. Peach mosaic: its identification and control. *US Dep. Agr. Circ.* 427:1–48
48. Jenkinson, J. G. 1955. The incidence and control of cauliflower mosaic in broccoli in south west England. *Ann. Appl. Biol.* 43:409–22
49. Jordovic, M. 1968. Effect of sources of infection on epidemiology of sărka (plum pox) virus disease. *VII Eur. Symp. viruskrankheiten Obstbäume Sunderdruck Tagungsber.* 97:301–8
50. Laird, E. F., Dickson, R. C. 1963. Tobacco etch virus and potato virus Y in pepper, their host plants and insect vectors in southern California. *Phytopathology* 53:48–57
51. Linford, M. 1932. Transmission of the pineapple yellow spot virus by *Thrips tabaci. Phytopathology* 22:301–24
52. Maramorosch, K. 1963. Arthropod transmission of plant viruses. *Ann. Rev. Entomol.* 8:369–414
53. Murumatsu, Y., Furuki, I., Kawaguchi, K., Sawaki, T., Takeshima, S. 1968. Forecasting rice dwarf disease. II Forecasting methods. *Rev. Plant Prot. Res. Japan,* 1:83
54. Nelson, M. R., Tuttle, D. M. 1969. The epidemiology of cucumber mosaic and watermelon mosaic 2 of cantaloups in an arid climate. *Phytopathology* 59: 849–56
55. Oldfield, G. N. 1970. Mite transmission of plant viruses. *Ann. Rev. Entomol.* 15:343–80
56. Ossiannilsson, F. 1966. Insects in the epidemiology of plant viruses. *Ann. Rev. Entomol.* 11:213–32
57. Pathak, M. D., Vea, E., John, V. T. 1967. Control of insect vectors to prevent virus infection of rice plants. *J. Econ. Entomol.* 60:218–25
58. Posnette, A. F. 1963. Control measures. In *Virus Diseases of Apples and Pears.* Technical Communication No. 30 Commonwealth Bureau of Horticulture and Plantation Crops, ed. A. F. Posnette, 121–25. Farnham Royal: Commonw. Agr. Bur. 141 pp.
59. Posnette, A. F., Cropley, R. 1954. Field studies on virus diseases of strawberries. II. Seasonal periods of virus spread.

*Ann. Rept. East Malling Res. Sta. 1953,* 154–57
60. Randles, J. W., Crowley, N. C. 1970. Epidemiology of lettuce necrotic yellows virus in S. Australia, I. Relationship between disease incidence and activity of *Hyperomyzus lactucae. Aust. J. Agr. Res.* 21:447–53
61. Rose, D. J. W. 1972. Times and sizes of dispersal flights by *Cicadulina* species (Homoptera: Cicadellidae), vectors of maize streak disease. *J. Anim. Ecol.* 41:495–506
62. Sakimura, K. 1962. The present status of thrips-borne viruses. In *Biological Aspects of Disease Transmission,* ed. K. Maramorosch, 33–40. New York: Academic. 192 pp.
63. Shands, W. A., Simpson, G. W., Wave, H. E. 1957. Some effects of two hurricanes upon populations of potato-infesting aphids in north-eastern Maine. *J. Econ. Entomol.* 49:252–53
64. Shepherd, R. J. 1972. Transmission of viruses through seed and pollen. In *Principles and Techniques in Plant Virology,* ed. C. I. Kado, H. O. Agrawal, 10:267–92. New York: Van Nostrand. 688 pp.
65. Shepherd, R. J., Hills, F. J. 1970. Dispersal of beet yellows and beet mosaic viruses in the inland valleys of California. *Phytopathology* 60:798–804
66. Simons, J. N. 1956. The pepper veinbanding mosaic virus in the Everglades area of South Florida. *Phytopathology* 46:53–57
67. Smith, H. C. 1963. Control of barley yellow dwarf virus in cereals. *N Z J Agr. Res.* 6:229–44
68. Strickland, A. H. 1951. The entomology of swollen shoot of cacao. II. The bionomics and ecology of the species involved. *Bull. Entomol. Res.* 42: 65–103
69. Stubbs, L. L., Grogan, R. G. 1963. Necrotic yellows: a newly recognised virus disease of lettuce. *Aust. J. Agr. Res.* 14:439–59
70. Stubbs, L. L., Guy, J. A. D., Stubbs, K. J. 1963. Control of lettuce necrotic yellows virus disease by the destruction of common sowthistle (*Sonchus oleraceus*). *Aust. J. Exp. Agr. Anim. Husb.* 3:215–18
71. Stubbs, L. L., O'Loughlin, G. T. 1962. Climatic elimination of mosaic spread in lettuce seed crops in the Swan Hill region of the Murray Valley. *Aust. J. Exp. Agr. Anim. Husb.* 2:16–19
72. Swarbrick, T., Thompson, S. R. 1932. Observations upon the incidence of 'reversion' and on the control of 'big bud' in black currant. *Ann. Rept. L. Ashton Res. Sta. 1931,* 101–11
73. Swenson, K. G. 1968. Role of aphids in the ecology of plant viruses. *Ann. Rev. Phytopathol.* 6:351–74
74. Swenson, K. G. 1969. Plant susceptibility to virus infections by insect transmission. See Ref. 27, pp. 143–57
75. Taylor, C. E., Cadman, C. H. 1969. Nematode vectors. See Ref. 27, pp. 55–94
76. Teakle, D. S. 1969. Fungi as vectors and hosts of viruses. See Ref. 27, pp. 23–54
77. Thresh, J. M. 1958. The spread of virus disease in cacao. *Tech. Bull. W. Afr. Cocoa Res. Inst.* 36 pp.
78. Thresh, J. M. 1967. Increased susceptibility of black currant bushes to the gall-mite vector (*Phytoptus ribis* Nal.) following infection with reversion virus. *Ann. Appl. Biol.* 60:455–67
79. Thresh, J. M. Unpublished
80. Valdez, R. B., McNamara, D. G., Ormerod, P. J., Pitcher, R. S., Thresh, J. M. 1974. Transmission of the hop strain of arabis mosaic virus by *Xiphinema diversicaudatum. Ann. Appl. Biol.* 76:113–22
81. Van der Plank, J. E. 1946. A method for estimating the number of random groups of adjacent diseased plants in a homogeneous field. *Trans. Roy. Soc. S. Afr.* 31:269–78
82. Van der Plank, J. E. 1959. Some epidemiological consequences of systemic infection. In *Plant Pathology: Problems and Progress 1908–1958,* ed. C. S. Holton, G. W. Fischer, R. W. Fulton, H. Hart, S. E. A. McCallan, 51: 566–73. Madison: Univ. Wisconsin Press. 588 pp.
83. Van der Plank, J. E. 1963. *Plant Diseases: Epidemics and Control.* New York: Academic. 349 pp.
84. Van der Plank, J. E., Anderssen, E. E. 1944. Kromnek disease of tobacco; a mathematical solution to a problem of disease. *Sci. Bull. No. 240 Union S. Afr. Dep. Agr. Forest. Bot. Plant Pathol. Ser. No 3,* 6 pp.
85. Van Hoof, H. A., Stubbs, R. W., Wouters, L. 1962. Beschouwingen over hoja blanca en zijn overbrenger *Sogata orizicola* Muir. *Surinamse Landb.* 10: 1–18
86. Wallin, J. R., Loonan, D. V. 1971. Low-level jet winds, aphid vectors, local weather and barley yellow dwarf virus outbreaks. *Phytopathology* 61:1068–70
87. Walters, H. J. 1969. Beetle transmission

of plant viruses. *Advan. Virus Res.* 15:339–64
88. Watson, M. A. 1966. The relation of annual incidence of beet yellowing viruses in sugar beet to variations in weather. *Plant Pathol.* 15:145–49
89. Watson, M. A., Healy, M. J. R. 1953. The spread of beet yellows and beet mosaic viruses in the sugar-beet root crop. II. The effects of aphid numbers on disease incidence. *Ann. Appl. Biol.* 40:38–59
90. Watson, M. A., Plumb, R. T. 1972. Transmission of plant-pathogenic viruses by aphids. *Ann. Rev. Entomol.* 17: 425–52
91. Watson, M. A., Serjeant, E. P. 1963. The effect of motley dwarf virus on yield of carrots and its transmission in the field by *Cavariella aegopodiae* Scop. *Ann. Appl. Biol.* 53:77–93
92. Whitcomb, R. F., Davis, R. E. 1970. Mycoplasma and phytarboviruses as plant pathogens persistently transmitted by insects. *Ann. Rev. Entomol.* 15:405–64
93. Wolf, F. A. 1935. *Tobacco Diseases and Decays.* Durham, N. Carolina: Duke Univ. Press. 454 pp.
94. Zinc, F. W., Grogan, R. G., Welch, J. E. 1956. The effect of the percentage of seed transmission upon subsequent spread of lettuce mosaic virus. *Phytopathology* 46:662–64

# SPOROGENESIS IN FUNGI

❖3592

*G. Turian*[1]
Department of Plant Biology, University of Geneva, 1211 Geneva 4, Switzerland

Fungal spores differ widely in their ontogeny (27), and comparative physiological studies are meaningful only when made between spores resulting from similar morphogenetic processes. For instance, phialoconidia of Aspergilli and Penicillia are more comparable to the basipetally budded microconidia than to the basifugally budded macroconidia of *Neurospora* (66). The field of sporogenesis is burgeoning (1, 7, 52, 63), but there is yet no one formula for fungal sporulation (13). Unifying concepts may be expected to be found only at the level of sporogenetic control mechanisms. These intervene at several developmental levels, primarily at the (*a*) genetic level, providing the sporogenic competence and ability to respond (*b*) to the environmental, physical, and chemiconutritional inducements mediated through (*c*) primary and then secondary metabolic pathways subjected to positive or negative regulations. It is from the harmonious cooperation of all these mechanisms that the specific ultrastructural patterns of the various spore types finally arise.

## GENETIC CONTROL

The apparent paradox of spore differentiation as opposed to differentiation in general is its occurrence in the presence of a constant genome. However, sequential transcription or expression of certain genes is needed to change hyphal chemistry and is a prerequisite to altering structures in order to place spores under the additional control of extrachromosomal elements (18). The most striking feature of Clutterbuck's (12) study of the spectrum of mutants affecting the conidial apparatus of *Aspergillus nidulans* is the paucity of mutants specifically and completely blocked in conidium formation, while the number of loci specifically involved in the whole conidiation process has been estimated at 45 to 150 (34). In 85% of the conidiation mutants the defect was not confined to conidiation (pleiotropic mutations). Mutants were classified into nine developmental stages and into asporogenous and oligosporogenous types. The largest group of mutants was blocked before conidiophore formation; only a few were blocked at the late stages, for example, "abacus," which

[1]The author is supported by the Swiss National Research Fund.

budded conidia basifugally in a *Monilia*-like way, and "bristle," which had its site of action at the continuity between the secondary wall of the conidiophore and the single wall of the sterigmata (42). Many of the asporogenous strains of *A. nidulans* had their mutations in loci with functions essential only at high temperatures (34). Such temperature-sensitive, conditional mutants have also been obtained in the cellular slime mold *Dictyostelium discoideum* (31). Mutants defective in their metabolic regulation can also be conditionally asporogenous, as shown by the "amycelial" strain of *Neurospora crassa* grown under sterile conditions with vesiculous hyphae on the surface of glucose-restrictive media, while recovering the capacity to conidiate on acetate-permissive media (67).

Conditional mutants must be used to study impairments in meiosis and sporulation in *Saccharomyces cerevisiae.* By use of adequate selection systems, mutants blocked in the whole sporogenic sequence (vegetative to sporogenic I–V) have been isolated, and it was calculated that 48 ± 27 loci are specifically involved in meiotic and sporulation function in *S. cerevisiae* (56). None of the Spo mutants have been found to be defective in protein or RNA synthesis at the restrictive temperature of 34°C, which blocked the formation of asci. Spo-1 and Spo-2 do not increase their DNA to the parental level during sporulation. While most of the Spo mutants of *S. cerevisiae* are concerned with defects in meiosis, the majority of the 300 asporogenic strains isolated from *Schizosaccharomyces pombe* are disturbed in their ascosporal maturation and wall synthesis (56).

"Analysis of mutants with known biochemical defects may be more informative than analysis of sporulation-specific mutants in which it is difficult to determine the primary effects of the respective mutations." This comment by Tingle et al (56) appears pertinent when applied to the relationship of enzymes to sporulation where a mutation in an enzyme involved in the endogenous metabolism can affect sporogenesis more than vegetative growth. This is the case of a yeast strain low in protease activity that failed to sporulate (56) and in the "cup" mutant of *Schizophyllum commune,* which has a disturbed glucans metabolism and forms abnormal sporocarps (41). This may also concern regulatory mutants such as amycelial of *N. crassa* found to be deficient in the gluconeogenic *c*-malate dehydrogenase when grown in aconidiogenic, sugar-restrictive conditions (45). A continuity appears, however, between biochemical asporogenous mutants and sporulation-specific mutants when we consider all sporulation genes as somehow concerned with the synthesis of new molecules that are characteristic of spores.

The complexity of genetic control of sporogenesis requires study of the regulation of a series of genes and the series of proteins derived from them. A first approach along this line has been made at the level of sequential DNA gene transcription using the selective inhibitor actinomycin D. By incorporating it at progressively later times during the developmental sequence, the transcriptive period of seven enzymes related to cell wall metabolism and therefore to sporogenesis was delineated in *Dictyostelium* (54). In *Allomyces,* spore induction and mitosporangium formation were insensitive to actinomycin D (6), as previously found for female (but not male) gametangia (63), suggesting that stable, preexisting mRNA is responsible for these developmental events. Inversely, in *Achlya,* actinomycin D inhibited the differentiation of sporangia and the incorporation of labeled precursors into RNA, thus indicating DNA-dependent RNA synthesis for that process (23). In *Neurospora,* a DNA-dependent RNA synthesis is required first for conidiophore formation (57) and then for conidiogenesis (57, 66).

## ENVIRONMENTAL-NUTRITIONAL CONTROL

The normal sequential pattern of expression of the genetic potential for sporogenesis depends on the environmental conditions. Light-induced sporulation has been described in numerous fungi of all classes (10, 39) but it is still little understood (63). Flavins appear to be involved in the process, as in *Alternaria solani* where they are photoinactivated during conidiogenesis (32), with the practical consequence that glycolate oxidase, a flavinic enzyme, is a good target for antisporulants (33). A substance having its peak of absorption at 310 mμ (P310) has been isolated from the water extracts of ultraviolet-treated mycelia from a wide range of fungi (30), and its sporogenic activity has been demonstrated (60).

Interactions of physical factors with morphogenetic effects are known. The size of the conidial heads in *Aspergillus giganteus* and the length of the conidiophores are affected by an interaction between temperature and light (59). Light enhanced conidial production in members of the *Aspergillus glaucus* group while darkness enhanced, and high osmotic pressures inhibited, cleistothecial formation (17).

Examples of inhibitory effect of insufficient aeration on sporulation are numerous (55). In *Neurospora,* the amount of oxygen required for vegetative hyphal growth of the substrate mycelium is lower than for macroconidiation occurring on aerial hyphae (50, 66); anaerobic mutants cannot conidiate (26). By contrast, *Neurospora* microconidiation is less aerophilic (63) and *A. nidulans* could still produce free phialoconidia, but without differentiated conidiophores, at low, dissolved oxygen tensions (11). In aquatic Phycomycetes, the gametangia of *Allomyces arbuscula,* the males more than the females can also differentiate at relatively low oxygen concentrations (29), while in subterrestrial forms as *Phytophthora* spp., sporangium formation was quickly reduced by decreasing $O_2$ concentration or increasing $CO_2$ levels from those in air (38). Volatile organic compounds are also implicated in the control processes of fungal sporulation (20) as well as temperature (25), pH (5), and humidity (58).

Starvation in buffer solution has been used in many fungi to induce sporulation (13) with a fair degree of synchrony as in the cases of *Achlya* zoosporogenesis (23) and *Neurospora* conidiation (66). It is only after complete exhaustion of sugar from the medium that the plasmodium of *Physarum* sporulates (49) or the pileus of *Schizophyllum* forms and expands (41). Lowering of the glucose content in the medium to a "maintenance ration" leads to early conidiation in *Penicillium chrysogenum* grown in continuous cultures (46). Inversely, high glucose concentration inhibits sporulation in yeast (37), *Schizophyllum* (41), and *Coprinus* (with cAMP reversal) (35, 70). In *Neurospora,* the anticonidial effect of glucose occurred only with an ammonium and not with a nitrate source (64, 65). In *Penicillium griseofulvum* and several related fungi in submerged culture, the absence or exhaustion of assimilable nitrogen, in the presence of residual carbohydrate, is the most general condition for the induction of conidiation (40). Organic nitrogen, especially glycine (66), most often favors sporulation, while an ammonium source is unfavorable to ascosporogenesis of *Saccharomyces cerevisiae* (37), conidiation of *Neurospora* spp. (62), and *Aspergillus niger* (52).

## METABOLIC CONTROL

The major problems in metabolic control are 1. complexity (i.e. how are the numerous enzymes that change during sporogenesis coordinated in a temporally and quantitatively regulated sequence) and 2. structure (i.e. how are the products of enzymatic reactions spatially laid down so as to form a structure of definite shape and size) (19). The metabolic processes at work at differentiation have been found, at least in the cellular slime molds (72), to depend not only on the enzymatic complement of the cell but also on factors such as compartmentalization, substrate availability, and enzyme activators and inhibitors. Wright (72) has built a computer model containing relevant biochemical data to identify the critical variables (bottlenecks) that limit the metabolism of trehalose and uridine diphosphoglucose during development of *Dictyostelium.* Among supporting data of the sporogenic role of substrate limitations are carbohydrate depletion, drop in amino acid pools, final catabolism of protein, and its use for polysaccharide synthesis (72). However, other physiological steps and further measurements of metabolic rates in vivo will have to be made to strengthen the predictive value of Wright's model, a provocative trial to cope with the complexity of biochemical changes at fungal differentiation, to say nothing of that of higher eukaryotic organisms. The orderly sequence of synthesis of enzymes concerned with amino acid and carbohydrate metabolism remains, however, a necessary, but not sufficient, part of the dynamics of sporogenesis in *D. discoideum* (21, 54).

Altered metabolism of carbohydrates and depletion of carbohydrate reserves also characterize the presporulation period induced by starvation in the acellular slime mold *Physarum polycephalum* (49). However, starvation in itself is not sufficient to induce sporulation; there must be DNA synthesis late in starvation and prior to the period of photoinduction (51).

Contrary to previous reports on the absence of most tricarboxylic acid cycle (TCA cycle) enzymes in the resistant sporangial (RS) cells of *Blastocladiella emersonii* (8), it has been found (28) that the TCA cycle enzymes are present in extracts from RS and OC (ordinary colorless) cells and that these enzymes have a higher activity per plant of RS cells. Moreover, and as shown by radioactive tracer experiments, the bicarbonate effect would depend on a stimulation of the TCA cycle by $CO_2$ fixation (28) rather than on induced loss of the TCA enzyme activities (8). The fact that aspartate received the greatest amount of label points to an increased synthetic function of the TCA cycle leading to protein synthesis. Such a bicarbonate-stimulated activity leads to self-regulation which, by progressively stopping growth, introduces an endotrophic differentiation of the thick-walled RS. Thus the trigger reaction concerned with establishing the point of no return in RS differentiation after 43 hr (8) might well be the conversion of protein synthesized in the initial $CO_2$-TCA cycle activated growth phase into the wall polysaccharides (chitin) and the accumulation of carotene-containing lipids in the cytoplasm of the RS. This suggestion reconciles the diverging results and has experimental support in Cantino's (8, 9) measurements of a decrease in the endogenous amino acid pool in the maturing RS, presumably following proteolysis.

In yeast, sporulation ability is in part governed by the shifting metabolism of a starving cell (56). As long as enough glucose is provided to the vegetative cells of *Saccharomyces cerevisiae,* they remain in their log phase growth as actively budding diploid cells, unable to oxidize acetate or to sporulate. As the concentration of glucose decreases either naturally or by induced starvation, the TCA cycle enzymes are derepressed and the cells in their late growth phase oxidize ethanol and acetate before starting early sporogenesis. Such a release of glucose inhibition of sporulation can also be achieved by exogenous cyclic AMP (61). The sporogenic shift from fermentation to oxidative respiration appears energetically essential to insure sporal syntheses from storage carbohydrates and lipids (56). The carbohydrate reserves then decrease when the ascospores start to differentiate, presumably to support carbon requirements for synthesis of ascospore wall polysaccharides (56). In yeast triggered to sporulation in the acetate medium, glycogen and fat start to disappear when endogenous respiration reaches a maximum in cells completing their nuclear divisions before ascosporogenesis. Extensive protein and RNA turnover and synthesis occur throughout sporulation; two maxima for protein synthesis occur: one before asci appear and a second during ascospore differentiation (56).

The systems of conidiogenesis in the Eu-Ascomycetes and Fungi imperfecti are devoid of the yeast duality of meiosis–sporogenesis but are heterogenous due to their preliminary period of vegetative filamentous growth. However, this hyphal phase can be short circuited by inducing straight or "microcycle" conidiation on germinating swollen conidia of *A. niger* (52, 53) and *N. crassa* (16) shifted down from elevated temperatures (44 and 46°C respectively). Normally, elongation of *Neurospora* hyphae continues freely as long as enough sugar is available from the medium (64). In subaerial conditions it stops when translocation of glucose to the tips becomes limiting and the aerobic stimulus is strong enough to induce a Pasteur effect (63), switching the predominantly alcoholic fermentation of the vegetative hyphae into an oxidative conidiogenic type of metabolism (71). The site of the anticonidiogenic effect of glucose appears to be the hyphal tip in which development of the mitochondria is restricted by catabolite repression (64). Premature conidiation could be obtained by increasing concentrations of acetate (62), lowering the glucose level in the ammonium medium, with parallel rise of the oxidative activity of alcohol dehydrogenase (65), or by adding cAMP (optimum concentration $10^{-6}M$) (67a). The derepressed oxidative activity in the budding proconidia relies in part at least on the TCA cycle (66) which provides increased energy for the new syntheses of RNA protein (69) and DNA of the dividing proconidial nuclei (68). Parallel stimulation of the glyoxylate shunt (62) then opens the gluconeogenic pathway necessary to build up the structural polysaccharides of the interconidial septa (68). In *Aspergillus niger,* a dependence of conidia formation on a functionally complete TCA cycle had been reported, but this has not yet been firmly established (53). Radiorespirometric studies with *Endothia parasitica* (36), *A. nidulans* (11), and *A. niger* (52) have indicated rather an enhanced contribution of the HMP pathway prior to and during phialoconidiation, with correlated changes in the levels of glycolytic intermediates (53). Interestingly, in *Neurospora* the iodoacetate inhibition of glycolysis, with presumed enhancement of HMP activity, selectively stimulated phialo (micro) conidium production (48). That lipids can also function as a source of carbon and energy during conidiation of *A. niger* is shown by the sharp increase in esterase activity prior to vesicle and phialide formation and its persistence in these structures (53). Net synthesis of RNA and protein occurs only at the induction of conidiation in *Penicillium chrysogenum* (46) , and the newly synthesized protoplasmic components seem to be translocated to the young phialides which are rich in RNA and other phosphorus compounds (52). New bands of proteins have been electrophoretically detected in conidiating compared with nonconidiating mycelia of *P. griseofulvum* (4).

## PROSPECTS

Following Klebs' first principle, Cochrane (13) had rightly stressed the concept that the major stimulus to sporulation is exhaustion of the medium, especially of its carbon source. The problem is to understand how starvation could cause a metabolic shift leading to spore differentiation. According to Wright (72), differentiation occurs in essentially endogenous self-sufficient systems that have to a greater or less degree cut themselves off from the environment; the limitation of exogenous substrate triggers the metabolic flux toward the accumulation of specialized, sporogenic endproducts.

However, it is sequential enzyme development following induced differential transcription-translation that commits the system to sporogenic processes (54). In such processes the pattern of cell wall construction is fundamental, as it is for the whole fungal morphogenesis (3). Profit can be drawn from an analogy of the many cases of apical sporulation, which involve uniformly dispersed wall construction in the spherical cells, with those of yeast-hyphal dimorphism (47), which involve apically restricted wall construction in the cylindrical cells (3). This alternate architectural design is reflected in the fibrillar organization of the dimorphic stages, as also found in the macroconidia of *Neurospora,* which are enclosed in a thick network of microfibrils devoid of the parallel bundles of chitinoglucan fibrils that strengthen the cylindrical portion of the hyphae (15). The structural polysaccharides result from a gluconeogenic reversal of the initially catabolic routes (66, 72), and strain-specific pigments (anthraquinones, melanins) produced by the secondary metabolism (52) finally impregnate spore walls.

Specific inductive factors of sporogenesis have been described as hormones indirectly involved in the differentiation of sexual spores like oospores of *Achlya* (2) and zygospores in Mucorales (22). Less or nonspecific ones from staling media from mature mycelia of Penicillia (24), *Fusarium, Geotrichum,* and other fungi were described as fungal morphogen(s) (44) and further characterized as bikaverin (14).

Natural antiglycolytic compounds such as quinones can also switch the hyphal tips to conidia in *Neurospora* (43). Most interesting, several fungi respond sporogenically to the addition of cAMP to glucose-containing media (see earlier sections). Considering both its glucose-controlled synthesis and faculty to promote transcription of the structural genes for sporulation in bacteria, cAMP could thus emerge also as a key sporogenic factor in fungi, filling the gap between the initial stimulus from the environment (starvation) and the commitment to sporulation.

*Literature Cited*


1. Ashworth, J. M., Smith, J. E. 1973. Microbial differentiation. *Symp. Soc. Gen. Microbiol., 23rd.* Cambridge, England: Cambridge Univ. Press. 450 pp.
2. Barksdale, A. W. 1969. Sexual hormones of *Achlya* and other fungi. *Science* 166:831–37
3. Bartnicki-Garcia, S. 1973. Fundamental aspects of hyphal morphogenesis. See Ref. 1, pp. 245–67
4. Bent, K. J. 1967. Electrophoresis of proteins of 3 *Penicillium* species on acrylamide gels. *J. Gen. Microbiol.* 49:195–200
5. Bergquist, R. P., Horst, R. K., Lorbeer, J. W. 1972. Influence of polychromatic light, carbohydrate source, and pH on conidiation of *Botryotinia squamosa. Phytopathology* 62:889–95
6. Burke, D. J., Seale, T. W., McCarthy, B. J. 1972. Protein and ribonucleic acid synthesis during the diploid life cycle of *Allomyces arbuscula. J. Bacteriol.* 110:1065–72
7. Burnett, J. H. 1968. *Fundamentals of Mycology.* London: Arnold. 546 pp.
8. Cantino, E. C. 1966. Morphogenesis in Aquatic Fungi. *The Fungi,* ed. G. C. Ainsworth, A. S. Sussman, II:283–337. New York: Academic. 805 pp.
9. Cantino, E. C. 1969. The $\gamma$ particle, satellite ribosome package, and spheroidal mitochondrion in the zoospore of *Blastocladiella emersonii. Phytopathology* 59:1071–76
10. Carlile, M. J. 1965. The photobiology of fungi. *Ann. Rev. Plant Physiol.* 16:175–202
11. Carter, B. L. A., Bull, A. T. 1971. The effect of oxygen tension in the medium on the morphology and growth kinetics of *Aspergillus nidulans. J. Gen. Microbiol.* 65:265–73
12. Clutterbuck, A. J. 1969. A mutational analysis of conidial development in *Aspergillus nidulans. Genetics* 63:317–27
13. Cochrane, V. W. 1958. *Physiology of Fungi.* New York: Wiley. 524 pp.
14. Cornforth, J. W., Ryback, G., Robinson, P. M., Park, D. 1971. Isolation and characterization of a fungal vacuolation factor (bikaverin). *J. Chem. Soc. C:* 2786–88
15. Coniordos, N., Turian, G. 1973. Recherches sur la différenciation conidienne de *Neurospora crassa.* IV. Modifications chimio-structurales de la paroi chez le type sauvage et chez deux mutants aconidiens. *Ann. Microbiol. Paris* 124:5–28
16. Cortat, M., Turian, G. 1974. Conidiation of *Neurospora crassa* in submerged culture without mycelial phase. *Arch. Microbiol.* 95:305–9
17. Curran, P. M. T. 1971. Sporulation in some members of the *Aspergillus glaucus* group in response to osmotic pressure, illumination and temperature. *Trans. Brit. Mycol. Soc.* 57:201–11
18. Fincham, J. R. S., Day, P. R. 1971. *Fungal Genetics.* Oxford: Blackwell. 402 pp.
19. Francis, D. 1973. Order in an anarchic field. *Science* 176:821–22
20. Fries, N. 1973. Effects of volatile organic compounds on the growth and development of fungi. *Trans. Brit. Mycol. Soc.* 60:1–21
21. Garrod, D., Ashworth, J. M. 1973. Development of the cellular slime mold *Dictyostelium discoideum.* See Ref. 1, pp. 407–35
22. Gooday, G. W. 1973. Differentiation in the Mucorales. See Ref. 1, pp. 269–94
23. Griffin, D. H., Breuker, C. 1969. Ribonucleic acid synthesis during the differentiation of sporangia in the water mold *Achlya. J. Bacteriol.* 98:689–96
24. Hadley, G., Harrold, C. E. 1958. The sporulation of *Penicillium notatum* Westling in submerged liquid culture. II. The initial sporulation phase. *J. Exp. Bot.* 9:418–25
25. Hawker, L. E. 1947. *The Physiology of Reproduction in Fungi.* Cambridge, England: Cambridge Univ. Press. 128 pp.

26. Howell, N., Zuiches, C. A., Munkres, K. D. 1971. Mitochondrial biogenesis in *Neurospora crassa.* I. An ultrastructural and biochemical investigation of the effects of anaerobiosis and chloramphenicol inhibition. *J. Cell. Biol.* 50: 721–36
27. Kendrick, B. 1971. *Taxonomy of Fungi Imperfecti.* Toronto: Univ. Toronto Press. 309 pp.
28. Khouw, B. T., McCurdy, H. D. 1969. Tricarboxylic acid cycle enzymes and morphogenesis in *Blastocladiella emersonii. J. Bacteriol.* 99:197–205
29. Kobr, M. J., Turian, G. 1967. Metabolic changes during sexual differentiation in *Allomyces. Arch. Mikrobiol.* 57:271–79
30. Leach, C. M. 1972. An action spectrum for light-induced sexual reproduction in the Ascomycete fungus *Leptosphaerulina trifolii. Mycologia* 64:475–90
31. Loomis, W. F. Jr. 1969. Temperature sensitive mutants of *Dictyostelium discoideum. J. Bacteriol.* 99:65–69
32. Lukens, R. J. 1963. Photo-inhibition of sporulation in *Alternaria solani. Am. J. Bot.* 50:720–24
33. Lukens, R. J., Horsfall, J. G. 1968. Glycolate oxidase, a target for antisporulants. *Phytopathology* 58:1671–76
34. Martinelli, S. D., Clutterbuck, A. J. 1971. A quantitative survey of conidiation mutants in *Aspergillus nidulans. J. Gen. Microbiol.* 69:261–68
35. Matthews, T. R., Niederpruem, D. J. 1972. Differentiation in *Coprinus lagopus.* I. Control of fruiting and cytology of initial events. *Arch. Mikrobiol.* 87:257–68
36. McDowell, L. L., De Hertogh, A. A. 1968. Metabolism of sporulation in filamentous fungi. I. Glucose and acetate oxidation in sporulating and nonsporulating cultures of *Endothia parasitica. Can. J. Bot.* 46:449–51
37. Miller, J. J., Hoffmann-Ostenhof, O. 1964. Spore formation and germination in *Saccharomyces. Z. Allg. Mikrobiol.* 4:273–95
38. Mitchell, D. J., Zentmyer, G. A. 1971. Effects of oxygen and carbon dioxide tensions on sporangium and oospore formation by *Phytophthora* spp. *Phytopathology* 61:807–12
39. Moore-Landecker, E. 1972. *Fundamentals of the Fungi.* Englewood Cliffs, NJ: Prentice-Hall. 482 pp.
40. Morton, A. G. 1967. Morphogenesis in Fungi. *Sci. Progr. Oxford* 55:579–611
41. Niederpruem, D. J., Wessels, J. G. H. 1969. Cytodifferentiation and morphogenesis in *Schizophyllum commune. Bacteriol. Rev.* 33:505–35
42. Oliver, P. T. P. 1972. Conidiophore and spore development in *Aspergillus nidulans. J. Gen. Microbiol.* 73:45–54
43. Oulevey-Matikian, N., Turian, G. 1968. Contrôle métabolique et aspects ultrastructuraux de la conidiation (macro-microconidies) de *Neurospora crassa. Arch. Mikrobiol.* 60:35–58
44. Park, D., Robinson, P. M. 1964. Isolation and bioassay of a fungal morphogen. *Nature* 203:988–89
45. Peduzzi, R., Turian, G. 1972. Recherches sur la différenciation conidienne de *Neurospora crassa.* III. Activité malicodéshydrogénasique de structures antigéniques et ses relations avec la compétence conidienne. *Ann. Inst. Pasteur Paris* 122:1081–97
46. Righelato, R. C., Trinci, A. P. J., Pirt, S. J., Peat, A. 1968. The influence of maintenance energy and growth rate on the metabolic activity, morphology and conidiation of *Penicillium chrysogenum. J. Gen. Microbiol.* 50:399–412
47. Romano, A. H. 1966. Dimorphism. See Ref. 8, II:181–209
48. Rossier, C., Oulevey, N., Turian, G. 1973. Electron microscopy of selectively stimulated microconidiogenesis in wild type *Neurospora crassa. Arch. Mikrobiol.* 91:345–53
49. Rusch, H. P. 1969. Some biochemical events in the growth cycle of *Physarum polycephalum. Fed. Proc.* 28:1761–70
50. Sargent, M. L., Kaltenborn, S. H. 1972. Effects of medium composition and carbon dioxide on circadian conidiation in *Neurospora. Plant Physiol.* 50:171–75
51. Sauer, H. W. 1973. Differentiation in *Physarum polycephalum.* See Ref. 1, pp. 375–405
52. Smith, J. E., Galbraith, J. C. 1971. Biochemical and physiological aspects of differentiation in the Fungi. *Advan. Microbiol. Physiol.* 5:45–134
53. Smith, J. E., Anderson, J. G. 1973. Differentiation in the *Aspergilli.* See Ref. 1, pp. 295–337
54. Sussman, M., Sussman, R. P. 1969. Patterns of RNA synthesis and of enzyme accumulation and disappearance during cellular slime mould cytodifferentiation. See Ref. 1, 19:403–35
55. Tabak, H. H., Cooke, B. W. 1968. The effects of gaseous environments on the growth and metabolism of fungi. *Bot. Rev.* 34:126–252
56. Tingle, M., Singh Klar, A. J., Henry, S. A., Halvorson, H. O. 1973. Ascospore

formation in yeast. See Ref. 1, pp. 209–43

57. Totten, R. E., Howe, H. B. Jr. 1971. Temperature-dependent actinomycin D effect on RNA synthesis during synchronous development in *Neurospora crassa*. *Biochem. Genet.* 5:521–32
58. Trinci, A. P. J., Banbury, G. H. 1967. A study of the growth of tall conidiophores of *Aspergillus giganteus* Wehmer. *Trans. Brit. Mycol. Soc.* 50:525–38
59. Trinci, A. P. J., Banbury, G. H. 1969. Effect of light on growth and carotenogenesis of the tall conidiophores of *Aspergillus giganteus*. *Trans. Brit. Mycol. Soc.* 52:73–86
60. Trione, E. J., Leach, C. M. 1969. Light-induced sporulation and sporogenic substances in fungi. *Phytopathology* 59:1077–83
61. Tsuboi, M., Kamisaka, S., Yanagishima, N. 1972. Effect of cyclic 3', 5'-adenosine monophosphate on the sporulation of *Saccharomyces cerevisiae*. *Plant Cell Physiol.* 13:585–88
62. Turian, G. 1966. The genesis of macroconidia of *Neurospora*. *The Fungus spore*, ed. M. F. Madelin. Colston Papers No. 18:61–66. London: Butterworth. 338 pp.
63. Turian, G. 1969. *Différenciation Fongique*. Paris: Masson. 144 pp.
64. Turian, G. 1972. Maintien de l'élongation végétative par répression catabolique dans l'apex hyphal de *Neurospora crassa*. *C. R. Acad. Sci. Ser. D* 275:1371–74
65. Turian, G. 1973. Induction of conidium formation in *Neurospora crassa* by lifting of catabolite repression. *J. Gen. Microbiol.* 79:347–50
66. Turian, G., Bianchi, D. E. 1972. Conidiation in *Neurospora*. *Bot. Rev.* 38:119–54
67. Turian, G., Oulevey, N., Coniordos, N. 1971. Recherches sur la différenciation conidienne de *Neurospora crassa*. I. Organisation chimio-structurale de la conidiation conditionnelle d'un mutant amycélien. *Ann. Inst. Pasteur Paris* 121:325–35

67a. Turian, G., Khandjian, E. 1973. Action biphasique de l'AMP-cyclique sur la conidiogenèse de *Neurospora* et la caroténogenèse en général. *Bull. Soc. Bot. Suisse* 83(4):352–57

68. Turian, G., Oulevey, N., Cortat, M. 1973. Recherches sur la différenciation conidienne de *Neurospora crassa*. V. Ultrastructure de la séquence macroconidiogène. *Ann. Microbiol. Paris* 124 A:443–58
69. Urey, J. C. 1971. Enzyme patterns and protein synthesis during synchronous conidiation in *Neurospora crassa*. *Develop. Biol.* 26:17–27
70. Uno, I., Ishikawa, T. 1973. Metabolism of adenosine 3', 5'-cyclic monophosphate and induction of fruiting bodies in *Coprinus macrorhizus*. *J. Bacteriol.* 113:1249–55
71. Weiss, B., Turian, G. 1966. A study of conidiation in *Neurospora crassa*. *J. Gen. Microbiol.* 44:407–18
72. Wright, B. 1973. *Critical Variables in Differentiation*. Englewood Cliffs, NJ. Prentice-Hall. 110 pp.

# NITROGEN FORM AND PLANT DISEASE[1,2]

◆3593

*D. M. Huber*
Department of Botany and Plant Pathology, Purdue University,
West Lafayette, Indiana 47907

*R. D. Watson*
Department of Plant and Soil Science, University of Idaho, Moscow, Idaho 83843

## INTRODUCTION

Nitrogen has been intensively studied in relation to host nutrition and disease severity for many years because of its essential requirement for plant growth, its limited availability in soil, and its effect on cell size and wall thickness. Because nitrogen frequently increases disease, disease control through nitrogen withdrawal by the use of carbohydrate amendments has been proposed (227). There seems to be little justification for starving the plant with this procedure (140), especially because no simple physiological pattern applies to all host-parasite interactions. Although a wide range of interactions of pathogens and their hosts are involved, it is generally the form of nitrogen available to the host or pathogen that affects disease severity or resistance rather than the amount of nitrogen. This has been observed for years; however, manipulation of the form of nitrogen available to the plant or soil microflora has only recently been possible. Enough information is now available so that developing patterns and relationships can be summarized relative to the form of nitrogen and disease severity. However, much of the reported data concerning the effect of nitrogen on plant disease is difficult to interpret because soil conditions, form of nitrogen, or rate and time of application are not published. Pot experiments

[1]Paper No. 5301 of the Journal Series of the Purdue University Agricultural Experiment Station, West Lafayette, Indiana, and Idaho Agricultural Experiment Station Research Paper No. 73718, Moscow, Idaho.

[2]The assistance of H. W. Warren, R. L. Nicholson, and G. E. Shaner with this manuscript is gratefully acknowledged.

with arbitrary rates of nitrogen and limited soil volume are equally difficult to assess relative to field conditions.

There are numerous reviews and discussions of factors that influence the form of nitrogen available to the host or pathogen (4, 9, 12, 17, 40, 91, 116, 140, 202, 224, 225, 243). This review illustrates the relationship between inorganic forms of nitrogen and plant disease, enumerates various factors influencing this relationship, and indicates possible mechanisms involved in disease control with specific forms of nitrogen.

## FACTORS INFLUENCING THE FORM OF NITROGEN IN SOIL

The biological mineralization of organic nitrogen to inorganic ammonium nitrogen ($NH_4$–N) and its subsequent nitrification to nitrate nitrogen ($NO_3$–N) are dynamic processes (210) resulting in the availability of several forms of nitrogen throughout plant growth. The rate of mineralization and nitrification are influenced by physical factors such as pH (26, 53, 202, 203, 209), soil type, (7, 53), nitrogen concentration (7, 113), source of nitrogen (92), oxygen tension (26), temperature (7, 68), salt concentration (113, 172), previous crop (43, 109, 126, 143), and agricultural chemicals (255).

Unlike the positively charged ammonium ion, which is relatively stationary because of its adsorption to organic matter or clay particles, the negatively charged nitrate ion is freely mobile in the soil solution (166). Thus, leaching and denitrification primarily involve a loss of $NO_3$–N (171). Inhibition or retardation of nitrification of applied $NH_4$–N can reduce nitrogen losses, increase efficiency of applied N, and establish a predominantly ammoniacal form of nitrogen available for plant uptake (105, 171, 222, 232). Nitrification is sometimes considered so universal and rapid that applications of $NH_4$–N are considered equivalent to $NO_3$–N. This is not true in many forest, orchard, and grassland soils. Even in areas where nitrification normally occurs, plants may take up large amounts of $NH_4$–N before nitrification is significant (40).

Crop sequence, root exudates, and organic amendments can profoundly affect the form of nitrogen predominating in soil. Crops such as alfalfa, corn, or peas, and organic amendments of manure, stimulate nitrification (42, 109, 154) while nitrification in many forest (36, 37, 126, 143) and grassland soils does not occur; thus $NO_3$–N is negligible while $NH_4$–N is usually measurable (43, 148, 151, 174, 216). Because considerable ammonification may occur in forest soils (143), reduced nitrification may be an advantage as some mycorrhizae use $NH_4$–N and simple organic compounds, while $NO_3$–N is a poor source of nitrogen if absorbed at all (36, 37). The level of available carbon and the form of nitrogen influence the rate of organic matter decomposition as well as the nitrogen balance in soil. Straw decomposition is much more rapid in the presence of $NH_4$–N than $NO_3$–N (39) probably because of a general increase in microbial activity and the induction of extracellular macerating enzymes. Nitrification is inhibited by soil fumigants, insecticides, fungicides, and chemical inhibitors such as 2-chloro-6-(trichloromethyl) pyridine (N–Serve ®) (70,

82, 90, 101, 105, 158, 171, 222, 232). Spore-forming ammonifiers that release ammonia from organic nitrogenous complexes return quickly after fumigation so that ammonification continues almost uninterrupted. Thus, $NH_4$–N accumulates from the decomposition of the organic fraction of soil, and added $NH_4$–N fertilizer will remain as $NH_4$–N for extended periods after the fumigant itself has dissipated (6, 53, 76, 81, 111, 254, 255). The specific nitrification inhibitor, N-Serve, is apparently toxic only to *Nitrosomonas* spp., which convert $HN_4$–N to $NO_2$–N, and does not affect *Nitrobacter* spp. which convert $NO_2$–N to $NO_3$–N (158).

## NITROGEN FORM AND PLANT DISEASE

The form of nitrogen influences the severity of many diseases (Table 1). Instead of an exhaustive discussion of each disease situation, a few diseases are discussed and similar situations noted. The effect of specific forms of nitrogen on disease severity depends on many factors and is not the same for all host-parasite associations. The greatest benefits from fertilizers have been observed on moderately susceptible or partially resistant varieties. Immune plants are not usually attacked regardless of the nitrogen form available, and resistance of many highly susceptible plants may not be improved by a form of nitrogen (140). The fact that a given form of nitrogen reduces one disease but favors another points up the need for detailed understanding of the consequences of manipulating soil factors to control specific diseases (100, 107, 108, 142). Inorganic mineral nutrients appear to influence disease potential more than inoculum potential, and the source of nitrogen appears more important than the rate (8). Organic amendments may decrease disease even though the population of the pathogen is increased (133, 149). Thus, the intricate relationship of plant pathogens with other microorganisms, environmental factors, and the host is dynamic as well as extremely complex to study. Nevertheless, studies of host nutrition in relation to disease development provide a basis for modifying current agricultural practices to reduce disease severity.

### *Seedling Diseases*

The effect of a specific form of nitrogen on soil-borne pathogens has been observed for many years. Afanasiev & Carlson (2) emphasized that the form of nitrogen, as well as the amount, is important in determining the severity of black root rot (*Rhizoctonia solani*) of sugar beets. In their studies, the number of diseased plants was doubled with $NH_4$–N compared with $NO_3$–N. This is consistent with increased damping-off of lettuce as rates of ammonium nitrate increased (51) and with the failure of varying rates of $NO_3$–N to affect damping-off of lettuce or cauliflower seedlings by *R. solani* (193). Because $NH_4$–N increases the level of glutamine and asparagine compared with $NO_3$–N (16, 257), the application of $NH_4$–N could be expected to increase damping off of beans similar to increasing levels of asparagine (244, 245). Growth of *R. solani* on stems, production of infection cushions, penetration, and lesion enlargement increased as asparagine levels increased (56). Yet, levels

**Table 1** Effect of inorganic forms of nitrogen on plant disease

| Disease | Host(s) | Pathogen | Nitrogen Form | | References |
|---|---|---|---|---|---|
| | | | Nitrate | Ammonium | |
| *Seedling and Damping-off Diseases:* | | | | | |
| Seedling disease | Sugar beets | *Rhizoctonia* | Decrease | Increase | 2 |
| *Root and Cortical Rots:* | | | | | |
| Root rot | Citrus | *Phytophthora* | Decrease | Increase | 119 |
| Root rot | Citrus | *Fusarium* | | Increase | 3 |
| Root rot | Bean | *Rhizoctonia* | Decrease | Increase | 52, 163–165 |
| Root rot | Bean | *Fusarium* | Decrease | Increase | 100, 107–109, 134, 208, 247 |
| Root rot | Pea, soybeans | *Aphanomyces* | Decrease | Increase | 35 |
| Root rot | Pea | *Pythium* | Increase | Decrease | 35 |
| Root rot | Corn, tomato | *Aphanomyces* | Decrease | Increase | 35 |
| Root rot | Corn | *Pythium* | Increase | Decrease | 35 |
| Stalk rot | Corn | *Fusarium* | Decrease | Increase | 162 |
| Stalk rot | Corn | *Diplodia* | Increase | Decrease | 152, 258 |
| Root rot | Pine | *Poria* | Decrease | Increase | 126 |
| Root rot | Pine | *Armillaria* | Decrease | Increase | 57, 126 |
| Root rot | Wheat | *Fusarium* | Decrease | Increase | 30, 78, 202, 205 |
| Root rot | Wheat | *Helminthosporium* | Decrease | | 197 |
| Eye spot | Wheat | *Cercosporella* | Decrease | Increase | 78, 100, 101, 107–109 |
| Root rot | Tobacco | *Thielaviopsis* | Decrease | Increase | 15, 221 |
| Root rot | Cotton | *Phymatotrichum* | Increase | Decrease | 114, 150 |
| Root & stem rot | Potatoes | *Rhizoctonia* | Decrease | Increase | 107, 184 |
| Take-all | Wheat | *Ophiobolus* | Increase | Decrease | 74, 78, 79, 107–109, 201–203, 223 |
| Sharp eye-spot | Wheat | *Rhizoctonia* | Decrease | Increase | 78, 107–109 |
| Black-root | Sugar beets | *Rhizoctonia* | Decrease | Increase | 2 |
| Patch | Turf grass | *Ophiobolus* | | Decrease | 206 |
| Southern blight | Tomato | *Sclerotium* | Decrease | Increase | 200 |
| Scab | Potato | *Streptomyces* | Increase | Decrease | 40, 107–109, 123, 170, 242 |
| *Vascular Diseases:* | | | | | |
| Wilt | Carnation | *Phialophora* | Increase | Decrease | 157 |
| Wilt | Cotton | *Fusarium* | Decrease | Increase | 140 |
| Wilt | Tomato | *Fusarium* | Decrease | Increase | 241 |
| Wilt | Potato, tomato | *Verticillium* | Increase | Decrease | 58, 87, 107–109, 242, 250, 251 |
| Yellows | Cabbage | *Fusarium* | Decrease | | 240 |
| Wilt | Tobacco, tomato | *Pseudomonas* | Increase | Decrease | 69, 207 |
| Canker | Tomato | *Corynebacterium* | Increase | | 241 |
| Stewart's wilt | Corn | *Erwinia* | Increase | | 141 |
| Ring rot | Potato | *Corynebacterium* | Increase | | 69, 241 |
| *Foliar Diseases:* | | | | | |
| Powdery mildew | Wheat | *Erysiphe* | Increase | | 124 |
| Fruit & root rot | Tomato | *Colletotrichum* | Increase | Decrease | 252 |
| Stripe rust | Wheat | *Puccinia* | Increase | Decrease | Huber (unpublished) |
| Stem rust | Wheat | *Puccinia* | Increase | Decrease | 50 |
| Northern leaf blight | Corn | *Helminthosporium* | Decrease | Increase | 152 |
| Blast | Rice | *Piricularia* | Decrease | Increase | 161, 220 |
| Chocolate spot | Broad bean | *Botrytis* | Decrease | Increase | 23, 209 |
| *Nematode, Gall, and Other Diseases:* | | | | | |
| Cyst nematode | Soybeans | *Heterodera* | Increase | Decrease | 10 |
| Root nematode | Tobacco | *Heterodera* | Increase | Decrease | 144 |
| Canker | Peaches, prunes | *Xanthomonas* | Decrease | | 140 |
| Storage rot | Sweet potato | *Rhizopus* | Decrease | Increase | 211 |
| *Virus-Like Diseases:* | | | | | |
| Virus | Tobacco | Potato virus Y | | Increase | 13, 14 |
| Virus | Potato | Potato virus Y | | Decrease | 13, 14 |
| Virus | *N. glutinosa* | Tomato aucuba mosaic | Increase | Decrease | 13, 14 |
| Virus | Tobacco | Tomato aucuba mosaic | Increase | Decrease | 13, 14 |
| Virus | *Nicotiana glutinosa* | Tobacco mosaic | | Decrease | 13, 14, 127 |
| Virus | Potato | Potato virus X | | Decrease | 13, 14 |

of carbon and nitrogen that reduced damping-off did not reduce the dry weight or colony diameter of inoculum pieces.

### Root Rots and Cortical Diseases

Cortical and root diseases caused by *Fusarium, Rhizoctonia, Aphanomyces, Cercosporella, Poria,* and *Armillaria* may be reduced by $NO_3$–N or increased by $NH_4$–N, while diseases caused by *Ophiobolus, Diplodia, Pythium,* and *Streptomyces* respond in an opposite manner. *Fusarium solani* f. sp. *phaseoli, R. solani,* and *Thielaviopsis basicola* are involved in the bean root complex (99, 208, 227, 228, 247). The nutritional environment of the infection court influences spore germination, mycelial growth, host penetration, and pathogenesis of *F. solani* f. sp. *phaseoli* (227, 228).

Root rot of bean is reduced with $NO_3$–N and increased with $NH_4$–N (45, 100, 107, 109, 134, 135, 228, 247). The reported relationship of the C:N ratio to root rot severity (208) was related to an effect of specific crop residues and C:N ratios on nitrification (109). Residues that stimulated the biological oxidation of $NH_4$–N to $NO_3$–N reduced the severity of root rot while those residues and chemicals that inhibited nitrification increased root rot when applied with $NH_4$–N (99, 100, 107–109, 134, 135). Root exudates stimulate *Fusarium* chlamydospores to germinate (187). However, Weinke (247) was unable to detect differences in hypocotyl exudation with different nitrogen sources even though root rot was increased much more with $NH_4$–N than urea or $NO_3$–N. By placing the nitrogen at various depths in the soil relative to the *Fusarium* inoculum and plant, Weinke (247) found that root rot increased significantly only when nitrogen was placed in the hypocotyl zone where the pathogen mainly invaded. Nitrogen applied to the root zone or to foliage that supplied adequate nitrogen for growth of bean plants did not result in increased disease. In this elaborate series of investigations, he also observed a more rapid development of a larger hyphal thallus, increased pathogenicity, and more rapid lesion coalescence with $NH_4$–N than $NO_3$–N. No pH changes were observed in either the sand or soil conditions employed, and nitrogen was the only nutrient tested that influenced bean root rot. Because neither $NO_3$–N nor $NH_4$–N influenced chlamydospore germination or affected the soil inoculum potential of *F. solani* f. *phaseoli,* the increase or decrease obtained with different forms of nitrogen could not be correlated with the inoculum potential in soil. Similar results have been reported by others (45, 99, 107, 109, 134, 135). Secondary infections resulting from growth and ramification of hyphae from the initial infection center take place abundantly with $NH_4$–N but not $NO_3$–N (45).

Populations of *Fusarium roseum* f. sp. *cerealis* "culmorum," *F. solani* f. sp. *pisi,* and saprophytic fusaria are drastically reduced by the fungicidal activity of $NH_3$ in the ammonia retention zone in field soils following application of anhydrous ammonia. However, the incidence and severity of *Fusarium* root rot and foot rot of winter wheat increased following fertilization with $NH_4$–N (204, 205) and diminished following $NO_3$–N fertilization (30, 78).

$NO_3$–N reduces *Rhizoctonia* root rot of bean compared with "slow-release" urea or $NH_4$–N fertilizers (52, 165). However, $NO_3$–N provides a better environment for saprophytic activity and survival of the pathogen (164). Thus, this pathogen may persist in soil and still cause no disease (184). $NO_3$–N also reduces infection of soybeans, sharp eyespot of wheat, and stem and stolon rot of potatoes by *R. solani* (38, 78, 107–109, 184). The severity of *R. solani* on potatoes is reduced in proportion to the rate of $NO_3$–N application, while $NH_4$–N markedly increases disease severity.

Simmonds (197) reported $NO_3$–N reduced *Helminthosporium sativum* root rot of wheat. Although chloropicrin soil fumigation inhibits nitrification it fails to reduce *Helminthosporium* root rot while favoring cereal-attacking species of *Fusarium* (28–30) and implies a form of nitrogen influence on foot and root rot severity. Butler's (30) conclusion that nitrogen fertilizers are generally ineffective in combating root rot caused by *H. sativum* failed to consider the effect of different forms of N.

Foot rot (eyespot) infection caused by *Cercosporella herpotrichoides* generally takes place during cool, moist periods in the early spring. Spring application of $NH_4$–N increases the incidence and severity of foot rot (78–80, 107–109) even though survival and germination of *Cercosporella* conidia is reduced by $NH_4$–N (33). Fall application of various nitrogen forms and rates, followed by nitrification, has no significant effect on foot rot. The increased foot rot encountered from spring fertilization in areas of high nitrogen loss can be avoided by stabilizing fall-applied $NH_4$–N with N-Serve (108). N-Serve prevents nitrogen loss, permits nitrification in the spring, and results in the lowest disease incidence (105, 108).

Interplanting red alder with conifers has been proposed for control of *Poria* and *Armillaria* root rots (126). Soil under a mixed stand of red alders and conifers contains markedly higher levels of $NO_3$–N than does soil under adjacent stands of pure conifers. Since *Poria weirii* and *Armillaria mellea* do not use $NO_3$–N but grow well with $NH_4$–N or amino–N, red alder has a potential for the biological control of these soil-borne pathogens by stimulating nitrification and increasing $NO_3$–N in the soil. Control was hypothesized to result from the inability of the pathogens to use $NO_3$–N for growth because of their lack of nitrate reductase (126) and because of the stimulation of antagonistic *Streptomyces* spp., which readily utilize $NO_3$–N.

Inconsistent results have been reported for the effects of nutrition on stalk rot of corn. Because several pathogens may cause stalk rot, it is important to designate the causal organism studied. *Diplodia zeae* stalk rot is common throughout the midwestern United States (1, 152), while *Fusarium moniliforme* is the primary cause of stalk rot in the irrigated western United States (162). Applications of potassium chloride, ammonium sulfate, or ammonium chloride decrease *Diplodia* stalk rot, while similar rates of potassium nitrate increase disease. The reduction in stalk rot with potassium chloride is due to the competitive inhibition of $NO_3$–N uptake by the chloride ion and is not dependent on potassium (152, 258). High rates of $NO_3$–N may overcome the competitive inhibition imposed by $Cl^-$. This explains why applications of $NO_3$–N without chloride increased stalk rot, while application of chloride without nitrogen had little effect on disease (1, 152). The omission of sulfate, potassium, or phosphrus had no effect on disease severity (152, 258). Resistant hybrids had low levels of stalk rot with either form of N, while intermediate and susceptible hybrids had more *Diplodia* stalk rot with $NO_3$–N and less with

$NH_4$–N (152). These findings indicate that resistance or susceptibility to stalk rot involves the nitrogen metabolism of the host.

In contrast to stalk rot caused by *Diplodia, F. moniliforme* stalk rot is reduced by $NO_3$–N. A marked increase in stalk rot of highly susceptible sweet corn lines occurred after fertilization with $NH_4$–N compared with $NO_3$–N (162). Disease severity was increased further by inhibiting nitrification with N-Serve or by late side-dressings of $NH_4$–N. Increasing rates of $NO_3$–N failed to increase stalk rot severity. Total nitrogen uptake was similar from all nitrogen sources; however, tissue nitrate levels were reduced 33% by inhibiting nitrification (162).

*Aphanomyces euteiches* causes a severe root rot of peas under cool, moist conditions or in poorly drained and compacted soil. This pathogen does not grow saprophytically in the soil but exists as oospores for prolonged periods (194). Infection is usually limited to cortical tissues macerated by pectolytic enzymes of the fungus (35). *Aphanomyces* root rot is suppressed by $NO_3$–N and increased by $NH_4$–N in nonsterile soil (35). Root rot resulted when $NH_4$–N was added to the soil but not when amounts adequate for plant growth were applied to foliage. Increased root rot from soil application of $NH_4$–N apparently results from an effect on the pathogen or soil microbial interactions, rather than increased host susceptibility which would also be affected by foliar absorption of $NH_4$–N. This lack of effect on host susceptibility was further demonstrated by dividing the root system so the same plant could be grown with different treatments. Again, only those roots exposed to $NH_4$–N developed severe root rot. *Aphanomyces* root rot was also more severe at lower inoculum levels when $NH_4$–N rather than $NO_3$–N was used. The occurrence of severe root rot with either form of N in sterilized soil, but suppression by $NO_3$–N in unsterilized soil, indicates that a competitive or antagonistic biological mechanism suppresses the activity of the pathogen in natural soils. This may be direct competition for nitrogen because *A. euteiches* utilizes $NO_3$–N poorly, or it may be a result of the increased population of antagonistic bacteria with $NO_3$–N (35, 126). Parasitism of corn and other crops by *A. euteiches* was also increased by the application of $NH_4$–N but not $NO_3$–N (35). Application of $NH_4$–N to corn increased the number of oospores in the soil 2.5 to 4.0 times compared with $NO_3$–N and no nitrogen applications, respectively. Thus, although peas are not generally fertilized with nitrogen, application of $NH_4$–N to corn may maintain a high level of *A. euteiches* inoculum in the soil between pea crops (35).

Susceptibility of citrus seedlings to *Phytophthora citrophthora* and *P. parasitica* was greater with ammonium sulfate or urea (47 and 59% roots infected, respectively), compared with fertilization with $NO_3$–N (3% infected roots) (119). Manure amendment increased the percentage of infected roots to 67 and 87% with $NH_4$–N and urea, compared with 23% with $NO_3$–N. In the absence of a high water table and organic amendment, only the $NH_4$–N fertilized trees had root rot. In all treatments, $NO_3$–N reduced root rot well below that on nonfertilized plants.

In the Pacific Northwest, there is a tendency to monocrop potatoes two or more years after clearing new land in order to recover development costs. The localized soil-sterilizing effect of heavy applications of anhydrous ammonia (204) used to obtain maximum yields in these desert soils, which are characteristically low in biological buffering capacity (107), and the accumulation of high residual $NO_3$–N are associated with a buildup of potato scab caused by *Streptomyces scabies.* $NO_3$–N increases scab and $NH_4$–N reduces it. Delaying nitrification of $NH_4$–N reduces scab more than does $NH_4$–N alone (40, 100, 107–109, 170, 242). Thus, the reduction of scab through the use of acid fertilizers may be the result of $NH_4$–N, which is stabilized by the low soil pH (107, 108). $NH_4$–N fertilization in the presence of high levels of residual $NO_3$–N was ineffective. However, planting potatoes after a previous cereal crop that removed the $NO_3$–N permitted control of scab with $NH_4$–N (108). Applying sulfur, calcium sulfate, potassium sulfate, or aluminum sulfate also failed to reduce scab in alkaline soils unless applied with $NH_4$–N.

Take-all of cereals, caused by *Ophiobolus graminis,* has been primarily associated with intensive wheat culture on low-fertility alkaline soils (30, 157). However, take-all has also been a serious problem following crops such as subterranean clover or alfalfa, which leave the soil at a high level of fertility and enhance nitrification (29, 30, 44, 106, 129, 130). Application of $NH_4$–N to field soils reduces take-all while $NO_3$–N may increase disease severity (73, 74, 78, 80, 106, 108, 183, 201–203). The effect of early rapid uptake of nitrogen by winter wheat on take-all is further indicated because $NH_4$–N (but not $NO_3$–N) applied in the fall or spring reduced take-all (78, 106, 108, 223) but increased the incidence of eyespot and lodging (78, 108, 183). Stabilization of $NH_4$–N with N–Serve provided more effective control of take-all than otherwise obtained (97, 201–203). Fall-applied $NH_4$–N which was rapidly nitrified increased take-all of irrigated spring wheat while spring-applied $NH_4$–N effectively controlled it (102, 108). Nitrogen can enhance or curtail survival of *Ophiobolus* (29, 30, 75), increase susceptibility or resistance of individual roots (74, 97, 203), and increase the host's ability to escape disease by producing new crown roots (74, 79, 157).

Although take-all is associated with alkaline soils, severe take-all infection also occurs in acid soils (157). *O. graminis* is capable of satisfactory growth over a wide pH range (3.2 to 9.6) so that adjustment of the soil reaction to provide an acid environment is unlikely to provide any appreciable control of take-all (30, 64). Thus, the reduction of take-all after acidification of the soil (78) may reflect a reduced rate of nitrification commonly encountered at acid pH levels so that mineralized nitrogen would persist longer as $NH_4$–N.

Genotype response to nitrogen may also be significant in evaluating take-all reactions. Glynne & Slope (80) reported high rates of nitrochalk reduced take-all of the cultivar Holdfast but not of Cappelle. We have selected lines for further study from the Purdue Small Grain Improvement Program with which take-all is reduced, increased, or not affected by nitrogen. Because cultivated field soils normally contain residual $NO_3$–N, disease reduction with $NH_4$–N may result from achieving a balance between the nitrogen forms (97). Control of take-all by $NH_4$–N under conditions of adequate nitrogen (223) supports this hypothesis but is in contrast to the frequently observed increase in take-all with $NO_3$–N in the field (73, 80, 106, 183, 201–203).

### *Vascular Wilts*

Wilt diseases caused by *Verticillium, Fusarium, Erwinia,* and *Corynebacterium* are influenced by the form of nitrogen. Infection and distribution of *Verticil-*

*lium* throughout the potato plant takes place early in the season, yet visible symptoms appear much later (87). $NH_4$–N delays plant maturity and inhibits *Verticillium* wilt while $NO_3$–N hastens maturity and increases disease severity (87, 100, 107, 108, 250). Stabilization of $NH_4$–N with N-Serve results in a further reduction in wilt and a corresponding increase in total yield and quality. Compounds that inhibit nitrification such as the nematicide 1,3-dichloropropene also improve *Verticillium* wilt control with $NH_4$–N (100–102, 107, 108). $NO_3$–N tends to nullify the beneficial effects of fumigation. $NH_4$–N enhanced, whereas $NO_3$–N nullified the reduction in *Verticillium* wilt of potato with the systemic insecticide Di-Syston® (101).

*Verticillium* wilt of hop (117), cotton (173), and tomato (241) increase with increasing levels of $NO_3$–N. Nitrogen-deficient hop plants generally failed to show symptoms even though *Verticillium* could be isolated from lower plant parts (117). On the other hand, *Fusarium* wilt of cotton (*F. oxysporum* f. sp. *vasinfectum*) (140), cabbage yellows (*F. oxysporum* f. sp. *conglutinans*), tomato wilt (*F. oxysporum* f. sp. *lycopersici*), and pea wilt (*F. oxysporum* f. sp. *pisi* race 1) decrease as concentration of $NO_3$–N increases (69, 240). Although the fusarial wilt pathogens are primarily xylem invaders and can use $NO_3$–N, increasing levels of $NO_3$–N generally reduce disease severity. The development and severity of *Fusarium* wilt of pea depend on temperature as well as on $NO_3$–N content (185). The different response at 21° versus 27°C may be related to nitrate reductase activity which is temperature sensitive (17).

Bacterial wilts are also influenced by forms of nitrogen. Corn seedlings deprived of nitrogen supported very poor growth of *Erwinia stewartii* (141). Because this pathogen is generally limited to the tracheal sap and is dependent on $NH_4$–N as an inorganic source of nitrogen, the suitability of sap as a nutrient substrate, rather than relative growth rate of the host, determines disease severity (141). Addition of $NH_4$–N to tracheal exudates low in nitrogen resulted in heavy growth of bacteria, while the addition of $NH_4NO_3$ supported heavy growth only at higher nitrogen levels. Increasing nitrogen levels in soil, which increases the nitrogen in tracheal sap, results in better growth of the pathogen and greater wilting (141). Potassium deficiency, which also results in a high nitrogen concentration in the sap, also increases disease severity (191).

Unlike *E. stewartii*, *Corynebacterium michiganense* (bacterial canker of tomato) is a phloem invader. Disease severity is increased by increasing levels of $NO_3$–N even though the bacterium requires an organic source of nitrogen (69). Ring rot of potato (*C. sepedonicum*) and bacterial wilt of tomato (*Pseudomonas solanacearum*) also increased as $NO_3$–N increased (69, 241), while high rates of urea reduced bacterial wilt of tobacco (*P. solanacearum*) (207).

### *Foliar Diseases*

Cereal rusts and mildews generally increase with $NO_3$–N and are reduced with $NH_4$–N. *Botrytis fabae* on broad bean and *Rhizoctonia solani* on bentgrass respond in an opposite manner. Daly (50) used resistant (Mindum), susceptible (Marquis), and mesothetic (Thatcher) wheat varieties in a study with race 56 of *P. graminis tritici* to evaluate the effect of various sources of nitrogen and temperature on rust development. The rust reaction of the resistant Mindum was not altered by nitrogen source or environment. Nonfertilized Thatcher plants and those supplied $NH_4$–N were resistant to rust, while plants supplied $NO_3$–N exhibited a mesothetic reaction which approached complete susceptibility. Application of $NH_4NO_3$ resulted in host reactions intermediate to those of the $NH_4$–N and $NO_3$–N treatments. At low temperatures, susceptible and mesothetic plants that received $NO_3$–N were moderately susceptible, while those receiving $NH_4$–N were completely resistant (50). Thus, temperature, nitrogen source, and the host/parasite combination all influenced rust reaction.

The more severe powdery mildew of wheat with increasing levels of $NO_3$–N was correlated with an increase in leaf area (124). However, leaves that had passed their normal susceptible stage and had become resistant to mildew were again susceptible when supplied more $NO_3$–N. Nitrogen-induced susceptibility was temporary in older leaves and appeared to be related to the onset of "internal nitrogen starvation." The percentage of conidial germination and appressorium formation on resistant and susceptible leaves did not differ, suggesting that nitrogen-induced susceptibility depends on physiological changes within the leaves rather than on a change in external environment (124).

Northern corn leaf blight, caused by *Helminthosporium turcicum,* was reduced on resistant but not on tolerant or susceptible hybrids by $NO_3$–N, while $NH_4$–N had no effect. This effect on the resistant hybrid was nullified if potassium chloride, which inhibits nitrate uptake, rather than potassium sulfate, was applied with the $NO_3$–N (152).

Rice blast (*Piricularia oryzae*) severity increased with increasing rates of $NH_4$–N applied to susceptible but not resistant varieties (161). Severity of blast was correlated with increasing amide levels in plant tissues (220) which were increased by applications of $NH_4$–N, low (20°C) night temperatures, or a combination of the two conditions. At night temperatures above 26°C, normally susceptible plants were resistant probably because of impaired amide synthesis from reduced nitrate reduction (248).

Anthracnose of cotton (*Colletotrichum gossypii*) was increased by ammonium sulfate but was not affected by $NO_3$–N or two other $NH_4$–N sources. Asparagine in the growth medium of resistant cotton seedlings, or low temperatures, which increase asparagine, induced complete susceptibility to anthracnose; however, asparagine-amended inoculum failed to increase disease over the unamended control (25). Tomato anthracnose (*Colletotrichum phomoides*) responded to nitrogen in a different manner. Increased virulence of *C. phomoides* after preconditioning on a $NO_3$–N instead of aspartate medium, rather than altered host resistance, appeared responsible for increased disease (252). Isolates transferred from aspartate to $NO_3$–N prior to inoculation were as virulent as isolates continuously cultured on $NO_3$–N.

Increased infection of broad bean by *B. fabae* was correlated with higher levels of sugars and amino acids in leaf exudates when roots were supplied $NH_4$–N as opposed to $NO_3$–N (23, 209). $NH_4$–N increased host-cell permeability and fungal spore germination, and increased infection 2.5 fold compared with $NO_3$–N.

Turf diseases caused by *Phialophora cinerescens* (156) and *Fusarium roseum* f. sp. *cerealis* "Culmorum" (*Fusarium* blight) (47) increased with high rates of $NO_3$–N, while dollar spot (*Sclerotinia homeocarpa*) (48, 59) and brown patch (*Rhizoctonia solani*) (24) decreased.

### *Nematode, Gall, and Canker Diseases*

The effect of nitrogen on the incidence and severity of nematode diseases appears directly related to nematode development rather than to host metabolism. $NH_4$–N compared with $NO_3$–N consistently depressed hatching, penetration, and cyst development of *Heterodera glycines* on soybean and thus reduced the damage caused by this nematode under field conditions (10, 125). The number of mature females and egg masses of *Meloidogyne incognita* per gram of bean root, and the root-gall index were much greater in plants receiving $NO_3$–N than in those receiving $NH_4$–N (160). Reduced invasion of eggplant roots by *Heterodera tabacum* after fertilizer or cellulose amendments was associated with high $NH_4$–N levels in soil and changes in soil acidity (144).

There are few studies on the effects of specific forms of nitrogen on other gall and canker-type diseases. However, the increased incidence of crown gall (*Agrobacterium tumefaciens*) in fumigated soil (6) and the reduction of peach canker (*Pseudomonas syringae*) following soil fumigation with materials inhibiting nitrification (54, 60) indicate a form of nitrogen effect.

### *Hyperparasitism*

Nitrogen source is an important factor in mycoparasitism of fungi. All evidence indicates that nitrogen sources affect the metabolism of the host, altering its susceptibility, rather than affecting the parasite directly (11). The mycoparasite *Calvarisporium parasiticum* depressed the growth of *Physalospora obtusa* more severely when the host was grown with $NO_3$–N than with casein hydrolysate. *Gonatobotryum fuscum* grew well on *Graphium* sp. supplied with $NH_4$–N but grew poorly when the host received urea or $NO_3$–N (195). In general, $NO_3$–N was also a poor nutrient for both *Piptocephilis virginiana* and its mucorales hosts, while $NH_4$–N supported moderate to good growth of the host but only poor growth of the parasite. This was reversed when *Syncephalastrum racemosum* was the host (21).

### *Virus-Like Diseases*

Much of the observed effect of nitrogen on virus infection or multiplication may be related to plant growth (13, 127). However, the intrinsic susceptibility of tobacco and potato plants to potato virus Y (PVY) is only slightly altered by differences in nutrition that have large effects on plant growth. Application of $NH_4$–N tended to increase the number of tobacco plants infected with PVY but reduced the number of infected potato plants. $NH_4$–N consistently reduced the concentration of tobacco mosaic virus (TMV) in sap of infected tobacco, the number of tomato aucuba mosaic virus (TAMV) lesions on tobacco, TMV lesions on *Nicotiana glutinosa*, and the concentration of potato virus X in potato. $NO_3$–N greatly increased the numbers of lesions caused by TAMV in both tobacco and *N. glutinosa* (13, 14).

## MECHANISMS OF CONTROL WITH NITROGEN FORMS

Disease control with a specific form of nitrogen may result from increased host resistance, altered virulence or growth of the pathogen, biological control through soil microfloral interactions, or a combination of these factors.

### *Altered Host Resistance*

Disease control as a result of modified host resistance may result from fewer penetrations (99, 106) or retarded pathogenesis after penetration (99, 100, 106). Evidence that nitrogen affected resistance of wheat to take-all was indicated by reduced infection and smaller lesions with $NH_4$–N compared with $NO_3$–N (97, 106, 203). Increased root growth permitting wheat to escape take-all has also been reported (30, 74, 97, 157). Neither $NH_4$–N nor $NO_3$–N prevents infection of peaches, prunes, and plums by *Xanthomonas pruni;* however, $NH_4$–N reduced bacterial canker of prunes by hastening periderm development so that cankers healed more promptly, while $NO_3$–N reduced defoliation of peaches and plums (140). Reduced disease from altered host resistance generally results from the influence of a specific form of nitrogen on metabolic pathways affecting growth, plant constituents, or exudates rather than a direct effect of the nitrogen per se.

EFFECT OF NITROGEN ON PLANT GROWTH The great majority of plants, fungi, and bacteria can utilize $NH_4$–N and $NO_3$–N as well as various organic forms of nitrogen, although some are better adapted to one form or another (243) (Table 2). In most well-aerated soils, $NO_3$–N predominates and plants adapted to such soils grow well with $NO_3$–N as the sole source of nitrogen. Crops that utilize $NH_4$–N grow well following fumigation, while crops requiring $NO_3$–N for optimum growth may be adversely affected (6, 53, 81, 254, 255).

**Table 2** "Preference" of plants for forms of nitrogen

| Plant | References | Plant | References |
|---|---|---|---|
| Preference for $NO_3$–N | | Preference for $NH_4$–N | |
| Bush bean | 81 | Bermuda grass | 147 |
| Celery | 254 | Conifer seedlings | 19 |
| Corn | 18 | Corn seedlings, oat seedlings | 243 |
| Cotton | 147 | Blueberry | 230, 231 |
| Grain sorghum | 147 | Mycorrhizas | 36 |
| Kale | 213, 214 | Orange trees | 115 |
| Onions | 254 | Rice | 111 |
| Pineapple | 139 | Ryegrass | 213, 214 |
| Potatoes | 139 | Sugar beets (pH 7.0) | 139 |
| Squash | 81 | Tea | 189 |
| Sugar beets (pH 5.0) | 139 | Wheat (seedlings) | 213, 214, 243 |
| Tobacco | 62 | Wheat (drought) | 213, 214 |
| Tomato | 81, 136, 256 | | |
| Wheat (older) | 213, 214 | | |

EFFECT OF NITROGEN ON PLANT CONSTITUENTS A great deal of research has compared the effects of $NH_4$–N and $NO_3$–N on plant metabolism (Table 3) (61, 66). Adsorption of nitrogen is more complicated than other essential elements because it is available both as a cation ($NH_4^+$) and an anion ($NO_3^-$) (139). Ammonium is rapidly assimilated, and it can immediately serve in the synthesis of amino acids or other compounds. This synthesis requires carbon skeletons that may deplete carbohydrates. As $NO_3$–N must be reduced before assimilation, organic acids may accumulate because the immediate demand for carbohydrate is lessened. Free $NO_3$–N may also accumulate in plant tissues (16). Age, plant species, root environment, and other factors are important in determining the form of nitrogen taken up (153). High pH, for example, favors the uptake of $NH_4$–N. Timing, intensity, and amount of rainfall also have a marked influence on the efficiency of nitrogen uptake and utilization from different sources (213, 214, 237).

Interactions of nitrogen with other nutrients are common. Potassium increases uptake of $NO_3$–N and promotes synthesis of organic nitrogen substances, while phosphate and chloride decrease $NO_3$–N uptake and increase $NH_4$–N uptake (258). Chloride also reduces amino acid and especially protein synthesis and promotes protein decomposition (233). $NO_3$ reduction requires molybdenum, magnesium, ferrous ion, and ascorbic acid. Mn is required for assimilation of $NO_3$–N by wheat roots as well as for protein synthesis (46, 139).

Many plant constituents altered by specific forms of nitrogen have been correlated with resistance or susceptibility to disease (198, 217, 235). Altered amino acid

**Table 3** Effect of nitrogen forms on plant constituents and exudation

| Constituent | $NH_4$-N | $NO_3$-N | References |
|---|---|---|---|
| Total nitrogen | equal | equal | 155, 162 |
| Total nitrogen | lower | higher | 159 |
| $NO_3$-N | lower | higher | 146, 155, 159, 213, 214 |
| Soluble carbohydrate | lower | higher | 139, 159 |
| Inorganic cations | lower | higher | 118, 139, 202 |
| Organic acids | lower | higher | 16, 118, 139 |
| NADH-NADPH nitrate reductase | lower | higher | 192 |
| Respiratory Quotient (*RQ*) | no change | higher | 16, 139, 257 |
| Photosynthate transport | inhibited | stimulated | 139 |
| Bleeding of cut stems | inhibited | increased | 139 |
| Amide-N (esp. asparagine) | higher | lower | 16, 66, 138, 159, 257 |
| Amino-nitrogen | higher | lower | 66, 139, 159 |
| Total carbohydrate | higher | lower | 155 |
| Phosphorus | higher | lower | 112, 176 |
| Soluble organic nitrogen | higher | lower | 138, 255 |
| Protein nitrogen | higher | lower | 20, 159 |
| $O_2$ requirement | higher | lower | 16, 139 |
| Leaf exudation | higher | lower | 23, 209 |
| Root exudation | higher | lower | 23 |

metabolism by specific forms of nitrogen is consistent with reported metabolic changes after infection, and the inhibition or enhancement of disease with specific amino acids (5, 49, 55, 63, 86, 98, 121, 169, 177, 190, 198, 199, 217, 218, 241). Total N, carbohydrate:nitrogen ratios (55, 67), silicon absorption (238), and sensitivity to toxins (182) have also been related to disease severity. Additional information in these areas would greatly facilitate the use of nitrogen for disease control and place our understanding on a firmer basis.

PLANT EXUDATES Nitrogen forms affect the composition of root and leaf exudates and bring about an effect similar to that on host constituents (Table 3) (23, 88, 180, 186–188). Environmental conditions, host nutrition, and genotype influence the release and availability of root and leaf exudates and in turn affect pathogenesis (23, 93). $NH_4$–N increased permeability and exudation of broad bean leaf surfaces resulting in higher levels of sugars and amino acids. *Botrytis cinerea* infected such leaves more frequently than leaves supplied $NO_3$–N (23, 209).

SOIL AND RHIZOSPHERE pH Root absorption of $NH_4$–N reduces the rhizosphere pH, while absorption of $NO_3$–N increases it (176, 201–203). The effect of nitrogen on rhizosphere pH ($pH_r$) is relatively short lived and is influenced by soil type (176, 201, 203). The difference between $NH_4$–N and $NO_3$–N treatments was as large as 1.9 units at an initial soil pH of 5.2, but it was only 0.2 units at a soil pH of 7.8 (176). The early depression of $pH_r$ by $NH_4$–N was temporary, and within 3 weeks the $pH_r$ was the same as the untreated controls (203). Reduced $pH_r$ has been recently proposed as the mechanism controlling take-all of wheat and other diseases (201–203). However, the $pH_r$ differences obtained with various nitrogen forms are insignificant relative to the level of control of take-all achieved with $NH_4$–N compared with $NO_3$–N. $NH_4$–N consistently reduced take-all, while disease severity was generally increased with $NO_3$–N regardless of pH or fumigation. Thus, the enhanced reduction in take-all at lower pH levels with $NH_4$–N may have resulted from inhibited nitrification of mineralized soil nitrogen providing a more persistent $NH_4$–N regime. $NH_4$–N also restricts infection and favors patch healing of turf grass infected by *Ophiobolus* even after liming acid turf (206).

Bulk soil pH has little relationship to the severity of other diseases influenced by forms of nitrogen (35, 203, 247), and rhizosphere pH has not generally been evaluated. Thus, a pH range of 4.0–9.0 failed to affect *Sclerotinia* dollar spot of turf grass which is influenced by levels of $NO_3$–N (48). Similar results are reported for bean root rot (247), *Aphanomyces* root rot of peas (35), and *Verticillium* wilt of various crops (87, 250). Application of acid fertilizers has been a standard recommendation for control of potato scab caused by *Streptomyces scabies* (202). This treatment may inhibit nitrification resulting in increased levels of $NH_4$–N which reduces scab severity (208).

### *Effects on the Pathogen*

Survival, germination, growth, or virulence of a pathogen may be influenced by the form of nitrogen (Table 4). Several workers have proposed using the fungicidal activity of ammonia for disease control (77, 204, 205). Propagules of *Fusarium roseum* f. sp. *cerealis* "culmorum" and *F. oxysporum* f. sp. *pisi* were rapidly destroyed under laboratory conditions and markedly reduced in the field after high rates of $NH_4$–N injection, yet foot rot of winter wheat was generally increased (205). Saprophytic survival of *Ophiobolus graminis* is reduced in low-fertility soils and promoted by high nitrogen fertility, especially high $NH_4$–N (29, 72). Butler proposed increased survival (29, 30) as an explanation for increased take-all after alfalfa (44, 106, 215) or ley farming (29, 30, 129). However, this proposal fails to explain why $NH_4$–N fertilization reduces the severity of take-all (30, 73, 78–80, 106, 108).

Survival is probably not the mechanism involved in control of many soil-borne pathogens because reduced disease frequently occurs without a reduction in population of the pathogen (99, 149, 247). Thus, presence of the pathogen may have little relationship to disease incidence or severity when the form of nitrogen is suitable for control, especially if the propagule is dormant or subjected to fungistasis (104, 128). Nutrients that overcome fungistasis increase germination of pathogen propagules and increase disease potential in the presence of a susceptible host. More spores of *Fusarium solani* f. sp. *phaseoli, F. oxysporum* f. sp. *pisi, Aspergillus niger,* and *Phytophthora cinnamomi* germinate with $NH_4$–N than with $NO_3$–N (32, 34, 45, 83, 84, 145, 167, 227). Increased spore germination of *F. solani* f. sp. *phaseoli* by $NH_4$–N may explain why bean root rot is more severe with $NH_4$–N, especially because each step in the pathogenesis of *F. solani* f. sp. *phaseoli* is sustained largely by nutrients surrounding the host and pathogen. It is only after parasitism is firmly established that this fungus derives most of its nutrition from within the host (227).

**Table 4** Inorganic nitrogen "preference" of various fungi

| Organism | References | Organism | References |
|---|---|---|---|
| Preference for $NH_4$–N | | Preference for $NO_3$–N | |
| *Aphanomyces euteiches* | 35 | *F. oxysporum* f. sp. *cubense* | 94 |
| *Armillaria mellea* | 126, 244 | *F. oxysporum* f. sp. *pisi* race 1 | 69 |
| *Aspergillus flavus* | 167 | *F. solani* f. sp. *cucurbitae* | 95 |
| *F. solani* f. sp. *phaseoli* | 33, 34, 45, 83, 84, 227, 247 | *Monilinia fructicola* | 89 |
| | | *Pythium aphanidermatum* | 85 |
| *Ophiobolus graminis* | 72 | *Rhizoctonia solani* | 226 |
| *Pythium aphanidermatum* | 120 | *Streptomyces* (*Actinomycetes*) | 239 |
| *Rhizoctonia solani* | 246 | *Verticillium albo-atrum* | 41 |
| *Rhizopus nigricans* | 178 | *Verticillium dahliae* | 41 |
| *Sclerotinia trifoliorum* | 253 | No Preference | |
| *Ustilago sphaerogena* | 212 | *Botrytis cinerea* | 229 |
| *Verticillium albo-atrum* | 131 | *Curvularia* | 182 |
| Preference for $NO_3$–N | | *F. oxysporum* f. sp. *lycopersici* | 69 |
| *Colletotrichum cocodes* | 122 | *Pythium ultimum* | 120 |
| *Colletotrichum falcatum* | 122 | *Sclerotinia gladiolis* | 242 |
| *Curvularia pallescens* | 182 | *Sclerotinia sclerotiorum* | 253 |
| *F. oxysporum* f. sp. *conglutinans* | 69 | | |

The inability of pathogens to use a specific form of nitrogen that reduces severity of diseases they cause has been found for *Aphanomyces* root rot of peas (35), *Rhizopus* rot of sweetpotato (178), *Poria* and *Armillaria* root rots of pine (126), *Streptomyces* scab of potato (246), Stewart's wilt of corn (57) , and *Fusarium* root rot of bean (247). This relationship does not include the *Fusarium* wilts, which are less severe with $NO_3$–N, even though the pathogens use this form of nitrogen (69, 94).

A form of nitrogen may affect virulence of a pathogen without affecting growth or germination, especially with facultative saprophytes. Thus, inhibition of enzyme synthesis or enzymatic activity, either directly or indirectly, could account for reduction in the severity of many diseases characterized by maceration of host tissues. Inhibition of pectolytic and cellulolytic enzymes by a specific form of nitrogen has been related to reduced *Rhizopus* fruit rot (211), cellulose decomposition in soil (234, 188), susceptibility of cotton to *Rhizoctonia solani* (244, 245), and *Verticillium* wilt of cotton (168). The specific crop, pathogen, and environmental conditions may influence disease control through ion inhibition of enzyme activity (211). Enzyme inhibition may result from a host response (99) or from microbial interactions (96, 107) which would prevent the production or activity of these enzymes in the infection court. Such mechanisms would not necessarily reduce the population of the pathogen while reducing disease severity.

The form of N may affect virulence in other ways than by affecting extracellular enzyme production. The thallus of *F. solani* f. sp. *phaseoli* was larger, developed earlier and was more extensive with $NH_4$–N than with $NO_3$–N (247). As long as virulent isolates of *Verticillium* had an adequate supply of $NO_3$–N in their own media, their effect on their hosts was relatively independent of the nitrogen supplied the host. Less virulent isolates were pathogenic only when their culture medium had a high $NO_3$–N content or when their hosts had an excessive supply of organic nitrogen (110). Increased pathogenicity of virulent isolates of *O. graminis* receiving $NO_3$–N more than nullified the $NO_3$–N-stimulated root production by the host so that take-all was most severe with $NO_3$–N (249).

The form of nitrogen has other effects on the physiology of pathogens that may or may not be correlated with virulence or increased disease severity. Thus, endogenous respiration of *Verticillium albo-atrum* conidia was stimulated by $NH_4$–N but not by $NO_3$–N. The content of glutamic acid, glycine, and alanine in weakly pathogenic isolates of *F. solani* f. sp. *phaseoli* was much greater than in highly pathogenic isolates when all isolates were grown on a $NO_3$ medium (137). Addition of glycine to *Agrobacterium tumefaciens* media increased the pathogenicity and virulence of this bacterium. Virulent isolates became avirulent in the absence of glycine (181). *Colletotrichum phomoides* was more virulent on tomato fruits when grown on a nitrate medium than on an aspartate medium. Virulence of isolates was restored when transferred from the aspartate to nitrate medium (252).

Effects such as the stimulation of perithecial production of *Fusarium solani* f. sp. *cucurbitae* by $NO_3$–N but not $NH_4$–N (95) and the reduction of saprophytic activity of *Rhizoctonia solani* with $NH_4$–N (164) may also influence survival and disease potential of these pathogens.

## *Microbial Interactions*

The soil is a common medium for many interacting microorganisms. Consequently, the effect of nutrients in the soil environment on a single organism is difficult to interpret without considering the likely indirect effects on other organisms (64). $NH_4$–N and $NO_3$–N had differential effects on *Fusarium* root rot of bean, *Aphanomyces* root rot of peas, take-all of cereals, and *Rhizoctonia* root rot of cotton only when in the zone occupied by the pathogen (35, 203, 245, 247). Disease severity was similar with either form of nitrogen in sterile soil, which suggests that disease control results from microbial interactions in the soil.

Suppression of pea root rot in natural soil by $NO_3$–N was correlated with increased numbers of bacteria and actinomycetes antagonistic to *A. euteiches* which were not observed with $NH_4$–N (35). A similar stimulation of antagonists (*Streptomyces*) under a $NO_3$–N regime has been postulated in the control of *Poria* and *Armillaria* root rot of pine (126) and for control of take-all of wheat with $NH_4$–N at pH levels above 5.0 (201–203). No differences in the kinds or numbers of microorganisms in *Ophiobolus*-infested or noninfested soils have been demonstrated (219), and antagonists could be isolated with equal frequency from either soil. Hypothesized inhibitory levels of rhizosphere $CO_2$ from increased microbial respiration as the mechanism controlling take-all of cereals (71, 72) was not confirmed by Louw (129, 130), who found rhizosphere respiration highest when legumes were included in cereal rotations which also resulted in the most severe disease.

Simultaneous inoculation of wheat seedlings with *Ophiobolus graminis* and *Didymella exitiales* (196) reduced the severity of take-all by direct mycoparasitic activity of *D. exitiales* in the rhizosphere, as well as from reduced pathogenicity of *O. graminis.* Excretion of aspartic acid, glutamic acid, alanine, leucine, valine, and two other ninhydrin-positive substances by *D. exitiales* reduced the pathogenicity but had no effect on runner hyphae of *O. graminis.* Vitamin $B_2$ (riboflavin) increased growth of runner hyphae of *O. graminis* but not mycelium and was largely responsible for the severity of take-all (65).

Foliar application of urea to peas increased root exudate, increased the number of Actinomycetes, decreased the population of *Fusarium oxysporum* f. sp. *pisi* in the rhizosphere of sprayed plants, and reduced wilt (179). Rhizosphere antagonists also appear responsible for reduced *Rhizoctonia* root rot of bean (8, 22, 163), *Phymatotrichum* root rot of watermelon (27), *Phytophthora* root rot of tea (189), and *Fusarium* root rot of bean (45, 163). However, increased populations of antagonists in the rhizosphere have not always been correlated with reduced disease (96, 99–103, 163, 259), and it is difficult to see how this mechanism could be functional without a reduction in pathogen numbers except through an effect on pathogenicity.

Specific bacterial-fungal associations have been correlated with control of bean root rot without a corresponding reduction in pathogen populations in soil (99, 103, 107). *F. solani* f. sp. *phaseoli* was not pathogenic when specific bacteria were present in the hyphosphere, but pathogenicity was restored in the absence of the bacteria. This hyphosphere association was enhanced by $NO_3$–N, conditions favoring nitrification, and specific crop rotations that reduced root rot but not by $NH_4$–N.

Competition between nonpathogenic *Fusarium roseum* and pathogenic *F. solani* f. sp. *phaseoli* was proposed as a mechanism for reducing bean root rot. Germination of *F. solani* conidia was reduced with $NO_3$–N, while *F. roseum* utilized either form of N (31, 32, 34, 134, 135). A change from the virulent conidial to the less virulent mycelial form of *F. solani* f. sp. *phaseoli* has been proposed to account for reducing bean root rot with specific crop rotations (132). Disease control may also be achieved through bacterial-stimulated chlamydospore formation and induced dormancy of *F. solani* f. sp. *phaseoli* (236).

### *Conclusions*

It is not surprising that the form of nitrogen plays an important role in plant disease. Of the soil-borne elements, nitrogen is required by the plant in the greatest quantity and it is an essential component of proteins. Assimilation of nitrogen, however, is more complicated than other essential elements, because it is assimilated both as a cation ($NH_4^+$) and as an anion ($NO_3^-$). $NH_4$–N is assimilated and rapidly converted to amino acids or other compounds. Water, pH, other elements, and temperature influence the uptake and utilization of specific forms of nitrogen by plants, pathogens, or interacting microorganisms.

Various factors influence the effect a specific form of nitrogen will have on disease. Because no one form of nitrogen controls all diseases or favors disease control on any group of plants, each disease must be considered individually. The disease control achieved with a specific form of nitrogen may also depend on host response or preference, previous cropping, nitrogen rate and stability, residual nitrogen, time of application, soil microflora present, ratio of $NH_4$–N to $NO_3$–N, or the disease complex present.

Because crops are fertilized to obtain maximum productivity and quality, the effects of nitrogen fertilizers on disease become an important consideration. The recent commercial availability of materials that stabilize the form of nitrogen make it possible to both control disease and improve nutrition of the plant. Many of the reported observations of "nonfungicidal, chemical control" of plant disease after herbicide, nematocide, or systemic insecticide applications may involve nitrogen transformations in soil or altered nitrogen constituents in host plants (101). The increased efficiency or activity of nitrate reductase in pea and rye seedlings after Simizine® (2-chloro-4,6-bisthylamino-*s*-triazine) applications (175) should affect diseases influenced by tissue $NO_3$–N levels.

Although disease control by a form of nitrogen is probably indirect, its practical significance through easily modified agricultural practices is worthy of serious consideration. Maximum disease control with a form of nitrogen will require reexamination of previous research and reevaluation of nitrogen as related to disease. It is hoped this review will stimulate this required research.

*Literature Cited*


1. Abney, T. S. 1967. *Influence of nutrition on stalk rot development of Zea mays L.* PhD thesis. Iowa State Univ., Ames. 99 pp.
2. Afanasiev, M. M., Carlson, W. E. 1942. The relation of phosphorus and nitrogen ratio to the amount of seedling diseases of sugar beets. *Ann. Am. Soc. Sugar Beet Tech.*, pp. 1–5
3. Allen, R. M. 1962. Dry root rot of citrus induced by ammonium excesses. *Phytopathology* 52:721 (Abstr.)
4. Allison, F. E. 1966. The fate of nitrogen applied to soils. *Advan. Agron.* 18: 219–58
5. Alten, F. 1940. Der Einfluss der Ernahrung auf den Stickstoffhaushalt der Pflanzen. Forschungsdienst Sonderheft 12. *Fortschr. Landwirt. Chem. Forsch.* 1939/40:45–59
6. Altman, J. 1970. Increased and decreased plant growth responses resulting from soil fumigation. In *Root diseases and Soil-borne Pathogens,* ed. T. A. Toussoun, R. V. Bega, P. E. Nelson, 216–21. Berkeley: Univ. Calif. Press. 252 pp.
7. Anderson, O. E., Boswell, F. C., Stacy, S. V. 1965. Effect of temperature on nitrification in Georgia soils. *Ga. Agr. Exptl. Sta. Bull.* 130
8. Baker, R., Martinson, C. A. 1970. Epidemiology of diseases caused by *Rhizoctonia solani. Rhizoctonia solani: Biology and Pathology,* ed. J. A. Parmeter, 172–88. Berkeley: Univ. Calif. Press. 255 pp.
9. Barber, D. A. 1968. Microorganisms and the inorganic nutrition of higher plants. *Ann. Rev. Plant Physiol.* 19: 71–88
10. Barker, K. R., Lehman, P. S., Huisingh, D. 1971. Influence of nitrogen and *Rhizobium japonicum* on the activity of *Heterodera glycines. Nematologia* 17: 377–85
11. Barnett, H. L. 1963. The nature of mycoparasitism by fungi. *Ann. Rev. Microbiol.* 17:1–14
12. Bartholomew, W. V., Clark, F. E. 1965. *Soil Nitrogen. Agron. Monogr. 10.* Madison, Wis.: Am. Soc. Agron. 615 pp.
13. Bawden, F. C., Kassanis, B. 1950. Some effects of host nutrition on the susceptibility of plants to infection by certain viruses. *Ann. Appl. Biol.* 37:46–57
14. Bawden, F. C., Kassanis, B. 1950. Some effects of host plant nutrition on the multiplication of viruses. *Ann. Appl. Biol.* 37:215–28
15. Beaumont, A. B. 1936. A hypothesis to explain brown root rot of Havana seed tobacco. *Science* 84:182–83
16. Beevers, H. 1961. *Respiratory Metabolism in Plants.* Evanston, Ill.: Row, Peterson. 232 pp.
17. Beevers, L., Hageman, R. H. 1969. Nitrate reduction in higher plants. *Ann. Rev. Plant Physiol.* 20:495–522
18. Bennett, W. A., Pesek, J., Hamway, J. J. 1964. Effect of nitrate and ammonium on growth of corn in nutrient solution sand culture. *Agron. J.* 56:342–45
19. Benzian, B. 1970. Nutrition of young conifers and soil fumigation. See Ref. 6, pp. 222–25
20. Berlier, Y., Guiraud, G. 1967. Absorption et utilisation par des graminees de l'azote nitrique on ammonical marque a l'azote –15. *Proc. Symp. Isotop. Plant Nutr. Physiol.,* pp. 155–57. Vienna, Austria: Int. At. Energy Agency
21. Berry, C. R. 1959. Factors affecting parasitism of *Piptocephilis virginiana* on other mucorales. *Mycologia* 51:824–32
22. Blair, I. D. 1943. Behavior of the fungus *Rhizoctonia solani* Kuhn in the soil. *Ann. Appl. Biol.* 30:118–27
23. Blakeman, J. P. 1971. The chemical environment of the leaf surface in relation to growth of pathogenic fungi. In *Ecology of Leaf Surface Micro-organisms,* ed. T. F. Preece, C. A. Dickenson, 255–68. New York: Academic. 640 pp.
24. Bloom, J. R., Couch, H. B. 1958. Influence of pH, nutrition, and soil moisture on the development of large brown patch. *Phytopathology* 48:260 (Abstr.)
25. Bollenbacher, K., Fulton, N. D. 1971. Effects of nitrogen compounds on resistance of *Gossypium arboreum* seedlings to *Colletotrichum gossypii. Phytopathology* 61:1394–95
26. Brandt, G. H., Wolcott, A. R., Erickson, A. E. 1964. Nitrogen transformations in the soil as related to structure, moisture, and oxygen diffusion rate. *Proc. Soil Sci. Soc. Am.* 28:71–75
27. Brown, J. G. 1933. Watermelon susceptible to Texas root rot. *Science* 78:509
28. Bruehl, G. W. 1952. Temporary sterilization of soil in the evaluation of root and crown rot losses with wheat and barley under field conditions. *Plant Dis. Reptr.* 36:234–37
29. Butler, F. C. 1959. Saprophytic behavior of some cereal root rot fungi. IV. Saprophytic survival in soils of high and low fertility. *Ann. Appl. Biol.* 47:28–36

30. Butler, F. C. 1961. Root and foot rot diseases of wheat. *Sci. Bull.* 77. Wagga Wagga, NSW, Australia: Agr. Res. Inst. 98 pp.
31. Byther, R. S. 1964. *Inorganic nitrogen utilization as a factor of competitive saprophytic ability.* MS thesis. Colorado State Univ., Fort Collins. 57 pp.
32. Byther, R. S. 1965. Ecology of plant pathogens in soils. V: Inorganic nitrogen utilization as a factor of competitive saprophytic ability of *Fusarium roseum* and *F. solani. Phytopathology* 55: 852–58
33. Byther, R. S. 1968. *Etiological studies on foot rot of wheat caused by Cercosporella herpotrichoides.* PhD thesis. Oregon State Univ., Corvalis. 134 pp.
34. Byther, R. S., Baker, R. 1964. Utilization of inorganic nitrogen by *Fusarium solani,* f. *phaseoli,* and *F. roseum* f. sp. *cerealis. Phytopathology* 54:889 (Abstr.)
35. Carley, H. E. 1969. *Factors affecting the epidemiology of pea* (*Pisum sativum* L.) *root rot caused by Aphanomyces euteiches* Drechs. PhD thesis. Univ. Minn., St. Paul. 120 pp.
36. Carrodus, B. B. 1966. Absorption of nitrogen by mycorrhizal roots of beech. I. Factors affecting the assimilation of nitrogen. *New Phytol.* 65:358–71
37. Carrodus, B. B. 1966. Absorption of nitrogen by mycorrhizal roots of beech. II. Ammonium and nitrate as sources of nitrogen. *New Phytol.* 66:1–4
38. Castano, J. J., Kernkamp, M. F. 1956. The influence of certain plant nutrients on infection of soybeans by *Rhizoctonia solani. Phytopathology* 46:326–28
39. Chandra, P., Bollen, W. B. 1960. Effect of wheat straw, nitrogenous fertilizers, and carbon-to-nitrogen ratio on organic decomposition in a subhumid soil. *J. Agr. Food Chem.* 8:19–24
40. Chase, F. E., Corke, C. T., Robinson, J. B. 1968. Nitrifying bacteria in soil. In *The Ecology of Soil Bacteria,* ed. T. R. G. Gray, D. Parkinson, 593–611. Toronto: Univ. Toronto Press. 681 pp.
41. Christie, T. 1966. Studies on the nitrogen metabolism of species of *Verticillium.* I: Amino acid nutrition of isolates of *Verticillium* pathogenic to tobacco in hops. *N. Z. J. Agr. Res.* 9:995–98
42. Clark, F. E. 1942. Experiments toward the control of the take-all disease of wheat and the *Phymatotrichum* root rot of cotton. *US Dep. Agr. Tech. Bull.* 835. 27 pp.
43. Clark, F. E., Paul, E. A. 1970. The microflora of grassland. *Advan. Agron.* 22:375–435
44. Cook, R. J., Huber, D. M., Powelson, R. L., Bruehl, G. W. 1968. Occurrence of take-all in wheat in the Pacific Northwest. *Plant Dis. Reptr.* 52:716–18
45. Cook, R. J., Schroth, M. N. 1965. Carbon and nitrogen containing compounds and germination of chlamydospores of *Fusarium solani* f. *phaseoli. Phytopathology* 55:254–56
46. Cordlus, R. L. 1938. The use of rapid chemical plant nutrient tests in fertilizer deficiency diagnosis and vegetable crop research. *Va. Truck Exp. Sta. Bull.* 89:1531–56
47. Couch, H. B., Bedford, E. R. 1966. *Fusarium* blight of turfgrasses. *Phytopathology* 56:781–86
48. Couch, H. B., Bloom, J. R. 1960. Influence of environment on diseases of turfgrasses. II: Effect of nutrition, pH, and soil moisture on *Sclerotinia* dollar spot. *Phytopathology* 50:761–63
49. Covey, R. P. Jr. 1971. The effect of methionine on the development of apple powdery mildew. *Phytopathology* 61: 346
50. Daly, J. M. 1949. The influence of nitrogen source on the development of stem rust of wheat. *Phytopathology* 39: 386–94
51. Das, A. C., Western, J. H. 1959. The effect of inorganic manures, moisture and inoculum on the incidence of root disease caused by *Rhizoctonia solani* Kuhn in cultivated soil. *Ann. Appl. Biol.* 47:37–48
52. Davey, C. B., Papavizas, G. C. 1960. Effect of dry mature plant materials and nitrogen on *Rhizoctonia solani* in soil. *Phytopathology* 50:522–25
53. Davidson, J. H., Thiegs, B. J. 1966. The effect of soil fumigation on nitrogen nutrition and crop response. *Down Earth* 22:7–12
54. Davis, J. R., English, H. 1969. Factors related to the development of bacterial canker in peach. *Phytopathology* 59: 588–95
55. Dickson, J. G. 1923. Influence of soil temperature and moisture on the development of the seedling-blight of wheat and corn caused by *Gibberella saubinetii. J. Agr. Res.* 23:837–70
56. Dodman, R. L. 1970. Factors affecting the prepenetration phase of infection by *Rhizoctonia solani.* See Ref. 6, pp. 116–21
57. dos Santos de Avevedo, N. F. 1963. Nitrogen utilization by four isolates of *Armillaria mellea. Trans. Brit. Mycol. Soc.* 46:281–84

58. Easton, G. D. 1964. The results of fumigating *Verticillium* and *Rhizoctonia* infested potato soils in Washington. *Am. Potato J.* 41:296
59. Endo, R. M. 1966. Control of dollar spot of turfgrass by nitrogen and its probable bases. *Phytopathology* 56:877 (Abstr.)
60. English, H., DeVay, J. E. 1964. Influence of soil fumigation on growth and canker resistance of young fruit trees in California. *Down Earth* 20:6–8
61. Epstein, E. 1972. *Mineral Nutrition of Plants; Principles and Perspectives.* New York: Wiley. 412 pp.
62. Evans, H. J., Weeks, M. E. 1947. The influence of nitrogen, potassium, and magnesium salts on the composition of burley tobacco. *Proc. Soil Sci. Soc. Am.* 12:315–22
63. Farkas, G. L., Kiraly, Z. 1961. Amide metabolism in wheat leaves infected with stem rust. *Physiol. Plant.* 14:344–53
64. Fellows, H. 1941. Effect of certain environmental conditions on the prevalence of *Ophiobolus graminis* in the soil. *J. Agr. Res.* 63:715–26
65. Fluck, V. 1955. Untersuchungen uber die Pathogenitat von Erregergemischen bei Getreidefusskrankheiten. *Phytopathol. Z.* 23:177–208
66. Fowden, L. 1965. Origins of amino acids. In *Plant Biochemistry,* ed. J. Bonner, J. E. Varner, 361–90. New York: Academic. 1054 pp.
67. Freeman, T. E. 1964. Influence of nitrogen on severity of *Piricularia grisea* infection of St. Augustine grass. *Phytopathology* 54:1187–89
68. Frederick, L. R. 1956. The formation of nitrate from ammonium nitrogen in soils. I. Effect of temperature. *Proc. Soil Sci. Soc. Am.* 20:496–500
69. Gallegly, M. E. Jr., Walker, J. C. 1949. Plant nutrition in relation to disease development. V: Bacterial wilt of tomato. *Am. J. Bot.* 36:613–23
70. Garretson, A. L., San Clemente, C. L. 1968. Inhibition of nitrifying chemolithotrophic bacteria by several insecticides. *J. Econ. Entomol.* 61:285–88
71. Garrett, S. D. 1934. Factors affecting the severity of take-all. *J. Agr. S. Aust.* 37:976–83
72. Garrett, S. D. 1938. Soil conditions and the take-all disease of wheat. III: Decomposition of the resting mycelium of *Ophibolus graminis* in infected wheat stubble buried in the soil. *Ann. Appl. Biol.* 25:742–66
73. Garrett, S. D. 1941. Soil conditions and the take-all disease of wheat. VI: The effect of plant nutrition upon disease resistance. *Ann. Appl. Biol.* 28:14–18
74. Garrett, S. D. 1948. Soil conditions and the take-all disease of wheat. IX. Interactions between host plant nutrition, disease escape, and disease resistance. *Ann. Appl. Biol.* 35:14–17
75. Garrett, S. D. 1967. The effect of nitrogen level on survival of *Ophiobolus graminis* in pure culture on cellulose. *Brit. Mycol. Soc.* 50:519–24
76. Gasser, J. K. R., Peachey, J. E. 1964. A note on the effects of some soil sterilants on the mineralization and nitrification of soil-nitrogen. *J. Sci. Agr.* 15:142–46
77. Gilpatrick, J. D. 1969. The role of ammonia in the control of avocado root rot with alfalfa meal soil amendment. *Phytopathology* 59:973–78
78. Glynne, M. D. 1951. Effects of cultural treatments on wheat and on the incidence of eyespot, lodging, take-all and weeds. *Ann. Appl. Biol.* 38:665–88
79. Glynne, M. D. 1953. Wheat yield and soil-borne diseases. *Ann. Appl. Biol.* 40:221–25
80. Glynne, M. D., Slope, D. B. 1959. Effects of previous wheat crops, seed rate and nitrogen on eyespot, take-all, weeds and yields of two varieties of winter wheat. *Ann. Appl. Biol.* 47:187–99
81. Good, J. M., Carter, R. L. 1965. Nitrification lag following soil fumigation. *Phytopathology* 55:1147–50
82. Goring, C. A. I. 1962. Control of nitrification by 2-chloro-6-(trichloromethyl) pyridine. *Soil Sci.* 93:211–18
83. Griffin, G. J. 1970. Carbon and nitrogen requirements for macroconidial germination of *Fusarium solani:* dependence on conidial density. *Can. J. Microbiol.* 16:733–40
84. Griffin, G. J. 1970. Exogenous carbon and nitrogen requirements for chlamydospore germination by *Fusarium solani:* dependence on spore density. *Can. J. Microbiol.* 16:1366–68
85. Grover, R. K., Sidher, J. S. 1965. Effect of nitrogen sources on the growth of *Pythium aphanidermatum* (Edson) Fitz. *Ann. Mycol.* 19:231–37
86. Guinn, G., Brinkerhoff, L. A. 1970. Effect of root aeration on amino acid levels in cotton plants. *Crop Sci.* 10:175–78
87. Guthrie, J. W. 1960. Early dying (*Verticillium* wilt) of potatoes in Idaho. *Idaho Agr. Exp. Sta. Res. Bull.* 45. 24 pp.

88. Hale, M. G., Foy, C. L., Shay, F. J. 1971. Factors affecting root exudation. *Advan. Agron.* 23:89–109
89. Hall, R. 1967. Carbon and nitrogen nutrition of *Monilinia fructicola. Aust. J. Biol. Sci.* 20:471–74
90. Hamaker, J. W., Kerlinger, H. O. 1967. Vapor pressure of pesticides. *Advan. Chem.* 86:39–54
91. Harmsen, G. W., Van Schreven, D. A. 1955. Mineralization of organic nitrogen in soil. *Advan. Agron.* 7:299–398
92. Hauck, R. D., Stephenson, H. F. 1965. Nitrification of nitrogen fertilizers: (1) Effect of N source, size and pH of the granule, and concentration. *J. Agr. Food Chem.* 13:486–92
93. Hayman, D. S. 1970. The influence of cotton seed exudate on seedling infection by *Rhizoctonia solani.* See Ref. 6, pp. 99–102
94. Hendrix, F. F. Jr., Toussoun, T. A. 1964. The influence of nutrition on sporulation of the banana wilt and bean root rot *Fusaria* on agar media. *Phytopathology* 54:389–92
95. Hix, S. M., Baker, R. 1964. Physiology of sexual reproduction in *Hypomyces solani* f. *cucurbitae.* I: Influence of carbon and nitrogen. *Phytopathology* 54: 584–86
96. Hooker, W. J. 1956. Survival of *Streptomyces scabies* in peat soil planted with various crops. *Phytopathology* 46: 677–81
97. Hornby, D., Goring, C. A. I. 1972. Effects of ammonium and nitrate nutrition on take-all disease of wheat in pots. *Ann. Appl. Biol.* 70:225–31
98. Howell, R. K., Krusberg, L. R. 1966. Changes in concentrations of nitrogen and free and bound amino acids in alfalfa and pea infected by *Ditylenchus dipsaci. Phytopathology* 56:1170–77
99. Huber, D. M. 1963. *Investigations on Root Rot of Beans.* PhD thesis. Michigan State Univ., E. Lansing. 97 pp.
100. Huber, D. M. 1966. How nitrogen affects soil-borne diseases. *Crops Soils* 18:10–11
101. Huber, D. M. 1967. Non-fungicidal, chemical control of soil-borne diseases. *Proc. 18th Ann. Fert. Conf. Pac. N. W.* pp. 1–6
102. Huber, D. M. 1972. Spring versus fall nitrogen fertilization and take-all of spring wheat. *Phytopathology* 62: 434–36
103. Huber, D. M., Andersen, A. L., Finley, A. M. 1966. Mechanisms of biological control in a bean root rot soil. *Phytopathology* 56:953–56
104. Huber, D. M., McKay, H. C. 1968. Effect of temperature, crop and depth of burial on the survival of *Typhula idahoensis* sclerotia. *Phytopathology* 58: 961–62
105. Huber, D. M., Murray, G. A., Crane, J. M. 1969. Inhibition of nitrification as a deterrent to nitrogen loss. *Proc. Soil Sci. Soc. Am.* 33:975–76
106. Huber, D. M., Painter, C. G., McKay, H. C., Peterson, D. L. 1968. Effects of nitrogen fertilization on take-all of winter wheat. *Phytopathology* 58:1470–72
107. Huber, D. M., Watson, R. D. 1970. Effect of organic amendment on soil-borne plant pathogens. *Phytopathology* 60:22—26
108. Huber, D. M., Watson, R. D. 1972. Nitrogen form and plant disease. *Down Earth* 27:14–15
109. Huber, D. M., Watson, R. D., Steiner, G. W. 1965. Crop residues, nitrogen and plant disease. *Soil Sci.* 100:302–8
110. Isaac, I. 1957. The effects of nitrogen supply upon the *Verticillium* wilt of *Antirrhinum. Ann. Appl. Biol.* 45:512–15
111. Iyatomi, K., Nishiaawa, T. 1970. Growth response of rice to soil fumigation. See Ref. 6, pp. 226–28
112. Jauert, R., Ansorge, H., Gorlitz, H. 1968. Influence of N-SERVE on nitrification, plant growth, and phosphorus absorption. *Thaer-Archives* 12:487–97
113. Johnson, D. D., Guenzi, W. D. 1963. Influence of salts on ammonium oxidation and carbon dioxide evolution from soil. *Proc. Soil Sci. Soc. Am.* 27:663–66
114. Jordan, H. V., Nelson, H. A., Adams, J. E. 1939. Relation of fertilizers, crop residues and tillage to yields of cotton and incidence of root rot. *Proc. Soil Sci. Soc. Am.* 4:325–28
115. Kadoya, K. 1966. Effects of different types of nitrogen fertilizers on growth and nutrient status of young Satsuma orange trees in sand culture. *Mem. Ehime Univ. Sect. 6,* 11:321–33
116. Keeney, D. R., Gardner, W. R. 1970. The dynamics of nitrogen transformations in the soil. *Symp. Global Eff. Environ. Pollut.,* ed. S. F. Singer, 96–103. New York: Springer-Verlag
117. Keyworth, W. G., Hewitt, E. J. 1948. *Verticillium* wilt of the hop (*Humulus lupulus*). V: The influence of nutrition on the reaction of the hop plant to infection with *Verticillium albo-atrum. J. Hort. Sci.* 24:219–27
118. Kirby, E. A., Mengel, K. 1967. Ionic balance in different tissues of the tomato plant in relation to nitrate, urea, or am-

monium nutrition. *Plant Physiol.* 42: 6–14
119. Klotz, L. J., DeWolfe, T. A., Wong, P. 1958. Decay of fibrous roots of citrus. *Phytopathology* 48:616–22
120. Kraft, J. M., Erwin, D. C. 1967. Effects of nitrogen sources on growth of *Pythium aphanidermatum* and *Pythium ultimum. Phytopathology* 57:374–76
121. Kuc, J., Barnes, E., Daftsios, A., Williams, E. B. 1959. The effect of amino acids on susceptibility of apple varieties to scab. *Phytopathology* 49:313–15
122. Kurtz, E. B., Fergus, C. L. 1964. Nitrogen nutrition of *Collectotrichum coccodes. Phytopathology* 54:691–92
123. Lapwood, D. H., Dyson, P. W. 1966. An effect of nitrogen on the formation of potato tubers and the incidence of common scab (*Streptomyces scabies*). *Plant Pathol.* 15:9–14
124. Last, F. T. 1953. Some effects of temperature and nitrogen supply on wheat powdery mildew. *Ann. Appl. Biol.* 40:312–22
125. Lehman, P. S., Barker, K. R., Huisingh, D. 1971. Effects of pH and inorganic ions on emergence of *Heterodera glycines Nematologia* 17:467–73
126. Li, C. Y., Lu, K. C., Trappe, J. M., Bollen, W. B. 1967. Selective nitrogen assimulation by *Poria weirii. Nature* 213:814
127. Ling, K. C. 1964. Effect of host nitrogen on the infective property of tobacco mosaic virus. *Diss. Abstr.* 25:2157–58
128. Lockwood, J. 1964. Soil fungistasis. *Ann. Rev. Phytopathol.* 2:341–62
129. Louw, H. A. 1957. Microbiological analysis of a Western Cape Province grain soil under various crop rotations. *S. Afr. Dep. Agr. Sci. Bull.* 378. 36 pp.
130. Louw, H. A. 1957. The effect of various crop rotations on the incidence of take-all (*Ophiobolus graminis* Sacc.) in wheat. *S. Afr. Dep. Agr. Sci. Bull.* 379. 12 pp.
131. Malca, I., Erwin, D. C., Moje, W., Jones, B. 1966. Effect of pH and carbon and nitrogen sources on the growth of *Verticillium albo-atrum. Phytopathology* 56:401–6
132. Maloy, O. C. Jr. 1959. Microbial associations in the *Fusarium* root rot of beans. *Diss. Abstr.* 19:2441–42
133. Martinson, C. A. 1959. *Inoculum potential studies of Rhizoctonia solani.* MS thesis. Colorado State Univ., Fort Collins. 132 pp.
134. Maurer, C. L. 1962. *Effect of carbon substrates and carbon to nitrogen ratio on bean root rot.* MS thesis. Colorado State Univ., Fort Collins. 50 pp.
135. Maurer, C. L., Baker, R. 1965. Ecology of Plant Pathogens in Soil. II: Influence of glucose, cellulose, and inorganic nitrogen amendments on development of bean root rot. *Phytopathology* 55:69–72
136. Maynard, D. N., Barker, A. V., Lachman, W. H. 1966. Ammonium-induced stem and leaf lesions of tomato plants. *Proc. Am. Soc. Hort. Sci.* 88:516–20
137. McAnelly, C. W. 1959. Free amino acids in the cultural media and cells of *Fusarium solani* f. *phaseoli. Phytopathology* 49:734–37
138. McKee, H. S. 1950. Studies on the nitrogen metabolism of the barley plant (*Hordeum sativum*). *Aust. J. Sci. Res. Ser. B* 3:474–86
139. McKee, H. S. 1962. *Nitrogen Metabolism in Plants.* Oxford: Clarendon
140. McNew, G. L. 1953. The effects of soil fertility. In *Plant Diseases,* 100–14. Washington DC: US Dep. Agr. Yearb. 973 pp.
141. McNew, G. L., Spencer, E. L. 1939. Effect of nitrogen supply of sweet corn on the wilt bacterium. *Phytopathology* 29:1051–67
142. Menzies, J. D. 1970. Factors affecting plant pathogen populations in soil. See Ref. 6, pp. 16–21
143. Messenger, A. S., Whiteside, E. P., Wolcott, A. R. 1972. Climate, time and organisms in relation to podzol development in Michigan sands. I: Site descriptions and microbiological observations. *Proc. Soil Sci. Soc. Am.* 36:633–38
144. Miller, P. M., Wihrheim, S. 1966. Invasion of roots by *Heterodera tabacum* reduced by cellulosic amendments or fertilizers in soil. *Phytopathology* 56:890 (Abstr.)
145. Mircetich, S. M., Zentmyer, G. A. 1970. Germination of chlamydospores of *Phytophthora.* See Ref. 6, pp. 112–15
146. Moore, F. D. III. 1973. N-serve nutrient stabilizer . . . a nitrogen management tool for leafy vegetables. *Down Earth* 28(4):4–8
147. Morris, H. D., Giddens, J. 1963. Response of several crops to ammonium and nitrate forms of nitrogen as influenced by soil fumigation and liming. *Agron. J.* 55:372–74
148. Munro, P. E. 1966. Inhibition of nitrifyers by grass root extracts. *J. Appl. Ecol.* 3:231–38
149. Nash, S. M., Snyder, W. C. 1962. Quantitative estimations by plate counts of propagules of the bean root rot

*Fusarium* in field soils. *Phytopathology* 52:567–72
150. Neal, D. C. 1935. Further studies on the effect of ammonium nitrogen on growth of the cotton root-rot fungus *Phymatotrichum omnivorum,* in field and laboratory experiments. *Phytopathology* 25: 967 (Abstr.)
151. Neal, J. L. Jr. 1968. Inhibition of nitrifying bacteria by grass and forb root extracts. *Can. J. Microbiol.* 15:633–35
152. Nelson, D. W. 1963. *The relationship between soil fertility and the incidence of Diplodia stalk rot and northern leaf blight in Zea mays.* MS thesis, Univ. Illinois, Urbana. 56 pp.
153. Nelson, L. B., Hauck, R. D. 1965. Nitrogen fertilizers: progress and problems. *Agr. Sci. Rev.* 3:38–47
154. Newton, J. D., Wyatt, F. A., Ignatieff, V., Ward, A. S. 1939. Nitrification under and after alfalfa, brome, timothy, and western rye grass. II: Soil microbiological activity. *Can. J. Bot.* 17:256–93
155. Nicholson, J. W. G., MacLeod, L. B. 1966. Effect of form of nitrogen fertilizer, a preservative, and a supplement on the value of high moisture grass silage. *Can. J. Anim. Sci.* 46(2):71–82
156. Nilsson, G. I., Nelson, P. V. 1964. Nitrogen nutrition and development of *Phialophora cinerescens* in carnation. *Phytopathology* 54:1172–73
157. Nilsson, H. E. 1969. Studies of root and foot rot diseases of cereals and grasses. *Ann. Agr. Coll. Swed.* 35:275–807
158. Norris, M. G. 1972. N-serve nitrogen stabilizers—a practical approach to better fertilizer nitrogen management. *Down Earth* 28:5–9
159. Nowakowski, T. Z., Cunningham, R. K., Nielsen, K. F. 1965. Nitrogen fractions and soluble carbohydrates in Italian ryegrass. I: Effects of soil temperature, form and level of nitrogen. *J. Sci. Food Agr.* 16(3):124–34
160. Oteifa, B. A. 1955. Nitrogen source of the host nutrition in relation to infection by a root-knot nematode, *Meloidogyne incognita. Plant Dis. Reptr.* 39: 902–3
161. Padmanabhan, S. Y. 1953. Effect of nitrogenous fertilization in the incidence of "blast" on rice varieties. *Curr. Sci.* 22:271–72
162. Painter, C. G., Simpson, W. R. 1969. Fertilizing sweet corn for seed production. *Idaho Agr. Exp. Sta. Res. Bull.* 501. 12 pp.
163. Papavizas, G. C. 1963. Microbial antagonism in bean rhizosphere as affected by oat straw and supplemental nitrogen. *Phytopathology* 53:1430–35
164. Papavizas, G. C. 1969. Survival of root-infecting fungi in soil. XI: Survival of *Rhizoctonia solani* as affected by inoculum concentration and various soil amendments. *Phytopathol. Z.* 64: 101–11
165. Papavizas, G. C., Davey, C. B. 1960. *Rhizoctonia* disease of bean as affected by decomposing green plant materials and associated microflora. *Phytopathology* 50:516–22
166. Parker, J. H. 1972. How fertilizer moves and reacts in soil. *Crops and Soils* 25:7–11
167. Pass, T., Griffin, G. J. 1972. Exogenous carbon and nitrogen requirement for conidial germination by *Aspergillus flavus. Can. J. Microbiol.* 18:1453–61
168. Patil, S. S., Dimond, A. E. 1967. Induction and repression of polygalaturonase synthesis in *Verticillium albo-atrum. Phytopathology* 57:825 (Abstr.)
169. Piening, L. J. 1972. Effects of leaf rust on nitrate in rye. *Can. J. Plant Sci.* 52:842–43
170. Potter, H. S., Norris, M. G., Lyons, C. E. 1971. Potato scab control studies in Michigan using N-Serve nitrogen stabilizer for nitrification inhibition. *Down Earth* 27(3):23–24
171. Prasad, R., Rajale, G. B., Lakhdive, B. A. 1971. Nitrification retarders and slow-release nitrogen fertilizers. *Advan. Agron.* 23:337–83
172. Rader, L. F. Jr., White, L. M., Whittaker, C. W. 1943. The salt index—a measure of the effect of fertilizers on the concentration of the soil solution. *Soil Sci.* 55:201–18
173. Ranney, C. D. 1962. Effects of nitrogen source and rate on the development of *Verticillium* wilt of cotton. *Phytopathology* 52:38–41
174. Rice, E. L. 1965. Inhibition of nitrogen fixing and nitrifying bacteria by seed plants. III: Comparison of three species of *Euphorbia. Proc. Okla. Acad. Sci.* 45:43–44
175. Ries, S. K., Chmiel, H., Dilley, D. R., Filner, P. 1967. The increase in nitrate reductase activity and protein content of plants treated with simazine. *Proc. Nat. Acad. Sci. USA* 58:526–32
176. Riley, D., Barber, S. A. 1971. Effect of ammonium and nitrate fertilization on phosphorus uptake as related to root-induced pH changes at the root soil interface. *Proc. Soil Sci. Soc. Am.* 35: 301–6

177. Rishbeth, J. 1970. The role of the basidiospores in stump infection by *Armillaria mellea.* See Ref. 6, pp. 141–46
178. Robbins, W. J. 1937. The assimilation by plants of various forms of nitrogen. *Am. J. Bot.* 24:243–50
179. Rothamsted Experimental Station. 1963. Effect of urea on pea wilt. *Rothamsted Exptl. Sta. Rept.,* 116–17
180. Rovira, A. D. 1965. Plant root exudates and their influence upon soil microorganisms. In *Ecology of Soil-borne Plant Pathogens,* ed. K. F. Baker, W. C. Snyder, 170–86. Berkeley: Univ. Calif. Press. 517 pp.
181. Rubio-Huertos, M., Beltra, R. 1962. Formas L. fijas del *Agrobacterium tumefaciens* obtenidas por medio de glicocola. *Microbiol. Espan.* 15:219–30
182. Sadasivan, T. S. 1965. Effect of mineral nutrients on soil microorganisms and plant disease. See Ref. 180, pp. 460–70
183. Salt, G. A. 1957. Effects of nitrogen applied at different dates, and of other cultural treatments on eyespot, take-all and yield of winter wheat (field experiment, 1953). *J. Agr. Sci.* 48:326–35
184. Sanford, G. B. 1947. Effect of various soil supplements on the virulence and persistence of *Rhizoctonia solani. Sci. Agr.* 27:533–44
185. Schroeder, W. T., Walker, J. C. 1942. Influence of controlled environment and nutrition on the resistance of garden peas to *Fusarium* wilt. *J. Agr. Res.* 65:221–48
186. Schroth, M. N., Hildebrand, D. C. 1964. Influence of plant exudates on root-infecting fungi. *Ann. Rev. Phytopathol.* 2:101–32
187. Schroth, M. N., Snyder, W. C. 1961. Effect of host exudates on chlamydospore germination of the bean root rot fungus, *Fusarium solani* f. *phaseoli. Phytopathology* 51:389–93
188. Schroth, M. N., Weinhold, A. R., Hayman, D. S. 1966. The effect of temperature on quantitative differences in exudates from germinating seeds of bean, pea, and cotton. *Can. J. Bot.* 44:1429–32
189. Shanmuganathan, N. 1970. Studies on the parasitism and control of tea root disease fungi in Ceylon. See Ref. 6, pp. 188–90
190. Shaw, M. 1963. The physiology and host parasite relations of the rusts. *Ann. Rev. Phytopathol.* 1:259–94
191. Shear, G. M., Wingard, S. A. 1944. Some ways by which nutrition may affect severity of disease in plants. *Phytopathology* 34:603–5
192. Shen, T. C. 1972. Nitrate reductase of rice seedlings and its induction by organic nitro-compounds. *Plant Physiol.* 49:546–49
193. Shephard, M. C., Wood, R. K. S. 1963. The effect of environment and nutrition of pathogen and host, in the damping-off of seedlings by *Rhizoctonia solani. Ann. Appl. Biol.* 51:389–402
194. Sherwood, R. T., Hagedorn, D. J. 1962. Studies on the biology of *Aphanomyces euteiches. Phytopathology* 52:150–54
195. Shigo, A. L. 1960. Parasitism of *Gonatobotryum fuscum* on species of *Ceratocystis. Mycologia* 52:583–98
196. Siegle, V. H. 1961. Uber mischinfektionen mit *Ophiobolus graminis* und *Didymella exitialis. Phytopathol. Z.* 42:305–48
197. Simmonds, P. M. 1960. Infection of wheat with *Helminthosporium sativum* in relation to the nitrogen content of the plant tissues. *Can. J. Plant Sci.* 40: 139–45
198. Singh, D., Brinkerhoff, L. A., Guinn, G. 1971. Effect of alanine on development of *Verticillium* wilt in cotton cultivars with different levels of resistance. *Phytopathology* 61:881–82
199. Singh, T. K. S., Subramanian, S. 1966. Nitrate reductase of rice (*Oryza sativa* L.) under foot rot disease. *Planta* 71 (2):125–29
200. Sitterly, W. R. 1962. Calcium nitrate for field control of tomato southern blight in South Carolina. *Plant Dis. Reptr.* 46:492–94
201. Smiley, R. W. 1972. *Relationship between rhizosphere pH changes induced by root absorption of ammonium versus nitrate-nitrogen and root diseases, with particular reference to take-all of wheat.* PhD thesis. Washington State Univ., Pullman. 78 pp.
202. Smiley, R. W. 1974. Forms of nitrogen and the pH in the root zone, and their importance to root infections. In *Biology and Control of Soil-borne Plant Pathogens,* ed. G. W. Bruehl. St. Paul, Minn.: Am. Phytopathol. Soc. In press
203. Smiley, R. W., Cook, R. J. 1973. Relationship between take-all of wheat and rhizosphere pH in soils fertilized with ammonium vs. nitrate-nitrogen. *Phytopathology* 63:882–89
204. Smiley, R. W., Cook, R. J., Pappendick, R. I. 1970. Anhydrous ammonia as a soil fungicide against *Fusarium* and fungicidal activity in the ammonia retention zone. *Phytopathology* 60: 1227–32

205. Smiley, R. W., Cook, R. J., Pappendick, R. I. 1972. *Fusarium* foot rot of wheat and peas as influenced by soil applications of anhydrous ammonia and ammonia potassium azide solutions. *Phytopathology* 62:86–91
206. Smith, J. D. 1956. Fungi and turf diseases. *Ophiobolus* patch disease. *J. Sports Turf Res. Inst.* 9:180–202
207. Smith, T. E. 1944. Control of bacterial wilt (*Bacterium solanacearum*) of tobacco as influenced by crop rotation and chemical treatment of the soil. *US Dep. Agr. Circ.* 692. 16 pp.
208. Snyder, W. C., Schroth, M. N., Christou, T. 1959. Effect of plant residues on root rot of bean. *Phytopathology* 49:755–56
209. Sol, H. H. 1967. *Meded. Landbouwhogesch. Opzoekingssta. Staat Gent.* 32:768–75 (Data cited by Blackman in Ref. 28).
210. Sommers, L. E., Biederbeck, V. O. 1973. Tillage management principles: soil microorganisms. *1973 Conserv. Tillage Conf., Des Moines, Iowa.* Soil Conserv. Soc. Am. pp. 87–108. 241 pp.
211. Spalding, D. H. 1969. Toxic effect of macerating action of extracts of sweet potatoes rotted by *Rhizopus stolonifer* and its inhibition by ions. *Phytopathology* 59:685–92
212. Spoerl, E., Sarachek, A., Smith, S. B. 1957. The effect of amino acids upon cell division in *Ustilago. Am. J. Bot.* 44:252–58
213. Spratt, E. D., Gasser, J. K. R. 1970. The effect of ammonium sulphate treated with a nitrification inhibitor, and calcium nitrate, on growth and N-uptake of spring wheat, ryegrass and kale. *J. Agr. Sci.* 74:111–17
214. Spratt, E. D., Gasser, J. K. R. 1970. Effect of ammonium and nitrate forms of nitrogen and restricted water supply on growth and nitrogen uptake of wheat. *Can. J. Soil Sci.* 50:263–73
215. Sprague, R. 1931. The distribution of cereal foot rots in the Pacific Northwest. *Northwest Sci.* 5:10–12
216. Stiven, G. 1952. Production of antibiotic substances by the roots of a grass [*Trachypogon plumosus* (H. B. K.) Nees] and of *Pentanisia variabilis* (E. Mey). Harv. (*Rubiaceae*). *Nature* 170:712
217. Stretch, A. W., Cappellini, R. A. 1965. Changes in free amino acids and reducing sugars in highbush blueberry fruit infected with *Glomerella cingulata. Phytopathology* 55:302–3
218. Strobel, G. A., Sharp, E. L. 1965. Proteins of wheat associated with infection type of *Puccinia striiformis. Phytopathology* 55:413–14
219. Stumbo, C. R., Gainey, T. L., Clark, F. E. 1942. Microbiological and nutritional factors in the take-all disease of wheat. *J. Agr. Res.* 64:653–65
220. Suryanarayanan, S. 1958. Role of nitrogen in host susceptibility to *Piricularia oryzae* cav. *Curr. Sci.* 27:447–48
221. Swanback, T. R., Anderson, P. J. 1947. Fertilizing connecticut tobacco. *Conn. Agr. Exp. Sta. New Haven Bull.* 503. 52 pp.
222. Swezey, A. W., Turner, G. O. 1962. Crop experiments on the effect of 2-chloro-6-(trichloromethyl)pyridine for the control of nitrification of ammonium and urea fertilizers. *Agron. J.* 54:532–35
223. Syme, J. R. 1966. Fertilizer and varietal effects on take-all in irrigated wheat. *Aust. J. Exp. Agr. Anim. Husb.* 6: 246–49
224. Thompson, L. M., Troeh, F. R. 1973. *Soil and Soil Fertility.* New York: McGraw. 495 pp.
225. Tisdale, S. L., Nelson, W. L. 1966. *Soil Fertility and Fertilizers.* New York: Macmillan. 2nd ed. 430 pp.
226. Tolmsoff, W. J. 1970. Metabolism of *Rhizoctonia solani.* See Ref. 8, pp. 93–107
227. Toussoun, T. A. 1970. Nutrition and pathogenesis of *Fusarium solani* f. sp. *phaseoli.* See Ref. 6, pp. 95–98
228. Toussoun, T. A., Nash, S. M., Snyder, W. C. 1960. The effect of nitrogen sources and glucose on the pathogenesis of *Fusarium solani* f. *phaseoli. Phytopathology* 50:137–40
229. Townsend, B. B. 1957. Nutritional factors influencing the production of sclerotia by certain fungi. *Ann. Bot.* (NS) 21:153–66
230. Townsend, L. R. 1966. Effect of nitrate and ammonium nitrogen on the growth of the lowbush blueberry. *Can. J. Plant Sci.* 46:209–10
231. Townsend, L. R. 1967. Effect of ammonium nitrogen and nitrate nitrogen, separately and in combination, on the growth of the highbush blueberry. *Can. J. Plant Sci.* 47:555–62
232. Turner, G. O., Goring, C. A. I. 1966. N-Serve—A status report. *Down Earth* 22(2):19–25
233. Udovenko, G. V., Min'ko, J. F. 1966. The nature of the influence of potassium and chlorine upon the nitrogen metabo-

lism of plants. *Fiziol. Rast.* 13(2): 214–37

234. Ulehlova, B. 1966. Wie verschiedene quellen des stickstoffes bei dem abbau der Zellulose in uterschiedlichen Bodentypen verwertet werden. *Preslia* 38(3): 308–11
235. van Andel, O. M. 1966. Amino acids and plant diseases. *Ann. Rev. Phytopathol.* 4:349–68
236. Venkata Ram, C. S. 1952. Soil bacteria and chlamydospore formation in *Fusarium solani. Nature* 170:889
237. Volk, G. N. 1966. Efficiency of fertilizer urea as affected by method of application, soil moisture, and lime. *Agron. J.* 58:249–52
238. Volk, R. J., Kahn, R. P., Weintraub, R. L. 1958. Silicon content of the rice plant as a factor influencing its resistance to infection by the blast fungus, *Piricularia oryzae. Phytopathology* 48: 179–84
239. Waksman, S. A. 1919. Cultural studies of species of *Actinomycetes. Soil Sci.* 8:71–216
240. Walker, J. C. 1946. Soil management and plant nutrition in relation to disease development. *Soil Sci.* 61:47–54
241. Walker, J. C., Gallegly, M. E., Bloom, J. R., Shepherd, R. D. 1954. Relation of plant nutrition to disease development. VIII: *Verticillium* wilt of tomato. *Am. J. Bot.* 41:760–62
242. Watson, R. D. 1966. Influence of nitrogen on potato scab in Idaho soils. *Phytopathology* 56:152 (Abstr.)
243. Webster, G. C. 1959. *Nitrogen Metabolism in Plants.* Evanston, Ill.: Row, Peterson. 152 pp.
244. Weinhold, A. R., Bowman, T., Dodman, R. L. 1969. Virulence of *Rhizoctonia solani* as affected by nutrition of the pathogen. *Phytopathology* 59: 1601–5
245. Weinhold, A. R., Dodman, R. L., Bowman, T. 1972. Influence of exogenous nutrition on virulence of *Rhizoctonia solani. Phytopathology* 62:278–81
246. Weinhold, A. R., Garraway, M. O. 1966. Nitrogen and carbon nutrition of *Armillaria mellea* in relation to growth-promoting effects of ethanol. *Phytopathology* 56:108–12
247. Weinke, K. E. 1962. *Influence of nitrogen on the root disease of bean caused by Fusarium solani f. phaseoli.* PhD thesis. Univ. Calif., Berkeley. 150 pp.
248. Went, F. W. 1957. *The Experimental Control of Plant Growth.* Waltham, Mass.: Chronica Botanica. 343 pp.
249. Weste, G., Thrower, L. B. 1971. The effect of added nitrate on the growth of *Ophiobolus graminis. Plant Soil* 35: 161–72
250. Wilhelm, S. 1950. The inoculum potential of *Verticillium albo-atrum* as affected by soil amendments. *Phytopathology* 40:970–74
251. Wilhelm, S. 1951. Is *Verticillium albo-atrum* a soil invader or a soil inhabitant. *Phytopathology* 41:944–45
252. Williams, F. J. 1965. Antecedent nitrogen sources affecting virulence of *Colletotrichum phomoides. Phytopathology* 55:333–35
253. Willis, C. B. 1968. Effect of various nitrogen sources on growth of *Sclerotinia. Can. J. Microbiol.* 14:1035–37
254. Wolcott, A. R., Maciak, F., Shepherd, L. N., Lucas, R. E. 1960. Effects of telone on nitrogen transformation and on growth of celery in organic soil. *Down Earth* 16:10–14
255. Wolcott, A. R., Liao, F. H., Kirkwood, J. I. 1967. Effects of fumigation, temperature, and level of nitrate on microbial numbers, $CO_2$ production and N-transformation in organic soil. *Soil Sci.* 103:131–38
256. Woolhouse, H. W., Hardwick, K. 1966. The growth of tomato seedlings in relation to the form of the nitrogen supply. *New Phytol.* 65.518–25
257. Yemm, E. W., Willis, A. J. 1956. The respiration of barley plants. IX: The metabolism of roots during the assimilation of nitrogen. *New Phytol.* 55:229–52
258. Younts, S. E., Musgrove, R. B. 1958. Chemical composition, nutrient absorption and stalk rot incidence of corn as affected by chloride in potassium fertilizer. *Agron. J.* 50:426–29
259. Zyngas, Z. P. 1963. The effect of plant nutrients and antagonistic microorganisms on the damping-off of cotton seedlings caused by *Rhizoctonia solani* Kuhn. *Diss. Abstr.* 23:3587

# CYTOPLASMIC SUSCEPTIBILITY IN PLANT DISEASE

❖3594

*A. L. Hooker*
Department of Plant Pathology, University of Illinois, Urbana, Illinois 61801

## INTRODUCTION

Biologists now generally accept the existence of extrachromosomal inheritance. In recent years cytoplasmic genes have been discovered and their role in cell heredity studied (38).

Several important traits are inherited through the cytoplasm. The United States corn blight epidemic of 1970 dramatically revealed the importance of plant cytoplasm in susceptibility to disease (19, 33, 48).

While the importance of extrachromosal inheritance (cytoplasmic genes) relative to chromosomal inheritance (nuclear genes) is not completely understood, far greater advances have been made in the knowledge of nuclear gene inheritance and its application to plant and animal improvement than has that of cytoplasm genes. The same can be said for reaction to disease in plants (21).

A brief discussion of what is known about the plant cytoplasm, cytoplasmic inheritance, and cytoplasmic male sterility is given as background information to aid the reader in understanding how the cytoplasm may play a role in the inheritance of disease susceptibility. This is followed by a review of selected reports on disease reaction and a discussion of possible mechanisms involved in cytoplasmic susceptibility in plant disease.

## THE CYTOPLASM

An idealized higher plant cell is composed of living protoplasm delimited by a plasma membrane and surrounded by a rigid cell wall. Within the plasma membrane are found the cell nucleus, the endoplasmic reticulum, ribosomes, vacuoles, plastids or proplastids, mitochondria, dictyosomes, and other membrane-enclosed organelles. Enzymes, soluble ribonucleic acid, and other macromolecules are contained in the aqueous medium of the cytoplasm. The living protoplast minus the nucleus is traditionally considered as the cytoplasm.

The cytoplasm constituents differ in composition and function. The plasma membrane is of tripartite lipoprotein structure. It is permeable to water and some molecules in solution but impermeable to others. The endoplasmic reticulum is continuous with the outer membrane surrounding the nucleus and subdivides the cytoplasm into irregular cavities. Ribosomes are both associated with the endoplasmic reticulum and free in the liquid phase of the cytoplasm. They associate with mRNA and play an integral part in protein synthesis. Among the plastids, the chloroplasts are perhaps the most noteworthy in leaves and other green tissue. The chloroplasts contain chlorophyll and are responsible for the production of carbohydrates during photosynthesis. Mitochondria are involved in respiration and perform the important function of generating adenosine triphosphate. Both chloroplasts and mitochondria are self-duplicating, have physical continuity from one cell generation to the next, and contain DNA (38).

## CYTOPLASMIC INHERITANCE

Several concepts can be advanced as to what constitutes inheritance. In its broadest aspects the transmission of a trait, through sexual or vegetative propagation or both, from parents to descendants constitutes inheritance. In a more restricted sense, inheritance is limited to the transmission of genetic material, not foreign to the organism, from parents to offspring. In higher plants the genetic material is DNA.

Higher plant development is controlled by genetic factors in the nucleus, the cytoplasm, and by the interaction of the two sets of factors. Although inheritance usually involves nuclear genes, several traits are mainly under the controlling influence of the cytoplasm.

### *Tests of Cytoplasmic Inheritance*

Several experimental tests are available to detect cytoplasmic inheritance. Some of these are breeding or genetic tests whereas others employ techniques of molecular biology.

The traditional implication of inheritance through the cytoplasm is that of maternal inheritance. The pattern of transmission is uniparental and non-Mendelian. The expression of the trait in the offspring is that of the maternal parent, while the paternal parent exerts little if any influence. In subsequent generations, segregation for the trait does not follow a pattern of chromosomal genes because, during sporogenesis and fertilization, the zygote receives its cytoplasm mainly from the egg or megagametophyte. In most higher plants, little cytoplasm, if any, is contributed by the male parent along with the two male nuclei from the pollen tube.

One preliminary test for cytoplasmic inheritance of a trait is to make reciprocal crosses between two parents that differ in the expression of that trait. The crosses may involve two individual plants or two homozygous and homogeneous inbred lines or varieties. In a pair of such reciprocal $F_1$ hybrids the nuclei should be genetically equivalent, whereas the cytoplasm could differ and would be that of the maternal parents. If the phenotypic expression of the trait differs in the reciprocal crosses and its expression is associated with the maternal parent, this is indicative but is not positive proof for cytoplasmic inheritance.

Another breeding test is to substitute the genome of one plant into the cytoplasm of another plant by backcrossing or to transfer by repeated backcrossing a series of genetically different nuclei from various plants to the common cytoplasm of one plant. If the trait in question continues to be expressed unchanged during backcrossing in a majority of the lines with the common cytoplasm, one can tentatively conclude that the trait is cytoplasmic in inheritance. Because of nuclear gene-cytoplasm interactions, some nuclei may modify or suppress completely traits that are inherited through the cytoplasm.

Using techniques of molecular biology, hybridization studies may reveal that rRNA, tRNA, and mRNA of the mitochondrion or of the chloroplast can be coded by the DNA of the mitochondria or chloroplasts, respectively.

### *DNA in Cytoplasmic Organelles*

The genetic material in living organisms is DNA; in viruses it is either RNA or DNA. In terms of structure, a gene may be considered a linear sequence of nucleotides. Genes replicate, recombine, and mutate. By means of an intermediate product, mRNA, a gene specifies the number and linear sequences of amino acids in a polypeptide chain of a protein. Such proteins are constructed at ribosomal surfaces in conjunction with tRNA, mRNA, and a battery of enzymes. Hence, in addition to replication, genetic information is transcribed and translated into macromolecules and eventually into more organized and complicated structures.

Because both chloroplasts and mitochondria contain DNA, they may be the most significant cytoplasmic constituents involved in extranuclear inheritance in plant cells (38, 46).

Chloroplast DNA (cDNA) and nuclear DNA differ in their properties. Replication of cDNA has been demonstrated. Different types of RNA form in the chloroplast, and the chloroplasts have been shown to possess an independent system for protein synthesis (38, 46, 47). Genes in the chloroplast recombine, and linkage maps have been constructed (38).

Although mitochondrial DNA (mDNA) of higher plants has not been studied extensively (5, 38), it is presently believed that mDNA replicates within the organelle. Furthermore, unique RNA can be synthesized in the mitochondria, and this organelle can incorporate amino acids into proteins. In addition, mDNA can mutate, and linkage maps have been established.

In both chloroplasts and mitochondria there is evidence for organelle DNA and the machinery for its transcription and translation (38, 46).

Conceivably, other cell constituents may eventually be found to have DNA and to be involved in extrachromosomal genetic systems.

## CYTOPLASMIC MALE STERILITY

In corn and other crops, first generation crosses between unrelated inbred lines or genetically different populations express hybrid vigor. This hybrid vigor or heterosis varies in degree, but plant size and grain yields are usually increased. Utilization of cytoplasmic male sterility facilitates exploitation of heterosis in the commercial production of hybrids. In sorghum and some other crops the wide-scale production of hybrid seed is not feasible without cytoplasmic male sterility. In the production of hybrid seed there is obviously a very practical advantage in a male-sterile parent that can be maintained and also used as a seed parent.

Cytoplasmic male sterility functions in much the same manner in most crops (8, 9, 51). Cytoplasms and nuclear factors interact so that pollen is produced by most plants, but in some plants nuclear factors interact with certain cytoplasms so that pollen abortion occurs. Other nuclear factors can interact with sterile cytoplasm to restore pollen production. The latter is the normal situation with sterile cytoplasms in nature. The "discovery" of a male-sterile cytoplasm comes about when, by accident or as a result of searching, a potentially male-sterile plant is crossed with another plant that does not have the needed nuclear factors to restore pollen production. Such crossed plants are sterile. Fertility-restoring genes for that cytoplasm are then usually found in the original male-sterile source but may also occur elsewhere.

To use the male-sterile system in seed production, the male-sterile cytoplasm is transferred to a seed parent line by repeated backcrossing. Sometimes nonrestoring nuclear genes must also be transferred. The male-sterile seed parent is maintained by crossing it with a genetically equivalent line that is fertile because it has normal cytoplasm. Hybrid seed is produced by crossing the male-sterile parent with another line. But such hybrid plants will be sterile unless the pollen parent carries nuclear genes restoring pollen production. Thus, fertility-restoring genes must be incorporated into the pollen parent line, or in the case of a cross-pollinated crop like corn it is possible to make blends of sterile and fertile plants to ensure seed production in farmers' fields. The fact that both cytoplasms and nuclear factors play a role in controlling male sterility and fertility makes possible a simple system for hybrid production and one easy for the plant breeder to manipulate.

Rhoades (36) first described a type of pollen abortion in corn inherited through the seed parent cytoplasm. Subsequently, other sources of male-sterile cytoplasm were discovered and studied in corn (8). Advances at this time were being made in the use of male sterility in onions and in other crops. In corn, the impetus for utilizing male sterility came with the discovery of the *cms*-T (Texas male sterile) cytoplasm by Rogers in Texas in 1944 (37). This cytoplasm came from the open-pollinated variety Golden June (Mexican June) which was then widely-grown in the southern United States. It was stable and reliable in many inbred backgrounds and environments. Furthermore, fertility was completely restored by the nuclear genes $Rf_1$ and $Rf_2$ that existed in many inbred lines (8). Subsequent testing in research plots and performance trials were favorable. Breeding programs during the 1950s and 1960s resulted in the rapid incorporation of *cms*-T cytoplasm into United States inbreds. By 1970 it was widely used in the production of hybrids in the United States and elsewhere.

Cytoplasmic male sterility is found in a wide variety of plants (8, 9). In addition to corn, it is used commercially in the production of hybrid sorghum, millet, onions, sugar beets, carrots, and other crops. Its use is also being contemplated in more crops.

## DISEASE REACTION

### *Southern Leaf Blight*

*Helminthosporium maydis* is a widespread pathogen of corn in tropical and subtropical areas of the world. In the United States it is usually limited to the warmer southern states and, hence, the disease it causes is commonly known as southern leaf blight.

While formerly of minor importance in the United States, the pathogen spread in the Corn Belt in 1970, developed in epidemic proportions, and caused major losses. Reviews describing the disease and this epidemic have been made by Hooker (19) and Ullstrup (48). Since 1970, the disease has increased in importance elsewhere in the world.

Southern leaf blight is unique in that the cytoplasm of the host plant determines in a major way susceptibility to a specific race of the pathogen. In breeding programs in the Philippine Islands between 1957 and 1961 (30) it was noticed that one cytoplasm type, *cms*-T cytoplasm for male sterility, conditions susceptibility. This was adequately studied and clearly interpreted as cytoplasmic susceptibility in subsequent reports (49). Following the Philippine report (30), the same inbred lines studied in the Philippines, i.e., F44 and F44T, with normal and *cms*-T cytoplasm, respectively, were tested against *H. maydis* in inoculated field plots in 1963 in Illinois (19, 44). In these tests and in following seasons no differences between cytoplasm types were seen in the corn pathology and genetics plots. Only in late 1969 were there indications that *cms*-T versions of corn plants were susceptible to *H. maydis* in the Corn Belt of the United States (19, 22, 42, 44, 48). Subsequent studies made during the winter and early spring of 1970 (22, 44) revealed that two races, race O and race T, of *H. maydis* were present in Illinois, that *cms*-T and *cms*-P cytoplasm plants were susceptible, that plants with *cms*-C or *cms*-S cytoplasms for male sterility or with normal cytoplasm were resistant, and that race T produced a pathotoxin specific to plants with susceptible cytoplasms. Our research program was expanded in the spring and summer of 1970 to include the following objectives (among others): (*a*) to compare races T and O in inoculated field plots; (*b*) to study a wide range of inbreds and hybrids in *cms*-T and normal cytoplasm versions for reaction to the two races; (*c*) to study other male-sterile cytoplasms for disease reaction; and (*d*) to seek sources of nuclear genes that might restore leaf blight resistance to plants with *cms*-T cytoplasm (19).

While it is not possible to cite all of the studies and observations made on southern leaf blight in recent years, an attempt is made to summarize the essential findings as they relate to cytoplasmic susceptibility. Because I am most familiar with the published and unpublished Illinois research, emphasis will be given to it.

Race T and race O of *H. maydis* differ in symptoms produced, cytoplasm specificity, toxin production, and other respects. Race O is primarily a leaf pathogen while race T infects the sheath, husk, and ear parts as well (6, 19). Only race T shows cytoplasm specificity. A few other cytoplasms such as *cms*-HA, *cms*-P, and *cms*-Q that are available as stocks in the collections of plant breeders also condition susceptibility (45). These cannot be distinguished clearly from *cms*-T cytoplasm (4). The fertile open-pollinated varieties, Golden June and Hastings Prolific, formerly widely grown in the southern United States, also are susceptible. A wide array of normal cytoplasms of different origins (18) and of other male-sterile cytoplasm are resistant to race T (45). An extensive survey of *H. maydis* isolates from the United States and Canada has failed to reveal a specific susceptibility of *cms*-C, *cms*-S, or one source of normal cytoplasm (10). This type of disease susceptibility has not been found in test plots exposed to natural infection. These tests, however, do not negate the possibility that quantitative or qualitative differences do exist or might be observed in the future. Such quantitative differences were reported recently by Julis (23).

Nuclear genes need to be considered in relation to cytoplasm reaction. Nuclear genes that restore full fertility to *cms*-T cytoplasm plants do not change disease reaction. Similarly, nuclear genes that modify the disease reaction conditioned by the cytoplasm do not alter pollen abortion characteristics. By breeding and selection, it has been possible to incorporate into plants with *cms*-T cytoplasm, nuclear genes that condition a high degree of resistance to race T. Some genotypes with normal cytoplasm are quite susceptible. Resistance is expressed in the form of lesion size, percentage of plant tissue infected, resistance to the pathotoxin, time of ear and plant death, and ability to yield well even though infected with *H. maydis* (19). Both monogenic and polygenic nuclear genetic systems that interact with the cytoplasm to condition disease reaction have been identified.

The unique susceptibility of plants with *cms*-T cytoplasm to *H. maydis* race T seems to be related to a pathotoxin produced by race T and its site of action at the mitochondria. Following our initial work and interpretation of a pathotoxin from race T and its specificity to plants with *cms*-T cytoplasm (22), the studies were extended to the development of a seed test (28), demonstration that the pathotoxin plays a role in disease (27), inheritance of toxin production (25), and determination of some of the properties of the toxin (26). That race T produces a host-specific toxin has now been confirmed by many others. Detailed studies of ultrastructural changes in plant cells (50), energy-linked processes of corn seedlings (1, 31), mitochondria swelling (14, 31), and membrane permeability (12, 15, 24) have also been made.

### *Yellow Leaf Blight*

Yellow leaf blight of corn caused by *Phyllosticta maydis* was observed in Ohio in 1965 (29) and in Wisconsin in 1967 (2). Subsequently, it has been found in much of the Corn Belt, the northeastern United States (3, 40, 41), adjacent parts of Canada, and perhaps elsewhere. McFeeley (29) described two types of isolates causing yellow leaf blight in Ohio. The isolates differed in morphology, cultural characteristics, and pathogenicity.

Various studies (2, 3, 29, 41) have shown that: (*a*) susceptibility of several corn inbreds and hybrids in *cms*-T cytoplasm is increased over their normal cytoplasm counterparts; (*b*) fertility-restoring genes do not influence disease reaction; (*c*) inbreds and hybrids in either *cms*-T or normal cytoplasm vary in reaction, and (*d*) some inbreds even in *cms*-T cytoplasm are quite resistant. Most cytoplasms for male sterility, other than *cms*-T and related types, react the same as normal cytoplasm (18, 34).

### *Other Diseases*

DIFFERENTIAL CYTOPLASM EFFECTS REPORTED Evidence suggesting the inheritance of virus-symptom expression through the cytoplasm was obtained in work done at the Central Potato Research Institute, Simla-1, India (32). When inoculated with potato virus X, *Capsicum annuum* develops systemic mosaic symptoms and *C. pendulum* develops necrotic local lesions. In the $F_1$ and $F_2$ generations of the cross *C. pendulum* X *C. annuum,* 44 and 201 plants, respectively, developed necrotic local lesions while none were observed with systemic mosaic mottling. While no seed was obtained from the cross *C. annuum* X *C. pendulum,* 33 plants were obtained from the backcross of the $F_1$ to *C. annuum.* All of these plants expressed the systemic mosaic mottling of the *C. annuum* parent. The absence of Mendelian segregation for the two reaction types in the $F_2$ and backcross generations is noteworthy.

In an abstract of some work done in England, Harland & King (16) suggested cytoplasmic effects of mildew reaction in strawberries. Reciprocal crosses, $F_2$ generations, and backcrosses were made between the commercial types of susceptible octoploid species *Fragaria grandiflora* and the resistant diploid *F. vesca.* Differences in mildew reaction were seen in the reciprocal crosses, and the authors indicated that the maternal plant influences appeared to persist in the advanced generations.

A cytoplasmic effect on the reaction of corn to common smut caused by *Ustilago maydis* has been reported (43). Plant-to-plant reciprocal crosses were made between foreign corn varieties and the Corn Belt single crosses WF9 X 38-11 and Oh43 X C103. The reciprocal crosses were evaluated for a number of agronomic and disease characters in replicated and randomized field plots. While maternal plant effects were rare and inconsistent, in 2 of 12 crosses studied, small but statistically significant differences were measured in a number of smutted plants. Plants with equal percentages of Corn Belt and exotic genotypes when in Oh43 cytoplasm had more smut than when in Peru cytoplasm from corn accession PI 186211, but less smut when in Guatemala cytoplasm from accession PI 163558. An interaction of genotype and cytoplasm was suggested.

Differences between reciprocal crosses in rice to two races of *Pyricularia oryzae* causing the blast disease were interpreted as cytoplasm effects (35). In this study variance values were obtained for disease reaction, lesion type, and lesion number for each of several crosses. The F-test was used to detect differences between reciprocal crosses. Among 69 F-tests made, 6 were at the 0.05 and 14 at the 0.01 level of statistical significance, and all others were nonsignificant. No consistent pattern of variance sizes or maternal effects was found for lesion number, lesion type, or reaction to the two pathogen races of the different crosses. The proportion of crosses where statistically significant differences were detected are more than one would expect due to chance alone, however.

Nagaich et al (32) cited reference to the work of Pohjakallio & Karhuvaara who observed maternal influence in $F_1$ clones from species crosses of *Solanum demissum*

X *S. tuberosum* in reaction to rugose mosaic incited by virus Y and to the work of Al'smik who reported that a potato clone 928-44 gave rise of up to 68% seedlings resistant to *Phytophthora infestans* when used as a seed parent but only 13% when used as a pollen parent.

Fleming (11) attributed different germination percentages and number of *Fusarium*-diseased seedling plants in double-cross corn hybrids to effects of the male cytoplasm. He also suggested a cytoplasm effect in reaction to another disease (presumably corn stunt) in a single inbred line in one test.

Specific resistance to two races of *Puccinia graminis* f. sp. *tritici* in wheats with alien cytoplasms revealed only a few cases where a cytoplasmic influence was detected (39). In eight cases, wheat varieties, susceptible in their own cytoplasm, became more resistant when the nuclei were transferred by backcrossing to cytoplasm of *Aegilops caudata, Ae. ovata,* and *Triticum timophievi.* In 13 other cases the effect was in the direction of increased susceptibility.

DIFFERENTIAL CYTOPLASM EFFECTS NOT SEEN Unless a specific purpose is served, negative data usually are not reported in the literature. In instances where it has been possible for many people to make many observations over a long period of time, however, the absence of certain data does have significance. Plant breeders have made many reciprocal hybrids in a wide array of plants. They have observed segregation patterns for nuclear genes in scores of cytoplasms. Furthermore, nuclear genes have been transferred intact by backcross breeding from one cytoplasm to another—usually to a male-sterile cytoplasm—and the two cytoplasm versions compared with each other in a wide array of environments. In some instances specific disease tests were made, but more commonly, various diseases were present in the environments where the performance tests were conducted. Had distinct cytoplasm effects for disease reaction been expressed they certainly would have been noted and reported. Hence, the absence of literature reports (21, 38, 51) and the experience of plant breeders and geneticists indicate that cytoplasm effects on disease reaction in plants are rare.

In Illinois, we have tested several corn cytoplasms for their reaction to disease (18). In this work, special efforts were made to maintain nuclear gene uniformity within the diversity of cytoplasms. This seems necessary because most characters where cytoplasm effects are possible may also be modified by genes in the nucleus. A set of 36 *cms* (male-sterile) cytoplasms and one *cmf* (normal or male-fertile) cytoplasm containing nuclei from the inbred 38-11 were available (4). Another set of 42 *cmf* cytoplasms containing nuclei from the inbred B14 had been developed by Sprague & Russell and were available. The set of *cmf* cytoplasms were diverse and distinguishable only in so far as the cytoplasms originated from different varieties or races of corn grown for many years in different areas of the world. In some instances the *cms* or *cmf* cytoplasm 38-11 inbreds were used as seed parents and crossed with another common inbred to make up single crosses.

Numerous disease tests were made in the greenhouse on seedlings and on plants in the field. While the tests were not exhaustive, no primary cytoplasm effects on disease reaction were seen with *Helminthosporium turcicum, H. maydis* race O, *H.*

*carbonum* race 1, *Puccinia sorghi, Diplodia maydis,* and two bacterial pathogens of corn (18). Cytoplasm effects were seen only to *Phyllosticta maydis* and *H. maydis* race T. Four *cms* cytoplasms (-HA, -P, -Q, -T) conditioned susceptibility to both pathogens. No cytoplasm effects were seen against common corn smut and other diseases naturally present in the field and to another leaf blight pathogen (20) in the greenhouse. Leaf blights predispose plants to root and stalk rots. Hence, secondary effects of plant cytoplasm may be experienced to these latter diseases.

## POSSIBLE MECHANISMS INVOLVED IN ACTUAL OR APPARENT CYTOPLASMIC SUSCEPTIBILITY TO DISEASE

### *Actual Cytoplasmic Inheritance*

Before the cytoplasm can be implicated in the inheritance of disease susceptibility, certain tests need to be made resulting in consistent, reproducible data. Care needs to be taken to compare plant types that differ in cytoplasms but that have identical nuclear genotypes. All plants should be free from viruses and other infectious agents and be unmodified in other ways by the maternal parent.

With the above requirements in mind only a few reports of cytoplasmic susceptibility to plant disease seem to be supported by ample evidence. The implication of cytoplasmic susceptibility of *cms*-T cytoplasm in corn to certain races or isolates of *H. maydis* and *P. maydis* seems justified. There may be others. In other instances, although maternal plant influences or differences between reciprocal crosses were seen, the conclusion that the effects were due to the cytoplasm seem unwarranted until more convincing data are presented.

In some instances an association of disease reaction with a cytoplasmic organelle is possible. The current view is that mDNA determines at least in part the structure of the inner mitochondrial membranes. Hence, an explanation for the cytoplasmic inheritance of disease reaction to *H. maydis* race T is provided by the association of a pathotoxin with race T and its site of action in a cytoplasmic organelle. The above hypothesis made earlier (22) has subsequently been supported by other studies (14, 31, 50). Furthermore, there is an exact correspondence between the reaction of corn cytoplasm to *H. maydis* race T and to *P. maydis* (18, 34). Both of these pathogens produce similar pathotoxins that affect in a similar way the functional processes of mitochondria from *cms*-T but not from normal cytoplasm corn (7, 13). Electron micrographs of isolated mitochondria previously treated with the pathotoxin (14) or of intact infected cells (50) show a disruption and vesiculation of the inner mitochondrial membrane.

Chloroplasts have not been implicated in specific susceptibility to disease. The same methods developed for the analysis of nuclear systems can be used to investigate cytoplasm genes (38). In a sense, disease susceptibility, or more precisely mitochondrial sensitivity to pathotoxin, provides a genetic marker to study mitochondrial genetics and gain new insight into the functioning of the cytoplasm. Some interesting issues can be raised concerning the distinctions between normal and male-sterile cytoplasms and between disease-resistant and susceptible cytoplasms. Is fertile normal cytoplasm always disease-resistant? Are sterility and susceptibility of *cms*-T cytoplasm mutually exclusive, or does each trait depend on a different mitochondrial gene? Are all male-sterile cytoplasms potentially vulnerable to disease?

## *Apparent Cytoplasmic Inheritance*

The maternal plant may influence seed composition or determine the structure of certain seed parts. When this occurs, reciprocal hybrids may differ in reaction to disease and the results interpreted incorrectly as cytoplasmic inheritance. Seed oil content of oats and soybeans for example, is strongly influenced by the maternal parent. One form of resistance in wheat to loose smut (*Ustilago tritici*) is determined by the ovary wall. Seed rot of corn kernels caused by several species of *Pythium* occurs when corn is planted in cold wet soils. The pericarp is diploid tissue of maternal parent origin and provides one of the barriers in the seed to *Pythium*. The seed endosperm is triploid with two sets of chromosomes from the maternal parent and one set from the paternal parent. Even the embroyo, which is diploid and truly hybrid in origin, is influenced by the seed parent during its development (17). Thus, it is not surprising that a strong maternal plant effect on reaction to certain diseases is seen.

In crosses between species or plants that differ in chromosome number, more chromosomes tend to be transmitted to the zygote through the egg than through the pollen. With supernumerary chromosomes, megagametophytes tend to be more functional than are microgametophytes. The maternal plant effect on mildew reaction mentioned earlier (16) in crosses between octoploid and diploid species of *Fragaria* could have been due to an unequal chromosome contribution from the two parents to the progeny.

An extreme case of unequal contribution of chromosomes from the female and male to the progeny was reported by Williams (51). In species belonging to the *Canina* section of the genus *Rosa,* only 14 of the 35 chromosomes pair at meiosis. The remaining 21 chromosomes are transmitted only through the megagametophyte. The somatic chromosome number of 35 is reconstituted at fertilization through the union of female and male nuclei that have 28 and 7 chromosomes, respectively. Because the female often contributes more chromosomes than the male, reciprocal crosses may differ, and for this reason the offspring may resemble the maternal parent.

Several types of apomixes occur in plants where reduction division and/or fertilization of the egg nucleus does not occur during seed reproduction. The embryo may develop from a diploid cell of the maternal parent. Hence, reciprocal "hybrids" are not crosses at all, and any reciprocal difference shown among individuals and similarities to the maternal parent will be quite independent of the cytoplasm.

Episomal systems are possible and merit consideration. This phenomenon has not been verified in higher plants, however.

Seed-borne viruses and other seed-borne pathogens could account for some maternal plant effects in reciprocal hybrids. Conceivably certain viruses might also be latent and brought into expression when nuclear genes change.

Incomplete nuclear recovery as a result of insufficient number of backcrosses to a different cytoplasm source could account for differences in disease reaction among backcross and original versions of an inbred line. Apparently unaware of the Philippine work (30, 49), Scheifele & Nelson (40) put forth this reason to explain why inbred lines converted to *cms*-T cytoplasm were more susceptible to yellow leaf blight than the original inbreds in normal cytoplasm.

Finally, when a large number of reciprocal hybrids are examined, some pairs will differ in disease reaction due to chance alone. Such differences are not biologically significant, although they are apparently statistically significant.

### *Future Prospects*

While genetic uniformity in a crop has long been recognized as inviting vulnerability to production hazards, disease susceptibility now adds the dimension of cytoplasmic uniformity.

Cytoplasmic uniformity is common in crops where agricultural technology is advanced. Most improved varieties resulting from the breeding of self-pollinated crops such as wheat, oats, barley, rice, soybeans, and others originate from a single seed. The cytoplasm of such varieties is uniform although it is of a normal type. Such varieties often are used as parents in the development of new varieties. Hence, a whole array of new varieties may all have the same cytoplasm of one of their parents. Other crop varieties are vegetatively propagated and their cytoplasm is uniform. It must also be remembered that some sterile cytoplasms go unrecognized and are regarded as so-called normal cytoplasms until nuclei that do not restore fertility are substituted for their own.

Where male-sterile cytoplasms are used in hybrid seed production, the breeder frequently has the choice of only one satisfactory type. Hence, as was the case in corn, only one source of cytoplasm is commonly used for the total commercial hybrid seed production of a crop. All United States sorghum hybrids, for example, carry a single source of sterile cytoplasm.

Cytoplasmic susceptibility to disease is rare but of importance where it occurs. The risks of uniformity are greatest when the cytoplasm determines reaction to a pathogen such as *H. maydis* that has the potential of developing in epidemic proportions and causing severe damage to the plant. Cytoplasmic uniformity seems undesirable, and methods of achieving cytoplasmic diversity have been proposed (19, 33).

Speculations as to where other instances of cytoplasmic susceptibility to disease may occur in the future are hazardous. But past experience would suggest that the following requirements would need to be met: (*a*) the crop has an economic character such as male sterility that is inherited through the cytoplasm; (*b*) this unusual and formerly rare type of cytoplasm is introduced by breeding and widely used in the crop; and (*c*) the crop is exposed to a pathogen capable of producing a pathotoxin with its site of action on membranes whose structure is determined by the DNA of a cytoplasmic organelle.

It seems impossible to predict when or where new pathogens or new races of a current pathogen with specificity to certain host cytoplasms will occur in nature.

Hopefully, plant pathologists and breeders will be vigilant in their observations of breeding materials, plants in performance test plots, and the commercial crop itself, so that new races will be detected early enough to avoid major epidemics. Fortunately, the shift from a susceptible to a resistant cytoplasm type is easily accomplished in most crops.

*Literature Cited*


1. Arntzen, C. J., Koeppe, D. E., Miller, R. J., Peverly, J. H. 1973. The effect of pathotoxin from *Helminthosporium maydis* (race T) on energy-linked processes of corn seedlings. *Physiol. Plant Pathol.* 3:79–89
2. Arny, D. C., Worf, G. L., Ahrens, R. W. Lindsey, M. F. 1970. Yellow leaf blight of maize in Wisconsin: Its history and the reactions of inbreds and crosses to the inciting fungus (*Phyllositcta* sp.). *Plant Dis. Reptr.* 54:281–85
3. Ayers, J. E., Nelson. R. R., Koons, C., Scheifele, G. L. 1970. Reactions of various maize inbreds and single crosses in normal and male-sterile cytoplasm to the yellow leaf blight organism (*Phyllosticta* sp.). *Plant Dis. Reptr.* 54: 277–80
4. Beckett, J. B. 1971. Classification of male-sterile cytoplasms in maize (*Zea mays* L.). *Crop Sci.* 11:724–27
5. Börst, P. 1970. Mitochondrial DNA: structure, information content, replication and transcription. *Symp. Soc. Exp. Biol.* 24:201–26
6. Calvert, O. H., Zuber, M. S. 1973. Ear-rotting potential of *Helminthosporium maydis* race T in corn. *Phytopathology* 63:769–72
7. Comstock, J. C., Martinson, C. A., Gengenbach, B. G. 1972. Characteristics of a host-specific toxin produced by *Phyllosticta maydis. Phytopathology* 62:1107 (Abstr.)
8. Duvick, D. 1959. The use of cytoplasmic male-sterility in hybrid seed production. *Econ. Bot.* 13:167–95
9. Edwardson, J. R. 1970. Cytoplasmic male sterility. *Bot. Rev.* 36:341–420
10. Fisher, D. E. 1972. *Pathogenicity of Helminthosporium isolates on corn plants representing four cytoplasm types.* MS thesis. Univ. Illinois, Urbana. 38 pp.
11. Fleming, A. A. 1972. Male cytoplasmic effect on reaction of maize to disease. *Plant Dis. Reptr.* 56:575–77
12. Garraway, M. O. 1973. Electrolyte and peroxidase leakage as indicators of susceptibility of various maize inbreds to *Helminthosporium maydis* races O and T. *Plant Dis. Reptr.* 57:518–22
13. Gengenbach, B., Koeppe, D. E., Miller, R. J. 1973. A comparison of mitochondria isolated from male-sterile and nonsterile cytoplasm etiolated corn seedlings. *Physiol. Plant.* 29:103–7
14. Gengenbach, B. G., Miller, R. J., Koeppe, D. E., Arntzen, C. J. 1973. The effect of toxin from *Helminthosporium maydis* (race T) on isolated corn mitochondria: swelling. *Can. J. Bot.* 51: 2119–25
15. Halloin, J. M., Comstock, J. C., Martinson, C. A., Tipton, C. L. 1973. Leakage from corn tissues induced by *Helminthosporium maydis* race T toxin. *Phytopathology* 63:640–42
16. Harland, S. C., King, E. 1957. Inheritance of mildew resistance in *Fragaria* with special reference to cytoplasmic effects. *Heredity* 11:287 (Abstr.)
17. Hooker, A. L. 1955. Additional seed factors affecting stands of corn in cold soils. *Agron. J.* 47:582–85
18. Hooker, A. L. 1972. Cytoplasmic effects on host-pathogen interactions of corn diseases. p. 11. American Society of Agronomy Abstracts
19. Hooker, A. L. 1972. Southern leaf blight of corn—present status and future prospects. *J. Environ. Qual.* 1:244–49
20. Hooker, A. L., Mesterhazy, A., Smith, D. R., Lim, S. M. 1973. A new Helminthosporium leaf blight of corn in the northern corn belt. *Plant Dis. Reptr.* 57:195–98
21. Hooker, A. L., Saxena, K. M. S. 1971. Genetics of disease resistance in plants. *Ann. Rev. Genet.* 5:407–24
22. Hooker, A. L., Smith, D. R., Lim, S. M., Beckett, J. B. 1970. Reaction of corn seedlings with male-sterile cytoplasm to *Helminthosporium maydis. Plant Dis. Reptr.* 54:708–12
23. Julius, A. J. 1973. *The effect of C and S group cytoplasms on resistance to southern corn leaf blight.* MS thesis. North Carolina State Univ., Raleigh, NC. 49 pp.

24. Keck, R. W., Hodges, T. K. 1973. Membrane permeability in plants: changes induced by host-specific pathotoxins. *Phytopathology* 63:226–30
25. Lim, S. M., Hooker, A. L. 1971. Southern corn leaf blight: genetic control of pathogenicity and toxin production in race T and race O of *Cochliobolus heterostrophus. Genetics* 69:115–17
26. Lim, S. M., Hooker, A. L. 1972. A preliminary characterization of *Helminthosporium maydis* toxins. *Plant Dis. Reptr.* 56:805–7
27. Lim, S. M., Hooker, A. L. 1972. Disease determinant of *Helminthosporium maydis* race T. *Phytopathology* 62:968–71
28. Lim, S. M., Hooker, A. L., Smith, D. R. 1971. Use of *Helminthosporium maydis* race T pathotoxin to determine disease reaction of germinating corn seed. *Agron. J.* 63:712–13
29. McFeeley, J. C. 1971. Comparison of isolates causing yellow leaf blight of corn in Ohio. *Plant Dis. Reptr.* 55:1064–68
30. Mercado, A. C. Jr., Lantican, R. M. 1961. The susceptibility of cytoplasmic male sterile lines of corn to *Helminthosporium maydis* Nish & Miy. *Philipp. Agr.* 45:235–43
31. Miller, R. J., Koeppe, D. E. 1971. Southern corn leaf blight: susceptible and resistant mitochondria. *Science* 173:67–69
32. Nagaich, B. B., Upadhya, M. D., Prakash, O., Singh, S. J. 1968. Cytoplasmically determined expression of symptoms of potato virus X crosses between species of *Capsicum. Nature London* 220:1341–42
33. National Academy of Sciences. 1972. Genetic Vulnerability of Major Crops. Washington DC. 307 pp.
34. Nelson, R. R., Ayers, J. E., Beckett, J. B. 1971. Reactions of various corn inbreds in normal and different male-sterile cytoplasms to the yellow leaf blight organism (*Phyllositcta* sp.). *Plant Dis. Reptr.* 55:401–3
35. Rath, G. C., Padmanabhan, S. Y. 1972. Cytoplasmic effects on the leaf blast reaction in rice. *Curr. Sci.* 41:338–39
36. Rhoades, M. M. 1931. Cytoplasmic inheritance of male sterility in *Zea mays. Science* 73:340–41
37. Rogers, J. S., Edwardson, J. R. 1952. The utilization of cytoplasmic male-sterile inbreds in the production of corn hybrids. *Agron. J.* 44:8–13
38. Sager, R. 1972. *Cytoplasmic Genes and Organelles.* New York and London: Academic. 405 pp.
39. Sanchez-Mange, E., Salazar, J., Branas, M. 1973. Cytoplasmic influence in specific wheat stem rust resistance. *Cereal Rusts Bull.* 1:16–18
40. Scheifele, G. L., Nelson, R. R. 1969. The occurrence of Phyllosticta leaf spot of corn in Pennsylvania. *Plant Dis. Reptr.* 53:186–89
41. Scheifele, G. L., Nelson, R. R., Koons, C. 1969. Male sterility cytoplasm conditioning susceptibility of resistant inbred lines of maize to yellow leaf blight caused by *Phyllosticta zeae. Plant Dis. Reptr.* 53:656–59
42. Scheifele, G. L., Whitehead, W., Rowe, C. 1970. Increased susceptibility to southern leaf spot (*Helminthosporium maydis*) in inbred lines and hybrids of maize with Texas male sterile cytoplasm. *Plant Dis. Reptr.* 54:501–3
43. Singh, M. 1966. Cytoplasmic effects on agronomic characters in maize. *Ind. J. Genet. Plant Breed.* 26:386–90
44. Smith, D. R., Hooker, A. L., Lim, S. M. 1970. Physiologic races of *Helminthosporium maydis. Plant Dis. Reptr.* 54:819–22
45. Smith, D. R., Hooker, A. L., Lim, S. M., Beckett, J. B. 1971. Disease reaction of thirty sources of cytoplasmic male-sterile corn to *Helminthosporium maydis* race T. *Crop Sci.* 11:772–73
46. Tewari, K. K. 1971. Genetic autonomy of extranuclear organelles. *Ann. Rev. Plant Physiol.* 22:141–68
47. Tewari, K. K., Wildman, S. G. 1970. Information content in the chloroplast DNA. *Symp. Soc. Exp. Biol.* 24:147–79
48. Ullstrup, A. J. 1972. The impacts of the southern corn leaf blight epidemic of 1970–1971. *Ann. Rev. Phytopathol.* 10:37–50
49. Villareal, R. L., Lantican, R. M. 1965. The cytoplasmic inheritance of susceptibility to Helminthosporium leaf spot in corn. *Philipp. Agr.* 49:294–300
50. White, J. A., Calvert, O. H., Brown, M. F. 1973. Ultrastructural changes in corn leaves after inoculation with *Helminthosporium maydis,* race T. *Phytopathology* 63:296–300
51. Williams, W. 1964. *Genetical Principles and Plant Breeding.* Philadelphia: Davis. 506 pp.

# SEED AND ROOT BACTERIZATION

❖3595

*Margaret E. Brown*
Soil Microbiology Department, Rothamsted Experimental Station, Harpenden, Hertfordshire, England

## BACTERIZATION TO INCREASE CROP YIELD

Seed or root bacterization usually means treatment of seeds or seedling roots with cultures of bacteria that will improve plant growth; such preparations are frequently called bacterial fertilizers. The classical example is treatment of legume seed with *Rhizobium,* whose value and mode of action is indisputable; preparations are on sale throughout the world. This review is not concerned with *Rhizobium,* but with the use and effects of other bacteria that have been the subject of controversy since the publication of experiments in the Soviet Union on the use of *Azotobacter chroococcum* and *Bacillus megaterium* as bacterial fertilizers, called "azotobacterin" and "phosphobacterin" respectively.

These bacteria were selected because *A. chroococcum* fixes atmospheric nitrogen, and *B. megaterium,* particularly the strain "phosphaticum" (59), mineralizes organic phosphorus compounds. It was thought that growth of these bacteria in soil around roots would supply nitrogen and phosphorus to the plant, but subsequent experiments (27, 65) showed this premise to be wrong. Crop growth was affected, but certainly not because of nitrogen fixation, and probably not because of phosphorus mineralization. Other bacteria, principally species of *Pseudomonas, Clostridium,* and *Bacillus* (other than *B. megaterium*), were also applied as inoculants and affected plant growth, but these have not been used as extensively as azotobacterin and phosphobacterin. By 1958 about $10^7$ ha in the Soviet Union were treated with these preparations, which were said to benefit 50–70% of the field crops, increasing yield 10–20% (27). Rubenchik (88) summarizes most of the work up to this period.

Mishustin & Naumova (65) then reviewed the subject critically, questioning whether the claims of large increases in yield were justified, for much of the early data had not been analyzed statistically. When analysis was possible they concluded that positive significant effects occurred in about one third of the trials, and an increase of 10% was almost within the limits of experimental error. Increase in yields of cereals rarely exceeded 10%, but some horticultural crops responded better with increases ranging from 13 to 39%, and best results were obtained in soils rich in organic matter with mineral fertilizers added. The different types of bacterial fertilizers produced similar effects on plant growth and thus probably had a common method of action. Mishustin & Naumova (65) listed 24 other bacteria, fungi, and actinomycetes, which in different experiments had stimulated plant growth, as a further demonstration of how widely this property was distributed in the microbial kingdom, suggesting a common mechanism.

Following this review a discussion ensued among Soviet scientists, and various points of view were published (10, 30, 51, 52, 60, 63, 89, 104). It was agreed that bacterial fertilizers would not replace mineral fertilizers, but that the correct combination of bacterial, mineral, and organic fertilizers would enhance plant growth more than any one treatment alone. Disagreement occurred over the mode of action of these fertilizers: one group (30, 63, 65, 89, 104) maintained that fungistatic properties and biologically active substances common to all the bacteria were responsible, and the other group (51, 52, 60) that phosphobacterin acted primarily by mineralizing phosphorus. It was also suggested (60, 89) that bacterial fertilizers changed the activity of other groups of microorganisms, thus improving nutritional conditions for the root. Microorganisms particularly affected were nitrifiers, ammonifiers, denitrifiers, anaerobic nitrogen fixers, and cellulose decomposers.

During the next ten years the experiments reported in journals from the Soviet Union and Eastern European countries further demonstrated that bacterial fertilizers slightly improved yields of a wide range of crop plants, especially vegetables, but gave little further information on the mode of action. The following yield increases are examples of such results: maize, 5–8% (38, 64); spring and winter wheat, 8–16% (64, 79); barley and oats, 9–12% (64); rice, 24–42% (91); soya beans, 10% (22); potatoes, 18–25% (64, 79, 103); cabbage, 19–33% (64, 79, 107); sugar beet, 7–10% (64, 79); tomatoes, 28–56% (64, 107); and the first harvest of cucumbers, 55% (64).

In this period Indian scientists also did pot and field experiments with azotobacterin, phosphobacterin, *Pseudomonas,* and *Beijerinckia,* using bacteria isolated from native soils and strains supplied by Russia (9, 48, 53, 58, 71, 95–98). The various inoculants improved growth of wheat, rice, *Eleusine corocana,* peas, cabbage, eggplant, and tomato. These effects increased when mineral fertilizers and farmyard manure were also added. The farmyard manure presumably acted as an energy source for bacterial growth, and in two experiments (98) adding an inoculum of cellulose decomposers with *Azotobacter* improved growth of peas more than *Azotobacter* alone. As Jensen (44) had shown that *Azotobacter* could use cellulose decomposition products as an energy source, such an association may have occurred in these experiments.

Besides yield effects, changes in the general growth of plants were recorded. Cuttings of grapevine and flowering shrubs rooted better after treatment with bacterial inoculants, and the number of shoots was increased (6, 62, 70, 84). *Dianthus* plants also had larger root systems and blooms after treatment (2). Sundara Rao et al (97) found that *Azotobacter* and phosphobacterin increased the height, number of tillers, and the rate of ear emergence of wheat. Such effects were indicative of the action of plant growth-regulating substances, but there was no experimental evidence of their direct involvement.

Up to 1962 scientists from other countries had dismissed the value of bacterial inoculants, for early experiments (3, 45, 93) had not produced significant increases in yield; but interest was renewed after the reviews by Cooper (27) and Rubenchik (88) were published. A series of comprehensive and statistically designed experiments with different crop plants were done by Brown, Burlingham, Jackson, and Patel (16, 18–21, 42, 43, 76, 77) using *A. chroococcum;* by Rovira and Ridge (81, 82, 86, 87) using species of *Azotobacter, Clostridium,* and *Bacillus;* by Tchan & Jackson (99) using *Azotobacter* species; by Denarié & Blachère (29) using species of *Azotobacter, Pseudomonas, Arthrobacter,* and *Bacillus circulans;* and by Eklund (32) using *Pseudomonas.*

These experiments confirmed many of the Soviet observations. Treatment with the different bacteria increased yields in some of the experiments in field soils. Greater increases were obtained in fertile soils with added minerals and with moisture content at field capacity than in poor soils or soils not at optimum moisture content. Trials in the controlled conditions of the glasshouse gave more consistent and reliable results than those in the field. Cereal crops responded less well than vegetables, and yields were of the same order as in the Soviet experiments. Changes in plant morphology and growth rates were also observed. In addition, these experiments answered many questions about behavior of the inoculum in the rhizosphere and the probable mechanisms which alter plant growth.

## PREPARATION OF THE BACTERIAL INOCULANTS

The effectiveness of bacterial fertilizers is determined by their quality, especially the number of viable organisms and their ability to multiply, once applied to seed, root, or soil (63). A minimum of 40 to 50 $\times$ $10^9$ viable cells (the number required in a standard bacterial fertilizer preparation in the Soviet Union) applied per hectare gives 15 to 20 bacteria per gram of soil, a number unlikely to affect plants or the microbial population unless multiplication occurs. Bacteria applied directly to seed coated with fungicide are liable to be killed; equally, bacteria on seeds combine-drilled with mineral fertilizers can be killed by osmotic shock (20). Hence bacteria can be directly applied only to seed not treated with fungicide and drilled separately from the mineral fertilizers.

To overcome these difficulties various peat preparations were made that could either be put on the soil or sprinkled onto moist seed at the time of sowing (88). Many of the Soviet preparations were of poor quality, mainly because of difficulties in keeping the cells viable and in preventing lysis by bacteriophages (89). Mishustin (64) used an improved peat-humus nutrient block treated with different mineral salts, formulated according to crop requirements, and wetted with a culture of *Azotobacter chroococcum.* These blocks were suitable for use only with vegetable crops whose seedlings were raised in them before transplanting. Ridge & Rovira (82) overcame the difficulties of combine-drilling *Azotobacter*-inoculated cereals with mineral fertilizers by coating the grain with a mixture of radiation-sterilized peat and cells grown on agar, or a suspension of peat-grown cells in aqueous gum arabic. The grain was free from fungicides.

In most recent experimental trials no attempt was made to use special preparations, but cultures of the various bacteria were applied directly to untreated seed by soaking or spraying (19, 20, 29, 32, 81, 82, 99). Roots of seedlings were usually dipped into cultures before transplanting to soil (16, 19, 20, 29, 43), and sometimes soil around seedling plants was watered with cultures (19, 20, 29).

## ESTABLISHMENT OF THE INOCULUM IN THE ROOT ZONE

Although it was assumed that survival of the inoculum in the soil was a prerequisite for an effect on crop growth, there was no supporting experimental data in the Soviet literature. Rakhno & Ryys (79) had pointed out that it was useless to add bacterial fertilizers to soils that would not support natural populations of the organisms; therefore, fertile soils rich in organic matter, with a pH near neutrality and moisture conditions favoring plant and microbial growth, would give best results. Preferably the bacteria should multiply in the rhizosphere, where their activity was most likely to influence plant growth. Under natural soil conditions neither *A. chroococcum* nor *B. megaterium* was regarded as true rhizosphere bacteria (19, 47, 65, 66). Some *Azotobacter* were present, depending on soil and the type and age of plant; a small proportion were on the root surface (42). *Pseudomonas* species, however, were abundant in this zone (85).

Brown, Burlingham & Jackson (20) inoculated wheat grain with *Azotobacter* cultures of different ages and found the bacteria established in the rhizosphere, the size of the population depending on culture age (14-day cultures were best) and number of cells originally on the grain. Inoculum survival tests made with young cultures of vegetative cells on wheat grain stored in a constant environment showed that viability decreased rapidly in one day and then declined slowly, until at 8 weeks no cells were recovered. Old cultures consisted entirely of encysted bacteria that were resistant to desiccation, which probably explains why these gave better rhizosphere establishment. Ridge (81), testing his peat inoculants for survival, observed a loss of inoculum when the peat dried on the grain, a disadvantage compared with liquid cultures where 75% inoculum is transferred to the grain. However, when grain is stored before sowing, cells from the peat inoculant have a half-life of 10–15 weeks compared with 8–10 weeks for cells from cultures.

Pot experiments with peas, wheat, and cabbage showed that inoculum of *Azotobacter* cysts germinated in 2 days and microcolonies developed only near roots (42). In a typical experiment with wheat, numbers of *Azotobacter* on inoculated grain decreased a hundredfold in 27 days, but increased tenfold on seedling roots. A mapping technique showed that *Azotobacter* was distributed continuously along young wheat roots, and cells were evenly distributed on older roots but less crowded. In field experiments *Azotobacter* became established throughout the root system of wheat with numbers more than thirty times those found in surrounding soil. All parts of the root were colonized including adventitious roots developing above the grain and primary system, numbers decreasing with distance from the grain. After initial multiplication the population remained static at the high level, but disappeared after harvest (19). Patel (76) also found this effect with wheat and tomato. Denarié & Blachère (29) established *Azotobacter* in the rhizospheres of rye grass, lettuce, and potatoes, the size of the inoculum influencing the population.

Pure culture studies by Rovira (86) in sand with lucerne, maize, tomato, and wheat indicated *Azotobacter* to be a poor colonizer of tomato and wheat, but *Clostridium, Bacillus,* and especially *Pseudomonas* multiplied. However, these results cannot be extrapolated to plants growing in soil, because in soil rhizosphere nutrients are modified by other microorganisms, and energy substrates may then be better suited to the inoculants. In fact, later experiments (87) in soil suggested that these bacteria must have multiplied on the seeds and roots during and following germination, using the soluble exudates as nutrients. When peat inoculants were used on six field sites, few *Azotobacter* were found in the root region. Naturally occurring *Azotobacter* could not be detected in these soils; it was not surprising, therefore, that the inoculants failed to become established, because inoculants are likely to survive only in favorable environments. Trials with rice (72, 91) showed that *Azotobacter* increased in the rhizosphere of young plants but later declined.

There is only one study on multiplication of *Pseudomonas* inoculum (32). This bacterium, as a rhizosphere inhabitant, would be expected to thrive, but, after multiplying initially to become predominant on young cucumber roots, the numbers decreased and cells were barely detectable by five weeks.

## CAUSES OF INOCULUM DECLINE

Different explanations for decline of the added inocula involved inhibitory root secretions and antagonistic microorganisms. Some root secretions contained polyphenols and gallotannins, and strongly inhibited *Azotobacter*. These substances were widely distributed in plants and could affect inoculants (33, 34). However, root excretions of cereals, plants frequently treated with inoculants, supported *Azotobacter* growth. The organic acid fraction acted as a better carbon source than either other fractions, such as pentoses or oligosaccharides, or unextracted root excretions (99, 101).

Microorganisms antagonistic to naturally occurring *Azotobacter* were abundant in rhizospheres of radish, onion, wheat, and red pepper, and numbers of *Azotobacter* and antagonists were directly related (40, 94). Patel & Brown (77) studied the possible role of antagonists in restricting multiplication of introduced *Azotobacter*. Inhibitory actinomycetes were usually as abundant in rhizospheres of inoculated wheat as in control plants, but significantly greater numbers of strongly antagonistic strains were present in the inoculated rhizosphere of 8-day-old seedlings. Root surface fungi were also tested, and production of antagonistic substances differed with the fungal growth medium. In dual inoculation tests in sterile soil, fungi had little effect on *Azotobacter* numbers, but some actinomycetes checked multiplication.

These experiments did not explain what happened in nonsterile soil where development of antibiotics would depend on adequate substrate in the microenvironment of the root. If such substrate were available, the actinomycetes abundant in the 8-day-old rhizosphere, for example, could become active, inhibiting *Azotobacter* and preventing further multiplication. Why should these actinomycetes be significantly increased in number? Possibly the multiplying population of *Azotobacter* directly supplied metabolites favoring actinomycete growth and antibiotic activity or, less directly, altered the quality of exudates to those beneficial to growth and antibiotic activity. Thus, for a brief period the *Azotobacter* population created an environment leading to self-destruction.

## MODE OF ACTION OF THE INOCULANTS

Decrease in numbers of the various inoculants did not detract from the fact that seed and root bacterization altered plant growth. Inoculants of *Azotobacter, B. megaterium, Clostridium pasteurianum, B. polymyxa, Pseudomonas fluorescens,* and other species were tested in soil and gave similar results. Crop yields were frequently increased, especially vegetables, and plant morphology and growth rates were altered. Increases were recorded in rates of germination, elongation of stems, enlargement of leaves, and earlier flowering and fruiting (20, 29, 65, 76, 82, 86, 87). Similar effects suggest similar causes, and four explanations deserve serious consideration, three being common to all inocula.

### *Changes in the Rhizosphere Microbial Population*

Inoculation with *Azotobacter* and *B. megaterium* was said to intensify the activity of rhizosphere bacteria involved in ammonification, anaerobic nitrogen fixation, nitrification, phosphate mineralization, and cellulose decomposition (10, 55, 70, 89). However, Rovira (86) did not find any consistent differences in the general heterotroph count of the rhizosphere microflora.

Patel (76), examining bacteria and root surface fungi from wheat, found that, compared with uninoculated plants, seed inoculation with *Azotobacter* significantly depressed colonization of emerging roots by bacteria and actinomycetes. After one week both groups of organisms increased rapidly to give larger populations in inoculated plants, but this difference later disappeared. Total rhizosphere populations and groups of organisms fermenting glucose, decomposing cellulose and chitin, growing anaerobically, or producing spores, all increased as plants aged but were unaffected by inoculation. Two separate experiments examining root surface fungi of wheat gave dissimilar and inconclusive results. Patel concluded that *Azotobacter* did not act by changing the rhizosphere microflora. However, suppression of bacteria colonizing emerging roots might be more important than previously thought, because when examining the ability of the general rhizosphere and rhizoplane microflora of wheat to produce plant growth regulators, Brown (14) found that bacteria producing inhibitory substances were significantly stimulated by roots of 6-day-old seedlings, particularly in the rhizosplane. Displacement of this population by *Azotobacter* could mean removal of metabolites potentially able to retard seedling development, particularly at a stage when primordia were differentiating and most vulnerable to modification.

## *Disease Suppression by the Inoculants*

Different experiments were quoted in the Soviet literature where azotobacterin and phosphobacterin decreased incidence of virus and bacterial diseases of potato (30, 89), of *Sclerotinia libertiana* on sunflower, of *Helminthosporium sativum* on grass, of smut on millet, and of brown rust on wheat (60). There was no evidence that this effect was one of direct antagonism of the pathogen; decreased disease could be related to general improvement in plant growth. Brown, Burlingham & Jackson (20) found that wheat infected with *Gaeumannomyces graminis* var. *tritici* grew better when inoculated with *Azotobacter,* but take-all disease was not decreased. In tube-culture experiments the improved wheat growth seemed to be related to a decrease in a subsidiary infection by *Fusarium* carried on the seed, rather than by an effect on *Gaeumannomyces* (15).

The effects of *Azotobacter* inoculum on development of pathogens borne naturally on different seeds were also examined. Infections of sugar beet by *Phoma betae,* of peas by *Mycosphaerella pinodes,* of barley by *Helminthosporium gramineum,* of oats by *H. avenae,* and of wheat by *Septoria nodosum* and *Fusarium nivale* were not suppressed. Only infection of *Medicago lupulina* by *Ascochyta imperfecta* was decreased, an effect directly related to the amount of pathogen on the seed (15). Mishustin (64), however, found that *Azotobacter* suppressed growth of fungi on germinating wheat. *Alternaria* infection of maize grown in sand or soil was also decreased and shoot growth accelerated, particularly if manure was added. Different strains of *Azotobacter* varied in their antagonism. The antibiotic was isolated: it belonged to the conactin group, resembled anisomycin, and was active against *Candida albicans* and *Monilia* (67). Seedlings of low vigor, which would be prone to infection, were reported to respond well to *Azotobacter* (82). The inoculant may give direct protection against pathogens or may stimulate seedling growth by another mechanism so that disease susceptibility was decreased.

Cultures of *Bacillus* and *Pseudomonas* frequently inhibit plant pathogens when tested in vitro and might give some disease control as inoculants; these effects are discussed later. However, as experience has shown that bacterization specifically for disease control is frequently unsuccessful (8), it seems unlikely that pathogen suppression should be a property common to all the bacterial fertilizers or should explain all the results.

## *Production of Plant Growth-Promoting Substances*

When describing the effects of inoculants, most authors reported changes in plant development that indicated activity of growth-regulating substances. Mishustin & Naumova (65) had suggested, but without definite evidence, that various vitamins, auxins, and gibberellin-like substances could be involved because these were found in *Azotobacter* cultures. Many soil organisms produced indolyl-3-acetic acid (IAA) in culture media with or without added tryptophan (14, 90), including the bacteria consistently used as inoculants (11, 17, 39, 41, 69). Root exudates contained tryptophan or related compounds that could act as precursors for IAA synthesis, and roots took up indoles (90). Microbial production of IAA in the root zone could therefore provide an exogenous source of hormone. Libbert & Manteuffel (54) showed that plants in the presence of epiphytic bacteria had more endogenous auxin than those maintained in sterile conditions. Exogenously produced auxin entered the tissues, and there were significant increases in the elongation and dry weights of maize.

Root growth is partly controlled by auxin, and external application usually causes stunting and thickening, but under some conditions, such as brief exposure to small doses, roots are elongated and growth of the whole plant may be promoted. Lateral root growth is also stimulated by auxin (90, 100, 106). Thus some of the plant growth effects observed after bacterization, such as improved rooting of cuttings and enlargement or stunting of primary roots, could partly be explained by IAA synthesis in the rhizosphere.

Effects on aerial parts of plants were much more indicative of gibberellin activity. Plants treated with gibberellin, even those normally unresponsive to growth stimulators, frequently developed elongated stems. Leaf area and shape were changed, and flowering and fruit set of some plants were enhanced, but yield was not necessarily increased. Treating seeds with gibberellin broke dormancy, and even in the absence of dormancy might have hastened extension of the radicle. The dormancy of underground organs like potato tubers was also broken by gibberellin. If gibberellin and IAA were both applied to plants their effect was often additive, and exogenous gibberellin increased the levels of endogenous IAA (75).

Involvement of gibberellin-like substances in the action of *Azotobacter* was shown by comparing the effects on tomato plants of inoculating with *Azotobacter* or treating with gibberellic acid ($GA_3$) (16, 21). Authentic $GA_3$ applied once to seeds or roots of tomatoes at the cotyledon stage of development caused elongation of stem internodes and expansion of leaves, and changed the number of flower buds and time of fruit formation. $GA_3$ and IAA added together did not affect development any differently than $GA_3$ alone. Quantities of growth substances used in these experiments were based on amounts found in 14-day-old cultures of *Azotobacter,* which had produced exactly similar effects on tomatoes. Three gibberellin-like substances were found in these cultures; that with the same *Rf* value as $GA_3$ or $GA_1$ was probably the most important and was found in amounts ranging from 0.01 to 0.1 $\mu$g $GA_3$ equivalents per ml, depending on the batch of culture. Although small, this quantity altered tomato development, possibly because it was taken up by seedlings at the critical stage when vegetative and reproductive primordia were differentiating.

Eklund (32) obtained effects on cucumber growth very similar to those on tomato after treating cucumber seed with *Pseudomonas* cultures. Germination of poor quality seed was improved. The *Pseudomonas* cultures contained gibberellin-like substances but no IAA. However, she did not consider that gibberellins alone produced all the observed effects, and suggested that quinones were also involved. *Pseudomonas* converted phenolic compounds present in the sphagnum peat plant medium to quinones. Quinones affect respiration and oxidative phosphorylation of plants and also give some protection against pathogens. Quinones could be further polymerized by the rhizoplane microflora to humus-like substances, thus increasing the water-soluble humic fraction around roots. It had been suggested that this fraction if added to a well-fertilized soil would improve vegetable growth (26) and that humic substances stimulated salt uptake (24) and influenced plant metabolism (35). Thus Eklund thought bacteria that combined gibberellin production with active stimulation of both quinone formation and humification processes would improve plant growth. *Azotobacter chroococcum* also produced dark colored humus precursors of quinones (83). The proposal that humus products are involved in the action of bacterial fertilizers is for the moment completely hypothetical. Considerably more experimental evidence is required, because humus decomposition products are very complex. In situations such as those used by Christewa (26) a number of undetermined interactions may have occurred among the humus, the manure, the minerals, and the soil, any one or all of which could have stimulated vegetable growth.

There is little doubt about the involvement of plant growth-regulating substances in the action of bacterial fertilizers. Gibberellin-like substances have also been identified in cultures of *Bacillus megaterium* (41, 69), *B. subtilis* (46), and *Pseudomonas* species (32, 41, 46, 69). Authentic growth-regulating substances applied once to seeds or seedlings affected subsequent plant development. Therefore bacterial inoculants supplying growth hormones and applied once could also cause changes in plant growth without necessarily continuously multiplying and producing more hormones in the rhizosphere. All inoculants multiplied around seedling roots and could have produced growth regulators at this critical stage. The magnitude of the plant response would undoubtedly be influenced by the amount of hormone produced and the environmental conditions, such as plant species, soil fertility, day length, light intensity, soil moisture, temperature, and length of growing season. Variations in any one of these could explain why plant responses to bacterization are unpredictable and unreliable.

### *Mineralization of Soil Phosphates*

Phosphorus occurs in soil as inorganic phosphate produced by weathering of parent rock, or as organic phosphate derived from plant, animal, and microbial residues. Organic phosphates accumulate in many soils and about half are present as inositol polyphosphates, especially *myo*-inositol hexaphosphate (4). Usually less than 5% of total soil phosphate is available to plants. Although many microorganisms can hydrolyze *myo*-inositol hexaphosphate, the inositol phosphates accumulate in soil, so there must be factors preventing hydrolysis (37).

*Bacillus megaterium* and *Pseudomonas fluorescens* are two of the bacteria decomposing organic phosphates, and several authors claimed that this activity was how inoculation improved crop yields (48, 51, 52, 60, 96, 97). Menkina (60) gave data from experiments in sand culture which showed that $P_2O_5$ was released from nucleic acid by *B. megaterium* but that most of the phosphate remained in the vessels and was not taken up by the plants.

Extrapolation to soil from results in sand is of doubtful validity, because Greaves & Webley (37) showed that soil inhibited organic phosphate breakdown. They could not detect release of inorganic phosphate when sodium *myo*-inositol hexaphosphate was incubated in soil or sand-soil mixture, nor did the bacterial population respond to the substrate. In contrast, on incubation in sand 78.6% of the substrate phosphorus was released as inorganic orthophosphate, and the bacteria multiplied. Sundara Rao et al (97) found that in soil with added farmyard manure an inoculum of *B. megaterium* solubilized bone meal, rock phosphate, superphosphate, and hydroxyapatite, and tomatoes were heavier and obtained significantly more total and fertilizer phosphorus than uninoculated plants. This was not proof that increased plant weight was a direct result of increased uptake of phosphorus released by bacterial action.

Bacterial metabolism of growth substances would yield large tomato plants which would inevitably take up more phosphate than small plants. Ramirez Martinez (80) concluded that direct participation of soil microorganisms in liberating inorganic phosphate was not proved. Martin (56) had no evidence that wheat-root microorganisms able to dephosphorylate *myo*-inositol hexaphosphate in culture did so in soil.

Further evidence against *B. megaterium* and *Pseudomonas* species influencing plant growth by mineralizing phosphate is that most significant effects occurred when mineral and organic fertilizers were added to the soil in amounts calculated to give optimum plant growth. If mineralization occurred at all, the amounts of phosphate released would be small and totally insignificant in the presence of available fertilizer. The only situation where microbial mineralization might be effective is in a phosphate-deficient soil.

It is interesting to examine the photographs published by Gerretsen (36) in the paper that stimulated much of the work on phosphate mineralization by soil microorganisms. All inoculated plants show symptoms of a type produced by treatment with a gibberellin. Could his results be explained on the basis that the different bacteria produced gibberellins that stimulated plant growth, and the larger plants therefore took up more phosphate than did the sterile controls?

## BACTERIZATION FOR BIOLOGICAL CONTROL OF PLANT DISEASES

Seed or root bacterization is also used specifically as a method for controlling root diseases. For success the inoculants must multiply in the rhizosphere and inhibit pathogens either by competition or antibiosis. If the inoculum multiplies around the emerging root but then declines or dies, it is unlikely to give permanent control, but may inhibit pathogens causing seedling diseases. To produce antibiotics the inoculants must have adequate food, especially carbon substrate, which may be provided by exudates from seed or root, and the quality of the substrate may affect the amount of antibiotic produced. Weinhold & Bowman (105) found that *Bacillus subtilis* produced more antibiotic when grown on water extract of soybean tissue than of barley tissue. These plant materials added to soil equally supported growth of *B. subtilis,* but difference in antibiotic activity could explain why a soybean cover crop and green manure prevented buildup of potato scab, whereas barley increased disease incidence. Similar effects could occur on a microscale in the rhizosphere where different substrates from plant roots could affect antibiotic activity of an inoculant and hence its ability to control disease. An antibiotic can be rapidly inactivated either by adsorption on soil particles or microbial degradation, but may be important for limited periods while the environment favors production.

Experience had shown that bacterization was not a very successful method of biological control in nonsterilized soils because of difficulties in establishing the bacteria. Most of the problems were discussed at the symposium on Ecology of Soil-borne Pathogens (7) and by Baker (8). When seed inoculation led to establishment in the wheat rhizosphere of bacteria antagonistic in vitro to *Cochliobolus sativus,* disease reactions were not affected (5). This demonstrated that organisms showing antibiosis in vitro were not necessarily those that gave biological control in soil, even when present around roots. However, there were some inoculation successes in nonsterile and partially sterilized soil using organisms antagonistic to pathogens in culture tests. It was suggested that such organisms were likely to operate only in a soil that already contained a natural population and that showed some inhibition (12). Natural activity of the antagonist in soil is a prerequisite rather similar to natural presence of the bacteria in soil treated with a bacterial fertilizer.

Seed inoculation with *Bacillus subtilis* has given promising results, significantly decreasing wilt of pigeon peas grown in sterile or nonsterile wilt-sick soil. The bacterial suspension was prepared in molasses and groundnut cake, materials that presumably enhanced antibiotic production around the seed coat and in the rhizosphere (92). Similarly, seedling blight of maize was controlled in pots of sterile or nonsterile soil by coating kernels with a suspension of *B. subtilis,* and in field trials inoculated kernels of three varieties gave stands comparable to those after treatment with captan or thiram. *B. subtilis* occurred naturally in the rhizosphere of maize, and inoculation augmented this population. In the sterile soil the bacterium was reisolated from 21 of 96 samples of seminal and adventitious roots. *Fusarium roseum* f sp *cerealis* colonized the pericarp, endosperm, and roots of maize; thus by protecting both pericarp and root, *B. subtilis* was a more effective controlling agent than chemicals that protected only the seed (25). Circumstantial evidence suggested that control was by antibiosis, but the treatment also significantly increased root emergence, root growth, and height of seedlings, so growth-regulating substances might also be involved.

Mitchell & Hurwitz (68) decreased tomato seedling mortality caused by *Pythium debaryanum* by seed inoculation with *Arthrobacter* species, but not by drenching the soil with a suspension of *Arthrobacter.* Vascular wilt caused by *Fusarium* was also decreased, but only in sterile soil. *Arthrobacter* lysed the mycelium of both pathogens. The organism was present in the seedling rhizosphere, but the population rapidly declined after two to three weeks, and normal microbial equilibrium was reestablished.

Rather than use seed treatment, Broadbent et al (13) added *B. subtilis* to soil treated by aerated steam. *Pythium ultimum* was controlled on *Antirrhinum*, possibly by hindering infection rather than by antibiosis. In the rhizosphere *Bacillus subtilis* multiplied at first but then the population became static. *B. subtilis* controlled *Rhizoctonia solani* in other experiments by lysing the mycelium, the degree of protection varying with soil and strains of fungi and bacteria (74). Undoubtedly more protection would occur in a soil favoring *B. subtilis* growth, and additions of manure and nitrogen improved control in nonsterile soil (31, 104).

Use of *B. subtilis* to control fusarium stem rot of carnation cuttings also seems promising. This disease is particularly destructive during propagation so that, provided an inoculant inhibits the pathogen during rooting, later control is not so essential. Control should also be achieved by dipping cuttings in the inoculant suspension before planting in the propagating medium. Usually this is a synthetic substance such as perlite, so problems of competition between inoculant and soil microflora do not arise. Aldrich & Baker (1) found that cuttings dipped in *B. subtilis* had fewer lesions, more extensive root development, and longer stems than controls. Changes in plant growth were another indication of hormone activity by this bacterium. Disease has also been controlled by dipping cuttings in suspensions of *Pseudomonas* and *Arthrobacter* (50, 61). *Arthrobacter* lysed the pathogen mycelium and remained a stable part of the total microflora in the propagating medium.

Recent pot and field experiments in Australia (78) to control *R. solani* by coating wheat grains with *B. subtilis* or *Streptomyces* gave decreased disease in pots, but not in the field. In both trials wheat growth was stimulated, tillering was increased, maturity hastened, dry matter increased by 24%, and grain yield by 40%. The same strain of *B. subtilis* applied to carrot seed in commercial pelleting increased yield by 40%. These experiments again indicate that bacterial inoculants alter plant growth by biological control of disease and action of growth-regulating substances, and that the two mechanisms are probably interrelated.

Other *Bacillus* species also inhibited pathogens. *B. polymyxa* added to soil decreased wilt of tomato seedlings (23), and a mixture of *B. cereus, B. mycoides,* and an unidentified species sprayed on leaves of *Pseudotsuga menziesii* controlled needle rust caused by *Melampsora medusae* (57). Although this last example of biological control is not achieved by seed or root bacterization, it demonstrates how many species of the genus *Bacillus* can inhibit pathogens.

Interestingly, the bacteria giving best disease control are those used at various times as bacterial fertilizers. Such results do much to substantiate the claim that bacterial fertilizers act as suppressors of disease in natural soil. While controlling disease some of the bacteria produce changes in plant growth indicative of hormone activity, possibly implicating these substances in the biological control mechanism. Davis & Dimond (28) found a significant decrease in disease of tomatoes caused by *Fusarium oxysporum* f. sp. *lycopersici* if the foliage was dipped into different auxins, including IAA, 4 to 10 days before inoculating with the pathogen. They suggested the hormones induced changes in the host metabolism that either regulated the parasite growth and toxin production or modified the host response. However, Volken (102) found that IAA increased, but $GA_3$ decreased *Fusarium* symptoms, and these effects were associated with changes in relations of total sugars and nitrogen due to aging in stems and leaves. In either case growth-regulating substances were involved in the expression of disease symptoms, and it can be postulated that they are involved, together with antibiosis and lysis, in control mechanisms following bacterization.

New & Kerr's (49, 73) success in controlling crown gall implicates another mechanism: competition for infection sites. Complete protection from galls was obtained by dipping peach seedling roots into a suspension of a nonpathogenic biotype of *Agrobacterium radiobacter* before planting in soil containing the pathogenic biotype. Seed inoculation also decreased galls on roots, especially in the crown region, but was less effective than dipping roots. In comparison, treatment with thiram was ineffective. The nonpathogenic biotype was predominant in the rhizosphere and was possibly replacing the pathogenic biotype at infection sites, probably root lenticels occurring at points of emergence of lateral roots. Seed rather than root inoculation might have been less effective because it led to an uneven distribution of the two forms along the root, and there were pockets on the root surface where the pathogenic form predominated.

## CONCLUSIONS

Bacterial inoculation of seeds or roots leads to changes in plant growth, sometimes to yield increases, and to biological control of some plant pathogens, effects that are probably interrelated. Changes in plant growth are caused by growth-regulating substances, especially those of the gibberellin type, and significant increases in yields of vegetables are almost certainly the result of gibberellin activity. Crops such as cabbage and lettuce are harvested in full leaf and are likely to benefit from a bacterial inoculant supplying small amounts of gibberellin, because this gives an increase in leaf area. Plants like carrots benefit from increased leaf area, giving a greater photosynthetic surface and hence more food to the storage organ, the crop. Although exogenous gibberellin causes earlier flowering and fruiting, plants have a built-in mechanism that compensates for the increased growth rate, and usually fruit ripening is delayed and yield is little affected. Hence cereals are less likely to benefit from bacterial inoculants.

Suppression of diseases, especially those affecting seedlings, can also lead to improved yields of crops sown from poor quality seed whose seedlings will be less vigorous and more prone to infection. An inoculant that either antagonizes or competes with the pathogen for colonization sites on the root, and at the same time produces stimulatory growth substances, will improve plant development and final yield.

Many environmental factors affect the ability of inoculants to produce growth regulators or antibiotics, and in field conditions these factors are uncontrollable so that beneficial effects are unpredictable. However, there are many situations in the world where even an unpredictable increase in yield of 5% would be an improvement. If treatment with bacterial inoculants together with mineral and organic fertilizers is economically possible, it should not be summarily dismissed. Development of peat inoculants should help in the economics.

Recent work on bacterization specifically to control plant diseases has shown promise in specialized environments. Antibiosis is not necessarily the only controlling mechanism; lysis or competition for infection sites or nutrients, and production of growth substances that either modify host or pathogen metabolism, may also be involved. Those organisms that antagonize in culture tests do not necessarily control disease in soil even when established in the rhizosphere. However, at the moment it is difficult to visualize other methods of selecting inoculants. Perhaps inoculation with mixtures of organisms combining antibiotic and lytic properties with an ability to produce growth regulators may prove more effective than the use of a single organism. Research into bacterization for biological control could profitably explore this possibility, as could development of techniques for modifying the rhizosphere environment to favor activity of the introduced organisms.

*Literature Cited*

1. Aldrich, J., Baker, R. 1970. Biological control of *Fusarium roseum* f. sp. *dianthi* by *Bacillus subtilis. Plant Disease Reptr.* 54:446–48
2. Alinicăi, N., Panait, T., Teşu, V. 1964. The effect of mineral and bacterial fertilizers in cultivation of *Dianthus caryophyllus* L. variety Chabaud. *Inst. agron. "I. Ionescu de la Brad" Luctari Ştiinţ.* 291–96
3. Allison, F. E., Gaddy, V. L., Pinck, L. A., Armiger, W. H. 1947. *Azotobacter* inoculation of crops. II. Effects on crops under greenhouse conditions. *Soil Sci.* 64:489–97
4. Anderson, G. 1967. Nucleic acids, derivatives, and organic phosphates. *Soil Biochemistry,* ed. A. D. McLaren, G. H. Peterson, 67–90. New York: Dekker. 509 pp.
5. Atkinson, T. G., Neal, J. L., Larson, R. I. 1974. Root-rot reaction in wheat: resistance not mediated by rhizosphere or laimosphere antagonists. *Phytopathology* 64:97–101
6. Avramov, L., Todorović, M., Lović, R. 1962. Possibility of applying bacterial fertilizers during planting of rooted cuttings of grapevine. *Arh. Poljopr. Nauke* 15:220–32
7. Baker, K. F., Snyder, W. C., Eds. 1965. *Ecology of Soil-borne Plant Pathogens.* Berkeley: Univ. California Press. 571 pp.
8. Baker, R. 1968. Mechanisms of biological control of soil-borne pathogens. *Ann. Rev. Phytopathol.* 6:263–94
9. Balasundaram, V. R., Sen, A. 1971. Effect of bacterization of rice (*Oryza sativa* L.) with *Beijerinckia. Ind. J. Agr. Sci.* 41:700–4
10. Berezova, E. P. 1963. The effectiveness of bacterial fertilizers. *Mikrobiologiya* 32:358–61
11. Brakel, J., Hilger, F. 1965. Étude qualitative et quantitative de la synthèse de substances de nature auxinique par *Azotobacter chroococcum in vitro. Bull. Inst. Agron. Sta. Rech. Gembloux* 33: 469–87
12. Broadbent, P., Baker, K. F. 1969. Bacteria and actinomycetes antagonistic to root pathogens in Australian soils. *Phytopathology* 59:1019
13. Broadbent, P., Baker, K. F., Waterworth, Y. 1971. Bacteria and actinomycetes antagonistic to fungal root pathogens in Australian soils. *Aust. J. Biol. Sci.* 24:925–44
14. Brown, M. E. 1972. Plant growth substances produced by micro-organisms of soil and rhizosphere. *J. Appl. Bacteriol.* 35:443–51
15. Brown, M. E., Burlingham, S. K. 1963. *Azotobacter* and plant disease. *Ann. Rept. Rothamsted Exp. Sta.,* 73
16. Brown, M. E., Burlingham, S. K. 1968. Production of plant growth substances by *Azotobacter chroococcum. J. Gen. Microbiol.* 53:135–44
17. Brown, M. E., Walker, N. 1970. Indolyl-3-acetic acid formation by *Azotobacter chroococcum. Plant Soil* 32:250–53
18. Brown, M. E., Burlingham, S. K., Jackson, R. M. 1962. Studies on *Azotobacter* species in soil. I. Comparison of media and techniques for counting *Azotobacter* in soil. *Plant Soil* 17:309–19
19. Brown, M. E., Burlingham, S. K., Jackson, R. M. 1962. II. Populations of *Azotobacter* in the rhizosphere and effects of artificial inoculation. *Plant Soil* 17:320–32
20. Brown, M. E., Burlingham, S. K., Jackson, R. M. 1964. III. Effects of artificial inoculation on crop yields. *Plant Soil* 20:194–214
21. Brown, M. E., Jackson, R. M., Burlingham, S. K. 1968. Effects produced on tomato plants, *Lycopersicum esculentum,* by seed or root treatment with gibberellic acid and indolyl-3-acetic acid. *J. Exp. Bot.* 19:544–52
22. Cantir, F., Comarovschi, G., Vasilică, C. 1962. The effect of bacterial inoculation on the yield of soya bean. *Inst. Agron. "I. Ionescu de la Brad" Luctari* Ştiinţ. 51–56
23. Celino, M. S., Gottlieb, D. 1952. Control of bacterial wilt of tomatoes by *Bacillus polymyxa. Phytopathology* 42:4
24. Chaminade, R. 1955. Le potassium et la

matière organique. *Potassium Symp.*, 203–14. Bern: Int. Potassium Inst.
25. Chang, I., Kommendahl, T. 1968. Biological control of seedling blight of corn by coating kernels with antagonistic microorganisms. *Phytopathology* 58:1395–1401
26. Christewa, L. A. 1958. Die stimulierende Wirkung der Pflanzen und die Effektivität der Humindünger in südlichen Gebieten der Ukrain S.S.R. *Int. Bodenk. Ges.* 11:46–50
27. Cooper, R. 1959. Bacterial fertilizers in the Soviet Union. *Soils Fertilizers* 22:327–33
28. Davis, D., Dimond, A. E. 1953. Inducing disease resistance with plant growth regulators. *Phytopathology* 43:137–40
29. Denarié, J., Blachère, H. 1966. Inoculation de graines de végétaux cultivés a l'aide de souches bactériennes. *Ann. Inst. Pasteur Paris* 111: Suppl. 3, 57–74
30. Dorosinskii, L. M. 1962. Some questions on the use of bacterial fertilizers. *Mikrobiologiya* 31:738–44
31. Dunleavy, J. 1955. Control of damping-off of sugar beet by *Bacillus subtilis*. *Phytopathology* 45:252–58
32. Eklund, E. 1970. Secondary effects of some pseudomonads in the rhizoplane of peat grown cucumber plants. *Acta Agr. Scand.* Suppl. 17:1–57
33. Elroy, R. L. 1964. Inhibition of nitrogen fixing and nitrifying bacteria by seed plants. *Ecology* 45:824–37
34. Elroy, R. L. 1965. Inhibition of nitrogen fixing and nitrifying bacteria by seed plants. Characterization and identification of inhibitors. *Physiol. Plant.* 18:255–68
35. Flaig, W., Söchtig, H. 1962. Einfluss organischer Stoffe auf die Aufnahme anorganischer Ionen. *Agrochimica* 6: 251–64
36. Gerretsen, F. C. 1948. The influence of microorganisms on the phosphate intake by the plant. *Plant Soil* 1:51–81
37. Greaves, M. P., Webley, D. M. 1969. The hydrolysis of *myo*inositol hexaphosphate by soil microorganisms. *Soil Biol. Biochem.* 1:37–43
38. Helmeczi, B. 1962. Possibilities of *Azotobacter* inoculation to maize. *Agrokem. Talajtan* 11:481–92
39. Hennequin, J. R., Blachère, H. 1966. Recherches sur la synthèse de phytohormones et de composés phénoliques par *Azotobacter* et des bactéries de la rhizosphère. *Ann. Inst. Pasteur Paris* 111: Suppl. 3, 89–102
40. Hovadík, A., Vančura, V., Vlček, F., Macura, J. 1965. Bacteria in the rhizosphere of red pepper. *Plant Microbes Relat. Proc. Symp. Relat. Soil Microorganisms Plant Roots*, ed. J. Macura, V. Vančura, 109–119. Prague: Czech. Acad. Sci. 333 pp.
41. Hussain, A., Vančura, V. 1970. Formation of biologically active substances by rhizosphere bacteria and their effect on plant growth. *Folia Microbiol. Prague* 11:468–78
42. Jackson, R. M., Brown, M. E. 1966. Behaviour of *Azotobacter chroococcum* introduced into the plant rhizosphere. *Ann. Inst. Pasteur Paris* 111: Suppl. 3, 103–12
43. Jackson, R. M., Brown, M. E., Burlingham, S. K. 1964. Similar effects on tomato plants of *Azotobacter* inoculation and application of gibberellins. *Nature London* 192:575
44. Jensen, H. L. 1940. Nitrogen fixation and cellulose decomposition by soil microorganisms II. The association between *Azotobacter* and facultative aerobic cellulose decomposers. *Proc. Linn. Soc. N. S. W.* 65:89–106
45. Jensen, H. L. 1942. Bacterial treatment of nonleguminous seeds as an agricultural practice. *Aust. J. Sci.* 4:117–20
46. Katznelson, H., Cole, S. E. 1965. Production of gibberellin-like substances by bacteria and actinomycetes. *Can. J. Microbiol.* 11:733–41
47. Katznelson, H., Strzelczyk, E. 1961. Studies on the interaction of plants and free-living nitrogen fixing microorganisms. I. Occurrence of *Azotobacter* in the rhizosphere of crop plants. *Can. J. Microbiol.* 7:437–46
48. Kavimandan, S. K., Gaur, A. C. 1971. Effect of seed inoculation with *Pseudomonas* sp. on phosphate uptake and yield of maize. *Curr. Sci.* 40:439–40
49. Kerr, A. 1972. Biological control of crown gall: seed inoculation. *J. Appl. Bacteriol.* 35:493–97
50. Koths, J. S., Gunner, H. B. 1967. Establishment of a rhizosphere microflora on carnations as a means of plant protection in steamed greenhouse soils. *Am. Soc. Hort. Sci.* 91:617–26
51. Kudzin, Yu. K., Yaroshevich, I. V. 1962. The use of phosphobacterin in the chernozem zone. *Mikrobiologiya* 31: 1098–1101
52. Kvaratskheliya, M. T. 1962. The advantages of using bacterial fertilizers. *Mikrobiologiya* 31:1102–6
53. Lehri, L. K., Mehrotra, C. L. 1972. Effect of *Azotobacter* inoculation on the yield of vegetable crops. *Indian J. Agr. Sci.* 6:201–4

54. Libbert, E., Manteuffel, R. 1970. Interactions between plants and epiphytic bacteria regarding their auxin metabolism. VII. The influence of the epiphytic bacteria on the amount of diffusable auxin from corn coleoptiles. *Physiol. Plant.* 23:93–98
55. Marendiak, D. 1964. Contribution to the question of dynamism and quantitative relations of some groups of microorganisms in the rhizosphere of maize. *Rostlinná Výroba* 10:137–44
56. Martin, J. K. 1973. The influence of rhizosphere microflora on the availability of $^{32}$P-*myo*inositol hexaphosphate phosphorus to wheat. *Soil Biol. Biochem.* 5:473–83
57. McBride, R. P. 1969. A microbiological control of *Melampsora medusae. Can. J. Bot.* 47:711–15
58. Mehrotra, C. L., Lehri, L. K. 1971. Effect of *Azotobacter* inoculation on crop yields. *J. Indian Soc. Soil Sci.* 19:243–48
59. Menkina, R. A. 1950. Bacteria mineralizing combined organic phosphorus. *Mikrobiologiya* 19:308–16
60. Menkina, R. A. 1963. Bacterial fertilizers and their importance for agricultural plants. *Mikrobiologiya* 32:352–58
61. Michael, A. H., Nelson, P. E. 1972. Antagonistic effect of soil bacteria on *Fusarium roseum* "Culmorum" from carnation. *Phytopathology* 62:1052–56
62. Minasyan, A. I., Nalbandyan, A. D. 1965. Effect of azotobacterin on rooting and growth of vine cuttings. *Dokl. Akad. Nauk Arm. SSR* 41:251–55
63. Mishustin, E. N. 1963. Bacterial fertilizers and their effectiveness. *Mikrobiologiya* 32:911–17
64. Mishustin, E. N. 1966. Action d'*Azotobacter* sur les végétaux supérieurs. *Ann. Inst. Pasteur Paris* 111: Suppl. 3, 121–35
65. Mishustin, E. N., Naumova, A. N. 1962. Bacterial fertilizers, their effectiveness and mode of action. *Mikrobiologiya* 31:543–55
66. Mishustin, E. N., Shil'nikova, V. K. 1971. *Biological fixation of atmospheric nitrogen.* London: Macmillan. 420 pp.
67. Mishustin, E. N., Naumova, A. N., Khokhlova, Yu. M., Ovshtoper, S. N., Smirnova, G. A. 1969. Antifungal antibiotic produced by *Azotobacter chroococcum. Mikrobiologiya* 38:87–90
68. Mitchell, R., Hurwitz, E. 1965. Suppression of *Pythium debaryanum* by lytic rhizosphere bacteria. *Phytopathology* 55:156–58
69. Montuelle, B. 1966. Synthèse bactérienne de substances de croissance intervenant dans le métabolisme des plantes. *Ann. Inst. Pasteur Paris* 111: Suppl. 3, 136–46
70. Mosiashvili, G. I., Topuridze, K. V., Kiriakova, N. G. 1963. Effectiveness of azotobacterin in vineyard soils. *Mikrobiologiya* 32:835–37
71. Nair, K. S., Ramaswamy, P. P., Perumal, R. 1972. Studies on *Azotobacter.* IV. Effect of *Azotobacter* inoculation on paddy. *Madras Agr. J.* 59:28–30
72. Neelakantan, S., Rangaswami, G. 1965. Bacterization of rice and okra seeds with *Azotobacter chroococcum* and establishment of the bacterium in the rhizospheres. *Curr. Sci.* 34:157–59
73. New, P. B., Kerr, A. 1972. Biological control of crown gall: Field measurements and glasshouse experiments. *J. Appl. Bacteriol.* 35:279–87
74. Olsen, C. M., Baker, K. F. 1968. Selective heat treatment of soil, and its effect on the inhibition of *Rhizoctonia solani* by *Bacillus subtilis. Phytopathology* 58:79–87
75. Paleg, L. G., West, C. A. 1972. The Gibberellins. In *Plant Physiology. VIB Physiology of Development: The Hormones,* ed. F. C. Steward, 146–81. New York: Academic. 365 pp.
76. Patel, J. J. 1969. Microorganisms in the rhizosphere of plants inoculated with *Azotobacter chroococcum. Plant Soil* 31:209–23
77. Patel, J. J., Brown, M. E. 1969. Interactions of *Azotobacter* with rhizosphere and root-surface microflora. *Plant Soil* 31:273–81
78. Price, R. D., Merriman, P. R., Kollmorgan, J. F. 1973. The effect of seed applications of selected soil organisms on the growth and yield of cereals and carrots. *2nd Int. Congr. Plant Pathol. Minneapolis,* Contrib. No. 0666
79. Rakhno, P. Kh., Ryys, O. O. 1963. The use of *Azotobacter* preparations. *Mikrobiologiya* 32:558–61
80. Ramirez-Martinez, J. R. 1968. Organic phosphorus mineralization and phosphatase activity in soils. *Folia Microbiol. Prague* 13:161–74
81. Ridge, E. H. 1970. Inoculation and survival of *Azotobacter chroococcum* on stored wheat seed. *J. Appl. Bacteriol.* 33:262–69
82. Ridge, E. H., Rovira, A. D. 1968. Microbial inoculation of wheat. *Trans. Int. Congr. Soil Sci. 9th* 111:473–81
83. Robert-Géro, M., Vidal, G., Hardisson, C., Le Borgue, L., Pochon, J. 1967.

Étude biogénétique des polymères humiques. Relation entre polymères humiques naturels, d'origine microbienne et lignine. *Ann. Inst. Pasteur Paris* 113:911–21

84. Roĭzin, M. B., Aleksandrova, N. M. 1966. Effect of azotobacterin on growth and development of decorative shrubs in the Murmansk region. *Izv. Akad. Nauk SSSR. Ser. Biol.* No. 3:452–56
85. Rouatt, J. W., Katznelson, H. 1961. A study of the bacteria on the root surface and in the rhizosphere soil of crop plants. *J. Appl. Bacteriol.* 24:164–71
86. Rovira, A. D. 1963. Microbial inoculation of plants. I. Establishment of free-living nitrogen-fixing bacteria in the rhizosphere and their effects on maize, tomato, and wheat. *Plant Soil* 19: 304–14
87. Rovira, A. D. 1965. Effects of *Azotobacter, Bacillus* and *Clostridium* on the growth of wheat. See Ref. 41, pp. 193–200
88. Rubenchik, L. I. 1963. *Azotobacter and Its Use in Agriculture.* Jerusalem: Israel Program for Scientific Translations. 278 pp.
89. Samtsevich, S. A. 1962. Preparation, use and effectiveness of bacterial fertilizers in the Ukrainian SSR. *Mikrobiologiya* 31:923–33
90. Scott, T. K. 1972. Auxins and roots. *Ann. Rev. Plant Physiol.* 23:235–58
91. Shende, S. T., Kokorina, L. M. 1964. Efficacy of the *Azotobacter* administered with mineral fertilizers in rice crop. *Mikrobiologiya* 33:467–71
92. Singh, P., Vasudeva, R. S., Bajaj, B. S. 1965. Seed bacterization and biological activity of bulbiformin. *Ann. Appl. Biol.* 55:89–97
93. Smith, J. H., Allison, F. E., Soulides, D. A. 1962. Phosphobacterin as a soil inoculant. Laboratory, greenhouse, and field evaluation. *US Dep. Agr. Tech. Bull.* 1263:1–22
94. Strzelczyk, E. 1961. Studies on the interaction of plants and free-living nitrogen fixing microorganisms. II. Development of antagonists of *Azotobacter* in the rhizosphere of plants at different stages of growth in 2 soils. *Can. J. Microbiol.* 7:507–13
95. Sulaiman, N. 1971. Non-symbiotic nitrogen fixation in rice plants. *Beitr. Trop. Subtrop. Landwirt. Tropenveterinaermed.* 9:139–71
96. Sundara Rao, W. V. B., Sinha, M. K. 1963. Phosphate dissolving microorganisms in the soil and rhizosphere. *Indian J. Agr. Sci.* 33:272–78
97. Sundara Rao, W. V. B., Bajpai, P. D., Sharma, J. P., Subbiah, B. V. 1963. Solubilization of phosphates by phosphorus solubilizing organisms using $P^{32}$ as tracer and the influence of seed bacterization on the uptake by the crop. *Indian Soc. Soil Sci.* 11:209–19
98. Sundara Rao, W. V. B., Mann, H. S., Paul, N. B., Mathur, S. P. 1963. Bacterial inoculation experiments with special reference to *Azotobacter. Indian J. Agr. Sci.* 33:279–90
99. Tchan, Y. T., Jackson, D. L. 1965. Studies of nitrogen fixing bacteria. IX. Study of inoculation of wheat with *Azotobacter* in laboratory and field experiments. *Proc. Linn. Soc. N. S. W.* 90:289–98
100. Thimann, K. V. 1972. *The Natural Plant Hormones* in *Plant Physiology.* See Ref. 75, pp. 3–145
101. Vančura, V., Macura, J. 1961. The effect of root excretions on *Azotobacter. Folia Microbiol. Prague* 6:250–59
102. Volken, P. 1972. Quelques aspects des relations Hôte-Parasite en fonction de traitements à l'acide indol-acétique et à l'acide gibberellique. *Phytopathol. Z.* 75:163–74
103. Volodin, V. 1969. Effect of treating seed potatoes with Mo and azotobacterin. *Kartofel Ovoshchi* 6:46
104. Voznyakovskaya, Yu. M. 1963. Choice of microorganisms for use in the composition of bacterial fertilizers. *Mikrobiologiya* 32:168–74
105. Weinhold, A. R., Bowman, T. 1968. Selective inhibition of the potato scab pathogen by antagonistic bacteria and substrate influence on antiobiotic production. *Plant Soil* 27:12–24
106. Wilkins, M. B. 1969. *Physiology of Plant Growth and Development.* London: McGraw-Hill. 695 pp.
107. Zak, G. A., Mukhutdinov, M. F. 1964. Effectiveness of bacterial fertilizers for vegetables. *Izv. Kuibyshev. Selsko-khoz. Inst.* 14:80–88

# SURVIVAL MECHANISMS OF PHYTOPATHOGENIC BACTERIA

❖3596

*M. L. Schuster and D. P. Coyne*
Departments of Plant Pathology and Horticulture-Forestry, respectively,
University of Nebraska, Lincoln, Nebraska 68503

## INTRODUCTION

Natural habitats usually do not provide bacteria the continuity of agricultural crops. With continuous culture, perpetuation of pathogen is no problem. Although agricultural practices provide some discontinuity between crops, it is less than that in nature. Uniformity of crop germ plasm also favors inoculum buildup and perhaps perpetuation of the pathogens.

The growth of most plant pathogens is discontinuous, because of the seasonal effect upon either the pathogen or the host. A successful pathogen must be able to bridge discontinuities, such as the gaps between successive crops and seasons. To reestablish when conditions are again favorable, inoculum must survive. Facultative saprophytes or facultative parasites are not as handicapped by discontinuous growth as are obligate parasites. Discontinuous growth of the pathogen mainly decreases the amount of inoculum (bacteria that cause disease when placed in suitable contact with the host).

The success of a bacterial plant pathogen depends in part on the amount of inoculum (bacterial cells) it produces. Because bacteria have a short generation time, a small amount of surviving primary inoculum can rapidly produce an epidemic. What is the minimum amount of inoculum necessary to initiate disease? Enough for mere survival is not necessarily the answer since transmission to a host is necessary. The source, exit, and transmission of primary inoculum are all necessary for occurrence of disease, for survival, and for continuity of the bacterial species. Establishment of a "curtain" between host and pathogen may be simpler and cheaper during the primary inoculum phase than thereafter.

The longevity of primary inoculum is important in the success of bacterial pathogens and depends upon its ability to escape or endure adverse environmental conditions (62). Survival varies with the form of primary inoculum and may depend on external factors as well as on the internal makeup of the pathogen. Because there are about 200 different species of bacterial plant pathogens, variations in modes of survival are certain to occur. The species included in this paper are limited to those that illustrate concepts and principles. Actimomycetes are excluded (14).

There are five genera of bacterial phytopathogens: *Agrobacterium,* (*A.*), *Corynebacterium* (*C.*), *Erwinia* (*E.*), *Pseudomonas* (*P.*), and *Xanthomonas* (*X.*). They are aerobic non-spore-forming rods. Only *Corynebacterium* is Gram-positive. Most are motile with polar or peritrichous flagella, but a few are atrichous.

Phytopathogenic bacteria do not form resting spores or structures comparable to fungi or nematodes; they remain dormant during the quiescent period in association with the following animate or inanimate agencies: (*a*) seeds, (*b*) perennial plant hosts or parts, (*c*) insects, (*d*) epiphytes, (*e*) plant residues, (*f*) soil and other nonhost materials. Longevity of pathogens in these agents under natural and artificial environmental conditions are discussed in this review, and selected references are provided.

## ASSOCIATION WITH SEED

"The flowers of all the tomorrows are in the seeds of today!" (22). The term *seed* is here used in a popular sense, and includes fruits, such as caryopses and achenes, but excludes vegetative propagules (e.g. seed potatoes). Seeds are an ideal agency for survival of plant pathogens when the growing host is lacking.

Orton (74) suggested in 1931 that any bacterial phytopathogen is likely to be transmitted by seed; he listed 59 such pathogens of the 128 species then described. In an annotated list of seed-borne pathogens, Noble & Richardson (73) included 95 species and varieties of bacteria, about half of the phytopathogenic species described. Baker & Smith (6) and Baker (5) outlined the essentials of the subject, referring to only a few bacterial species.

Many seeds, especially those of temperate zone plants, have a period of dormancy. Although vegetative cells of bacteria are not subject to dormancy per se, it is important that their survival correspond to that of seed viability. The occurrence of a virulent bacterium on, in, or accompanying a prescribed amount of seed does not assure transmission; various environmental and inherent seed factors could affect transmission. Transport of a bacterial pathogen in seed is an important means of survival and transfer in time and space. Another factor concerning effectiveness of seed transfer is the time of survival of the bacterium on or in seed. Most pathogens survive as long as the seed is viable.

Some bacterial pathogens die before seed loses its viability, while many survive even beyond the time of germinability. Those that infect *Phaseolus vulgaris* and other legumes are excellent examples, and *C. flaccumfaciens* was viable after 5 to 24 years (139). Schuster & Sayre (105) isolated virulent *C. flaccumfaciens* var. *aurantiacum, X. phaseoli,* and *X. phaseoli* var. *fuscans* from 15-year-old bean seed, and *C. flaccumfaciens* var. *violaceum* from 8-year-old seed kept at 10°C. Basu & Wallen (7) found viable *X. phaseoli* in seeds after 3 years at 20–35°C and nonviable in other 2-year-old seed. Although aging of seed was suggested as a possible control (13), it is too variable and unreliable for use as a control method. The discrepancies in longevity of bacteria in beans apparently are due to storage temperatures and moisture, bacterial species and strains, and length of experiments.

The development of the seed-bean industry in the semiarid West was based on relative freedom from several bacterial pathogens (15): *P. phaseolicola, X. phaseoli, X. phaseoli* var. *fuscans, C. flaccumfaciens,* and *P. syringae.* However, the bean industry sustained disastrous economic losses during 1963–1966 from severe outbreaks of bacterial diseases. Halo blight occurred in commercial fields in Wisconsin (132) and Nebraska (101) planted with western-grown seed, as well as at Twin Falls, Idaho (41, 77, 132) where seed production is concentrated. A pilot 600-acre snap-bean acreage, for example, in McCook, Nebraska was devastated in 1964 by *P. phaseolicola* (101).

The importance of bacterial diseases to the bean industry demands the use of pathogen-free seed. Recent outbreaks of bacterial diseases of snap and dry edible beans may be attributed in part to weather favorable to halo blight (16, 41) and to changes in races of the pathogen (41) in Idaho during the 1963–1966 growing seasons. The occurrence of three bacterial pathogens in bean plants from certified seed emphasizes the problem of control of the bacterial diseases (93, 127). The explanation offered (40, 41) is that seed contamination might occur during harvesting and processing even though the disease is rare and difficult to detect by field inspection. This explanation may seem unlikely since the extensive epidemics in the 1960s imply more than rare amounts of seed infestation. Nevertheless a few infested seeds may result in considerable losses in humid bean production areas; for example, in Wisconsin a dozen seed infected with *P. phaseolicola* per acre (0.02%), distributed at random, promoted a general epidemic (132). Epidemiological work, especially in Canada, has shown that 0.5% seed infection with *X. phaseoli* var. *fuscans* can be disastrous (133). Sabet & Ishag (91) considered survival of *X. phaseoli* in *Dolichos* seeds of little practical significance even though 2% of seed had viable bacteria after storage for 549 days. They claimed that the symptoms occurring on seedlings were atypical for secondary infections, but typical for systemic infections. Nevertheless, secondary spread should be possible from systemically infected seedlings.

Sutton & Wallen (121) thought that *X. phaseoli* var. *fuscans* assumed epidemic proportions in Ontario, Canada shortly after the introduction of a new bean cultivar, Sanilac, emphasizing that a new pathogen may be introduced into an environmentally suitable region and become a serious problem. Widespread epidemics can occur when seed for large areas is grown in one locality when favorable conditions for disease occur. *P. syringae* has been isolated from Wisconsin- and Idaho-grown bean seed; Idaho beans did not exhibit the brown-spot disease in the field. Wisconsin beans recently sustained losses from this ubiquitous pathogen (47).

Taylor (125) found that a high proportion of bean seeds infected with *P. phaseolicola* failed to produce infected plants; this may account for very low levels of primary infection reported in the field. Taylor devised a means of quantitative estimation of seed infection, but no data are available on the level of seed infection necessary for primary infection. Ten percent of infected plants arose from infected seeds (125). What value is the detection of 3 bacterial cells per seed if it requires

100,000 to initiate seedling infection? Workers (47, 120, 125, 135) have painstakingly devised procedures to detect one infected seed in 10,000 to 40,000 or more, but the basic question regarding number of cells required to initiate seedling infection remains unanswered. Removal of infected seed based on discoloration did not eliminate all infected seed (125); a point not determined by Taylor was whether so-called clean seed would give rise to infected plants. The degree of success of transfer depends on the inoculum load on the seed; a heavily infested seed may not germinate or may give rise to a weak seedling that dies before emergence. *P. phaseolicola* is more apt to fall in this category than are other bean pathogens.

Most investigators agree that *X. malvacearum* overwinters principally in cotton seed fuzz which remains on the seed coat after delinting. The internal infection by the organism in seed rarely attains 24% in artificially inoculated bolls. On the other hand, Massey (66) reported that internal infection of seed occurs in nature, but this could not be demonstrated artificially. Brinkerhoff & Hunter (11), Hunter & Brinkerhoff (49), and others also reported internal infection. Stoughton (116) found no evidence of the bacterium within the seed; infections of the cotyledon margins occur during germination, and this process is influenced by environmental conditions. Fallen bolls containing seed furnish a source of overwintering bacteria, and volunteer seedlings from such bolls in the spring become a source of infection for the planted crop. Brinkerhoff (9) and Schnathorst (94) recovered the viable pathogen on the seed after several years, although it lost viability more rapidly than the seed. A relatively small amount of inoculum is required to initiate an epidemic (11). Conflicting results of survival in seed may be attributed to systemic vs nonsystemic infection. The preponderance of evidence indicates that angular leaf spot is nonsystemic; however, Massey (66) thinks it is systemic and that cotyledons are infected while still enclosed in the seed coat. According to him, the pathogen is never truly vascular but passes through the tissue external to the vascular strands or in the intercellular spaces of the cortical parenchyma. Wickens (136) claims the bacterium is on the seed as a surface contaminant of the micropyle. Control through acid delinting suggests that the pathogen is principally lint-borne (95, 124).

Bacterial leaf blight is recognized as one of the most important diseases of rice in Asian countries. *X. oryzae* overwinters on diseased grains stored in farmhouses as well as in rice straw. It normally is found in husk tissues (123, 131), and has not been detected in unhulled rice grains in Japan. It has been located in glumes and occasionally in the endosperm in severely infected rice in China (28). One of the most important sources of inoculum in India is infected seed, and seedlings from infected seed are usually diseased (24, 112).

Two bacterial diseases of corn deserve consideration from the standpoint of kernel infection and survival mechanisms. These are Stewart's bacterial wilt caused by *E. stewartii* and the recently discovered Nebraska leaf freckle and wilt induced by *C. nebraskense* (104). Both invade the kernels via the vascular system and contaminate the seed surface from bacterial exudate on the inner husks (51, 104, 109). The organisms are present in the old vascular tissue of the chalazal region of the endosperm (51). There is no evidence of *E. stewartii* in the embryo region. On occasion *C. nebraskense* is found around the embryo, but under these circumstances the kernels are small and do not germinate (104). The vascular elements of the pedicel terminate in the chalazal region; further progression into the kernel is by dissolution of the chalazal areas, resulting in lysigenous cavities filled with bacterial ooze (51, 104). Both bacterial species were still viable and pathogenic in one-year-old kernels. Low seed transmission (about 2%) of *E. stewartii* was reported when kernels were planted in sterilized soil (80). Ivanoff (51), on the other hand, found no transmission to seedlings from infected seed in autoclaved or in field soil. Schuster et al (104) found that kernels infected with *C. nebraskense* gave less than 1% transmission when planted in autoclaved or in nontreated field soil. Explanations for the low amount of seedling infection from internally infected kernels are that both *E. stewartii* and *C. nebraskense* require wounding of host tissues for infection, and that there is a small percentage of infected embryos, which are the biological equivalents of infected seedlings. As in most seeds, the vascularization of corn kernels does not reach the embryo.

Shelled corn kernels usually retain the pedicels, providing a protective cover for their vascular elements and the infected chalazal regions. Because of this characteristic, 5-min treatments with tetracycline (100 and 200 ppm) or Clorox® (15%) were ineffective, whereas 10-min soak in mercuric chloride (1:1000) was effective in controlling *C. nebraskense* in the seed, based on plate culturing. Infection was decreased by 90% after 10 to 15-min soaking of *E. stewartii*-infected seed in mercuric chloride 1:1000, but chlorophol 0.25%, or several organic mercuries in up to 24 hr treatments were ineffective in decreasing the amount of infected plants from such treated seed (85). The percentage of transmission, which was at times over 50%, is surprising in light of reported average transmission of 2%. Since the experiments were field-conducted and infection data were recorded at the canning stage, insect transmission of *E. stewartii* might have been overlooked.

A few pathogens infect the seeds through the vascular elements (*X. campestris, X. incanae*). *C. michiganense* invades the tomato fruit via the vascular elements, but seed contamination results during extraction. Thus the tomato-canker bacterium accompanies the seed but apparently is external to it. Rates of transmission to seedlings from infested seed ranged from 1–4%. Strider (118) reviewed the variable data on rates of transmission of *C. michiganense* to seedlings from infested seed. A few diseased plants in a seed bed can result in a high percentage of diseased plants. Infested seed in a seed bed contaminate the soil as they decompose, resulting in root infection of adjacent plants. Schuster & Wagner (106) found that *C. michiganense* can infect unwounded tomato roots, resulting in 100% infected plants.

That seed anatomy is related to seed infection and transmission has already been demonstrated for two pathogens of corn. The testa of legume seed contains vascular elements that may be extensive in large seeds. In bean, cotton, cucurbits, and tomato the vascular tissue (raphe), a continuation of the funiculus, provides favorable sites for internal transmission of pathogens. The parasites are retained in the raphe and may also invade the embryo, e.g., *P. pisi* (108), *P. phaseolicola, X. phaseoli,* and other bean pathogens (139). The bean bacteria may gain entrance through seed openings, such as hilum, micropyle, and threshing injuries (40, 41, 139). The natural openings through tomato seed coats into the space between the testa and endosperm cuticle is easily penetrated. In most annual plants bacterial pathogens may survive nonpathogenically on or inside the testa of dry seeds in a manner comparable to epiphytic saprophytic bacteria on plant parts.

The manner of germination may affect transmission of seed-borne bacteria. Seeds with hypogeal germination (Gramineae, pea, sweet pea) could affect infection by *C. fascians* (4), *P. pisi* (108), *X. translucens* (134), *C. nebraskense* (104), and others. This type of germination could limit transmission of bacteria that infect only aerial parts. Epigeal germination (beans, alfalfa, beet, cabbage, carrot, cotton, garden stock, lettuce, pepper, castor bean, and tomato) could favor transmission of seed-borne bacteria that infect the aerial parts (exemplified by *X. campestris* in cabbage, *P. phaseolicola* of bean, *X. incanae* of stock, *X. carotae* of carrot, and *C. michiganense* of tomato), but may limit pathogens infecting subterranean plant organs. These characteristics for five classes of seed-borne pathogens were compared by Baker (5). The success of seed-borne bacteria is dependent on their location on the seed, anatomical structure and germination type of seed, survivability, and the bacterial species itself.

*C. sepedonicum,* the cause of the ring rot of potatoes, can infect tomato seed, aiding its persistence and spread (58). The bacteria are restricted to the vessels of the xylem. Alfalfa seed grown in areas infested with *C. insidiosum* can provide an important means of survival and spread of this important pathogen (21). However, Cormack & Moffat (21) made no transmission studies to seedlings but thought the seed-borne organism could be introduced into the soil.

Leppik (64) made a plea to the First International Congress of Plant Pathology for post-entry surveillance of introduced seed as a safeguard against importation of seed-borne pathogens. New strains of *X. phaseoli* recently recovered from bean seeds from Colombia and Uganda were much more virulent than Nebraska strains (100). Localized production of pathogen-free-seed has an inherent disadvantage in that diseases developing in these areas quickly spread through the distribution area, as shown by the bean seed infections in USA in the 1960s. Bean varieties with a narrow genetic base are vulnerable to a pathogen. This is especially true for snap beans because of the specific demands concerning the type and quality of beans (138). Through centralization of seed production, the producer may eliminate or minimize important problems for his entire clientele through distributing clean seed, if other infection sources are eliminated. Some seed-borne pathogens persist in soil in infested plant residues and serve as an infection reservoir for succeeding crops. Nevertheless, such other sources for seed-borne pathogens are not requisite for perpetuation of the diseases.

## ASSOCIATION WITH PLANT RESIDUES

Although the primary inoculum source of *Xanthomonas malvacearum* is infested seed, the pathogen persists in dry plant tissues for years (9, 66). The bacterium in infested debris buried in soil was infective until debris was thoroughly decomposed (10). In the arid climate of Sudan, infested debris is a threat to the next cotton crop.

In moister Tanganyika, *X. malvacearum* barely survived between cotton crops in refuse on the soil surface.

Although seed free of *Xanthomonas phaseoli* is commonly planted, common blight occurs where crop rotation is not practiced (139). *X. phaseoli* survives at least one winter in infested bean straw (96, 102). *P. phaseolicola* can live overwinter in stems, pods, and leaves on the soil surface in New York (71). Infested bean straw is important in the overwintering of *P. phaseolicola* (Race 1, Race 2, and Nebr. 16), *X. phaseoli, X. phaseoli* var. *fuscans, C. flaccumfaciens, C. flaccumfaciens* var. *aurantiacum,* and *P. syringae* (98, 99, 101). Bacteria in infested straw on the soil surface survives better than in debris buried 8 inches in the soil. Recovery 22 months later was successful.

Bacteria survived best in dry bean debris (91, 98, 99). Brinkerhoff & Fink (10) found that *X. malvacearum* lost its viability sooner in nonsterilized soil than autoclaved soil. Under furrow irrigation in the semiarid Western United States, infestation with dried bean dust may be an important seed contamination, even when halo blight is rare or difficult to detect by field inspection (40, 41).

Rice stubble and straw from infested fields are sources of inoculum of *X. oryzae* in Japan (123, 131). *X. oryzae* survives until spring in dried rice straw in farmhouses, but only one or two months when straw is plowed into soil. Bacteria can survive in rice stubble in the field in the warm areas, but not in northern Japan (69).

*X. vesicatoria* overwinters in infested debris in tomato fields (1, 81).

*Erwinia stewartii* was not recovered from soil under a diseased corn crop, but was from overwintered corn stubble (51, 80, 87). *Corynebacterium nebraskense* (103, 104) and *E. stewartii* invade the vascular elements and parenchyma, producing profuse amounts of bacterial slime in the stalks and leaves. *C. nebraskense* also overwinters in the field in corn residues, particularly on the soil surface.

## ASSOCIATION WITH PERENNIAL HOST

Survival of pathogenic bacteria in perennial plants requires conditions different from those in annuals. Bacteria remain in the living, but perhaps dormant tissues of the perennial host during the off-season in temperate climates. In the tropics or where the intervals between crops is brief, the dry or dormant period may be brief or absent. Certain pathogens, such as *X. juglandis, P. mors-prunorum, X. citri, E. amylovora,* and *X. pruni* (23) overwinter principally in holdover cankers or blighted twigs, which may produce a bacterial exudate furnishing initial spring inoculum. Varieties most subject to fire-blight blossom infection, have, as a rule, the most overwintering sources of primary inoculum.

Although small percentages of *E. amylovora*-induced cankers contain viable bacteria, it is generally agreed that under favorable conditions a small number of holdover cankers is sufficient to initiate a spring epidemic of blight. Percentages of blight cankers harboring viable *E. amylovora* ranged from 2.0 (89) to over 50 in California (128). Viable *E. amylovora* in a blight canker is limited to the region immediately surrounding the periphery of the discolored margin. Rosen's (89) report that cankers with well-defined, suberized, and cracked margins maintain the parasite over winter is in error.

Survival in holdover cankers does not adequately explain severe fire blight in orchards where the disease has not been noted for one to several years (53). Virulent *E. amylovora* may exist as a natural resident in healthy stems, shoots, and buds of apple and pear (54) and move into newly developed shoots, with or without producing blight symptoms. A severe outbreak of fire blight of resistant pears in northern Arkansas was associated with a severe hailstorm suggesting that the causal organism was internal (52, 129). Epidemics in orchards where the disease has not been observed may also result from wind dissemination of aerial bacterial strands. In a comparable situation, walnut catkins provide a perennial source of infection, and when mature they shed contaminated pollen; this explains the epidemic occurrence of *X. juglandis* when only a few cankers are evident (3).

Two alfalfa pathogens, *Xanthomonas alfalfae* and *Corynebacterium insidiosum,* overwinter in the host (19). Wounds are necessary for infection by the wilt bacterium (*C. insidiosum*) but not for the other pathogen. Winter frosts and mowing provide injuries for entry of the wilt pathogen. Using the invaded vessels as indicators, it was found that some active infections had occurred 17 years before. Infected plants usually succumb after the second year, but apparently spring inoculum can arise for many years from persistent infected alfalfa plants.

## EPIPHYTES AS INOCULUM SOURCES

Although considered to be poor saprophytes in natural soils, phytopathogenic bacteria manifest a diversity of facultative combinations with higher plants. Evidence is accumulating that different plant organs and plant species have characteristic epiphytic bacterial floras (61, 82). These epiphytes have been found on roots (rhizoplane), buds (gemmiplane), and leaves (phylloplane). Epiphytic organisms may also survive on seed surfaces. Recent investigations have stressed the occurrence of phylloplane bacteria.

Epiphytic bacteria may be a source of primary inoculum. Hagedorn et al. (43) recovered *P. syringae* throughout the year from leaf surfaces of healthy *Vicia villosa,* and associated natural outbreaks of bean brown spot with the epiphytes on nonsusceptible hairy vetch. *P. syringae* was also isolated from bean debris overwintering in Wisconsin until April, but not May (47). Schuster (99) recovered *P. syringae* in Nebraska in May and June from overwintered bean debris. The discrepancy may be due to sensitivity of isolation methods, or differences in strains of the pathogen or in environment. That this ubiquitous pathogen resorted to resistant nonhost plant species for its survival was unexpected.

Haas (42) found under artificial inoculations that *X.phaseoli* var. *fuscans* survived on the phylloplane of primary leaves of *Phaseolus vulgaris* cultivar Sanilac, but disappeared quickly from the unifoliate leaves. Mew & Kennedy (67) found that varieties of *Glycine max* differentially supported strains of *P. glycinea* epiphytically; on susceptible leaves the bacterium increased 1000-fold within 2 weeks, but it remained unchanged or declined on the leaf surface of resistant cultivars. Race specificity of *P. glycinea* was correlated with the resident phase of the pathogen on the phylloplane of soybeans. Leben (61) and co-workers showed that three pathogens have a resident phase on host plants, namely *X. vesicatoria* on tomato, *P. glycinea* on soybean, and *P. syringae* on beans.

The phylloplane growth of *P. mors-prunorum* was suggested as a primary source of inoculum for infection of branches of stone fruit trees in the autumn (23). This and another important pathogen of stone fruits, *X. pruni*, have a foliar stage that alternates with an active winter stage in the branches or stems. Although the role of leaf epiphytes in the annual cycle of the disease is still obscure, fewer cells of *P. mors-prunorum* resided on leaves of resistant cherry variety than on susceptible ones. This was substantiated, using two species of bacteria on cherries and pears, with survival on the natural host in each instance (23). *X. citri* survived overwinter as a leaf epiphyte on 17 nonsusceptible weeds in citrus groves (39).

Bud epiphytes may serve as a primary inoculum source. Although they can be found on annuals, they would be especially important in perennials (61). Goodman (35) recovered overwintering *Erwinia amylovora* and *E. herbicola* from healthy apple buds; *E. herbicola* was inhibitory for *E. amylovora* (36).

Research of the aerosphere flora is limited, but that of the rhizophere and the vicinity of roots is more extensive. Epiphytic bacteria migrate from seed and may be found on mature plants in the field. Certain phytopathogenic bacteria are able to survive in the rhizoplane of nonhost plants. *Pseudomonas tabaci* and *P. angulata* colonize root surfaces of wheat, clover, vetch, and certain weeds, which thus might be their overwintering sites; these organisms may persist in the soil indefinitely, apparently in association with roots. *X. vesicatoria* was found to overwinter on wheat roots but *P. phaseolicola* and *X. phaseoli* var. *sojense* did not (25).

Stanek & Lasik (113) discovered *X. phaseoli* var. *fuscans* colonizing bean roots, but it disappeared after two weeks; root exudates retarded the pathogen, and seed exudates stimulated its growth. Apparently plant metabolism is altered during the transition from the cotyledonary stage to the photosynthetic assimilation stage. Diachun & Valleau (25) initiated isolation studies of bacteria from wheat roots one month after seeding; that bean and soybean bacteria survived during the early development of the wheat was demonstrated in gnotobiotic tests with plants less than a week old. *X. malvacearum* isolated from roots of 14 different weed species in blighted cotton fields were thought to be unimportant in overwinter survival of the pathogen because samples collected in winter were nonpathogenic. Perhaps *X. citri* found in low numbers the year round on rhizomes and roots of *Zoysia japonica* could be important in survival, although the different strains on citrus and *Z. japonica* differ physiologically (39). Outdoors *X. citri* survived 6 months on the surface of *Calystegia japonica* rhizomes and for 5 months in the soil. Proof is still lacking that the rhizoplane bacteria on this common weed can provide primary inoculum for citrus infections. It is questionable whether the rhizoplane bacteria represent a partial and possibly transitory extension of the phylloplane phase. This might depend on the interactions of the pathogenic bacterium, host, and other microorganisms involved (33, 42, 113). Gibbins (33) thought that there was a paucity of data on relationships between pathogenic and nonpathogenic bacterial epiphytes on leaves, and that it was necessary to investigate mixed populations of microorganisms in the different aspects of their biology.

## ASSOCIATION WITH NONHOST MATERIALS

Many plant parasitic bacteria may reside in nonhost materials such as soil, or in plant parts. We should determine whether the bacteria survive in host parts or are free living in the soil. A case in point is the assumption that *X. phaseoli* (97) survived in the soil with the exact inoculum source not ascertained.

Buddenhagen (12) categorized phytopathogenic bacteria in soil into three groups: transient visitors, resident visitors, and residents. In the first group the soil phase is a rapidly declining one, commonly not contributing to the perpetuation of the pathogens. Perhaps most phytopathogenic bacteria fall in this category, although data are incomplete: *E. stewartii, E. amylovora, E. tracheiphila, X. citri, X. vesicatoria, X. vasculorum,* and *E. rubrifaciens* (92). Several other xanthomonads were found to overwinter in natural soil, probably in debris of diseased plants: *X. campestris, X. malvacearum, X. juglandis, X. vesicatoria* (1), *X. oryzae* (69), *X. translucens* (134), and *X. phaseoli,* and *X. phaseoli* var. *fuscans* (96).

*X. citri* underwent a rapid and continuous population decline in different nonsterile soil types tested, and vanished in about two weeks (31, 63, 79). Similar results were obtained with infested leaf debris.

Although Peltier & Frederich (79) and Goto (39) thought that *X. citri* gained entrance into the outer bark tissue through lenticels and remained dormant through the winter months, the latter worker contended that the pathogen also overwintered in soil. Peltier & Frederich (79) attempted direct soil isolations by germinating citrus seeds in infested soil. Goto (39) claimed that seedlings grown in infested soil with high *X. citri* numbers would not become infected, but that populations of $10^2$ cells/ml could easily be detected by a leaf-infiltration technique. The extent to which results of the workers from USA and Japan reflected the sensitivity of the isolation method should be examined. Both Lee (63) and Fulton (31) used a very sensitive method involving inoculation of punctured leaves with soil infusion, not unlike the multineedle inoculator used by Goto. Both Lee and Fulton employed varied conditions that were much more severe than would occur under the most favorable natural conditions for spread of the bacterium from soil to plants. Isolation of *X. citri* from soil under severely infected grapefruit trees or after removal of infected trees was negative (79). Goto (39) had no direct evidence that *X. citri* at low concentrations in soil or nonhost plants served as a source of primary inoculum in the orchards under natural conditions. The work of Lee (63), Fulton (31), and Peltier & Frederich (79) is substantiated by eradication of citrus canker in citrus areas in United States through systemic destruction of diseased grove and nursery trees and by use of strict sanitation. Despite claims by Goto and co-workers, *X. citri* may not possess sufficient survivability to long maintain populations in the soil phase. Because of the contagious nature of the pathogen it seems improbable that survival in soil and nonhost plants could serve as primary inoculum sources.

Corynebacteria do not survive very long in the free state, but may overwinter in soil. *C. sepedonicum* (110), *C. insidiosum* (72), *C. nebraskense* (104), and *C. flaccumfaciens* (13) have poor survival in natural soil. Strider (118) thought that *C. michiganense* could persist in soil for a long period of time. The pathogen kept in tubes of air-dried soil persisted outdoors for five years. Circumstantial evidence of survival for several years in field soil is commonly reported without experimental proof. *C. fascians* is common in soil (4), but there is some uncertainty regarding its biology (12) and taxonomic position there.

Certain pseudomonads are incapable of persisting in the free state in natural soils for extended periods: *P. syringae, P. pisi, P. phaseolicola, P. solanacearum* race 2, *P. tabaci, P. glycinea, P. lachrymans, P. mors-prunorum, P. savastanoi, X. pruni,* and others.

Buddenhagen's second category is characterized by organisms whose numbers gradually decline in soil; their long-term occurrence there is host-dependent, and their populations increase or gradually decrease according to cropping practices. Pathogens with an extended soil phase include *A. tumefaciens, P. solanacearum* race 1, and *E. carotovora.* The pathogens *P. solanacearum* and *Agrobacterium* spp. can be considered true soil-borne pathogens, yet both are wound parasites. Host wounding may be common.

Few bacteria survive in soil in a free state. One of the most successful pathogens in this regard is the common and important *P. solanacearum* race 1. This pathogen survives four to six years under bare fallow or up to 10 years in soils cropped to nonsusceptible plants (55). The disease may occur in the first planting of a susceptible crop on virgin land; this has been attributed to the presence of susceptible weed hosts in the natural flora (107). Because of the pathogen's low tolerance of desiccation, increase in microbial antagonism, exposure to sunlight, and absence of weed hosts, it gradually decreases in some soils under cultivation. Since these reports have been based on the occurrence of brown rot in crops replanted in the field, the differences may be due to soil factors and bacterial strains (12).

The brown-rot pathogen enters the plant through wounds in the roots and invades the xylem, inducing wilt and possible death of the plant (56). Although races 2 and 3 are transmitted independently of the soil, soil is the principal source of inoculum. Loss of viability in infested plant residues is associated with desiccation rate of tissue. Persistence in potato tubers was comparable to that in cultures exposed to cold-storage temperature. This is not in agreement with Smith's (109) report that the bacterium can survive in vitro exposure to –77°C.

Some aspects of ecology of *P. solanacearum* are still obscure (23). It has a host range of 200 plant species, wider than that of any other bacterial plant pathogen (55). Its host range, occurrence as pathogen in indigenous plants, and adaptability are consistent with root inhabitants (32). Records of the distribution of *P. solanacearum* are based more on the occurrence of disease than on the occurrence of the species. It may be widely distributed as a harmless rhizoplane organism.

*A. tumefaciens,* the crown-gall bacterium, is universally distributed. This wound parasite commonly survives in soil long enough to infect crops the following year. The bacteria can be reisolated after 669 days in artificially infested nonsterile soil without plant cover, but they gradually decrease in numbers in natural soils. They are favored by moderate temperature, alkaline, and moist soil (26). Field observations of *A. tumefaciens* suggest it is capable of long survival in soil, but, as with *P. solanacearum*, this interpretation is confused by the extensive host range, survival in plant residue, and independently distributed strains in plant material. There is, however, some evidence of prolonged persistence in soil in warm moist climates. The *Erwinia* soft-rot bacteria may be included in the category with *P. solanacearum* and *A. tumefaciens*. However, research might show them to be root epiphytes.

Buddenhagen's third ecological group of phytopathogens is typified by bacteria that reproduce in soil and by root epiphytes whose relation to plant disease is ephemeral. These include green-fluorescent *Pseudomonas* spp. that produce soft rot of plants in or near the soil (27), as well as rhizoplane bacteria and true soil saprophytes. Some of the *Erwinia* (*Pectobacterium*) soft-rot bacteria also might be included here. Vorokevich (130) thought that *Erwinia* soft-rot bacteria were not capable of long persistence in soil but survive in plant residue until it is decomposed. Evidence on soil relations of soft-rot bacteria of *Erwinia* is conflicting (12).

A frequent explanation for poor survivability of plant pathogenic bacteria in soil is that they are inhibited by antagonistic microflora. Brian (8) discussed antibiotic production in this respect. Patrick (78) demonstrated that many soil actinomycetes, bacteria, and fungi are antibiotic to phytopathogenic bacteria in culture. Of 1200 microorganisms, 120 produced large inhibition zones with 28 species of plant-pathogenic bacteria. Addition of organic matter to soil increases antagonistic microorganisms and decreases *A. tumefaciens* (26). In mixed cultures, antagonistic bacteria decreased infection by several pathogens (13, 126).

Bacteriophages are common in soil in association with diseased plants (23) and have been thought to reduce bacterial pathogens in soil (66), but this seems improbable. The requisite for control are high bacteriophage concentration and a low concentration of sensitive bacterial cells. Anderson (2) reported that, under favorable situations in vitro, the lowest initial concentration of a virulent bacteriophage required to eliminate a single cell of *Salmonella typhi* was about $10^7$ particles/ml. He concluded that change in natural bacterial numbers due to elimination of sensitive cells was improbable. Crosse (23) found that soils enriched with concentrated suspensions of *P. syringae* and *P. mors-prunorum* have rarely exceeded $10^2$ phage particles/ml of soil suspension. This yield would only lyse two or three cells, and the chance of absorption onto sensitive cells would be remote. Sutton & Wallen (120) failed to detect *X. phaseoli* in soil from infested fields or to isolate phages of this pathogen without phage enrichment. It is improbable that bacteriophage significantly affects the occurrence of *X. phaseoli* strains. Obligately parasitic bdellovibrios (114, 115) or predacious protozoans and free-living nematodes should be studied with respect to bacterial survival.

Certain bacteria can survive on nonhost materials other than soil. An example is *C. sepedonicum* which persists on planting equipment, harvesting and grading machinery, sacks, and storage bins, and is resistant to heat and desiccation (86). Planting and harvesting equipment can also be survival sites for *X. malvacearum* (94) and *P. phaseolicola* (40). The importance of nonhost agencies should not be minimized for contagious organisms such as *C. sepedonicum, P. phaseolicola,* and *X. malvacearum.*

## ASSOCIATION WITH INSECTS

Insects are *socius criminis* with certain bacterial pathogens in inciting plant diseases. A few bacteria transmitted primarily by insects are capable of survival within the bodies of the insects vectors. Specialized relationships between the bacterial pathogens and the transmitting insects have been summarized (17, 59).

Because seed and soil transmission have failed to explain the prevalence and dissemination of Stewart's bacterial wilt of corn, insect vectors have been extensively studied. The geographical distribution of this disease coincides with the prevalence of the 12-spotted cucumber beetle (*Diabrotica undecimpunctata howardi*), the corn flea beetle (*Chaetocnema pulicaria*), and the toothed flea beetle (*C. denticulata*). Dissemination of the wilt organism by adult 12-spotted cucumber beetles is not important, although the insect harbors the organism in its alimentary tract for considerable periods of time. Field observations and tests in Maryland demonstrated that most of the late spring and summer infection occurred from bacteria carried by corn flea beetles.

Of 40 insects studied, *E. stewartii* was isolated only from the intestinal tract of overwintered adults of the corn flea beetle *C. pulicaria*. Out of every 100 of these insects from different hosts and localities, 75 contained the wilt bacterium. Robert (87) found that 10–20% of the beetles emerging from hibernation carried *E. stewartii;* up to 75% of the beetles feeding on corn in midsummer may be carriers. Root-feeding insects (*Phyllophaga* spp., *Diabrotica longicornis, Hylemya platura*) expedite invasion by *E. stewartii* of seedlings from infected seed due to root injury. Since 160 of the more than 350 species of insects attacking corn cause noticeable damage, it is possible that other insects can also act in transmission and survival of the bacteria.

There is an apparent relation between winter temperatures and prevalence of Stewart's bacterial wilt. Low temperatures decrease the number of pathogen-harboring insects that overwinter. Forecasting of this disease is based on the sum of the mean temperatures for December, January, and February. In winters with mean temperatures above 37–38°C, large numbers of beetles survive. Cold winters (sum of mean temperatures below 32°C) reduce the populations and limit disease development (80).

*C. nebraskense,* causal agent of leaf freckles and wilt (LFW) of corn, is very similar in symptomatology to Stewart's disease. It does not overwinter in the corn flea beetle, *C. pulicaria* (104). LFW could become a problem in other corn-growing areas if a relationship similar to that in Stewart's disease were established with insects such as the corn flea beetle.

The cucurbit-wilt bacterium, *E. tracheiphila,* is completely dependent on cucumber beetles for its survival between seasons. Smith (109) thought that the striped cucumber beetle (*Acalymma vittatum*) was the only disseminator of the pathogen. Careful research showed that primary infection was not associated with soil or seed, but that the hibernating adult striped cucumber beetle (*A. vittatum*) and the 12-spotted cucumber beetle (*D. undecimpunctata howardi*) harbor the pathogen overwinter in their intestinal tracts and transmit it. Primary infection in the spring originates from feeding punctures of such overwintered beetles. The bacterium was recovered from a relatively small percentage of the overwintered beetles tested, but this is adequate to establish infection centers for secondary spread. The dependence of the pathogen on the insects is therefore complete. Feces from infective beetles contain virulent bacteria and may serve as primary inoculum if dropped into fresh wounds. Overwintered adults of *A. vittatum* feed on plants of wild cucumber in the spring before migration to cucurbit seedlings, and may obtain fresh inoculum. Infection of cucurbits therefore may be from secondary spread, a primary inoculum source, or both. Weather conditions that affect the abundance of cucumber beetles also influence the prevalence of bacterial wilt of cucurbits.

The potato blackleg organism, *E. atroseptica,* can live in all stages of the seed-corn maggot (*Hylemya platura*), and may persist in this insect, even though it can also persist in tubers and soil (59). The insect increases disease incidence by transporting the pathogen and placing it in ideal infection courts. Because *E. atroseptica* is a facultative anaerobe, and cork formation by the host is diminished in wet soils, infection is increased by wet, poorly drained soils. Leach (59) demonstrated that the bacterium is present in the intestinal tracts of both adult flies and larvae. Since the pathogen survives pupation, the emerged adult may contaminate eggs as they are laid. *E. carotovora* survives in the castout linings of the fore and hind intestines and in the lumen of the pupal midintestine. Survival in the fore and hind intestines must be entirely saprophytic. Leach (59) believed that soft rot of crucifers associated with cabbage maggot (*H. brassicae*) was similar to potato blackleg.

Survival of bacterial pathogens in the alimentary canals of insects is not, however, an assurance of transmission. Beetles in the subfamily Chrysomilidae commonly regurgitate their food and have been shown to transmit plant viruses (30). It is logical that, in representatives (*A. vittatum, D. undecimpunctata howardi*) of this subfamily, the mechanism of transmission could be by regurgitation. Other related beetles (*C. pulicaria, C. denticulata*) in the subfamily Galerucinae also regurgitate and presumably thus transmit internally borne bacteria. Feces, regurgitated fluids, and crushed bodies of the insects were found to carry the bacteria.

It is unlikely that symbiosis is involved in the insect-pathogen relationship of bacterial wilt of cucurbits and Stewart's bacterial wilt of corn since the vectors of both are chewing insects, and the bacteria in plant tissues on which they feed enter and pass through the intestinal tract. The *E. atroseptica* seed-corn maggot association suggests symbiosis; the association cannot be parasitic because the bacteria develop on castoff tissues. No specificity is associated with the pupation-surviving bacteria.

The relationship between *Daucus oleae,* the olive fly, and *P. savastanoi,* cause of olive-knot disease, is in a different category from the above cases. The pathogen enters the egg through the micropyle. The diverticulum is a reservoir for the pathogen, from which the alimentary tract of the fly becomes contaminated. These specialized organs indicate a highly developed association. The olive fly is not, however, essential to survival and transmission of *P. savastanoi.*

A number of aspects of insect-bacterial associations need investigation. Strains of *E. stewartii* of differing virulence should be investigated to ascertain the effect of virulence on survival in the intestinal tract and in passage through wild host plants. A similar situation may occur in cucurbit wilt. Indiscriminate feeders, such as the cucumber beetles, should also be studied. Other areas for study include determining the minimal bacterial dosage, and the effect of pathogen retention and mode of inoculation on efficiency of transmission. Does the pathogen multiply in the vector, and is it there symbiotic or pathogenic? Aposymbiosis is an antibiotic method to demonstrate symbiosis in insect vectors. For example, antibiotics eliminated pseudomonads in maggots, especially *P. savastanoi,* the microsymbiote of *D. oleae.* Streptomycin applied to adult flies prevented larval development in olives. Thus it is possible to treat plants simultaneously against hibernating pathogens and symbiote-dependent insects (57). What is the relationship of plant-pathogenic Rickettsia-like bacteria and insects in the epidemiology of a disease such as Pierce's disease of grapes? Could the leafhopper be a winter carrier of the Rickettsia-like bacterium (34)?

## DISCUSSION

Without means of distribution and the ability to live through the winter in the temperate zone or the dry periods in the tropics, a plant pathogenic bacterium would not long survive. The common pathogens are perhaps universal because they have succeeded in both dissemination and survival.

The mechanisms pathogens adopt in overwintering are few and uncomplicated. Survival through unfavorable conditions can be in association with animate or inanimate agencies. Any survival site is not mutually exclusive for individual species. In survival in host plants some pathogens have become so dependent that they become quite vulnerable when this association fails.

Although plant-pathogenic bacteria are non-sporeforming, many are tolerant of desiccation and survive for relatively long periods under dry conditions. Are these bacteria protected by some agency or material? Bacterial exudate, ooze, or slime has been considered as offering protection. This substance is commonly found in infected seed, cankers, or in living or dead plant parts. *C. flaccumfaciens,* for example, was shown to survive about five years in "dried bacterial ooze." In fact the viability of this and other bean bacterial pathogens exceeds that of the bean seed.

Workers have found, for the following bacteria, that the production of bacterial exudate, ooze, or polysaccharides in culture is comparable with that in infested host plants: *X. campestris* (122); *E. amylovora* (29, 45, 70); *X. phaseoli* (20, 59, 137); *E. stewartii* (29, 37); *E. carotovora* (20); *A. tumefaciens* (20, 46); *C. michiganense* (83); *C. insidiosum* (111); *C. sepedonicum* (111); *E. rubrifaciens* (92); and *P. solanacearum* (50). Hedrick (44) reported that 16 species of pathogens in five genera

(*Bacterium, Corynebacterium, Erwinia, Pseudomonas,* and *Xanthomonas*) produced exudate on media containing sucrose or glucose in basal casein hydrolysate.

The ooze produced by *E. amylovora* has been given much attention with respect to bacterial survival. It was believed that *E. amylovora* was very susceptible to desiccation, although the fire blight bacterium was known to survive a long time in dried exudate. Rosen (89, 90) found that the pathogen thus survived for more than 300 days at different temperatures and relative humidities. Hildebrand (45) recovered virulent cultures of *E. amylovora* from dried exudate after 15 and 25 months, but the organism survived only 13 days in moist exudate. When the exudate becomes moist in the spring the opportunity for infection is short indeed. Bacteria survived only 13 days in nutrient media in which capsules or slimy layers around the bacterial cells were absent. *E. amylovora* remains viable up to 12 months in aerial strands, another form of ooze. Both aerial strands and ooze are composed of about four-fifths matrix and one-fifth bacterial cells. Because of their ease of spread by wind the aerial strands may explain severe outbreaks of fire blight not explainable by survival in holdover cankers.

Hedrick (44), Leach et al (60), Feder & Ark (29), Corey & Starr (20), and others thought that bacterial exudates behave as a hydrophilic colloid. Its high water-holding capacity may aid bacterial survival during unfavorable conditions and seasons. This has been reported for *E. amylovora.* Leach et al (60) found that appreciable numbers of *X. phaseoli* cells survived in exudate for as long as 1325 days under a variety of conditions.

Maintenance under dry conditions commonly favors bacterial survival in plant residues. Residues on or near the soil surface are more favorable to bacterial survival than those incorporated in soil. By inference, the pathogens in dry undecayed residues are protected from antagonistic microflora (78) and from voracious protozoa (117) and free-living nematodes (75). Moisture is necessary for movement of these microfauna. Protozoa and nematodes were estimated to consume $9 \times 10^{14}$ bacterial cells/$m^2$ and about one ton of bacteria per hectare/year, respectively. These microfauna therefore may be involved in bacterial survival. Decreasing moisture is one factor that causes bacteria to become dormant. Control may thus be effected by maintaining the survival site in a moist condition. This is possible under irrigation systems. *P. solanacearum* and *A. tumefaciens* both survive better under moist conditions, and for this reason are true soil pathogens.

Living plant tissue and the debris of dead plants provide more favorable loci for bacteria than does the soil matrix. Lucas (65) showed that the morphology of an individual piece of straw may influence the species colonizing it. After some decomposition of colonized materials in soil some species disappear, while other fungi and bacteria (119) continue in sites in humus specially favorable for survival. This may be due in part to protection afforded by organic matter (84). The rate of decomposition of infested plant parts may have a bearing on longevity. For example, virulent *C. nebraskense* was recovered for a longer time from parts of corn plants resistant to decay. Organic matter in tropical soils is decomposed very rapidly, whereas in temperate zones it reaches a fairly stable equilibrium level; repeated addition of organic manures causes no permanent increase (23). Bacterial pathogens in dead residues obviously will be differentially reduced in the two climatic zones in a given period of time.

Presumably some bacterial phytopathogens may once have been soil inhabitants. Changes due to mutations may have caused free-living types to assume a parasitic habit and in due time become dependent on the plant for their nutritional requirements (68). Under natural conditions these bacteria are essentially obligate parasites, since they presumably lost their saprophytic capability.

"Bacterial plant pathogens are not soil-inhabiting organisms and are apparently unable to stand the competition in nature" (13). Many pathogens have a soil phase, even if only one of a rapidly declining population. Soil matrix may be protective in function during survival because of its colloidal properties. Chen & Alexander (18) and Robinson et al (88) found that certain asporogenous soil bacteria unprotected by colloids survived in extremely dry conditions for long periods. They found that a higher percentage of drought-tolerant than drought-sensitive bacteria were able to grow under dry conditions. They also found that the bacteria remained viable under dry conditions if collected in the stationary phase, but became nonviable if harvested in the exponential phase. It was thought that the effect of age resulted from differences in internal osmotic pressure.

Many refined techniques have been developed to detect surviving bacteria, but we question their relation to natural conditions. For example, the vacuum method of seed inoculation (38) or the use of additives are artificial. The water-soaking method or its modifications tend to duplicate nature in detecting viable pathogenic cells (25, 96). Leaf clipping or root injury duplicate natural conditions for certain diseases. Infested residues, nonhost materials, vegetative propagules, and seed lend themselves to field tests of survival. In greenhouse tests it is nearly impossible to simulate population levels and environmental conditions that occur in the field. Some pathogenic bacteria have the ability to persist on the root surface or in leaves of nonhost crops and weeds. Perhaps true soil-borne pathogens tend to survive in the rhizoplane rather than saprophytically. Pathogens that survive poorly in soil may have lost their saprophytic ability during the change from the saprophytic to parasitic habit or they could be sensitive to antagonistic microorganisms. We commonly assume that soil bacteria usually change in the direction of parasitism. Virulent isolates become avirulent in culture; why not in nature? Plant pathologists cannot readily recognize avirulent isolates. Differential selection pressures in nature affect persistence of bacterial strains and species. The greater vulnerability to antibiotics of root pathogens than soil saprophytes is in part responsible for their low competitive saprophytic ability (8). Differential persistence has been noted for races of *X. malvacearum* on different host varieties (9). Buddenhagen (12) suggested that races of pathogens best adapted to saprophytic survival are not necessarily the most virulent. However, because of their population buildup during their pathogenic phase these strains might effect saprophytic survival. We have found that the more virulent strains of *P. phaseolicola* and *C. flaccumfaciens* are better adapted for survival, and that two equally virulent strains of *C. nebraskense* differed in overwinter survival.

Phytopathogenic bacteria must have evolved successful survival mechanisms or they would have become extinct. The best example of man's attempts to eliminate primary inoculum sources is the eradication of citrus canker between 1914–1927. Recently this method also proved successful in California for *X. malvacearum* (95). Other general controls of primary inoculum consist of the use of pathogen-free seed, tubers, and cuttings, cultural practices, and resistant cultivars.

Hollis (48) discussed the origin of different disease types, and Buddenhagen (12) suggested ways in which competitive saprophytic ability of bacterial pathogens may have been altered. Crosse (23) postulated a localized origin of a bacterial pathogen. A recent example, the new bacterial pathogen *C. nebraskense,* was first noted in two locations in south central Nebraska in 1969, apparently of local origin. The primary source of inoculum is corn stubble on or near the soil surface; seed transmission is negligible. Minimum tillage and corn monoculture had been widely practiced in south central Nebraska for a decade before the sudden and widespread appearance of this new disease in Nebraska. Were the corn roots exposed to the bacteria long enough for buildup of inoculum, or for soil bacteria or another vascular parasite to adapt to corn? It is improbable that the new bacterium appeared de novo and assumed serious proportions in less than a decade. Because of the narrow corn germ-plasm base, the pathogen was favored. Perhaps the organism was present as an epiphyte or as a parasite of green foxtail (*Setaria viridis*), shatter cane (*Sorghum* spp.), or common weeds.

That nonpathogenic bacteria predominate over pathogenic in association with higher plant relations is well established. During the evolution of plants, roots have been subjected to bacteria so long that the survivors are tolerant of most soil-borne bacteria. The effect of host and nonsusceptible crops on survival needs clarification. For example, *X. phaseoli* and *C. flaccumfaciens* var. *aurantiacum* overwintered inside weeds (99). Do different bacterial strains possess the ability to colonize different plant parts? What are the active and inactive stages of survival in soil? We need to determine the environmental conditions in which activity and survival occur. What effect does cultivation have on survival in contrast to noncultivation? Debris may differ from mineral soil matrix as a site for survival. Factors inimical to survival might differentially affect bacteria in their active and inactive phases; the periods of active and inactive phases are perhaps not the same for all pathogens. Nutrient deficiencies and competition with other organisms could operate only during the active phases since the inactive stage has no demands on nutrients. Antibiosis could operate during the active stage and possibly to some extent during the dormant stage. Senescence of inactive stages may be expected to occur in soil as it does in culture and may account for decline in populations. Survival in the free state in soil may be misinterpreted; bacteria may occur not as single cells but as microcolonies in a mucilaginous matrix. The bacteria also may assume L-forms or resting cells induced by antibiosis or unfavorable conditions (76).

Acknowledgments

This work was done under Nebraska Agricultural Experiment Station Project No. 20–016. The authors gratefully acknowledge assistance of Kathryn Radant and Kathie Bragg for their patience, kindness, and diligence in translating longhand to legible type.

*Literature Cited*


1. Allington, W. B. 1961. Plant Pathology. Report of progress. *Nebr. Agr. Exp. Sta. Quart.,* 7(4):7–10
2. Anderson, E. S. 1957. The relations of bacteriophages to bacterial ecology. *Symp. Soc. Gen. Microbiol.* 7:189–217
3. Ark, P. A. 1944. Pollen as a source of walnut bacterial blight infection. *Phytopathology* 34:330–34
4. Baker, K. F. 1950. Bacterial fasciation disease of ornamental plants in California. *Plant Dis. Reptr.* 34:121–26
5. Baker, K. F. 1972. Seed Pathology. In *Seed Biology.* ed, T. T. Kozlowski, 1:317–416. New York: Academic
6. Baker, K. F., Smith, S. H. 1966. Dynamics of seed transmission of plant pathogens. *Ann. Rev. Phytopathol.* 4: 311–34
7. Basu, P. K., Wallen, V. R. 1966. Influence of temperature on the viability, virulence, and physiologic characteristics of *Xanthomonas phaseoli. Can. J. Bot.* 44:1239–45
8. Brian, P. W. 1957. The ecological significance of antibiotic production. *Symp. Soc. Gen. Microbiol.* 7:168–88
9. Brinkerhoff, L. A. 1970. Variation in *Xanthomonas malvacearum* and its relation to control. *Ann. Rev. Phytopathol.* 8:85–110
10. Brinkerhoff, L. A., Fink, G. B. 1964. Survival and infectivity of *Xanthomonas malvacearum* in cotton plant debris and soil. *Phytopathology* 54: 1198–1201
11. Brinkerhoff, L. A., Hunter, R. E. 1963. Internally infected seed as a source of inoculum for the primary cycle of bacterial blight of cotton. *Phytopathology* 53:1397–1401
12. Buddenhagen, I. W. 1965. The relation of plant-pathogenic bacteria to the soil. In *Ecology of Soil-borne Plant Pathogens.* ed. K. F. Baker, W. C. Synder. pp. 269–284. Berkeley: Univ. Calif. Press 571 pp.
13. Burkholder, W. H. 1948. Bacteria as plant pathogens. *Ann. Rev. Microbiol.* 2:389–412
14. Burkholder, W. H. 1959. Present-day problems pertaining to the nomenclature and taxonomy of the phytopathogenic bacteria. In *Omagiu lui Traian Săvulescu,* 119–27. Bucharest: Ed. Acad. Repub. Pop. Romîne
15. Butcher, C. L., Dean, L. L., Guthrie, J. W. 1969. Effectiveness of halo blight control in Idaho bean seed crops. *Plant Dis. Reptr.* 53:894–96
16. Butcher, C. L., Dean, L. L., Laferriere, L. 1967. Incidence of halo blight in Idaho. *Plant Dis. Reptr.* 51:310–11
17. Carter, W. 1962. *Insects in Relation to Plant Disease.* New York: Interscience 705 pp.
18. Chen, M., Alexander, M. 1973. Survival of soil bacteria during prolonged desiccation. *Soil Biol. Biochem.* 5:213–21
19. Claflin, L. E., Stuteville, D. L. 1973. Survival of *Xanthomonas alfalfae* in alfalfa debris and soil. *Plant Dis. Reptr.* 57:52–53
20. Corey, R. R., Starr, M. P. 1957. Colony types of *Xanthomonas phaseoli. J. Bacteriol.* 74:137–40
21. Cormack, M. W., Moffat, J. E. 1956. Occurrence of the bacterial wilt organism in alfalfa seed. *Phytopathology* 46: 407–9
22. Cowan, J. R. 1973. The seed. *Agron. J.* 65:1–5
23. Crosse, J. E. 1968. Plant pathogenic bacteria in soil. In *Ecology of Soil Bacteria,* ed. T. R. Gray, D. Parkinson. pp. 552–72. Toronto: Univ. Toronto Press. 681 pp.
24. Devadath, S. 1969. *Studies on Xanthomonas oryzae (causal organism of bacterial blight) occurring on rice.* PhD thesis. Utkal Univ., Cuttack, India
25. Diachun, S., Valleau, W. D. 1946. Growth and overwintering of *Xanthomonas vesicatoria* in association with wheat roots. *Phytopathology* 36:277–80
26. Dickey, R. S. 1961. Relation of some edaphic factors to *Agrobacterium tumefaciens. Phytopathology* 51:607–14
27. Dowson, W. J. 1958. The present position of bacterial plant diseases, and subjects for future research. *Commonw. Phytopathol. News* 4:33–35
28. Fang, C. T., Lin, C. F., Chu, C. L. 1956. A preliminary study on the disease cycle of the bacterial leaf blight of rice. *Acta Phytopathol. Sinica* 2:173–85 (In Chinese with English summary)
29. Feder, W. A., Ark, P. A. 1951. Wilt-inducing polysaccharides derived from crown-gall, bean blight, and soft-rot bacteria. *Phytopathology* 41:804–8
30. Freitag, J. H. 1956. Beetle transmission, host range, and properties of squash mosaic virus. *Phytopathology* 46:73–81
31. Fulton, H. R. 1920. Decline of *Pseudomonas citri* in the soil. *J. Agr. Res.* 19:207–23
32. Garrett, S. D. 1950. *Ecology of Root Inhabiting Fungi.* London: Cambridge Univ. Press 292 pp.

33. Gibbins, L. N. 1971. Relationships between pathogenic and non-pathogenic bacterial inhabitants of aerial plant surfaces. In *Plant Pathogenic Bacteria.* Geesteranus, H. P. M., Ed. pp. 15–24. *Proc. 3rd Int. Conf. Plant Pathog. Bact. Wageningen.* 365 pp.
34. Goheen, A. C., Nyland, G., Lowe, S. K. 1973. Association of a rickettsialike organism with Pierce's disease of grapevines and alfalfa dwarf and heat therapy of the disease in grapevines. *Phytopathology* 63:341–45
35. Goodman, R. N. 1965. In vitro and in vivo interaction between components of mixed bacterial cultures isolated from apple buds. *Phytopathology* 55:217–21
36. Goodman, R. N. 1967. Protection of apple stem tissue against *Erwinia amylovora* infection by avirulent strains of three other bacterial species. *Phytopathology* 57:22–24
37. Gorin, P. A. J., Spencer, J. F. T. 1961. Structural relationship of extracellular polysaccharides from phytopathogenic *Xanthomonas* spp. Part I. Structure of the extracellular polysaccharides from *Xanthomonas stewartii. Can. J. Chem.* 39:2282–89
38. Goth, R. W. 1966. The use of partial vacuum to inoculate bean seeds with phytopathogenic bacteria. *Plant Dis. Reptr.* 50:110–11
39. Goto, M. 1971. The significance of the vegetation for the survival of pathogenic bacteria. See Ref. 33, pp. 39–53
40. Grogan, R. G., Kimble, K. A. 1967. The role of seed contamination in the transmission of *Pseudomonas phaseolicola* in *Phaseolus vulgaris. Phytopathology* 57:28–31
41. Guthrie, J. W., Fenwick, H. S. 1967. Bacterial pathogens of beans. *Idaho Agr. Res. Progr. Rept.* 121:1–36
42. Haas, J. H. 1972. Epiphytic populations of *Xanthomonas phaseoli* var. *fuscans* on *Phaseolus vulgaris* "Sanilac." *Proc. 38th Can. Phytopathol. Soc.* 39:31
43. Hagedorn, D. J., Rand, R. E., Ercolani, G. L. 1972. Survival of *Pseudomonas syringae* on hairy vetch in relation to epidemiology of bacterial brown spot of bean. *Phytopathology* 62:672
44. Hedrick, H. G. 1956. Exudates produced by phytopathogenic bacteria. *Phytopathology* 46:14–15
45. Hildebrand, E. M. 1939. Studies on fire-blight ooze. *Phytopathology* 29: 142–55
46. Hodgson, R., Riker, A. J., Peterson, W. H. 1945. Polysaccharide production by virulent and attenuated crown gall bacteria. *J. Biol. Chem.* 158:89–100
47. Hoitink, H. A. J., Hagedorn, D. J., McCoy, E. 1968. Survival, transmission, and taxonomy of *Pseudomonas syringae* van Hall, the causal organism of bacterial brown spot of bean (*Phaseolus vulgaris* L.) *Can. J. Microbiol.* 14: 437–41
48. Hollis, J. P. 1952. On the origin of diseases in plants. *Plant Dis. Reptr.* 36: 319–27
49. Hunter, R. E., Brinkerhoff, L. A. 1964. Longevity of *Xanthomonas malvacearum* on and in cotton seed. *Phytopathology* 54:617
50. Husain, A., Kelman, A. 1958. Relation of slime production to mechanism of wilting and pathogenicity of *Pseudomonas solanacearum. Phytopathology* 48: 155–65
51. Ivanoff, S. S. 1933. Stewart's wilt disease of corn, with emphasis on the life history of *Phytomonas stewartii* in relation to pathogenesis. *J. Agr. Res.* 47:749–70
52. Keil, H. L., Imale, B. C., Wilson, R. A. 1964. Longevity of fire blight organism, *Erwinia amylovora* on pear trees in the greenhouse as demonstrated by infection after sandblast injury. *Phytopathology* 54:747
53. Keil, H. L., van der Zwet, T. 1972. Aerial strands of *Erwinia amylovora:* structure and enhanced production by pesticide oil. *Phytopathology* 62:355–61
54. Keil, H. L., van der Zwet, T. 1972. Recovery of *Erwinia amylovora* from symptomless stems and shoots of Jonathan apple and Bartlett pear trees. *Phytopathology* 62:39–42
55. Kelman, A. 1953. The bacterial wilt caused by *Pseudomonas solanacearum. N.C. Agr. Exp. Sta. Tech. Bull.* 99: 1–194
56. Kelman, A., Sequeira, L. 1965. Root-to-root spread of *Pseudomonas solanacearum. Phytopathology* 55:304–9
57. Krieg, A. 1971. Aposymbiosis, a possible method of antimicrobial control of arthropods. In *Microbial Control of Insects and Mites,* ed. H. D. Burges, N. W. Hussey, pp. 673–77. London New York: Academic 861 pp.
58. Larson, R. H. 1944. The ring rot bacterium in relation to tomato and eggplant. *J. Agr. Res.* 69:309–25
59. Leach, J. G. 1940. *Insect Transmission of Plant Diseases.* New York London: McGraw-Hill. 615 pp.
60. Leach, J. G., Lilly, V. G., Wilson, H. A., Purvis, M. R. Jr. 1957. Bacterial

polysaccharides: The nature and function of the exudate produced by *Xanthomonas phaseoli*. *Phytopathology* 47: 113–20

61. Leben, C. 1965. Epiphytic microorganisms in relation to plant disease. *Ann. Rev. Phytopathol.* 3:209–30
62. Leben, C. 1973. Survival of plant pathogenic bacteria. Abstr. 0326, *2nd Int. Congr. Plant Pathol. Minneapolis*
63. Lee, H. A. 1920. Behavior of the citrus-canker organism in the soil. *J. Agr. Res.* 19:189–206
64. Leppik, E. E. 1968. Quarantine and seed pathology. *1st Int. Congr. Plant Pathol. London* Abstr. p. 114
65. Lucas, R. L. 1955. A comparative study of *Ophiobolus graminis* and *Fusarium culmorum* in saprophytic colonization of wheat straw. *Ann. Appl. Biol.* 43: 134–43
66. Massey, R. E. 1931. Studies on black-arm disease of cotton. II. *Emp. Cotton Grow. Rev.* 8:187–213
67. Mew, T. W., Kennedy, B. W. 1971. Growth of *Pseudomonas glycinea* on the surface of soybean leaves. *Phytopathology* 61:715–16
68. Misaghi, I., Grogan, R. G. 1969. Nutritional and biochemical comparisons of plant-pathogenic and saprophytic fluorescent pseudomonads. *Phytopathology* 59:1436–50
69. Mizukami, T., Wakimoto, S. 1969. Epidemiology and control of bacterial leaf blight of rice. *Ann. Rev. Phytopathol.* 7:51–72
70. Moore, L. W., Hildebrand, D. C. 1966. Electron microscopy of *Erwinia amylovora* and *Pseudomonas phaseolicola* in bacterial ooze. *Phytopathology* 56:891
71. Natti, J. J. 1967. Overwinter survival of *Pseudomonas phaseolicola* in New York. *Phytopathology* 57:343
72. Nelson, G. A., Semeniuk, G. 1963. Persistence of *Corynebacterium insidiosum* in the soil. *Phytopathology* 53:1167–69
73. Noble, M., Richardson, M. J. 1968. An annotated list of seed-borne diseases. *Commonw. Mycol. Inst.* Phytopathol. Papers 8:1–191
74. Orton, C. R. 1931. Seed-borne parasites. A bibliography. *W. V. Agr. Exp. Sta. Bull.* 245:1–47
75. Overgaard-Nielsen, C. 1949. Studies on the soil microfauna. 2. The soil inhabiting nematodes. *Natura Jutlandica* 2: 1–131
76. Park, D. 1965. Survival of microorganisms in soil. See Ref. 12, pp. 82–98
77. Patel, P. N., Walker, J. C. 1965. Resistance in Phaseolus to halo blight. *Phytopathology* 55:889–94
78. Patrick, Z. A. 1954. The antibiotic activity of soil microorganisms as related to bacterial plant pathogens. *Can. J. Bot.* 32:705–35
79. Peltier, G. L., Frederich, W. J. 1926. Further studies on the overwintering of *Pseudomonas citri*. *J. Agr. Res.* 32: 335–45
80. Pepper, E. H. 1967. Stewart's bacterial wilt of corn. *Am. Phytopathol. Soc. Monogr.* 4. St. Paul, Minn: Am. Phytopathol. Soc. 36 pp.
81. Peterson, G. H. 1963. Survival of *Xanthomonas vesicatoria* in soil and diseased tomato plants. *Phytopathology* 53:765–67
82. Preece, T. F., Dickinson, C. H. Ed. 1971. *Ecology of Leaf Surface Microorganisms*. London New York: Academic 640 pp.
83. Rai, P. V., Strobel G. A. 1969. Phytotoxic glycopeptides produced by *Corynebacterium michiganense* II. Biological properties. *Phytopathology* 59: 53–57
84. Record, B. R., Taylor, R. 1953. Some factors influencing the survival of *Bacterium coli* on freeze-drying. *J. Gen. Microbiol.* 9:475–84
85. Reddy, C. S., Holbert, J. R., Erwin, A. T. 1926. Seed treatments for sweet-corn diseases. *J. Agr. Res.* 33:769–79
86. Richardson, L. T. 1957. Quantitative determination of viability of potato ring rot bacteria following storage, heat, and gas treatments. *Can. J. Bot.* 35:647–56
87. Robert, A. L. 1955. Bacterial wilt and Stewart's leaf blight of corn. *US Dep. Agr. Farmers Bull.* 2092:1–13
88. Robinson, J. B., Salonius, P. O., Chase, F. E. 1965. A note on the differential response of *Arthrobacter* spp. and *Pseudomonas* spp. to drying in soil. *Can. J. Microbiol.* 11:746–48
89. Rosen, H. R. 1929. The life history of the fire blight pathogen, *Bacillus amylovorus*, as related to the means of overwintering and dissemination. *Ark. Agr. Exp. Sta. Bull.* 244:1–96
90. Rosen, H. R. 1938. Life span and morphology of the fire blight bacteria as influenced by relative humidity, temperature and nutrition. *J. Agr. Res.* 56:239–58
91. Sabet, K. A., Ishag, F. 1969. Studies on the bacterial diseases of Sudan crops. VIII. Survival and dissemination of *Xanthomonas phaseoli* (E. F. Smith) Dowson. *Ann. Appl. Biol.* 64:65–74

92. Schaad, N. W., Wilson, E. E. 1971. The ecology of *Erwinia rubrifaciens* in the development of phloem canker of Persian walnut. *Ann. Appl. Biol.* 69:125–36
93. Schnathorst, W. C. 1954. Bacteria and fungi in seeds and plants of certified bean varieties. *Phytopathology* 44: 588–92
94. Schnathorst, W. C. 1964. Longevity of *Xanthomonas malvacearum* in dried cotton plants and its significance in dissemination of the pathogen on seed. Phytopathology 54:1009–11
95. Schnathorst, W. C. 1966. Eradication of *Xanthomonas malvacearum* from California through sanitation. *Plant Dis. Reptr.* 50:168–71
96. Schuster, M. L. 1955. A method for testing resistance of beans to bacterial blights. *Phytopathology* 45:519–20
97. Schuster, M. L. 1955. Bean blight survives winter in soil. *Nebr. Agr. Exp. Sta. Quart.,* 3(4):3
98. Schuster, M. L. 1967. Survival of bean bacterial pathogens in the field and greenhouse under different environmental conditions. *Phytopathology* 57:830
99. Schuster, M. L. 1970. Survival of bacterial pathogens of beans. *Bean Impr. Coop.* 13:68–70
100. Schuster, M. L., Coyne, D. P., Hoff, B. 1973. Comparative virulence of *Xanthomonas phaseoli* strains from Uganda, Colombia, and Nebraska. *Plant Dis. Reptr.* 57:74–75
101. Schuster, M. L., Coyne, D. P., Kerr, E. D. 1965. New virulent strains of halo blight bacterium overwinters in the field. *Phytopathology* 55:1075
102. Schuster, M. L., Harris, L. 1957. Find new ground for your 1958 bean crop. *Nebr. Agr. Exp. Sta. Quart.* 5(1):3–4
103. Schuster, M. L., Hoff, B. 1973. Epidemiology of leaf freckles and wilt of corn. *2nd Int. Congr. Plant Pathol. Minneapolis* Abstr. 0817
104. Schuster, M. L., Hoff, B., Mandel, M., Lazar, I. 1973. Leaf freckles and wilt, a new corn disease. *Proc. Ann. Corn Sorghum Res. Conf. Chicago* 27:176–91
105. Schuster, M. L., Sayre, R. M. 1967. A coryneform bacterium induces purple-colored seed and leaf hypertrophy of *Phaseolus vulgaris* and other Leguminosae. *Phytopathology* 57:1064–66
106. Schuster, M. L., Wagner, L. J. 1972. Control of bacterial canker and root knot of hydroponic tomatoes. *Plant Dis. Reptr.* 56:139–40
107. Sequeira, L. 1962. Control of bacterial wilt of bananas by crop rotation and fallowing. *Trop. Agr. London* 39: 211–17
108. Skoric, V. 1927. Bacterial blight of peas; overwintering, dissemination, and pathological histology. *Phytopathology* 17: 611–27
109. Smith, E. F. 1911. *Bacteria in Relation to Plant Diseases.* Carnegie Inst. Wash. Publ. 27, Vol. 2. 368 pp.
110. Snieszko, S. F., Bonde, R. 1943. Studies on the morphology, physiology, serology, longevity and pathogenicity of *Corynebacterium sepedonicum. Phytopathology* 33:1032–44
111. Spencer, J. F. T., Gorin, P. A. J. 1961. The occurrence in the host of physiologically active gums produced by *Corynebacterium insidiosum* and *Corynebacterium sepedonicum. Can. J. Microbiol.* 7:185–88
112. Srivastava, D. N., Rao, Y. P. 1964. Seed transmission and epidemiology of bacterial blight disease of rice in North India. *Indian Phytopathol.* 17:77–78
113. Stanek, M., Lasík, J. 1965. The occurrence of microorganisms parasitizing on the over-ground parts of plants in the rhizosphere. pp. 300–307. In *Plant Microbes Relationships,* ed. J. Macura, V. Vančura. Prague: Publ. House Czech. Acad. Sci. 333 pp.
114. Starr, M. P., Seidler, R. J. 1971. The bdellovibrios. *Ann. Rev. Microbiol.* 25:649–78
115. Stolp, H., Petzold, H. 1962. Untersuchungen über einen obligat parasitischen Mikroorganismus mit lytischer Aktivität für *Pseudomonas-Bakterien. Phytopathol. Z.* 45:364–90
116. Stoughton, R. H. 1930. Angular leaf spot disease of cotton. *Nature* 125: 350–51
117. Stout, J. D., Heal, O. W. 1967. Protozoa. In *Soil Biology,* ed. A. Burges, F. Raw. pp. 149–95. New York: Academic 532 pp.
118. Strider, D. L. 1969. Bacterial canker of tomato caused by *Corynebacterium michiganense. N.C. Agr. Exp. Sta. Tech. Bull.* 193:1–110
119. Strugger, S. 1948. Fluorescence microscope examination of bacteria in soil. *Can. J. Res.,* Sec. C., 26:188–93
120. Sutton, M. D., Wallen, V. R. 1967. Phage types of *Xanthomonas phaseoli* isolated from beans. *Can. J. Bot.* 45: 267–80
121. Sutton, M. D., Wallen, V. R. 1970. Epidemiological and ecological relations of *Xanthomonas phaseoli* and *X. phaseoli* var. *fuscans* on beans in south-

from culture and from infested cabbage leaves. *Can. J. Bot.* 48:645–52
123. Tagami, Y. et al 1964. Epidemiological studies on the bacterial leaf blight of rice, *Xanthomonas oryzae* (Uyeda et Ishiyama) Dowson. I. The overwintering of the pathogen. *Bull. Kyushu Agr. Exp. Sta.* 9:89–122
124. Tarr, S. A. J. 1953. Seed treatment against blackarm disease of cotton in the Anglo-Egyptian Sudan. I. Dry dressings, laboratory evaluation of bactericides and the toxicity of mercuric chlorida-iodide powders to cotton seed. *Emp. Cotton Grow. Rev.* 30:19–33
125. Taylor, J. D. 1970. The quantitative estimation of the infection of bean seed with *Pseudomonas phaseolicola* (Burkh.) Dowson. *Ann. Appl. Biol.* 66:29–36
126. Teliz-Ortiz, M., Burkholder, W. H. 1960. A strain of *Pseudomonas fluorescens* antagonistic to *Pseudomonas phaseolicola* and other bacterial plant pathogens. *Phytopathology* 50:119–23
127. Thomas, W. D. Jr., Graham, R. W. 1952. Bacteria in apparently healthy pinto beans. *Phytopathology* 42:214
128. Van der Zwet, T. 1969. Study of fire blight cankers and associated bacteria in pear. *Phytopathology* 59:607–13
129. Van der Zwet, T., Keil, H. L. Smale, B. C. 1969. Fire blight in the Magness pear cultivar in north central Arkansas. *Plant Dis. Reptr.* 53:686–89
130. Vorokevich, I. W. 1960. On the survival in the soil of bacteria of the genus *Erwinia*, causal agents of soft rots in plants. *Bull. Soc. Nat. Moscow. Ser. Biol.* 65:95–105
131. Wakimoto, S. 1955. Overwintering of *Xanthomonas oryzae* on unhulled grains of rice. *Agr. Hort.* 30:1501
132. Walker, J. C., Patel, P. N. 1964. Splash dispersal and wind as factors in epidemiology of halo blight of bean. *Phytopathology* 54:140–41
133. Wallen, V. R., Sutton, M. D. 1965. *Xanthomonas phaseoli* var. *fuscans* (Burkh.) Starr & Burkh. on field bean in Ontario. *Can. J. Bot.* 43:437–46
134. Wallin, J. R. 1946. Seed and seedling infection of barley, bromegrass, and wheat by *Xanthomonas translucens* var. *cerealis. Phytopathology* 36:446–59
135. Wharton, A. L. 1967. Detection of infection by *Pseudomonas phaseolicola* (Burkh.) Dows. in white-seeded dwarf bean seed stocks. *Ann. Appl. Biol.* 60:305–12
136. Wickens, G. M. 1953. Bacterial blight of cotton. *Emp. Cotton Grow. Rev.* 30:81–103
137. Wilson, H. A., Lilly, V. G., Leach, J. G. 1965. Bacterial polysaccharides. IV. Longevity of *Xanthomonas phaseoli* and *Serratia marcescens* in bacterial exudates. *Phytopathology* 55:1135–38
138. Zaumeyer, W. J. 1972. Genetic vulnerability in snap beans. *Bean Impr. Coop.* 15:37–40
139. Zaumeyer, W. J., Thomas, H. R. 1957. A monographic study of bean diseases and methods for their control. *US Dep. Agr. Tech. Bull.* 868:1–255

# REPLICATION OF PLANT VIRUSES

❖3597

*R. I. Hamilton*
Agriculture Canada, Research Station, Vancouver, Canada

## INTRODUCTION

For more than just practical reasons it is instructive to consider viruses in regard to the messenger capacity of the nucleic acid encapsidated in the virion. With this in mind, we can classify the plant viruses that have been characterized into four groups. This may do violence to some of the groupings of viruses that have been erected primarily on the basis of the morphology of the virion and the physicochemical properties of its protein and nucleic acid components, but it does serve a purpose in illustrating some aspects of the biology of the viral nucleic acids. One such scheme, which is really a modification of Wildy (130), is shown in Table 1.

Viruses in group 1 are characterized by viral RNA, which is believed to function directly as mRNA. Group 1*a* viruses, exemplified by tobacco mosaic virus (TMV) and turnip yellow mosaic virus (TYMV), are representative of a large number of

**Table 1** Virus groups

| Group | Nucleic Acid | Virion-Associated Enzyme | Example |
|---|---|---|---|
| 1. | Infectious ssRNA (+)[a] | | |
| | (*a*) single genome | | tobacco mosaic virus |
| | | | turnip yellow mosaic virus |
| | (*b*) split genome | | |
| | similar virions | | brome mosaic virus |
| | dissimilar virions | | alfalfa mosaic virus |
| | | | tobacco rattle virus |
| 2. | Infectious dsDNA (±)[b] | | cauliflower mosaic virus |
| 3. | Noninfectious ssRNA (–) | transcriptase | lettuce necrotic yellows virus |
| 4. | Noninfectious dsRNA (±) | transcriptase | wound tumor virus |

[a] + signifies the translatable nucleic acid strand.
[b] – signifies the strand that serves as template for transcriptase.

the plant viruses. The isolated RNA of each virus has been extensively characterized; it is infectious, and although there seems to be evidence of size heterogeneity (30), these viruses are considered to be comprised of single genomes. Group 1*b* viruses are characterized by the distribution of their genetic information over several genomes encapsidated in two or more virions of similar or different properties, depending on the virus. The RNA of brome mosaic virus (BMV) is comprised of 4 species (1.09, 0.99, 0.75, and 0.28 $\times$ $10^6$d), and these are encapsidated in virions that differ only slightly in density. On the other hand, alfalfa mosaic virus (AMV) RNA, which also consists of 4 species (1.27, 1.0, 0.76, and 0.33 $\times$ $10^6$d), is encapsidated in virions of different shape (isometric and bacilliform) while tobacco rattle virus (TRV) RNA (2.5 and 1.3–0.6 $\times$ $10^6$d) is encapsidated in tubular virions of different lengths. Infectivity of AMV-RNA inocula requires all four species; infectivity of BMV-RNA requires only the three largest RNAs; the 2.5 $\times$ $10^6$d RNA of TRV is infectious but no TRV coat protein is produced in the absence of the smaller (1.3–0.6 $\times$ $10^6$d) RNA. The cistron for coat protein is located on the smallest RNAs of each of these viruses, although there is evidence that some of the genetic information in these genomes is duplicated on the next larger genome in AMV and BMV (16, 124).

Group 2 viruses is a minor group comprised of cauliflower mosaic virus and related viruses. The DNA is double stranded (4.7 $\times$ $10^6$d) and is a single genome. Presumably one strand of the genome is transcribed by host RNA polymerase to mRNA.

Group 3 viruses are comprised of those plant-infecting rhabdoviruses such as lettuce necrotic yellows (LNYV) and other related viruses. The isolated RNA (4.0 $\times$ $10^6$d) is not infectious because it is not directly translatable. Consequently the RNA must first be transcribed by the virion-associated RNA-dependent RNA polymerase (transcriptase) to translatable complimentary RNAs of low molecular weight (57).

Group 4 viruses comprise those viruses with double-stranded RNA genomes, which are also segmented but encapsidated in a single virion. The isolated RNA is not infectious, and a virion-associated transcriptase is involved in the translation of one of the two RNA strands (132).

This review is organized with a view to examining the replicative cycles of representative viruses in each of the above groups. Most of the research in plant virology has employed viruses in group 1 (TMV and TYMV), and it is becoming apparent that other members of this group (AMV, BMV etc) are extremely useful for studies on plant virus genetics. Much of this review deals with this group.

## UNCOATING OF THE VIRAL GENOME

The events subsequent to the association of virions with the cell have been studied in abraded leaves and more recently in protoplasts. In the abraded leaf system, extracts of leaf tissue are examined by sucrose gradient centrifugation at intervals after inoculation with labeled virus, usually TMV. The evidence to date indicates that some virions undergo disassembly within minutes of inoculation (109).

The disassembly involves stripping of protein subunits in an apolar fashion (111) with the attendant release of RNA, some of which is native but most of which is degraded (65, 110). Similar studies with BMV and TYMV indicate that uncoating of their genomes is by a disassembly process in the case of BMV and by a rearrangement of the nucleocapsid to form top component in the case of TYMV (60). The uncoating of TMV-RNA is not inhibited by cycloheximide in its early stages but it appears to be affected at 4–5 hr postinoculation (109). Release of viral RNA occurs in leaves of hosts and nonhosts alike as evidenced by the more rapid uncoating of TYMV-RNA in barley leaves than in cabbage leaves (60) and by the detection of input $^{32}$P-labeled TMV-RNA in the polysomal fraction of nonhosts (54).

Protoplasts prepared from tobacco mesophyll cells can be infected with a number of plant viruses or their separated RNAs (2, 72, 78, 108, 119). The mode of virion entry appears to be by pinocytosis (20, 79) as evidenced by the appearance of invaginations, formation of intraprotoplastic vesicles containing virions, and the absolute dependence of infectivity upon the presence of poly-L-ornithine, a compound that stimulates pinocytosis in mammalian cells (99). No uncoating studies per se have been reported for virion-protoplast interactions, but it has been suggested that TMV undergoes disassembly in intraprotoplastic vesicles (20).

The site of viral RNA release has not been elucidated in these studies, although Machida & Kiho (65) suggest a two-step process in which TMV virions, partially uncoated in the cytoplasm, are completely stripped in a particulate fraction, probably in the nucleus. The possibility of some cellular membrane system serving as the site of virion disassembly or reorganization has been mentioned (60, 109), and it would appear that the protoplast system offers an excellent opportunity to test this hypothesis.

## VIRAL NUCLEIC ACID REPLICATION

### *Group 1 (Infectious Single-Stranded RNA)*

The replication of viral RNA in this group involves a double-stranded RNA (dsRNA) and an associated RNA-dependent RNA polymerase (replicase) as an intermediate in the cycle between parental and progeny genomes; a similar situation obtains for some animal and bacterial RNA viruses. Research on the replication of plant viruses has concentrated on the detection and characterization of the dsRNA, demonstration of replicase activity in vitro and characterization of the products of this enzymatic activity, and the attempted isolation and characterization of the replicase itself.

DETECTION AND CHARACTERIZATION OF dsRNA Extraction of the cellular nucleic acids from virus-infected plants several days after inoculation followed by fractionation of the nucleic acid preparation by chromatography or by treatment with DNase and RNase has resulted in the demonstration of an RNA species that is RNase-resistant and double stranded (88). If suitable precautions (i.e. blocking of RNase activity in the extraction procedure) are used, this dsRNA can be resolved into two species, one of which is completely double stranded (replicative form, RF) and the other of which is partially double stranded (replicative intermediate, RI). Recent papers have demonstrated the isolation of dsRNA from TMV-infected tobacco (42, 76), from barley stripe mosaic virus (BSMV)-infected barley (86), from BMV-infected barley (40), from AMV-infected tobacco (74), and from cowpea mosaic virus (CPMV)-infected cowpea (122, 123). The dsRNAs associated with several virus infections have been characterized, with most of the data coming from studies on TMV, CPMV, and TYMV.

The dsRNA of TMV-infected plants, purified from the crude extract by chromatography on cellulose, constitutes about 0.2% of the total nucleic acid isolated from leaf tissue (42). Nilsson-Tillgren (76) fractionated the dsRNA of TMV-infected tobacco into RF and RI by differential solubility in 1 *M* NaCl; the precipitated RI and the soluble RF were purified by three cycles of chromatography before characterization of their properties. RF sedimented at 15–16 S, the value expected for a double-stranded molecule corresponding to TMV-RNA and thermal denaturation (100°C, 1 min, 1 m*M* EDTA, pH 5.5) converted its sedimentation properties to those of TMV-RNA indicating that it was composed of two single strands with the molecular weight of TMV-RNA. RI exhibited pronounced centrifugal heterogeneity, sedimenting at 16–30 S. Treatment of RI with low concentration of RNase for 30 min at 25°C prior to centrifugation converted its sedimentation properties to those of RF, suggesting that the centrifugal heterogeneity was due to the nascent tails of newly synthesized single-stranded RNA associated with RF. Further evidence in support of the isolation of RI was obtained by comparing the relative resistance of RF and RI to RNase. Under conditions whereby $^3$H-TMV-RNA was completely susceptible to RNase, $^3$H-TMV-RF and $^3$H-TMV-RI were 90 and 42% resistant to RNase, respectively.

The observations and conclusions of Nilsson-Tillgren were confirmed and extended by Jackson et al (42) who further characterized RF and RI from TMV-infected tobacco. The major portion of the dsRNA was associated with the mitochondrial fraction, a result that confirmed previous findings (91).

Isopycnic centrifugation in $Cs_2SO_4$ indicated the presence of two components of $\rho = 1.615$g/ml (major) and $\rho = 1.630$g/ml (minor). The minor component could be eliminated by RNase treatment, suggesting that the dsRNA from TMV-infected tobacco contains RF and RI. Isopycnic centrifugation of dsRNA that had been enriched in RI by precipitation of RI in 2 *M* LiCl showed that the major species had a density of 1.630 g/ml but that it could be converted to RF ($\rho = 1.615$ g/ml) by treatment with RNase. Thermal denaturation studies of dsRNA (RF + RI) and dsRNA treated with RNase (RF) indicate a denaturation temperature ($T_m$) of 68°. Annealing of heat-denatured dsRNA in the presence of $^3$H TMV-RNA rendered most of the labeled RNA resistant to RNase, clearly suggesting that dsRNA contains TMV-RNA and its complement. Heat-denatured dsRNA was infectious, confirming the presence of TMV-RNA. The molecular weights of RF and RI were estimated at $4.0 \times 10^6$d and $5.0 \times 10^6$d respectively by polyacrylamide gel electrophoresis using *E. coli* rRNA ($5.5 \times 10^5$d) and TMV-RNA ($2.05 \times 10^6$d) as standards. The values of 4.0 and $5.0 \times 10^6$d are the expected values for native TMV-RNA in completely and partially double-stranded structures.

In a subsequent paper (43) suspensions of separated leaf cells were prepared from tobacco leaves inoculated with TMV 3–7 days previously and the synthesis of viral-related RNAs was followed in the cells by isotope ($^3$H uridine) labeling experiments. RNAs, corresponding to TMV-RNA, RF, and RI were detected by polyacrylamide gel electrophoresis; in addition, a high molecular weight RNA ($2.3 \times 10^6$d) similar to that observed by Fowlks & Young (30) and a low molecular weight RNA, ($3.5 \times 10^4$d) which was later characterized by Siegel et al (115) as a segment of the TMV genome, were detected. Experiments designed to determine the time after infection at which the labeled RNAs could be detected showed that no label could be detected in any species at 24 hr postinoculation but that labeled RF and TMV-RNA were detected in 48 hr infections. At 3–4 days postinoculation, all five species of RNA were labeled. Pulse-chase experiments suggested that RF and RI were precursors in the synthesis of TMV-RNA. Pulse labeling for 0.5–1.0 hr with $^3$H uridine showed that the label was preferentially incorporated into RF and RI and that the label could be chased from these structures into TMV-RNA during the chase period (3.4–15 hr). Similar preferential incorporation of label into RF was observed by Nilsson-Tillgren (75) in TMV-infected tobacco. An interesting observation in the work of Jackson et al (43) is that some label is retained by RF even after a prolonged chase (15 hr), suggesting that not all of the RF present in an infected cell functions directly as a progenitor of TMV-RNA. Similar observations were made by others (131) as well as by those investigating viral RNA replication in bacteria and animal cells.

An RNase-resistant RNA specific to TYMV-infected chinese cabbage leaves was isolated by chromatography of nucleic acid extracts on methylated albumin kieselguhr (88). The RNA could be detected prior to detection of infective virus. Its location in subcellular fractions was a matter of controversy for some time, but Ralph et al (89) conclusively showed that it sedimented with the chloroplasts in cell fractionation experiments rather than with the nucleoli as had been suggested by Bové et al (13).

The dsRNA of CPMV-infected cowpea was isolated by phenol-treatment of a chloroplast-nucleolar fraction followed by treatment of the partially purified dsRNA by DNase and RNase and chromatography through Sephadex (122). The average yield of dsRNA, expressed as a percentage of the total RNA of fresh leaves was in the order of 0.05%; expressed as a percentage of the chloroplast-nucleolar RNA it was 1%. The dsRNA consisted only of RF because of the use of RNase in its purification, but the dsRNA sedimented as two species with sedimentation coefficients of 18 S and 15 S corresponding to the values expected for RFs of the two RNA species (2.5 and $1.4 \times 10^6$d) of this virus.

Further characterization of the dsRNA from CPMV-infected tissue shows it to have the usual properties of dsRNA: hybridization with CPMV-RNA upon heat denaturation of the dsRNA, resistance to RNase in high salt but susceptibility to the enzyme in low salt, and a sharp heat denaturation profile with $T_m$ dependent upon the ionic strength of the buffer (123). Electron microscopy of dsRNA using a modification of the Kleinschmidt technique showed that a majority of the molecules were arranged in two size groups of 1.2–4 $\mu$m and 2.0–2.4 $\mu$m. It was assumed that the shorter dsRNA corresponded to middle component RNA and that the longer corresponded to the RNA of bottom component.

The dsRNA of AMV-infected tobacco has been recently characterized (85) after earlier work had indicated its presence (74). Four RNA species have been isolated from the four nucleoprotein components, and the molecular weights have been estimated at 1.3, 1.0, 0.7, and 0.34 $\times$ $10^6$d (85). The replicative RNA of AMV-infected tobacco labeled with $^{32}P$, purified by chromatography on hydroxylapatite and by treatment with DNase, was found to consist of three major and two minor components by polyacrylamide gel electrophoresis. Co-electrophoresis of DMSO-denatured $^{32}P$-labeled dsRNA and $^{14}C$-labeled AMV-RNA extracted from purified virions resulted in perfect coincidence of the two labels, clearly indicating that each of the AMV-RNA components had a corresponding double-stranded replicative RNA.

The dsRNA of BSMV-infected barley apparently consisted only of RF, probably because RI was discarded in the 2.0 *M* LiCl precipitation step of the purification of RF (86). Two RFs sedimenting at 11.2 and 12.7 S were detected in dsRNA preparations from barley infected with a North Dakota isolate (ND 18) of BSMV. This isolate was also characterized by the presence of two single-stranded RNA species, which sedimented at 19.5 S and 21.3 S. Further work by other investigators (41) has shown BSMV to be a split genome virus; the number of genomes varies according to the strain but RFs have been associated with each of the single-stranded RNA components investigated to date.

A similar situation was observed in virions of brome mosaic virus (BMV) which contain 4 RNA species (1.09, 0.99, 0.75, and 0.28 $\times$ $10^6$d); a fifth species with a molecular weight of 0.54 $\times$ $10^6$d was also detected occasionally (63). The dsRNA (10–50 $\mu$g/100 g tissue) isolated from BMV-infected plants was resolved by polyacrylamide gel electrophoresis into 5 species, each corresponding to one of the BMV-RNA components. Double-stranded RNA specific to BMV-infected barley had been isolated earlier (40), but the method of collecting the dsRNA (collecting material from the 12–16 S region of a sucrose gradient after centrifugation of a leaf nucleic acid preparation) precluded resolving the dsRNA into fractions of different molecular weight.

An interesting anomaly in the replication of viral RNA is a recent observation that the replication of tobacco ringspot virus (TRSV)-RNA is associated with dsRNA of very low molecular weight (94). No structures analogous to RF or RI could be detected in infected cucumber or bean. Heat denatured dsRNA hybridized with TRSV-RNA. The function of this dsRNA in the replication of TRSV is unknown.

IN VITRO SYNTHESIS The examples cited above provide the bulk of the recent evidence for the presence of a dsRNA structure in nucleic acid preparations of virus-infected plants. An integral part of the replicative apparatus is the enzyme, which using template RNA synthesizes the complementary strand in the duplex and then using the complementary strand as template synthesizes the progeny viral RNA strand. The presence of the enzyme (RNA-dependent RNA polyermase, replicase) has been inferred from a number of experiments using cell-free extracts of virus-infected plants at early stages (2–10 days postinoculation). The criterion for the activity of the enzyme has been the demonstration that labeled isotope (usually $^3$H uridine triphosphate) is incorporated into an acid-insoluble precipitate which resists degradation by RNase. Although this criterion suggests a double-stranded RNA structure, it has also become necessary to determine that the structure is in fact a double-stranded duplex containing complementary strands and that this structure is involved in the synthesis of progeny RNA. This can only be convincingly done by pulse-chase experiments which are difficult to do with whole plant tissues. Consequently, much of the research effort has concentrated on cell-free preparations from infected leaves, tissue slices, or, more recently, cell suspensions, where it is easier to effect isotope incorporation. Suppression of host-directed RNA synthesis (DNA-dependent RNA synthesis) has usually been necessary in order to detect the small amounts of viral RNA-directed synthesis; this is routinely done by incorporating actinomycin D (AMD) into the RNA synthesizing system. Virus infections involving BMV, TMV, and TYMV have provided most of the data on this aspect of viral RNA replication.

A cell-free system capable of incorporating $^3$H-UTP into a trichloroacetic acid (TCA) insoluble precipitate was used to study RNA synthesis in BMV-infected barley (106). The cell-free extract (prepared by differential centrifugation to yield a fraction pelleted at 10,000 g in 15 min) was obtained from the symptomless leaf (leaf 2) above the inoculated leaf (leaf 1) of barley plants inoculated 3–4 days previously with BMV. It was necessary to incorporate AMD (20 $\mu$g/ml of incorporating medium) in order to repress DNA-dependent RNA synthesis; under the conditions of incorporation, RNA synthesis in cell-free extracts from BMV-infected barley was largely resistant to AMD while that in cell-free extracts from healthy plants was sensitive to the drug.

The product obtained in the incorporating system from BMV-infected tissue was largely resistant to RNase in high salt buffer, but it became sensitive to the enzyme upon heating at 100°C in low salt buffer prior to incubation with the enzyme; the product in the incorporation system from healthy barley was susceptible to RNase at either ionic strength. These results indicate that the product in the incorporating system from BMV-infected tissues is double stranded. The RNase-resistant RNA from BMV-infected tissue sedimented at 12–14 S, the value reported for Rfs of several RNA viruses, while the majority of the RNA from healthy tissue sedimented at 4 S, and no peak was detectable at the position of 12–14 S RNA. Three of the four BMV-RNA species could be detected as products of the cell-free RNA synthesizing system provided exogenous RNA (yeast RNA or liver ribosomal RNA) was added to the system, ostensibly to protect the nascent RNA (59). The radioactive RNA products obtained after a 2 min pulse of $^3$H-UTP or after a 2 min pulse followed by a 2 min chase were fractionated in 2 *M* LiCl to yield RF (soluble) and RI (insoluble), and then analyzed by gradient centrifugation. Most of the LiCl-soluble RNA produced by the incorporation system in a 2 min pulse with $^3$H-UTP was in the form of RF as evidenced by its resistance to RNase and by its sedimentation coefficient (14 S); the LiCl-insoluble radioactive RNA was essentially RI and single-stranded RNA because it showed partial resistance to RNase and exhibited sedimentation heterogeneity (14–28 S). After a two minute chase with unlabeled UTP, the amount of labeled RF decreased, but this was associated with an increase in slowly sedimenting single-stranded RNA. On the other hand, the LiCl insoluble product (RI and single-stranded RNA) was composed primarily of single-stranded RNA exhibiting the sedimentation properties of BMV-RNAs 1, 2, and 3 (sedimentation coefficients of 27, 22, and 14 S respectively). These results clearly indicate that labeled nascent BMV-RNA was released from RI during a chase with unlabeled UTP and presents good evidence for synthesis of BMV-RNA in an in vitro RNA-synthesizing system. Similar results were obtained in an incorporating system prepared from broad bean leaves inoculated with broad bean mosaic virus (44), a virus with properties similar to those of BMV.

A virus-specific RNA synthesizing system was prepared from the leaves of Samsun tobacco inoculated 4 days previously with the U1 strain of TMV (14). The system was isolated by differential centrifugation (1000 g, 20,000 g) to yield a fraction that pelleted at 20,000 g. Incorporation of $^3$H-UTP into a TCA-precipitable product was the criterion of replicase activity; the products of the replicase system were assayed for RNase resistance, and estimates were made of their molecular weight by polyacrylamide gel electrophoresis. Replicase activity was resistant to AMD and DNase. No single-stranded TMV-RNA was synthesized in this system, but labeled RF and RI with estimated molecular weights of $4 \times 10^6$d and $5 \times 10^6$d were detected.

A particulate fraction (10,000 g pellet) from TYMV-infected chinese cabbage was shown to incorporate $^3$H-ATP into an RNase-resistant product which co-chromatographed with the labeled dsRNA product isolated directly from TYMV-infected leaves (90). Subsequent work by Bové and his colleagues (62) showed that the bulk (70%) of the labeled product sedimented at 17 S, it had a molecular weight of 4.2 $\times 10^6$d, twice that of TYMV-RNA, and it hybridized with unlabeled TYMV-RNA. The product had to be released from a cellular fraction sedimentable at 10,000 g by several cycles of freezing and thawing, and subsequent experiments using high resolution autoradiography of thin sections of the pellet showed that the label was associated with the chloroplast envelope. Comparative estimates of the specific activity of the replicase (cpm/mg protein) in incorporating systems containing complete chloroplasts compared to those enriched in chloroplast envelopes showed that the membrane-bound replicase had a specific activity six times that of the replicase in whole chloroplasts.

Infection of cucumber cotyledons with cucumber mosaic virus (CMV) induces the activity of an RNA-dependent RNA polymerase which can be detected in the soluble fraction (16,000 g supernatant) or in several particulate fractions (67). The particulate fraction contained two RNA polymerases, a major one dependent on added RNA for activity and a minor one independent of added RNA. Treatment of the 16,000 g pellet with 50% saturated $(NH_4)_2SO_4$ or with 30–70 m*M* $MgSO_4$ resulted in the release (about 90%) of the enzyme activity dependent upon added RNA, and its properties were shown to be the same as those reported for the soluble enzyme. The product of the particulate enzyme was 80–90% resistant to RNase, had a $T_m$ of 92°, and sedimented mainly at 8 S with a minor fraction at 14 S. Denaturation of the dsRNA product followed by sedimentation yielded an RNase-sensitive species which sedimented at 5 S and had a molecular weight of about 0.5 X $10^6$d. Hybridization of heat-denatured dsRNA with CMV-RNA or TYMV-RNA did not affect the self-reannealing properties of the dsRNA, thus making it impossible to conclude that the product contained viral RNA.

Studies with cell-free extracts from cucumber cotyledons infected with tobacco ringspot virus (TRSV) showed that this virus also induced the activity of an RNA polymerase with properties similar but not identical to those of the CMV-induced RNA polymerase (80). The replicase activity of the TRSV-induced enzyme was maximal at 3 days postinoculation, whereas the activity of the CMV-induced replicase did not reach a plateau until 10 days after inoculation. It was concluded that the enzymes were not identical because the two enzymes had different solubilities in 50–100 m*M* $MgSO_4$, as less than 10% of the TRSV-induced replicase was solubilized under conditions where 90% of the CMV-induced enzyme was released.

CHARACTERIZATION OF PARTIALLY PURIFIED VIRUS-INDUCED REPLICASES The replicases induced in tissue infected with BMV, CMV, and TMV have been partially purified by a variety of techniques, and some of their properties have been determined. Purification of the replicases essentially involves their release from the particulate fraction and template with which they are usually associated in the isolation of the crude enzyme. In general, the soluble replicases are dependent upon added RNA as template, but the specificity of the enzyme is markedly decreased. The CMV-induced replicase shows no specificity for CMV-RNA or several other viral RNAs (68), and neither does the TMV-induced replicase (134). The replicase from BMV-infected barley showed maximum activity with RNA from BMV and CCMV but no activity with TMV-RNA (34). These observations indicate that purification procedures may remove some fractions that control the specificity of an RNA polymerase. It has been shown, for example, that the RNA polymerase activity of bacteriophage QB replicase requires at least two host proteins obtained from either uninfected or QB-infected bacteria and which are involved in the early steps of the QB-RNA: replicase interaction prior to actual synthesis on the QB-RNA template (33).

The molecular weights of the replicases have been estimated at between 120 X $10^3$d and 180 X $10^3$d (34, 80, 134), but Brishammar (15) reported that a TMV-induced replicase had a molecular weight of about 70 X $10^3$d. A low molecular weight protein (34.5 X $10^3$d), detected only in extracts of BMV-infected barley and which is distinct from the coat protein (20 X $10^3$d) has been suggested as a possible component of the BMV-RNA replicase, because it is associated with cell fractions containing BMV-induced replicase activity (35). This is an important observation because it suggests that for the BMV-RNA replicase, at least, the enzyme may be constituted of subunits that are coded in part from the host genome as well as from the viral genome. The QB replicase has been shown to consist of four polypeptides with molecular weights of about 70, 65, 45, and 35 X $10^3$d; the 65 X $10^3$d protein is coded by QB, while the others preexist in the uninfected cell (50, 58). The role of each subunit has not been defined at present, but it is possible that the role of the virus-coded protein is to confer specificity of the complex replicase for transcription of QB-RNA.

The recent demonstration of an RNA-dependent RNA polymerase in apparently healthy chinese cabbage (3) and in cucumber (4), the synthesis of a low molecular weight RNase-resistant RNA in the presence of actinomycin D by cell-free extracts of apparently healthy as well as TMV-infected tobacco (134), and the oft-noted actinomycin D-resistant RNA polymerase activity in cell-free extracts of uninoculated plants strongly suggest the existence of the essential components of an RNA polymerase system. This system could be modified by translational products of the viral genome so that the complex becomes essentially dependent upon the viral genome for template. In this connection, it is interesting to note that the host protein subunits in the QB replicase are apparently distinct from the proteins that comprise the bacterial DNA-dependent RNA polymerase (50, 58). Thus, a plant virus genome may be able to subvert the host's RNA transcriptional apparatus by modifying the specificity of some of the enzymes involved in RNA transcription, a concept which is attractive because of the resultant economy of genetic information it affords for the virus. The extent of this economy can be appreciated if we calculate the amount of viral RNA required to code for a replicase (120–180 $\times$ $10^3$d) and a coat protein (20 $\times$ $10^3$d); the total amount required is approximately 1.4–2.0 $\times$ $10^6$d. For a virus such as TMV, this would require 70–100% of the genetic information of its RNA. On the other hand, if we assume that the replicase is in the order of 70 $\times$ $10^6$d, then less than 50% of the genome is required to code for these two proteins.

OTHER REPLICASE PRODUCTS Evidence to date shows that the principal products of the replication of plant virus RNA genomes are the replicative apparatus (RF and RI) and the progeny viral genomes. For both single and multiple genome viruses, each of the RNA genomes is replicated from a corresponding replicative apparatus. Recent studies have shown that there are additional minor RNA components detectable in nucleic acid preparations from purified virions or from virus-infected tissue, but the role of these components is unclear. The best characterized of these is the low molecular weight component (LMC) of TMV (115), but similar RNAs may have been detected in other infections (63, 97).

The minor RNA species (LMC) associated with TMV-RNA replication was detected by gel-electrophoresis of the nucleic acids synthesized in separated leaf cells or leaf strips in the presence of $^3$H uridine (43). It has a molecular weight of 350,000d, it is single stranded, and it hybridizes with TMV-dsRNA but not with TMV-RNA (115). The RNA is thus a fragment of the TMV genome. Moreover, it is apparently one of a number of minor RNA species that can be detected in infected cells, but it is unique in that it does not assemble with TMV coat protein to form a nucleoprotein in a reconstitution experiment. Previous reconstitution experiments (17, 77) have shown that a nucleotide sequence at the 5' end of TMV-RNA is the initiation site for the in vitro assembly of TMV coat protein with TMV-RNA. The implication of the failure of LMC to assemble with TMV coat protein is that LMC may represent a polynucleotide derived from some part of the genome other than the 5' end; no data are yet available on the terminal nucleotide sequences of this RNA. A similar species may be the RNA that is rapidly labeled and able to be dissociated from the 40 S ribosomal subunit in ribosomal fractions of TMV-infected cells (7). A recent observation (T. J. Morris, personal communication) suggests that a legume strain of TMV induces the synthesis of a similar RNA component in cowpea which, unlike that of the strain studied by Siegel, is encapsidated in TMV protein. When the legume strain is inoculated to tobacco, only one nucleoprotein (the usual 3000 Å rod) is obtained, but the two RNA species are synthesized, suggesting that the minor RNA species is not encapsidated in tobacco.

Other virus infections appear to evidence low molecular weight RNAs whose functions are unknown. A slowly sedimenting highly labeled virus-specific RNA, which is apparently not encapsidated, was detected in tissues infected with broad bean mottle virus (BBMV) (97). This RNA is single stranded, its molecular weight has been estimated at slightly less than $0.3 \times 10^6$d by polyacrylamide gel-electrophoresis, and it hybridizes with BBMV-dsRNA. The molecular weight of the RNA is very close to that of the lowest molecular weight component of BBMV-RNA ($0.3 \times 10^6$d), and possibly the two RNAs are identical. A similar RNA component has been detected in barley infected with brome mosaic virus (63, 84). This RNA, with molecular weight of $0.5 \times 10^6$d, is not always detected in the RNA extracted from virions but its dsRNA has been detected in BMV-infected tissue (63), suggesting that it is replicated by its own replicative apparatus. The component can be eliminated from virus stocks by passage through several local lesion transfers (63), which suggests that the component may be a satellite RNA associated with BMV infection.

A unifying feature of these minor RNA species is their low molecular weight ($0.3$–$0.5 \times 10^6$d). From genetic analysis of alfalfa mosaic virus (127), brome mosaic virus (63), and cowpea chlorotic mottle virus (9), it has been established that the coat protein cistron is located on the smallest genome required for infectivity using RNA inocula (i.e. top b RNA for AMV; RNA 3 for BMV and CCMV). However, it has also been established that top a RNA of AMV and RNA 4 of BMV are able to direct the synthesis of proteins similar to their respective coat proteins in an *E. coli* cell-free amino acid incorporation system (118, 126), which implies that the genetic information in these RNA components is duplicated in top b RNA and RNA 3 respectively. That top a RNA of AMV and RNA 4 of BMV can be synthesized in infected plants even though they were not included in the RNA inocula (12, 25, 63) and that the nucleotide sequence of BMV-RNA 4 is present in RNA 3 but not in RNA 1 or 2 (113) clearly shows that these minor RNA components are derived as a consequence of the transcription of viral genomes.

The coincidence of the appearance of these minor RNA species in the transcription of viral RNA and their property of coding for proteins similar to coat protein strongly suggests that these RNA components are amplifications of the coat protein cistron. Such gene amplification may be necessary when one considers the large amount of coat protein subunits required to encapsulate the progeny viral RNA.

The shorter length of the amplified region should allow for a competitive advantage with the viral replicase for rapid transcription and, by the same token, with host ribosomes for rapid translation into coat protein. Whether or not these minor RNAs are themselves encapsulated is apparently of no consequence to the biology of their parental viruses and may be explainable, at least in the case of TMV, by the presence or absence of a specific sequence of bases at the 5' end of the RNA which initiates assembly with TMV protein. There is a parallel between the possible gene amplification in plant viruses and the derivation of short replicating RNAs derived from QB phage RNA; such mutants contain less than 10% of the original QB genome (46). A competitive advantage in the replication of short RNAs has been demonstrated in QB (69), tobacco rattle virus (70), the satellite virus systems of tobacco ringspot virus (104), and tobacco necrosis virus (45). In the tobacco necrosis system the consequence of the replication of the satellite RNA is the marked depression of tobacco necrosis coat protein synthesis.

These results suggest that it should be possible to derive shorter replicating RNAs that can be translated; the small RNA species of some plant viruses may have evolved for that purpose.

### *Group 2 (DNA)*

Relatively little is known of the replicative process per se of cauliflower mosaic virus and other related viruses. Presumably, transcription and translation would occur by conventional mechanisms. Preferential incorporation of $^3H$ thymidine into the characteristic cytoplasmic inclusions caused by virus infection has been observed (49), suggesting that these inclusions represent the site of virus-directed DNA synthesis. A later study (29) indicated that the label is first incorporated into host nuclei during a three hour labeling but that the label then accumulates in the cytoplasmic inclusion. No distinction could be made between virus-directed DNA synthesis in the nucleus and subsequent transport to the cytoplasmic inclusion and the possibility of synthesis in the inclusion. The results are probably explained on the basis of the length of the labeling period; in the work of Kamei et al (49), labeling was for 48 hr, while labeling was for 3 hr in the experiments of Favali et al (29). The possibility of the host DNA participating in the synthesis of cauliflower mosaic virus DNA is suggested by the recent report that viral DNA and cauliflower host DNA have similar nucleotide doublet frequencies (98), suggesting that the two nucleic acids may have similarities in base sequences at certain points of the polynucleotides.

### *Group 3 (Non-Infectious Single-Stranded RNA)*

Several plant viruses including lettuce necrotic yellows virus (LNYV), potato yellow dwarf virus, and wheat striate mosaic virus have been classified as rhabdoviruses. Rhabdoviruses are generally bullet shaped or bacilliform. They contain single-stranded RNA, which is assembled with protein to form a loosely helical ribonucleoprotein which in turn is associated with a matrix protein that may stabilize the shape of the ribonucleoprotein; the resulting nucleocapsid is enveloped in a lipoprotein envelope acquired by a budding process through the host-cell membrane (57).

Recent studies have shown that the animal-infecting rhabdoviruses have an RNA-dependent RNA polymerase associated with the nucleocapsid, possibly the ribonucleoprotein; the same enzyme activity has been shown for the nucleocapsid of LNYV (31). Purified virus, treated with the non-ionic detergent Nonidet P-40 to remove the lipoprotein envelope and suspended in an enzyme assay medium containing the four ribonucleoside triphosphates including $^{14}$C-UTP, directs the synthesis of a TCA-insoluble product that is predominately single stranded (>90%) and complimentary to the RNA isolated from the virus. The nucleocapsid enzyme is thus a transcriptase rather than a replicase that would be expected to produce a product homologous to the viral RNA. The product RNA has been further characterized (32) with respect to its molecular size, and the major portion of the product is of low molecular weight, sedimenting at about 4 S with a minor fraction sedimenting at 9–12 S. LNYV-RNA, on the other hand, sediments at about 43 S; estimates of its molecular weight range from 3.5–4.5 $\times$ $10^6$d, depending on the method (sedimentation or electrophoresis) used to estimate this parameter. Transcription of rhabdovirus RNAs into short pieces of RNA is a unifying feature of this group of viruses.

Vesicular stomatitis virus (VSV), an animal rhabdovirus, is basically similar to LNYV; its RNA has a molecular weight of 4 $\times$ $10^6$d and in vitro transcription of its RNA results in the formation of several RNAs with molecular weights of 0.2–1.0 $\times$ $10^6$d. The current view is that these short pieces of transcribed viral RNA represent mRNA, and this is somewhat strengthened by the observation that there is a correspondence between the molecular weight of these RNAs (0.35, 0.59, and 0.75 $\times$ $10^6$d) and the molecular weights of the three most abundant virus proteins which have molecular weights of 0.33, 0.57, and 0.76 $\times$ $10^5$d (102). The detection of two apparently virus-specific RNAs in the polyribosomal fraction of LNYV-infected *Nicotiana glutinosa* (92) suggests that LNYV messenger RNAs may be similar in molecular size to those reported for VSV.

No information is currently available on the replication of plant rhabdovirus RNA, but from studies on VSV-RNA replication there is evidence for the presence of double-stranded RNAs of 7.0–7.8 $\times$ $10^6$d that would correspond to the replicative apparatus for the VSV-RNA and also for the presence of single-stranded RNA of 3.8 $\times$ $10^6$d corresponding to the RNA encapsulated in the ribonucleoprotein (102).

The synthesis of rhabdovirus RNA can thus be distinguished from the scheme proposed for most RNA viruses. Whereas the viral genome of most RNA viruses is mRNA, the viral genome of the rhabdoviruses must first be transcribed by the virus RNA-dependent RNA polymerase (transcriptase) to complementary messenger-sized RNAs which then code for the viral proteins including replicase.

### *Group 4 (Double-Stranded RNA)*

There is little direct information available on the details of the replication of the double-stranded RNA viruses that infect plants (132). The best characterized of these viruses is wound tumor virus (WTV), which contains an RNA-dependent RNA polymerase (11). The polymerase is associated with complete virions in contrast to the same enzyme in reovirus which is activated when reovirus is partially uncoated in vivo or by chymotrypsin in vitro. The product of the WTV-RNA polymerase is single-stranded RNA which hybridizes with heat-denatured WTV-RNA, thus indicating that the enzyme is a transcriptase. Based on what is known of the replication of reovirus, it is likely that the process in WTV-infected cells is similar. In vitro studies indicate that transcriptase of reovirus transcribes one strand (minus strand) of each dsRNA genome segment to a complimentary single-stranded (plus strand) RNA. This RNA can also be found in the ribosomal fraction of infected cells, and in vitro translation of the RNA in cell-free extracts from mammalian cells resulted in the synthesis of 8 of the 10 known viral proteins (73), indicating that it can function as mRNA. A study of dsRNA synthesis in vivo indicates that the plus strand accumulates during the early phase of virus replication and that this RNA serves as the template for the synthesis of the complimentary minus stand. More recent studies suggest that the nascent plus strand condenses with viral coat protein, including a dsRNA synthetase, and the synthesis of the complimentary minus strand is completed ir this complex. Characterization of the complex by isopycnic centrifugation in CsCl indicates properties similar to that of the virion, suggesting that genomic RNA is synthesized within particles that resemble mature virions (135). The coincidence in the association of the dsRNA synthetase and the RNA transcriptase within the complex has resulted in the suggestion that the two enzymes may exhibit alternate activities, depending on the template (1).

The establishment of insect cell tissue cultures and the successful infection of these cell cultures by WTV (55) will hopefully lead to an understanding of the replication of this virus.

## TRANSLATION OF VIRAL RNA

The main translational product of plant viral RNA appears to be the coat protein, although it is obvious that other proteins such as replicase and gene products that regulate symptom expression must also be produced. If one considers the total coding capacity of an RNA genome of $2 \times 10^6$d, it is obvious that coding for a coat protein of $20 \times 10^3$d would require about 10% of the coding capacity, clearly indicating that significant portions of most plant virus genomes probably are coding for proteins that we have failed to detect. Recent advances in the genetics of plant viruses have assigned some of the biological characteristics of these viruses to particular sections of the genome. The genes for coat protein and the local lesion reaction in *Nicotiana sylvestris* have been assigned to the half of the TMV genome terminating at the 5' end (47, 48), and the genes for symptom expression and coat protein have been located on the smallest of the RNAs required for infection in alfalfa mosaic virus (25), brome mosaic and cowpea chlorotic mottle viruses (10), and for raspberry ringspot virus (36). The characterization of the gene products, on the other hand, is still in its infancy and represents a formidable task. Initial studies are being done in two ways: 1. by analyzing the proteins produced in an in vitro amino acid incorporating system primed with viral RNA as messenger and comparing these proteins with known virus proteins, and 2. analyzing the virus-induced proteins produced in infected cells.

## IN VITRO TRANSLATION

Addition of the small RNA components of alfalfa mosaic virus (126) and brome mosaic virus (118) to an in vitro amino acid incorporation system derived from *E. coli* and the necessary factors for protein synthesis leads to enhanced amino acid incorporation and the detection of proteins similar to the respective coat proteins based on comparison of the tryptic digests of the in vitro products and authentic coat protein. The RNA of the satellite virus of tobacco necrosis virus directs the synthesis of its coat protein and other uncharacterized proteins in cell-free systems from prokaryotic (*E. coli*) and eukaryotic (wheat embryo) sources (56). Recently the wheat embryo system has been used to translate TMV-RNA (27) and BMV-RNA (112a) into proteins, some of which closely resemble the authentic coat proteins of these viruses. The BMV-RNA wheat embryo system is of considerable interest because of the different messenger functions of the RNAs. RNA 4 ($0.3 \times 10^6$d) functions as the messenger for coat protein while RNA 3 ($0.75 \times 10^6$d) contains cistrons for coat protein and a larger protein (3a) of unknown function. RNAs 1 and 2 (1.09 and $0.99 \times 10^6$d, respectively) cannot be adequately separated for studies on messenger function, but their mixture is translated to a heterogeneous mixture of proteins. Interestingly, the translation product of unfractionated BMV-RNA is coat protein, and the significance of this observation to the overall problem of controlling virus-induced protein synthesis is obvious. The use of the wheat embryo system hopefully portends the exploitation of other eukaryotic sources, and the possibility for real progress in characterizing the proteins encoded in plant virus genomes appears to be at hand.

## VIRUS-INDUCED PROTEIN SYNTHESIS IN VIVO

The products of the in vivo replication and translation of viral RNAs have been characterized by several workers using TMV as the infecting virus. The basic approach has been to compare the proteins produced in TMV-infected leaves, separated cells, or protoplasts with those produced in uninoculated controls. Plants or cells are labeled with the appropriate isotope, usually $^{14}$C leucine or $^{14}$C arginine; in some instances a double labeling technique is used in which uninfected tissue is labeled with one isotope and infected tissue with another, and the ratio of the two labels is used to detect virus-induced proteins. Analysis is usually by polyacrylamide electrophoresis of the protein extracts. The results of these investigations indicate that the major protein synthesized in tobacco varieties that support systemic infection is TMV coat protein (116, 125); in plants containing genes for the hypersensitivity response, several low molecular weight proteins unrelated to TMV coat protein were detected. High molecular weight proteins, specific to infected leaves, with molecular weights of 245, 195, 155, and $37 \times 10^3$d were also detected in addition to TMV coat protein in tobacco (133). One of these proteins ($155 \times 10^3$d) is similar to a noncoat protein ($140 \times 10^3$d) detected in TMV-infected protoplasts (100). In tobacco (Xanthi n.c.) infected with tobacco necrosis virus, the major protein synthesized was the coat protein (mol wt $32 \times 10^3$d), but other proteins with molecular weights of 64, 41, 23, 15, and $12 \times 10^3$d were detected in minor amounts (45). In plants infected with tobacco necrosis virus and its satellite, the only satellite virus-induced protein was the coat protein with a molecular weight of $20 \times 10^3$d.

Recent studies have shown that the cellular inclusions characteristic of infections by the potato virus Y group are antigenically unrelated to the causal virus (112). Subsequent work has established that the inclusions are composed of protein subunits with molecular weights of about $70 \times 10^3$d, twice those of the respective virus coat proteins (39), and that they are antigenically unique although related in some instances (87). These results suggest that the inclusion proteins are translational products of the viral genome; the inclusion may represent a "sink" for these protein subunits.

Aside from the coat and inclusion proteins as the major virus-induced proteins in infected cells, none of the proteins have been characterized. Zaitlin & Hariharasubramanian (133) suggested that the $155 \times 10^3$d protein in TMV-infected cells was involved in replicase activity because it is predominantly associated with a particulate fraction that contains TMV-RNA replicase activity (14). However, possibly the same protein as well as the $195 \times 10^3$d protein were also detected in extracts of tomato infected with potato spindle tuber viroid, a low molecular weight ($80 \times 10^3$d) RNA devoid of coat protein (23, 24). The coding capacity of this RNA is sufficient only to code for a protein of about $10 \times 10^3$d. It is more likely therefore that some if not most of the proteins detected in virus-infected plants are of host origin and that their synthesis is a consequence of virus infection. Their synthesis may arise as the consequence of the derepression of host DNA by the virus, or they may be normal host components whose synthesis is enhanced as a consequence of virus infection. Considering the size of the genomes of most plant viruses, it is more likely that virus-coded proteins will be of relatively low molecular weight and their detection in protein preparations from infected plants correspondingly difficult.

## VIRION ASSEMBLY

Our understanding of the specific steps in the assembly of virus coat protein(s) and viral nucleic acid in the infected cell is limited, for the time being, to the observations made on the assembly of virions in vitro. The reconstitution of TMV laid the basis for subsequent investigations by others which led to the assembly of other viruses (8, 107). The details of the in vitro assembly process are perhaps better understood with the tubular virions of TMV and tobacco rattle virus than with the isometric viruses. The critical step in the assembly of tubular virions is the association of a recognition region at the 5' end of the viral RNA with an aggregate (2-turn disc) of coat protein (17, 71). Nucleoprotein elongation is then accomplished either by sequential addition of protein subunits (77, 95) or by the addition of protein discs, followed by local depolymerization to allow the subunits within the discs to fit to the end of the growing rod (17). Mixed reconstitution between the RNA and protein of unrelated viruses has also been reported (5, 6, 128, 129), some of which were infective, although the mixed reconstitution of BMV and CCMV (129) must now be interpreted as reconstitution between related viruses (105).

Mixed assembly of virions as a consequence of the interaction between two viruses in the same host has been reported. Encapsidation of the genome of one virus by the protein coat of the other (genomic masking) has been reported for mixed infections involving two apparently unrelated but structurally similar viruses (96) or in infections with a mixture of TMV strains, usually one of which is unable to produce its own coat protein under the conditions of the experiment (51–53, 101). Genomic masking of barley stripe mosaic RNA in the coat protein of BMV was suggested by Peterson & Brakke (81), who reported that BSMV virions could be detected in barley plants inoculated with BMV isolated from doubly infected plants. The RNA of TMV is encapsidated in BSMV coat protein as a consequence of a double infection in barley (26), and further experiments have demonstrated the physical presence of infective TMV-RNA in the RNA isolated from virions specifically precipitated from extracts of doubly infected plants by BSMV antibody (J. A. Dodds and R. I. Hamilton, mansucript submitted). In the same immunoprecipitate from which TMV-RNA could be isolated, no TMV coat protein could be detected, clearly indicating encapsidation of TMV-RNA in BSMV protein. A novel encapsidation of tobacco host nucleic acid complementary to tobacco chloroplast and nuclear DNA in TMV protein (pseudovirions) has been reported (114). The significance of this phenomenon as well as a similar one involving encapsidation of host DNA by simian virus 40 (120) is that pseudovirions may afford an opportunity for genetic exchange between eukaryotic organisms in a way similar to transduction in bacteria.

The site of virion assembly has not been convincingly demonstrated for any plant virus (103). The TMV pseudovirions may offer an insight into the site of TMV virion assembly in the infected cell. The majority of the encapsidated host nucleic acid, presumably RNA, is complementary to chloroplast DNA, suggesting that pseudovirions, and by inference TMV virions, are assembled in the chloroplast. Evidence in support of this is the report that the chloroplast, or membranes associated with this organelle, is involved in the assembly of TMV (117). In short labeling periods using $^{14}C$ leucine, most of the label is associated with the chloroplast fraction, but treatment of the chloroplast with the detergent Triton X-100 removes the label which can then be pelleted under conditions that would pellet TMV but not protein subunits. On the other hand, other investigators suggest that the chloroplast is not the site of virion assembly (19, 93).

The chloroplast vesicles characteristic of TYMV infection appear to be involved in TYMV-RNA synthesis and possibly in the assembly of virions. These double-membraned vesicles are thought to arise by an invagination of the chloroplast envelope (61), but recent studies of the vesicle membranes by the freeze-fracture method (37) reveal that the inner membrane does not contain the protein particles characteristic of cellular membranes on either of its two fracture faces, while the outer chloroplast membrane, from which the inner vesicle membrane is thought to be derived (61), contains the normal complement of these protein particles. These results suggest that the chloroplast membrane structure is altered as a consequence of TYMV infection. In a series of papers, Bové and his colleagues have established by high resolution autoradiography that the vesiculated chloroplast is the most likely site of TYMV dsRNA synthesis (61, 62). The vesicles contain fibrillar material suggestive of a double-stranded nucleic acid (121), and they border on cytoplasmic pockets filled with viruses. It has been suggested that the assembly of TYMV occurs as the TYMV-RNA leaves the vesicle on its way to the cytoplasm (66).

It appears that vesiculation is a common feature of plant virus infection. Peripheral chloroplastic vesicles have also been observed in barley infected with barley stripe mosaic virus (18); other viruses induce similar structures in nuclei (22, 28) and in mitochondria (38). Vesicles have also been reported in mitochondria of *Begonia* infected with a virus-like disease (83). The involvement of these vesiculated membrane systems in the synthesis and assembly of virus components, although demonstrated in TYMV infections, is not known. de Zoeten et al (22) have shown that the fibrillar material in nuclear vesicles induced by pea enation mosaic virus is susceptible to DNase but not to RNase. These vesicles may be specialized sites of virus-induced DNA-dependent RNA synthesis, and it would be interesting to determine whether their formation and possible role in DNA synthesis is affected by AMD. Of even more interest would be studies to determine whether such vesicles occur in cowpea infected with cowpea mosaic virus and other viruses, because it has been demonstrated that viral RNA synthesis in this host is severely inhibited by applications of AMD at the time of inoculation or shortly afterwards (64).

## DISCUSSION

This review has concentrated on recent research concerning plant viral RNA transcription and translation. I think it should be apparent that membrane systems play key roles in the replicative cycle of a plant virus. Entrance to the cell presumably involves association of the virion with a membrane, and virions may be associated with vesicles in the early stages of the infective process. The association of template-specific replicases with cellular membrane systems and the relatively nonspecific activity of these enzymes upon their dissociation from the bound state further points to the involvement of membranes in viral nucleic acid synthesis. The increasing record of vesiculated membranes associated with diverse virus infections will hopefully spur a concentrated research effort on the role that these may play in the synthesis and assembly of virus components. The recent implication of membrane systems in the synthesis of polio virus RNA and protein and their role in the assembly of protein shells (82) suggests an approach that plant virologists might employ. The establishment of synchronous infection would be a distinct advantage in this type of research, and the use of protoplasts or synchronous systemic infections (21) may prove useful in this respect.

### Acknowledgments

The author thanks Dr. T. J. Morris and J. A. Dodds for the many stimulating discussions in the preparation of this review.

*Literature Cited*


1. Acs, G. et al 1971. Mechanism of reovirus double-stranded ribonucleic acid synthesis *in vivo* and *in vitro*. *J. Virol.* 8:684–89
2. Aoki, S., Takebe, I. 1969. Infection of tobacco mesophyll protoplasts by tobacco mosaic virus ribonucleic acid. *Virology* 39:439–48
3. Astier-Manifacier, S., Cornuet, P. 1971. RNA-dependent RNA polymerase in Chinese cabbage. *Biochim. Biophys. Acta* 232:484–93
4. Astier-Manifacier, S., Cornuet, P. 1973. *RNA Dependent RNA Polymerase in Healthy and Virus-Infected Plants.* Presented at 2nd Int. Congr. Plant Pathol., Minneapolis, Minn.
5. Atabekov, J. G., Novikov, V. K., Vishnichenko, V. K., Javakhia, V. G. 1970. A study of the mechanisms controlling the host range of plant viruses: II The host range of hybrid viruses reconstituted in vitro and of free viral RNA. *Virology* 41:108–15
6. Atabekov, J. G., Novikov, V. K., Vishnichenko, V. K., Kaftanova, A. S. 1970. Some properties of hybrid viruses reassembled *in vitro*. *Virology* 41:519–32
7. Babos, P. 1971. TMV-RNA associated with ribosomes of tobacco leaves infected with TMV. *Virology* 43:597–606
8. Bancroft, J. B. 1970. The self-assembly of spherical plant viruses. *Advan. Virus Res.* 16:99–134
9. Bancroft, J. B. 1972. A virus made from parts of the genomes of brome mosaic and cowpea chlorotic mottle viruses. *J. Gen. Virol.* 14:223–28
10. Bancroft, J. B., Lane, L. C. 1973. Genetic analysis of cowpea chlorotic mottle and brome mosaic viruses. *J. Gen. Virol.* 19:381–89
11. Black, D. R., Knight, C. A. 1970. Ribonucleic acid transcriptase activity in purified wound tumor virus. *J. Virol.* 6: 194–98
12. Bol, J. F., van Vloten-Doting, L., Jaspars, E. M. J. 1971. A functional equivalence of top component *a* RNA and coat protein in the initiation of infection of alfalfa mosaic virus. *Virology* 46: 73–85
13. Bové, J. M., Bové, C., Rondot, M. J., Morel, G. 1967. Chloroplasts and virus RNA synthesis. In *Biochemistry of Chloroplasts,* ed. T. W. Goodwin, 2: 329. New York: Academic
14. Bradley, D. W., Zaitlin, M. 1971. Replication of tobacco mosaic virus. II. The *in vitro* synthesis of high molecular weight virus-specific RNA's. *Virology* 45:192–99
15. Brishammar, S. 1970. Identification and characterization of an RNA replicase from TMV-infected tobacco leaves. *Biochem. Biophys. Res. Commun.* 41: 506–11
16. Brown, F., Hull, R. 1973. Comparative virology of the small RNA viruses. *J. Gen. Virol.* 20: Suppl., 1–130
17. Butler, P. J. G., Klug, A. 1971. Assembly of the particle of tobacco mosaic virus from RNA and disks of protein. *Nature New Biol.* 229:47–50
18. Carroll, T. W. 1970. Relation of barley stripe mosaic virus to plastids. *Virology* 42:1015–22
19. Cech, M. 1967. On the role of chloroplasts in tobacco mosaic virus reproduction. *Phytopathol. Z.* 59:72–82
20. Cocking, E. C., Pojnar, E. 1969. An electron microscope study of the infection of isolated tomato fruit protoplasts by tobacco mosaic virus. *J. Gen. Virol.* 4:305–12
21. Dawson, W. O., Schlegel, D. E. 1973. Differential temperature treatment of plants greatly enhances multiplication rates. *Virology* 53:476–78
22. de Zoeten, G. A., Gaard, G., Diez, F. B. 1972. Nuclear vesiculation associated with pea enation mosaic virus-infected plant tissue. *Virology* 48:638–47
23. Diener, T. O. 1972. Viroids. *Advan. Virus Res.* 17:295–313
24. Diener, T. O., Smith, D. R. 1973. Potato spindle tuber viroid. IX. Molecular weight determination by gel electrophoresis of formylated RNA. *Virology* 53:359–65
25. Dingjan-Versteegh, A., van Vloten-Doting, L., Jaspars, E. M. J. 1972. Alfalfa mosaic virus hybrids constructed by exchanging nucleoprotein components. *Virology* 49:716–22
26. Dodds, J. A., Hamilton, R. I. 1971. Evidence for possible genomic masking between two unrelated plant viruses. *Phytopathology* 61:889 (Abstr.)
27. Efron, D., Marcus, A. 1973. Translation of TMV-RNA in a cell-free wheat embryo system. *Virology* 53:343–48
28. Esau, K., Hoefert, L. L. 1972. Development of infection with beet western yellows virus in the sugar beet. *Virology* 48:724–38
29. Favali, M. A., Bassi, M., Conti, G. G. 1973. A quantitative autoradiographic study of intracellular sites for replica-

tion of cauliflower mosaic virus. *Virology* 53:115–19
30. Fowlks, E., Young, R. J. 1970. Detection of heterogeneity in plant viral RNA by polyacrylamide gel electrophoresis. *Virology* 42:548–50
31. Francki, R. I. B., Randles, J. W. 1972. RNA-dependent RNA polymerase associated with particles of lettuce necrotic yellows virus. *Virology* 47:270–75
32. Francki, R. I. B., Randles, J. W. 1973. Some properties of lettuce necrotic yellows virus RNA and its *in vitro* transcription by virion-associated transcriptase. *Virology* 54:359–68
33. Franze de Fernandez, M. T., Hayward, W. S., August, J. T. 1972. Bacterial proteins required for replication of phage QB ribonucleic acid. Purification of host factor I, a ribonucleic acid binding protein. *J. Biol. Chem.* 247:824–31
34. Hadidi, A., Fraenkel-Conrat, H. 1973. Characterization and specificity of soluble RNA polymerase of brome mosaic virus. *Virology* 52:363–72
35. Hariharasubramanian, V., Hadidi, A., Singer, B., Fraenkel-Conrat, H. 1973. Possible identification of a protein in brome mosaic virus infected barley as a component of viral RNA polymerase. *Virology* 54:190–98
36. Harrison, B. D., Murant, A. F., Mayo, M. A. 1972. Two properties of raspberry ringspot virus determined by its smaller RNA. *J. Gen. Virol.* 17:137–41
37. Hatta, T., Bullivant, S., Matthews, R. E. F. 1973. Fine structure of vesicles induced in chloroplasts of chinese cabbage leaves by infection with turnip yellow mosaic virus. *J. Gen. Virol.* 20: 37–50
38. Hatta, T., Nakamoto, T., Takagi, Y., Ushiyama, R. 1971. Cytological abnormalities of mitochondria induced by infection with cucumber green mottle mosaic virus. *Virology* 45:292–97
39. Hiebert, E., McDonald, J. G. 1973. Characterization of some proteins associated with viruses in the potato virus Y group. *Virology* 56:349–61
40. Hiruki, C. 1969. Properties of single- and double-stranded ribonucleic acid from barley plants infected with bromegrass mosaic virus. *J. Virol.* 3:498–505
41. Jackson, A. O., Brakke, M. K. 1973. Multicomponent properties of barley stripe mosaic virus ribonucleic acid. *Virology* 55:483–94
42. Jackson, A. O., Mitchell, D. M., Siegel, A. 1971. Replication of tobacco mosaic virus. I. Isolation and characterization of double-stranded forms of ribonucleic acid. *Virology* 45:182–91
43. Jackson, A., Zaitlin, M., Francki, R. I. B. 1972. Replication of tobacco mosaic virus. II. Viral RNA metabolism in separated leaf cells. *Virology* 48:655–65
44. Jacquemin, J. M. 1972. *In vitro* product of an RNA polymerase induced in broadbean by infection with broadbean mottle virus. *Virology* 49:379–84
45. Jones, I. M., Reichmann, M. E. 1973. The proteins synthesized in tobacco leaves infected with tobacco necrosis virus and satellite tobacco necrosis virus. *Virology* 52:49–56
46. Kacian, D. L., Mills, D. R., Kramer, F. R., Spiegelman, S. 1972. A replicating RNA molecule suitable for a detailed analysis of extracellular evolution and replication. *Proc. Nat. Acad. Sci. USA* 69:3038–42
47. Kado, C. I., Knight, C. A. 1966. Location of a local lesion gene in tobacco mosaic virus RNA. *Proc. Nat. Acad. Sci. USA* 55:1276–83
48. Kado, C. I., Knight, C. A. 1968. The coat protein gene of tobacco mosaic virus. I. Location of the gene by mixed infection. *J. Mol. Biol.* 36:15–23
49. Kamei, T., Rubio-Huertos, M., Matsui, C. 1969. Thymidine-$^3$H uptake by X-bodies associated with cauliflower mosaic virus infection. *Virology* 37:506–8
50. Kamen, R. 1970. Characterization of the subunits of QB replicase. *Nature New Biol.* 228:527–33
51. Kassanis, B., Bastow, C. 1971. In vivo phenotypic mixing between two strains of tobacco mosaic virus. *J. Gen. Virol.* 10:95–98
52. Kassanis, B., Bastow, C. 1971. Phenotypic mixing between two strains of tobacco mosaic virus. *J. Gen. Virol.* 11:171–76
53. Kassanis, B., Conti, M. 1971. Defective strains and phenotypic mixing. *J. Gen. Virol.* 13:361–65
54. Kiho, Y., Machida, H., Oshima, N. 1972. Mechanism determining the host specificity of tobacco mosaic virus. I. Formation of polysomes containing infecting viral genome in various plants. *Jap. J. Microbiol.* 16:451–59
55. Kimura, I., Black, L. M. 1972. Growth of wound tumor virus in vector cell monolayers. *Virology* 48:852–54
56. Klein, W. H., Nolan, C., Lazar, J. M., Clark, J. M. Jr. 1972. Translation of tobacco necrosis virus ribonucleic acid. I. Characterization of *in vitro* procaryotic and eucaryotic translation products. *Biochemistry* 11:2009–14

57. Knudson, D. L. 1973. Rhabdoviruses. *J. Gen. Virol.* 20: Suppl., 1–130
58. Kondo, M., Gallerani, R., Weissmann, C. 1970. Subunit structure of QB replicase. *Nature New Biol.* 228:525–27
59. Kummert, J., Semal, J. 1972. Properties of single-stranded RNA synthesized by a crude RNA polymerase fraction from barley leaves infected with brome mosaic virus. *J. Gen. Virol.* 16:11–20
60. Kurtz-Fritsch, C., Hirth, L. 1972. Uncoating of two spherical plant viruses. *Virology* 47:385–96
61. Laflèche, D., Bové, J. M. 1969. Development of double membrane vesicles in chloroplasts from turnip yellow mosaic virus-infected cells. *Progr. Photosynth. Res.* 1:74–83
62. Laflèche, D. et al 1972. Site of viral RNA replication in the cells of higher plants: TYMV-RNA synthesis on the chloroplast outer membrane system. In *RNA Viruses: Structure and Function. FEBS Amsterdam Symp.* 27:43–65
63. Lane, L. C., Kaesberg, P. 1971. Multiple genetic components in bromegrass mosaic virus. *Nature New Biol.* 232: 40–43
64. Lockhart, B. E. L., Semancik, J. S. 1969. Differential effect of actinomycin D on plant virus replication. *Virology* 39:362–65
65. Machida, H., Kiho, Y. 1970. *In vivo* uncoating of tobacco mosaic virus. II. Complete uncoating to TMV-RNA. *Jap. J. Microbiol.* 14:441–49
66. Matthews, R. E. F. 1973. Induction of disease by viruses, with special reference to turnip yellow mosaic virus. *Ann. Rev. Phytopathol.* 11:147–70
67. May, J. T., Gilliland, J. M., Symons, R. H. 1970. Properties of a plant virus-induced RNA polymerase in particulate fractions of cucumbers infected with cucumber mosaic virus. *Virology* 41: 653–64
68. May, J. T., Symons, R. H. 1971. Specificity of the cucumber mosaic virus-induced RNA polymerase for RNA and polynucleotide templates. *Virology* 44: 517–26
69. Mills, D. R., Peterson, R. L., Spiegelman, S. 1967. An extracellular Darwinian experiment with a self-duplicating nucleic acid molecule. *Proc. Nat. Acad. Sci. USA* 58:217–24
70. Morris, T. J., Semancik, J. S. 1973. Homologous interference between the components of tobacco rattle virus. *Virology* 52:314–17
71. Morris, T. J., Semancik, J. S. 1973. In vitro protein polymerization and nucleoprotein reconstitution of tobacco rattle virus. *Virology* 53:215–24
72. Motoyoshi, F., Bancroft, J. B., Watts, J. W., Burgess, J. 1973. The infection of tobacco protoplasts with cowpea chlorotic mottle virus and its RNA. *J. Gen. Virol.* 20:177–93
73. McDowell, M. J., Joklik, W. K., Villa-Komaroff, L., Lodish, H. F. 1972. Translation of reovirus messenger RNA synthesized in vitro into reovirus polypeptides by several mammalian cell-free systems. *Proc. Nat. Acad. Sci. USA* 69:2649–53
74. Nicolaieff, A. L., Pinck, L., Koenig-Nikes, A. M., Hirth, L. 1972. Electron microscopy of replicative form and single-stranded RNA of alfalfa mosaic virus. *J. Gen. Virol.* 16:47–59
75. Nilsson-Tillgren, T. 1969. Studies on the biosynthesis of TMV. II. On the RNA-synthesis of infected cells. *Mol. Gen. Genet.* 105:191–202
76. Nilsson-Tillgren, T. 1970. Studies on the biosynthesis on TMV. III. Isolation and characterization of the replicative form and the replicative intermediate RNA. *Mol. Gen. Genet.* 109:246–56
77. Okada, Y., Ohno, T. 1972. Assembly mechanism of tobacco mosaic virus particle from its ribonucleic acid and protein. *Mol. Gen. Genet.* 114:205–13
78. Otsuki, Y., Takebe, I. 1973. Infection of tobacco mesophyll protoplasts by cucumber mosaic virus. *Virology* 52: 433–38
79. Otsuki, Y., Takebe, I., Matsui, C., Honda, Y. 1972. Ultrastructure of infection of tobacco protoplasts by tobacco mosaic virus. *Virology* 49:188–94
80. Peden, K. W. C., May J. T., Symons, R. H. 1972. A comparison of two plant virus-induced RNA polymerases. *Virology* 47:498–501
81. Peterson, J. F., Brakke, M. K. 1973. Genomic masking in mixed infections with brome mosaic and barley stripe mosaic viruses. *Virology* 44:54–66
82. Perlin, M., Phillips, B. A. 1973. *In vitro* assembly of polioviruses III. Assembly of 14S particles into empty capsids by polio virus-infected Hela cell membranes. *Virology* 53:107–14
83. Petzold, H. 1972. Elektronmikroskopische Beobachtungen über eine auf *Begonia semperflorens* beschränkte und ubertragbare Erkrankung der Mitochondrien. *Phytopathol. Z.* 74:249–62
84. Phillips, G., Gigot, C., Hirth, L. 1972. Rapid labeling of a nonencapsulated RNA of bromegrass mosaic virus. *FEBS Lett.* 25:165–69

85. Pinck, L., Hirth, L. 1972. The replicative RNA and the viral RNA synthesis rate in tobacco infected with alfalfa mosaic virus. *Virology* 49:413–25
86. Pring, D. R. 1972. Barley stripe mosaic virus replicative form RNA: preparation and characterization. *Virology* 48: 22–29
87. Purcifull, D. E., Hiebert, E., McDonald, J. G. 1973. Immunochemical specificity of cytoplasmic inclusions induced by viruses in the potato Y group. *Virology* 55:275–79
88. Ralph, R. K. 1969. Double stranded RNA. *Advan. Virus Res.* 15:61–158
89. Ralph, R. K., Bullivant, S., Wojcik, S. J. 1971. Evidence for the intracellular site of double-stranded turnip yellow mosaic virus RNA. *Virology* 44:473–79
90. Ralph, R. K., Clark, M. F. 1966. Intracellular location of double stranded plant viral ribonucleic acid. *Biochim. Biophys. Acta* 119:29–36
91. Ralph, R. K., Wojcik, S. J. 1969. Double stranded tobacco mosaic virus RNA. *Virology* 37:276–82
92. Randles, J. W., Coleman, D. 1972. Changes in polysomes in *Nicotiana glutinosa* leaves infected with lettuce necrotic yellows virus. *Physiol. Plant Pathol.* 2:247–58
93. Reddi, K. K. 1972. Tobacco mosaic virus with emphasis on the events within the host cell following infection. *Advan. Virus Res.* 17:51–94
94. Rezaian, M. A., Francki, R. I. B. 1973. Replication of tobacco ringspot virus. I. Detection of low molecular weight double stranded RNA from infected plants. *Virology* 56:238–49
95. Richards, K. E., Williams, R. C. 1972. Assembly of tobacco mosaic virus *in vitro:* effect of state of polymerization of the protein component. *Proc. Nat. Acad. Sci. USA* 69:1121–24
96. Rochow, W. F. 1972. The role of mixed infections in the transmission of plant viruses by aphids. *Ann. Rev. Phytopathol.* 10:101–24
97. Romero, J. 1973. Properties of a slow-sedimenting RNA synthesized by broadbean leaf tissue infected with broadbean mottle virus. *Virology* 55:224–30
98. Russell, G. J., Follett, E. A. C., Subak-Sharpe, J. H., Harrison, B. D. 1971. The double stranded DNA of cauliflower mosaic virus. *J. Gen. Virol.* 11:129–38
99. Ryser, H. J. -P. 1968. Uptake of protein by mammalian cells; an underdeveloped area. *Science* 159:390–96
100. Sakai, F., Takebe, I. 1972. A non-coat protein synthesized in tobacco mesophyll protoplasts by tobacco mosaic virus. *Mol. Gen. Genet.* 118:93–96
101. Sarkar, S. 1969. Evidence of phenotypic mixing between two strains of tobacco mosaic virus. *Mol. Gen. Genet.* 105: 87–90
102. Schincariol, A. L., Howatson, A. F. 1972. Replication of vesicular stomatitis virus. II. Separation and characterization of virus-specific RNA species. *Virology* 49:766–83
103. Schlegel, D. E., Smith, S. H., de Zoeten, G. A. 1967. Sites of virus synthesis within cells. *Ann. Rev. Phytopathol.* 5:223–46
104. Schneider, I. R. 1971. Characteristics of a satellite-like virus of tobacco ringspot virus. *Virology* 45:108–22
105. Scott, H. A., Slack, S. A. 1971. Serological relationship of brome mosaic and cowpea chlorotic mottle viruses. *Virology* 46:490–92
106. Semal, J., Hamilton, R. I. 1968. RNA synthesis in cell-free extracts of barley leaves infected with bromegrass mosaic virus. *Virology* 36:293–302
107. Semancik, J. S., Reynolds, D. A. 1969. Assembly of protein and nucleo-protein particles from extracted tobacco rattle virus protein and RNA. *Science* 164: 559–60
108. Shalla, T. A., Peterson, L. J. 1973. Infection of isolated plant protoplasts with potato virus X. *Phytopathology* 63:1125–30
109. Shaw, J. G. 1969. *In vivo* removal of protein from tobacco mosaic virus after inoculation of tobacco leaves. II. Some characteristics of the reaction. *Virology* 37:109–16
110. Shaw, J. G. 1970. Uncoating of tobacco mosaic virus ribonucleic acid after inoculation of tobacco leaves. *Virology* 42:41–48
111. Shaw, J. G. 1973. In vivo removal of protein from tobacco mosaic virus after inoculation of tobacco leaves. III. Studies on the location in virus particles for the initial removal of protein. *Virology* 53:337–42
112. Shepard, J. F., Shalla, T. A. 1969. Tobacco etch virus cylindrical inclusions: antigenically unrelated to the causal virus. *Virology* 38:185–88
112a. Shih, D. -S., Kaesberg, P. 1973. Translation of brome mosaic viral ribonucleic acid in a cell-free system derived from wheat embryo. *Proc. Nat. Acad. Sci. USA* 70:1799–1803

113. Shih, D-S., Lane, L. C., Kaesberg, P. 1972. Origin of the small component of brome mosaic virus RNA. *J. Mol. Biol.* 64:353–62
114. Siegel, A. 1971. Pseudovirions of tobacco mosaic virus. *Virology* 46:50–59
115. Siegel, A., Zaitlin, M., Duda, C. 1973. Replication of tobacco mosaic virus. IV. Further characterization of viral-related RNA's. *Virology* 53:75–83
116. Singer, B. 1971. Protein synthesis in virus-infected plants. I. The number and nature of TMV-directed proteins detected on polyacrylamide gels. *Virology* 46:247–55
117. Singer, B. 1972. Protein synthesis in virus-infected plants. II. The synthesis and accumulation of TMV and TMV coat protein in subcellular fractions of TMV-infected tobacco. *Virology* 47: 397–404
118. Stubbs, J. D., Kaesberg, P. 1967. Amino acid incorporation in an *Escherichia coli* cell-free system directed by bromegrass mosaic ribonucleic acid. *Virology* 33:385–97
119. Takebe, I., Otsuki, Y. 1969. Infection of tobacco mesophyll protoplasts by tobacco mosaic virus. *Proc. Nat. Acad. Sci. USA* 64:843–48
120. Trilling, D., Axelrod, D. 1972. Analysis of the three components of the simian virus 40: pseudo-, mature and defective viruses. *Virology* 47:360–69
121. Ushiyama, R., Matthews, R. E. F. 1970. The significance of chloroplast abnormalities associated with infection by turnip yellow mosaic virus. *Virology* 42:293–303
122. van Griensven, L. J. L. D., van Kammen, A. 1969. The isolation of a ribonuclease-resistant RNA induced by cowpea mosaic virus: evidence for two double-stranded RNA components. *J. Gen. Virol.* 4:423–28
123. van Griensven, L. J. L. D., van Kammen, A., Rezelman, G. 1973. Characterization of the double-stranded RNA isolated from cowpea mosaic virus-infected *Vigna* leaves. *J. Gen. Virol.* 18:359–67
124. van Kammen, A. 1972. Plant viruses with a divided genome. *Ann. Rev. Phytopathol.* 10:125–50
125. van Loon, L. C., van Kammen, A. 1970. Polyacrylamide disc electrophoresis of the soluble leaf proteins from *Nicotiana tabacum* var. Samsun and Samsun NN. II. Changes in protein constitution after infected with tobacco mosaic virus. *Virology* 40:199–211
126. van Ravenswaay, Claasen, J. C., van Leeuwen, A. B. J., Duijts, G. A. H., Bosch, L. 1967. *In vitro* translation of alfalfa mosaic virus RNA. *J. Mol. Biol.* 23:535–44
127. van Vloten-Doting, L., Kruseman, J., Jaspars, E. M. J. 1968. The biological functions and mutual dependence of bottom component and top component *a* of alfalfa mosaic virus. *Virology* 34: 728–37
128. Verduin, B. J. M., Bancroft, J. B. 1969. The infectivity of tobacco mosaic virus RNA in coat proteins from spherical viruses. *Virology* 37:501–6
129. Wagner, G. W., Bancroft, J. B. 1968. The self-assembly of spherical viruses with mixed coat proteins. *Virology* 34: 748–56
130. Wildy, P. 1973. Unity and variety in virology. *J. Gen. Virol.* 20: Suppl., 1–130
131. Wollum, J. C., Shearer, G. B., Commoner, B. 1967. The synthesis of tobacco mosaic virus RNA: relationships to the biosynthesis of virus-specific ribonuclease-resistant RNA. *Proc. Nat. Acad. Sci. USA* 58:1197–1204
132. Wood, H. A. 1973. Viruses with double-stranded RNA genomes. *J. Gen. Virol.* 20: Suppl., 1–130
133. Zaitlin, M., Hariharasubramanian, V. 1972. A gel electrophoretic analysis of proteins from plants infected with tobacco mosaic and potato spindle tuber viruses. *Virology* 47:296–305
134. Zaitlin, M., Duda, C. T., Petti, M. A. 1973. Replication of tobacco mosaic virus. V. Properties of the bound and solubilized replicase. *Virology* 53: 300–11
135. Zweerink, H. J., Yoshiaki, I., Matsuhisha, T. 1972. Synthesis of reovirus double-stranded RNA within virion-like particles. *Virology* 50:349–58

# CUTINOLYTIC ENZYMES IN RELATION TO PATHOGENESIS

❖3598

*G. van den Ende and H. F. Linskens*
Department of Botany, Catholic University, Nijmegen, The Netherlands

## INTRODUCTION

Cutin is the basic structural unit of the cuticle. The arrangement of cutin also determines the physical structure of the epicuticular wax layer. The structure of the cuticle varies from one plant to another (9, 44, 46, 47, 51, 62, 65); Figure 1 illustrates the usual arrangement (19, 44, 46, 51, 60, 62). The outer layer of epidermal cells consists of a pectic, a cutinic, and a waxy layer. Cellulose is found in highest concentration at the inner side of the cutin layer, whereas wax is concentrated at the outer side. The specific structure of the cuticle is influenced not only by the genetic background and ploidy of the plant (36, 51) but also by climate (44, 47), humidity (3, 44), and other environmental factors (45).

A cuticular layer is found not only on surfaces of epidermal cells of aerial parts of plants but also in substomatal areas where it exists as a boundary between the subepidermal cells and the substomatal cavity (46, 62). A cuticular layer also is present on the mesophyll and palisade cells (9, 62). Marked variations exist in the thickness of the cuticle and in the degree of wax formation among leaves of different

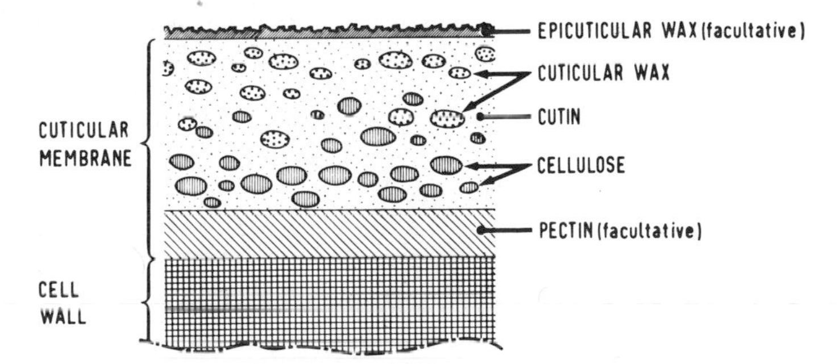


*Figure 1* Schematic representation of the plant cuticle.

species (51, 62) and varieties (49) of plants, between the upper and lower surfaces of the leaves (10, 45), and also among leaves, fruits, stems, seeds (47), and pistils (39). There is also a great difference between the cuticles of plants cultivated under glass and those in the field (36). The thickness and composition of the cuticle changes during plant development (45, 46).

## THE CHEMICAL AND PHYSICAL NATURE OF CUTIN

The structural framework of the cuticle is formed by the cutin. Cutin usually is defined (16, 18, 43, 44, 60) as a polymer of crosswise esterified mono-, di-, and trihydroxy fatty acids (FA) with peroxide and ether bridges between them. The chief components of the cuticle are cutin, waxes, cellulose, and pectic compounds. The cuticular membrane of leaves varies in both weight and cutin content (51). The basic chemistry of cutin is fairly well understood (5, 18–24, 46). Its poor solubility in most organic solvents has made direct chemical analysis of this substance difficult. Microbial enzymes have been useful in elucidating the chemical nature of cutin (18).

The main components of cutin are hydroxy acids (cutinic acids), especially 9,10,18-trihydroxyoctadecanoic acid (9,10,18-THOA) and 10,16-dihydroxyhexadecanoic acid (10,16-DHHA) (6, 12, 47, 51, 60, 61). An acidic function from one molecular chain may esterify with an alcohol function from another molecular chain, the linked chains forming an "estolide." The FA components of cutin from various plant species are described by many authors (6, 12, 44–46, 51, 60, 61). No significant differences in cutin components between mono- and dicotyledons were found in investigations of cutin from 24 species of angiosperms (4). The physically strongest cutinized structures are found in membranes containing large amounts of 9,10,18-THOA and 10,16-DHHA. Thus, the degree of hydroxylation has an influence on the structure and thickness of the cuticle. The thin cuticular membranes of leaves of deciduous herbs contain large quantities of FA but no 9,10,18-THOA and only very small amounts of 10,16-DHHA. The main FA component in leaves of woody plants is 10,16-DHHA; it makes up 69% of total FA in *Citrus aurantifolia,* 60% in *Coffea,* and 21% in *Vitis vinifera.* The leaves of xerophytes have a cutin membrane that contains 9,10,18-THOA as the major component together with significant quantities of 10,16-DHHA.

The FA components in the cuticle membrane of fruits are the same as those found in the leaves (47), although the cuticle in fruits is usually thicker and more complex in structure. The main component in tomato cutin is dihydroxyeicosanoic acid (60).

The chemical and physical properties of the cutin layer have been studied intensively during the last 50 years (46, 47, 51). In cutin biosynthesis, precursors migrate through the cellulosic layer, and polymerize into cutin under the influence of oxygen. The basic structure of cutin, according to Heinen (23), is shown in Figure 2.

## CUTIN AS A BARRIER AGAINST PENETRATION

Because cutin is one of the main components of the cuticle of plants, it is logical to think that it could play an important role in preventing penetration of microor-

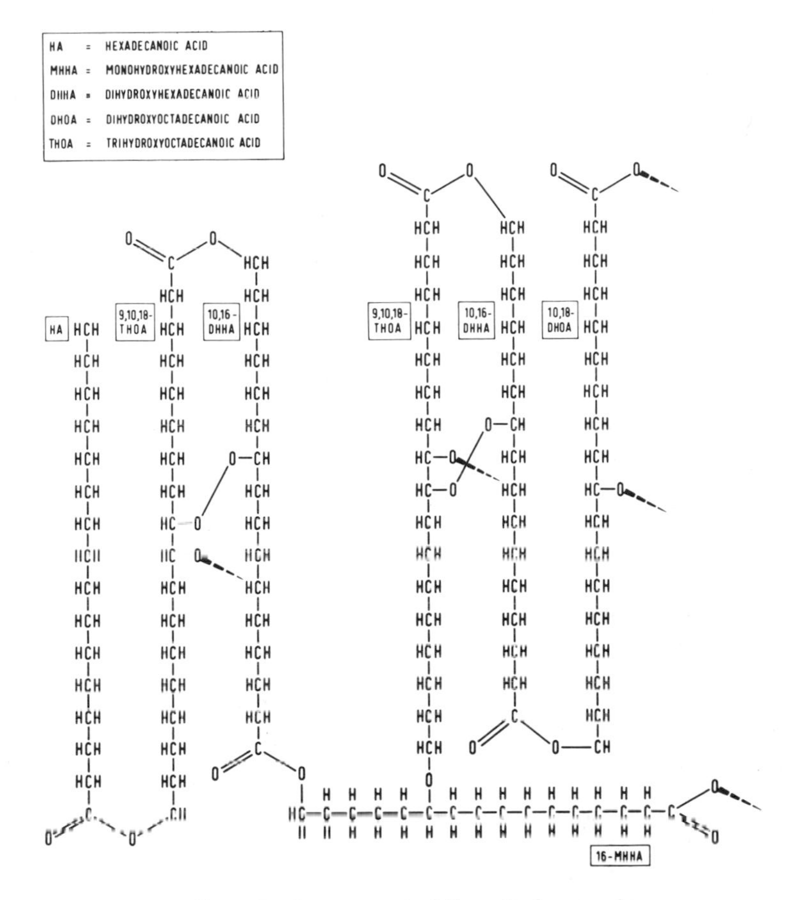


*Figure 2* Arrangement of the cutin framework.

ganisms. The cormosphere[1] and the microhabitat within it play an important role in relation to cutinolytic enzymes and penetration of the plant pathogen. Microorganisms present in the aerial cormosphere have a typical composition. Most research is done on the phyllosphere (51, 54). Stimulation of penetration has been demonstrated with pollen extracts of rye on *Phoma betae* spores present on leaves of *Beta vulgaris* (65, 66) and on fungi in the phyllosphere of rye (51). Germinating pollen of different plants produce an enzyme, probably the same as cutin esterase,

[1]The *cormosphere* is the microecosystem consisting of the plant surface and its immediate environment; it is the region of exchange between biotic and abiotic components. It is composed of the rhizosphere, caulosphere, calusphere, phyllosphere, anthosphere, and carposphere. Caulo-, calu-, antho- and carposphere refer respectively to the surface of the stem, bud, flower, and fruit.

which can play a role in the breakdown of cutin and penetration of plant pathogens into the cuticle.

Many substances are exuded by the plant through the cuticle (7, 11, 13, 28, 37, 47, 51, 54, 63) including: cations such as K, Na, Mg, Ca; amino acids; ammonia; free sugars; pectic substances; organic acids; sugar alcohols; vitamins and growth-regulating substances (gibberellin). Attention should be focused on exchange reactions and the influence of epicuticular wax and cutin structure upon attachment or adhesion of spores, infection hyphae, or appressoria on the aerial surface. The roughness of the leaf and its wettability (36) are strongly influenced by the epicuticular wax and the structure of cutin in the cuticle. The thickness of the cuticle also affects infection and penetration (29, 46). The presence of aliphatic hydrocarbons in spores of some phytopathogenic fungi (67) suggests that attention should be given to organisms, with the ability to improve adhesion between plant surface and the spores or their germ tubes. Yeasts and bacteria which live on plant surfaces and excrete lipids (51, 54) may contribute to the capacity of other organisms for attachment and penetration of plants.

## CUTINOLYTIC ENZYMES

Enzymes are known which can break down the cellulosic part of the cuticle (2, 8, 10, 19, 57). The pectic layer of the cuticle is readily attacked by pectin esterase and pectin polygalacturonidase (2, 8, 10, 18, 19, 34, 51, 57).

In the past 13 years research on cutinolytic enzymes has been performed mainly by Heinen and his co-workers (18–26), usually utilizing an isolate of the fungus *Penicillium spinulosum* Thom isolated from rotting leaves. This isolate is able to grow on a culture medium with cutin as the only carbon source. Hence, the fungus contains enzymes able to destroy cutin. A cutinolytic enzyme was purified by treating a protamine-sulfate extract of the fungus with ammonium sulfate at 70% salt concentration. In this way the cutinolytic enzyme was separated from other "disturbing enzymes." The enzyme reaction was independent of temperature between 15–35°C, the pH optimum was at 6.0, and the optimum substrate concentration was 1.0–1.5 mg cutin/ml enzyme solution. This enzyme, cutin esterase, is stable against H ion concentration, temperature, aging, and dialysis, and needs no dialyzable cofactors. The enzyme is inactivated by shaking. In addition to the cutin esterase mentioned above, the "cutinase complex" contains three oxidizing enzymes that are active on stearic, oleic, and linoleic acid. The amount of stearic oxidase activity does not seem to be correlated with that of the cutin esterase, but oleic and linoleic oxidase activity is correlated with that of cutin esterase. Cutin esterase is able to split off FA from cutin by hydrolytic cleavage of ester links and therefore delivers substrates for the oxidases. Oleic acid is the first substrate acted on by oleic oxidase, delivering the substrate for linoleic oxidase. An enzyme solution obtained from crude extracts of the fungus at 20% saturation with ammonium sulfate contains a peroxidase in addition to FA-oxidizing enzymes. This enzyme, carboxycutin peroxidase, catalyzes the release of oxygen from carboxycutin by loosening the peroxide bridges as shown in the following sequence of reactions in which R and R' are FA:

$$\text{cutin} \xrightarrow{\text{cutin esterase}} \text{R–O–O–R' carboxycutin}$$

$$\text{R–O–O–R'} + \text{Enzyme}\begin{smallmatrix}\text{OH}\\ \text{H}\end{smallmatrix} \xrightarrow{\text{peroxidase}} \text{R–OH} + \text{Enzyme}\begin{smallmatrix}\text{H}\\ \text{OOR'}\end{smallmatrix}$$

$$\text{Enzyme}\begin{smallmatrix}\text{H}\\ \text{OOR'}\end{smallmatrix} \longrightarrow \text{Enzyme}\begin{smallmatrix}\text{H}\\ \text{OOH}\end{smallmatrix} + \text{R'–OH}$$

$$\text{Enzyme}\begin{smallmatrix}\text{H}\\ \text{OOH}\end{smallmatrix} + H_2O \longrightarrow \text{Enzyme}\begin{smallmatrix}\text{H}\\ \text{OH}\end{smallmatrix} + H_2O + O$$

In this manner the cutin esterase produces open FA chains, which are still linked by peroxide bridges (= carboxycutin). The peroxide bridges are exposed and accessible to the carboxycutin peroxidase. Endproducts of the reactions are 2 hydroxy-FA and hydrogen peroxide, which is then split by catalase from the fungus into water and oxygen.

## BIOSYNTHESIS OF CUTIN

The precursors of cutin migrate through the cellulose layer and then polymerize into cutin (12, 40, 46). The following cutin synthetic scheme can be postulated (23): After synthesis of stearic acid from "$C_2$ units" by the acetyl-CoA pathway, which also occurs in wax synthesis (31), this acid is dehydrated to oleic acid by stearic acid oxidase. An equal concentration of oleic acid oxidase accounts for the oxidation of this substrate to linoleic acid. Further dehydration to linolenic acid is assumed but not yet proved. At this stage lipoxidase converts linoleic and possibly the linolenic acids to their hydroperoxides, linoleic-9-hydroperoxide and linolenic-9-hydroperoxide. The following reactions may involve stearic or palmitic acid (27, 32), oleic, linoleic, and linolenic acid, and also the hydroperoxide forms of the two last mentioned FA. Secondary reactions produce many saturated and unsaturated mono- and polyhydroxy-FA (38), which polymerize to cutin via peroxide bridges, followed by the esterification of the hydroxyl and carbonyl groups of the individual FA chains to cutin.

## PATHOGENIC ORGANISMS AND THEIR CUTINASE COMPLEX

Cutin is degraded by many soil microorganisms (16, 25). Data dealing with cutinolytic enzymes of phytopathogenic organisms can be divided into three categories: direct observation, formation of cutinolytic enzymes obtained in pure cultures, and information obtained by histochemical methods.

### *Direct Observation*

*Synchytrium endobioticum* was not able to infect potato stems from which the epidermis was removed but plants with an intact cuticle became infected (15). When zoospores of this fungus come into contact with the cuticle, a fat droplet within the zoospore is liberated and wets and dissolves the cuticle, permitting the fungus to penetrate the tissue. A similar phenomenon has been observed for *Pellicularia filamentosa* and *Raphanus sativus.* Infection cushions can only develop on intact epidermal cells plus cuticle. The presence of the cuticle seems to be important for attachment and penetration (29). Germ tubes of *Pyricularia oryzae* as well as of appressoria form lipid droplets, and the cutin just beneath them is dissolved (17). The cuticle of apple breaks down just at the penetration point of *Venturia inaequalis* (43). In the appressoria a membrane-bound "infection sac" was observed in contact with the hostplant cuticle at the penetration point. The view that a cutin-dissolving enzyme is present is supported by the observation that organisms which do not form an appressorium are able to pass through the cuticle and infect the host plants (30).

### *Enzymes in Pure Culture*

In 1960 Heinen (18, 24) demonstrated that two phytopathogenic fungi, *Fusarium moniliforme* and *Rhodotorula glutinis* var. *rubescens,* excreted cutinolytic enzymes. *R. glutinis* showed 1.7 X greater cutinolytic activity than *P. spinulosum.* The cutin used in these tests was from *Gasteria verrucosa.* The cutinolytic activity of *Rhizoctonia solani, Botrytis cinerea, Cladosporium cucumerinum, Pyrenophora graminea,* and *P. spinulosum* was studied (38) on cutin of the specific "host cutin" and on "standard cutin," a preparation from leaves of *Gasteria verrucosa.* All fungi used showed cutinase activity. For host cutin the level of cutinase activity was 2–3 times greater when compared with standard cutin. The activity of the enzymes with regard to the host cutin was highly variable. The amount of activity was correlated with the specificity of the pathogen used and with the degree of susceptibility of the variety used.

The non-host-specific pathogen, *Botrytis cinerea* showed less activity on the cutin of potato leaves than on standard cutin. *Pyrenophora graminea,* a parasite of the *Gramineae,* showed greater cutinolytic activity on the typical host cutin from leaves of *Triticum aestivum,* while the activity on standard cutin was much lower (one fourth to one fifth). Greater "cutinase" activity was found in congenial than in noncongenial host-parasite relationships; for example, *Rhizoctonia solani/Solanum tuberosum* showed an activity of 2.26 ml, whereas *Penicillium spinulosum/S. tuberosum* has an activity of 0.37 ml (the activity expressed in milliliters of free FA 0.01 n/10 mg cutin/mg dry weight). *Cladosporium cucumerinum* showed little activity on the cutin of a fruit from a resistant *Cucumis* variety. The cutin of the susceptible variety was very thin and its degree of polymerization was low. Therefore, breakdown of the cutin of this variety is more easily accomplished.

The cutinolytic activity of *B. cinerea* was confirmed by Shishiyama and coworkers (60, 61). The network structure of the cuticular membrane was changed during fungal attack. The specific activity of the enzyme of *B. cinerea,* measured spectrophotometrically by Duncombe's method, was much greater than that found by Heinen with *P. spinulosum,* using the titration method. Shishiyama mentioned that his own method was better; in a later paper, Heinen & de Vries (26) discussed problems with the Duncombe method and arrived at a better spectrophotometric method. Six different free FA were found in hydrolysates of the tomato cutin. The main component of tomato cutin, dihydroxyeicosanoic acid, was isolated only in small amounts. Therefore the enzyme should be an exotype cutin esterase that hydrolyzes minor FA side chains bound to the major cutin structure.

### *Histochemical Data*

Loprieno & Tenerini (41) demonstrated enzymatic breakdown of cutin in leaves infected by *Spilocaea oleagina.* The enzymes split off free FA, yielding a dark spot around the penetration site of the parasite. *Sphaerotheca pannosa* var. *rosae,* cause of a mildew of rose, attacked the cuticula enzymatically (10) at precisely the same locations where the appressorium was attached to the epidermis. A discoloration developed in this area of the cuticle when stained with Sudan III, a selective stain for cutin. Congo red and Ruthenium red did not stain the appressorial contact region; therefore, all substances of the epidermal wall are successively attacked. Observations on barley leaves infected with *Erysiphe graminis hordei* (1, 34) showed that around the penetration peg a halo is formed which is not colored by Sudan III. This halo is caused by the degradation of the epidermal cell wall by fungal enzymes including cutinase and cellulase. Esterase activity was demonstrated (50) in the pregermination, the "just germinating," and the appressorial stages of *Venturia inaequalis* on hypocotyls of *Malus silvestris* and *Phaseolus vulgaris.* The substrate for this enzyme is possibly cutin.

In summary, cutin is attacked enzymatically by phytopathogenic fungi. The amount of this activity is different for the various parasites and is also determined by the host specificity of the pathogen. This enzymatic activity is not limited to a special taxon of fungi, but is found in different families.

## SIGNIFICANCE OF CUTINOLYTIC ENZYMES IN PHYTOPATHOGENESIS

Penetration of phytopathogenic microorganisms into the aerial parts of host plants takes place through the cuticle of epidermal cells or of cutinized cells below a natural aperture. Penetration through the cuticle and cell wall, may take place by mechanical pressure, by enzymatic dissolution, or both. Opinions regarding these two modes of penetration have been in flux during the past forty years (Table 1). As early as 1915 Wiltshire mentioned the possibility of a cuticle-dissolving enzyme produced by *Venturia inaequalis.* Many plant pathologists believe in an enzymatic breakdown of cell wall material during penetration and in a mechanical penetration of the cuticle. This view can have an important influence on the interpretation of experimental results. For example, Roberts, Martin & Peries (53) could have given a more logical explanation of their results if they had considered the possibility of enzymatic breakdown of cutin.

Indented and irregular edges of the cuticle around a penetration hole and enlargement of the penetration peg after passing the cuticle are sometimes cited (51) in arguments for mechanical penetration of the cuticle; but enzymatic breakdown of cutin also may produce irregular, rough surfaces (18, 64). The fact that the penetration peg enlarges after passing the cuticle indicates only that the cuticle is more resistant to enzymatic degradation than cell wall components such as cellulose and pectin. Many authors distinguish between penetration of the epidermis and penetration of natural openings, while in both circumstances the cuticle must be penetrated. In the latter case of natural openings, however, the cuticle is usually thinner and may have a lower degree of polymerization. The time during which the parasite spore rests on the aerial part of the plant until the penetration peg is formed is of great importance (14).

**Table 1** Experiments concerning enzymatic breakdown of the plant cell wall plus cuticle and of the cuticle alone

| Year | Author(s) | Cell wall plus cuticle | | Cuticle alone | |
|---|---|---|---|---|---|
| | | Mechanical | Enzymatic | Mechanical | Enzymatic |
| 1936 | Brown (7) | + | – | + | – |
| 1951 | Gretschuschnikow & Jakowlewa (15) | – | + | – | + |
| 1954 | Pristou & Gallegly (52) | + | – | + | – |
| 1956 | Lupton (42) | – | – | + | – |
| 1957 | Chakravarty (11) | – | – | + | – |
| 1960 | Wood (68) | – | – | + | – |
| 1963 | Linskens & Haage (38) | – | + | – | + |
| 1964 | Meredith (48) | – | – | + | – |
| 1965 | Linskens, Heinen & Stoffers (40) | – | + | – | + |
| 1968 | Ellingboe (14) | + | + | + | + |
| 1968 | Hashioka (17) | – | – | + | + |
| 1968 | Akai c.s. (1) | – | + | – | + |
| 1969 | Kunoh c.s. (34) | – | + | – | + |
| 1970 | Shishiyama (61) | – | – | + | + |
| 1971 | Stanbridge[a] | – | – | + | – |
| 1971 | Mercer[a] | – | – | + | – |
| 1972 | Schroeder (55) | – | – | + | – |
| 1972 | Nicholson (50) | – | + | – | + |

[a]Cited from 51.

This period can be divided into four stages: 1. pregermination, 2. germination, 3. appressorial development or attachment, and 4. development of the penetration peg. During stages 1 and 2 the inoculum unit can be influenced by the cormosphere and must maintain itself on the plant surface. In stage 1 the inoculum unit can form cutinolytic exoenzymes (15, 30, 43, 50). During this period, organisms of the aerial cormosphere, e.g. yeasts, bacteria, or pollen may release exoenzymes (39, 54) and thus facilitate penetration by the pathogen. During the germination stage, growth and development of the germ tube also can be influenced by exudates of the leaf or by substances which are excreted by cormosphere organisms and may have a positive influence on enzymatic activity (25). Certain wax substances have been proven to exert a positive and selective influence on specific pathogens (56, 58, 59).

The development of appressoria or infection cushions is influenced by the physical nature of the cuticle (2). Esterase activity is also found in the appressoria (43, 50). But nothing is known about the influence of cormosphere microorganisms on the development of the appressoria, the direct influence of nutritive excretion products on appressorial development, or the indirect influence of cutinolytic enzymes on this process. On the other hand, in some cases where the cuticle is not intact, no characteristic appressoria or infection cushions are formed. Exudates of the plants are sometimes required for successful penetration. Histochemical studies have clearly demonstrated the formation of cutinolytic enzymes in the halo around the infection peg on the cuticle (1, 10, 34, 41, 43, 50). Only Akai and co-workers (1) mentioned an occasional crack in the center of a halo.

The most crucial question in this entire chapter is whether the cutin in the cuticle is a real barrier in the phytopathogenesis. Experiments (38) have shown the following: 1. a considerable difference in enzyme activity by pathogenic fungi on standard cutin and host cutin, 2. greater cutinolytic activity in a congenial host-parasite relationship than in a noncongenial relationship, and 3. a difference in pathogen enzyme activity on the cutin of resistant and susceptible varieties (37). In the above experiments the authors used isolated cutin. The results of Martin (44) are in partial agreement with the above-mentioned results. On the other hand, many authors (3, 28, 33, 35, 52, 53) found penetration of the cuticle by pathogens in both resistant and susceptible varieties or species. These authors looked to the end result, namely the growth of the pathogen under the cuticle or the epidermis, without focusing on the actual penetration process and the time required for it. There are several examples where this difference in penetration rate must be very small.

Plant pathogenic bacteria mainly utilize natural openings of the plants or wounded tissue for infection sites. The cutin layer in such areas is often much thinner and possibly less complex. On the other hand, many nonpathogenic bacteria possess cutinolytic enzymes.

The cutin meshwork is interwoven with the wax and pectic layers which border it (5, 25). Aliphatic hydrocarbons (67) in phytopathogenic fungi may change the adhesion forces between the inoculum unit and the cuticle; but these compounds can also form chemical links between the spore, the hyphe or the appressoria, and the cuticle to achieve a stronger attachment of the inoculum unit. For *Pyricularia oryzae* Hashioka and co-workers (17) clearly showed dissolution of the cuticle just beneath a germ tube and beneath an appressorium strongly attached to the cuticle with mucilaginous strands, so that the cuticle is raised locally around the appressorium. This need not be in contradiction with enzymatic breakdown of the cuticle, but suggests that the adhesion forces and eventually the chemical links play an important role.

The significance of cutinolytic enzymes in relation to pathogenesis must not be overemphasized; it is still too early to be confident about its significance. More research must be done concerning host specificity of the enzymes and their activity levels, host/parasite relations, the influence of the degree of polymerization of cutin on its biodegradability, and finally the effects of organisms and various substances in cormosphere on the breakdown of cutin and penetration of parasites. Further attention also must be focused on two other aspects: 1. the role of cormospheric organisms in preventing cutin formation—Ruinen (54) has suggested such a role for the microorganisms of the phyllosphere; and 2. the use of the cuticle as a food source. Some phytopathogenic fungi have saprophytic or residental stages prior to penetration, others possess subcuticular mycelia, and still others directly penetrate the plant. It will be interesting to know more about the cutinolytic activity of those organisms. It is possible that in organisms like *Spilocaea oleagina,* which live sub- or intracuticularly, the cuticle is also used as a food source.

*Literature Cited*


1. Akai, S., Kunoh, H., Fukutomi, M. 1968. Histochemical changes of the epidermal cell wall of barley leaves infected by *Erysiphe graminis hordei. Mycopathol. Mycol. Appl.* 35:175–80
2. Albersheim, P., Jones, T. M., English, P. D. 1969. Biochemistry of the cell wall in relation to infective processes. *Ann. Rev. Phytopathol.* 7:171–94
3. Ayres, P. G., Owen, H. 1971. Resistance of barley varieties to establishment of subcuticular mycelia by *Rhynchosporium secalis. Trans. Brit. Mycol. Soc.* 57:233–40
4. Baker, E. A., Holloway, P. J. 1970. The constituent acids of angiosperm cutins. *Phytochemistry* 9:1557–62
5. Bredemeijer, G., Heinen, W. 1968. Cutin synthesis in plants I. Free fatty acid movement during cutin synthesis in injured *Gasteria verrucosa* leaves. *Acta Bot. Neerl.* 17:15–25
6. Brieskorn, C. H., Kabelitz, L. 1971. Hydroxyfettsäuren aus dem Cutin des Blattes von *Rosmarinus officinalis. Phytochemistry* 10:3195–3204
7. Brown, W. 1936. The physiology of hostparasite relations. *Bot. Rev.* 2: 236–81
8. Brown, W. 1965. Toxins and cell wall dissolving enzymes in relation to plant disease. *Ann. Rev. Phytopathol.* 3:1–18
9. Calvin, C. L. 1970. Anatomy of the aerial epidermis of the mistletoe *Phoradendron flavescens. Bot. Gaz. Chicago* 131:62–74
10. Caporali, L. 1960. Sur la formation des suçoirs de *Sphaerotheca pannosa* (Wallr.) Lév. var. *rosae* dans les cellules épidermiques des folioles de Rosa pouzini Tratt. *C.R. Acad. Sci. Paris* 250:2415–17
11. Chakravarty, T. 1957. Anthracnose of banana (*Gloeosporium musarum* Cke & Massee) with special reference to latent infection in storage. *Trans. Brit. Mycol. Soc.* 40:337–45
12. Croteau, R., Fagerson, I. S. 1972. The constituent cutin acids of cranberry cuticle. *Phytochemistry* 11:353–63
13. Cruickshank, I. A. M. 1963. Phytoalexins. *Ann. Rev. Phytopathol.* 1:351–74
14. Ellingboe, A. H. 1968. Inoculum production and infection by foliage pathogens. *Ann. Rev. Phytopathol.* 6:317–30
15. Gretschuschnikow, A. J., Jakowlewa, N. N. 1951. Cited from Suchorukow, K. T. 1958. *Beiträge zur Physiologie der pflanzlichen Resistenz.* p. 26. Berlin: Akad.-Verlag
16. Hankin, L., Kolattukudy, P. E. 1971. Utilization of cutin by a Pseudomonad isolated from soil. *Plant Soil* 34:525–29
17. Hashioka, Y., Ikegami, H., Horino, O., Kamei, T. 1967. Fine structure of rice blast II Electronmicrographs of the initial infection. *Res. Bull. Fac. Agr. Gifu Univ.* 24:78–90
18. Heinen, W. 1960. Über den enzymatischen Cutin-Abbau I. Mitteilung: Nachweis eines "Cutinase"-Systems. *Acta Bot. Neerl.* 9:167–90
19. Heinen, W. 1962. Über den enzymatischen Cutin-Abbau III. Mitteilung: Die enzymatische Ausrüstung von *Penicillium spinulosum* zum Abbau der Cuticularbestandteile. *Arch. Mikrobiol.* 41:268–81

20. Heinen, W. 1963. Über den enzymatischen Cutin-Abbau IV. Mitteilung: Trennung der Cutinase von oxydativen Begleitfermenten. *Enzymologia* 25:281–91
21. Heinen, W. 1963. Über den enzymatischen Cutin-Abbau V. Mitteilung: Die Lyse von Peroxyd-Brücken im Cutin durch eine Peroxydase aus *Penicillium spinulosum* Thom. *Acta Bot. Neerl.* 12:51–57
22. Heinen, W., van den Brand, I. 1961. Über den enzymatischen Cutin-Abbau II. Mitteilung: Eigenschaften eines cutinolytischen Enzyms aus *Penicillium spinulosum* Thom. *Acta Bot. Neerl.* 10:171–89
23. Heinen, W., van den Brand, I. 1963. Enzymatische Aspekte zur Biosynthese des Blatt-Cutins bei *Gasteria verrucosa*-Blättern nach Verletzung. *Z. Naturforsch.* 18b:67–79
24. Heinen, W., Linskens, H. F. 1960. Cutinabbau durch Pilzenzyme. *Naturwissenschaften* 47:18–20
25. Heinen, W., de Vries, H. 1966. Stages during the breakdown of plant cutin by soil microorganisms. *Arch. Mikrobiol.* 54:331–38
26. Heinen, W., de Vries, H. 1966. A combined micro and semi-micro colorimetric determination of long-chain fatty acids from plant cutin. *Arch. Mikrobiol.* 54:339–49
27. Holloway, P. J., Deas, H. B. 1971. Occurrence of positional isomers of dihydroxyhexadecanoic acid in plant cutins and suberins. *Phytochemistry* 10:2781–85
28. Jones, P., Ayres, P. G. 1972. The nutrition of the subcuticular mycelium of *Rynchosporium secalis:* permeability changes induced in the host. *Physiol. Plant Pathol.* 2:383–92
29. Kerr, A., Flentje, N. T. 1957. Host infection in *Pellicularia filamentosa* controlled by chemical stimuli. *Nature* 179:204–5
30. Kilpatrick, R. A. 1959. A disease of Ladino white clover caused by a yeast, *Rhodotorula glutinis* var. *rubescens. Phytopathology* 49:148–51
31. Kolattukudy, P. E. 1967. Biosynthesis of paraffins in *Brassica oleracea:* Fatty acid elongation-decarboxylation as a plausible pathway. *Phytochemistry* 6: 963–75
32. Kolattukudy, P. E. 1970. Cutin biosynthesis in *Vicia faba* leaves. *Plant Physiol.* 46:759–60
33. Kreber, H., Pethold, H. 1972. Licht- und elektronenmikroskopische Untersuchungen über Wirt-Parasit-Beziehungen bei anfälligen und gegen *Perenospora* spp. resistent gezüchteten Sorten von Tabak und Spinat. *Phytopathol. Z.* 74:296–313
34. Kunoh, H., Akai, S. Y. 1969. Histochemical observation of the halo on epidermal cell wall of barley leaves attacked by *Erysiphe graminis hordei. Mycopathol. Mycol. Appl.* 37:113–18
35. Lapwood, D. H., McKee R. K. 1966. Dose-response relationships for infection of potato leaves by zoospores of *Phytophthora infestans. Trans. Brit. Mycol. Soc.* 49:679–86
36. Linskens, H. F. 1952. Über die Änderung der Benetzbarkeit von Blattoberflächen und deren Ursache. *Planta* 41:40–51
37. Linskens, H. F. 1955. Der Einfluss der toxigenen Welke auf die Blattausscheidungen der Tomatenpflanze. *Phytopathol. Z.* 23:89–106
38. Linskens, H. F., Haage, P. 1963. Cutinase-Nachweis in phytopathogenen Pilzen. *Phytopathol. Z.* 48:306–11
39. Linskens, H. F., Heinen, W. 1962. Cutinase-Nachweis in Pollen. *Z. Bot.* 50:338–47
40. Linskens, H. F., Heinen, W., Stoffers, A. L. 1965. Cuticula of leaves and the residue problem. *Residue Rev.* 8:136–78
41. Loprieno, N., Tenerini, I. 1959. Methodo per la diagnosi precoce dell' "Occhio di pavone" dell'olivo. *Phytopathol. Z.* 34:385–92
42. Lupton, F. G. H. 1956. Resistance mechanisms of species of Triticum and Aegilops and of amphidiploids between them to *Erysiphe graminis* DC. *Trans. Brit. Mycol. Soc.* 39:51–59
43. Maeda, K. M. 1970. *An ultrastructural study of Venturia inaequalis (Cke) Wint. infection of Malus hosts.* MS thesis. Purdue Univ., 112 pp. Cited in Ref. 50
44. Martin, J. T. 1964. Role of cuticle in the defense against plant disease. *Ann. Rev. Phytopathol.* 2:81–100
45. Martin, J. T., Batt, R. F. 1958. Studies on plant cuticle. I. The waxy coverings of leaves. *Ann. Appl. Biol.* 46:375–87
46. Martin, J. T., Juniper, D. E. 1970. The Cuticles of Plants, pp. 1–10, 59–117, 144–47, 174–78, 254–77, 287–91. London: Edward Arnold
47. Mazliak, P. 1968. Chemistry of plant cuticles. In *Progress in Phytochemistry,* ed. L. Reinhold, Y. Liwschitz, 1:49–111. London/New York/Sydney. Wiley-Interscience

48. Meredith, D. S. 1964. Appressoria of *Gloeosporium musarum* Cke and Massee on banana fruits. *Nature* 201: 214–15
49. Morozova, N. P., Sal'kova, E. C. 1970. The composition of paraffin hydrocarbons of the plant cuticular wax. *Prikl. Biokhim. Microbiol.* 6:697–702
50. Nicholson, R. L., Kuć, J., Williams, E. B. 1972. Histochemical demonstration of transitory esterase activity in *Venturia inaequalis*. *Phytopathology* 62: 1242–47
51. Preece, T. F., Dickinson, C. H. 1971. *Ecology of Leaf Surface Micro-Organisms.*, especially Sect. I:1–4; II:4, 5; III:5–7, 13, 14; V:1–10. London/New York: Academic
52. Pristou, R., Gallegly, M. E. 1954. Leaf penetration by *Phytophthora infestans*. *Phytopathology* 44:81–86
53. Roberts, M. F., Martin, J. T., Peries, O. S. 1960. Studies on plant cuticle IV. The leaf cuticle in relation to invasion by fungi. *Ann. Rept. Agr. Hort. Res. Sta. Long Ashton* 102–10
54. Ruinen, J. 1966. The phyllosphere IV Cuticle decomposition by micro-organisms in the phyllosphere. *Ann. Inst. Pasteur* 111:342–46
55. Schroeder, C. 1972. Untersuchungen zum Wirt-Parasit Verhältnis von Tulpe und *Botrytis* spp. *Z. Pflanzenkrankh.* 79:94–104
56. Schuck, H. J. 1972. Composition of needle wax of *Pinus silvestris* in relation to the provenience origin and needle age and its significance in the susceptibility to *Lophodermium pinastri*. *Flora Jena* 161:604–22
57. Schulz, F. A. 1972. Production of pectinolytic and cellulolytic enzymes by *Gloeosporium perennans*. *Phytopathol. Z.* 74:97–108
58. Schütt, P. 1971. Untersuchungen über den Einfluss von Cuticularwachsen auf die Infektionsfähigkeit pathogener Pilze. 1. *Lophodermium pinastri* und *Botrytis cinerea*. *Eur. J. Forest Pathol.* 1:32–50.
59. Schütt, P. 1972. Untersuchungen über den Einfluss von Cuticularwachsen auf die Infektionsfähigkeit pathogener Pilze. 2. *Rhytisma acerinum, Microsphaera alphitoides* und *Fusarium oxysporum*. *Eur. J. Forest Pathol.* 2:43–59
60. Shishiyama, J., Araki, F., Akai, S. 1970. Studies on cutin esterase: I. Preparation of cutin and its fatty acid component from tomato fruit peel. *Plant Cell Physiol.* 11:323–34
61. Shishiyama, J., Araki, F., Akai, S. 1970. Studies on cutin esterase. II Characteristics of cutin esterase from *Botrytis cinerea* and its activity on tomato-cutin. *Plant Cell Physiol.* 11:937–45
62. Sitholey, R. V. 1971. Observations on the three dimensional structure of the leaf cuticle in certain plants. *Ann. Bot. London* 35:637–39
63. Topps, J. H., Wain, R. L. 1957. Fungistatic properties of leaf exudates. *Nature* 179:652–53
64. de Vries, H., Bredemeijer, G., Heinen, W. 1967. The decay of cutin and cuticular components by soil micro-organisms in their natural environment. *Acta Bot. Neerl.* 16:102–10
65. Warren, R. C. 1972. The effect of pollen on the fungal leaf microflora of *Beta vulgaris* L. and on infection of leaves by *Phoma betae*. *Neth. J. Plant Pathol.* 78:89–98
66. Warren, R. C. 1972. Attempts to define and mimic the effects of pollen on the development of lesions caused by *Phoma betae* inoculated into sugarbeet leaves. *Ann. Appl. Biol.* 71:193–200
67. Weete, J. D. 1972. Review article aliphatic hydrocarbons of the fungi. *Phytochemistry* 11:1201–5
68. Wood, R. K. S. 1960. Chemical ability to breach the host barriers. In *Plant Pathology*, ed. J. G. Horsfall, A. E. Dimond, II: 239. New York/London: Academic

# TOXINS PRODUCED BY PHYTOPATHOGENIC BACTERIA

❖3599

*Suresh S. Patil*
Department of Plant Pathology, University of Hawaii, Honolulu, Hawaii 96822

In memorium: Albert E. Dimond, teacher, colleague, and friend

## INTRODUCTION

The following conversation was overheard at a recent international gathering of phytopathologists:

Q: What is your special research interest?
A: Bacterial phytotoxins.
Q: They're not host specific, are they?
A: No.
Q: But aren't the bacteria which produce them host specific?
A: Yes, but . . .
Q: Then how can studies of nonspecific toxins tell us anything about the nature of host specificity of the pathogens which produce such toxins?

The point made by the questioner may be indicative of the general view among host-parasite physiologists that the ultimate aim of biochemical studies of host-parasite interaction is to elucidate the basis of pathogenicity and specificity of parasites for their host plants. The fact of nonspecificity of isolated toxins of phytopathogenic bacteria (21), as contrasted to the host specificity (81) exhibited by toxins of some fungal pathogens, has led many pathologists to believe that non-specific toxins are not involved in the host specificity of pathogens which produce them. I argue later in this discussion that such a belief may not be justified and could have been responsible for the historical lack of greater interest in bacterial phytotoxins. Whatever the reason, little is known about the role of these toxins in the host specificity of pathogens which elaborate them.

The primary purpose of this paper is to review the chemical structure of a few bacterial phytotoxins and the biochemical changes they induce in affected host tissues. Such information is critical to understanding the role of these toxins in determining the host specificities of phytobacteria. It also has practical implications in terms of control. Mention is made of some recently discovered toxins and of a new class of toxins that appears to be involved in the induction of the hypersensitive reaction in plants. A complete survey of the occurrence of bacterial phytotoxins has been made recently (15). Table 1 contains a tabulation of the major types of bacterial phytotoxins and a few of their properties.

## CHEMICAL COMPOSITION AND BIOCHEMICAL EFFECTS

### *Tabtoxin(s) and Related Inactive Products*

CHEMISTRY The toxin produced by *Pseudomonas tabaci* (Wolf and Foster) Stevens, traditionally has been referred to as the "wildfire" toxin. Recently it was found that the so-called wildfire toxin is a mixture of two analogs (102) and that at least two other species of *Pseudomonas* produce the same mixture of compounds (88). For these reasons the name wildfire is considered inappropriate (100) and the new nomenclature described at the end of this section is preferred.

The production of an extracellular toxin by *P. tabaci,* the causal agent of the wildfire disease of tobacco (*Nicotiana tabacum*), was first reported in 1925 (34). Clayton (8) later tested its host range and showed that the toxin was not host specific. In spitc of this early work, attempts to characterize the toxin did not begin in earnest until the early 1950s. Woolley, Pringle & Braun (104) devised a purification method and a biological assay to study it. A structure for *P. tabaci* toxin was

**Table 1** Sources and properties of toxins produced by phytopathogenic bacteria

| Bacterial Pathogen | Toxins | Elemental Composition | Chemical Nature | Molecular Weight | Mode of Action |
|---|---|---|---|---|---|
| *P. tabaci* (94)<br>*P. coronafaciens* (88)<br>*Pseudomonas* Spp. (Timothy) (102) | Tabtoxin and<br>2-serine-tabtoxin | $C_{11}H_{19}N_3O_6$<br>$C_{10}H_{17}N_3O_6$ | β-lactamthreonine<br>β-lactamserine | 289<br>275 | ?<br>? |
| *P. phaseolicola* (66, 70) | phaseotoxin | — | peptide? | 500–1000 | inhibition of OCT? |
| *P. syringae* (85) | syringomycin | — | polypeptide | low | membrane disruption |
| *R. japonicum* (61) | rhizobitoxin | $C_7H_{14}N_2O_4$ | enol ether amino acid | 190 | inhibition of β-cystathionase? |
| *C. sepedonicum* (96) | — | $C_{48}H_{96}O_{48}N$ | glycopeptide | 21,450 | membrane disruption |
| *C. insidiosum* (96) | — | $C_{108}H_{126}O_{132}N$ | glycopeptide | 5,000,000 | ? |
| *C. michiganense* I (73) | — | $C_{42}H_{82}O_{40}N$ | glycopeptide | > 200,000 | membrane disruption |
| *C. michiganense* II | — | $C_{473}H_{940}O_{449}N$ | glycopeptide | 129,700 | membrane disruption |
| *C. michiganense* III | — | $C_{27}H_{52}O_{25}N$ | glycopeptide | 35,280 | membrane disruption |

proposed in 1955 (105): it was then thought to be a $\alpha$-lactylamino-$\beta$-hydroxy-($\epsilon$-aminopimelic acid) lactone (Figure 1a). Complete acid hydrolysis (6 *M* HCl, 121°C, 60 min) of the toxin resulted in the production of tabtoxinine ($\alpha,\epsilon$-diamino-$\beta$-hydroxy pimelic acid) and lactic acid. The structure of tabtoxinine proposed by Woolley and his colleagues was later withdrawn (93).

Active work on the *P. tabaci* toxin did not resume until more than a decade after these early findings. Durbin and co-workers (87, 88), while working on a disease of oats caused by *P. coronafaciens* (Elliott) Stevens, found that an extracellular toxin was involved in the production of chlorotic symptoms in infected plants. Despite some differences in the environmental conditions under which the toxins of *P. coronafaciens* and *P. tabaci* produced symptoms in host tissues, (15), it appeared that both toxins were chemically similar.

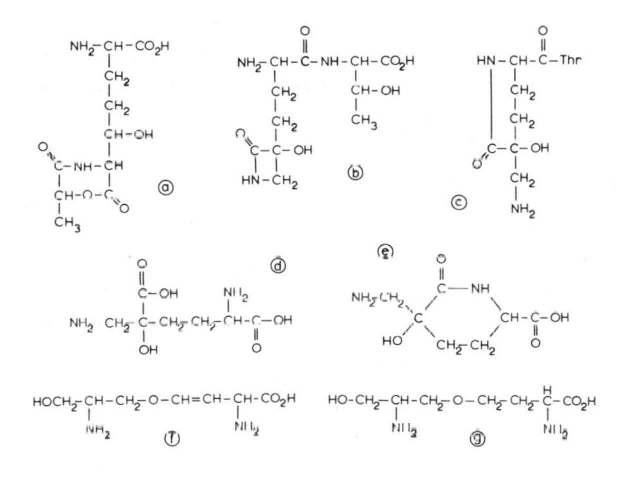

*Figure 1* Structures of toxic and nontoxic metabolites of *P. tabaci* and *R. japonicum*. (*a*) original inaccurate structure of tabtoxin (105), (*b*) tabtoxin (102), (*c*) isotabtoxin, (*d*) tabtoxinine, (*e*) tabtoxinine-$\delta$-lactam, (*f*) rhizobitoxine (61), (*g*) dihydrorhizobitoxine (62).

Further detailed comparative studies (88) using ion exchange chromatography revealed the presence of an identical compound in culture filtrates of both pathogens. Acid hydrolysis of this biologically active compound from either source yielded equal molar ratios of threonine and tabtoxinine. Serine was also detected in these hydrolysates but at a much lower molar ratio with respect to tabtoxinine. No lactic acid was found in hydrolysates of either toxin.

Alkali treatment (0.4 *M* $NaHCO_3$) or heating (110°C, 20 min) of *P. tabaci* or *P. coronafaciens* toxin produced two inactive products, as opposed to only one observed by Woolley and co-workers (105) after alkali treatment of wildfire toxin. These comparative studies indicated that the two pathogenic *Pseudomonas* species produce identical toxins. They also revealed the possibility that the active toxin may actually be a mixture of two closely related compounds: one containing L-threonine and the other containing L-serine. Subsequent mass spectrometric studies confirmed the existence of the two analogs (102). This and other evidence made the originally proposed structure of wildfire toxin untenable.

Stewart (94) examined the validity of the original structure of *P. tabaci* toxin. Purification of the toxin by ion exchange chromatography resolved it into a major and minor peak. The major peak contained 99% activity. Hydrolysis of the compound from this peak produced tabtoxinine and L-threonine. After similar treatment of the compound in the minor peak, tabtoxinine and L-serine were detected. This compound was not investigated further. Boiling the threonine analog at 100°C (pH 3.1) for 60 min produced an inactive compound with the same chemical properties as those reported by Sinden & Durbin (88) for their inactivation product.

Stewart deduced the structure of the purified threonine analog by infrared, NMR, and mass spectrometry. It shows a tabtoxinine-$\beta$-lactam residue linked to the amino group of L-threonine (Figure 1b). The inactive compound is a tabtoxinine-$\delta$ lactam residue linked to the amino group of L-threonine (isotabtoxin) (Figure 1c). Finally, the structure of tabtoxinine was deciphered. It is 5-amino-2-amino methyl-2-hydroxyadipic acid (Figure 1d).

Taylor, Schnoes & Durbin (101, 102) reported that yet another *Pseudomonas* species pathogenic to timothy (*Phleum pratense* L.) produces the same two toxic analogs as those produced by *P. tabaci* and *P. coronafaciens.* They confirmed the structures proposed by Stewart (94) for the threonine and serine analogs, their corresponding inactivation products (isotabtoxins), and tabtoxinine. These workers characterized tabtoxinine-$\delta$-lactam (Figure 1e), a hitherto uncharacterized product produced either by acid hydrolysis of the active toxins and their inactivation products (isotabtoxins) or by heating tabtoxinine at 100°C for 60 min at pH 3.1.

Because all three bacteria produce the same toxins and related products, the original name wildfire toxin became inapproporate. A new nomenclature was therefore proposed, according to which the threonine-containing active compound is called tabtoxin, its serine analog is referred to as [2-serine]-tabtoxin, and the corresponding inactive isomers are designated isotabtoxin and [2-serine]-isotabtoxin. The original name tabtoxinine is retained for the hydrolysis product of tabtoxins and isotabtoxins (all four products are degraded to tabtoxinine). Finally, the newly characterized degradation product of tabtoxinine is designated tabtoxinine-$\delta$-lactam. To avoid confusion I refer to the biologically active mixture containing tabtoxin and [2-serine]-tabtoxin in the following section as tabtoxin. This is appropriate because most of the work on mode of action has been done with this mixture, whose individual components have equal biological potency on a molar basis (100).

MODE OF ACTION Although the structures of tabtoxins and related inactive products have been neatly deciphered, their mode of action in plant tissues remains a mystery. Almost two decades ago Braun (5) suggested that tabtoxin acted as an antagonist of L-methionine. This suggestion was based on several lines of evidence: (*a*) Toxin inhibited growth of the test organism *Chlorella vulgaris,* which could be prevented by adding liver or yeast extracts. Experiments showed that L-methionine in the extracts was responsible for preventing inhibition of growth of the alga by the toxin. (*b*) Once methionine was implicated, experiments were conducted with its structural analog methionine sulfoximine (MSO). MSO inhibited growth of *Chlorella vulgaris,* which was also prevented by adding methionine, and induced chlorosis in tobacco leaves. (*c*) Mutants of *Chlorella* selected for resistance to MSO were also resistant to the toxin. Carlson (7) recently showed that MSO-resistant mutants of tobacco are also resistant to chlorosis induced by *P. tabaci* and that several mutants show a specific increase in amount of free methionine. Thus, the close similarities between the behavior of MSO and tabtoxin indicated that, like MSO, the toxin acts as a structural analog of methionine.

But several problems stood in the way of the acceptance of this hypothesis: (*a*) The chemical structure of tabtoxin proposed by Woolley (105) did not resemble L-methionine. (*b*) Neither MSO nor toxin affected reactions involving L-methionine (14). The formation of S-adenosylmethionine or incorporation of charged methionine into a polypeptide fraction was not affected by MSO or toxin. (*c*) MSO inhibition of bacterial growth was reversed not only by L-methionine but also by glutamine (5). The similar behavior of the two disparate amino acids could not be explained on grounds of shared affinity for a common metabolic site within the bacterial cell (50) and suggested the possibility that both glutamine and methionine prevent the uptake of MSO (51). Later, it was shown that L-methionine prevents uptake of MSO in animal cells (41) and in *C. vulgaris* (14); this may also occur in Carlson's tobacco mutants. It was therefore surprising that L-methionine did not protect tobacco leaves from the effects of either toxin or MSO (86), which suggests that it does not act as a toxin analog.

Since the original work of Braun (4, 5) evidence had begun to mount that in both animals and plants MSO interferes with glutamine rather than methionine metabolism (14, 26) and that this occurs via the inhibition of glutamine synthetase (L-glutamate: ammonia ligase, EC 6.3.1.2), the enzyme that catalyzes the conversion of glutamic acid to glutamine (41, 49).

Taking a cue from this information, Durbin and colleagues reopened the investigations of the mode of action of tabtoxin in tobacco tissues. They found (86) that partially purified tabtoxin inhibits semipurified glutamine synthetase (GS) of pea plants. A further discovery that purified tabtoxin, like MSO, causes convulsions in rats (89) and inhibits GS of rat brain in vitro (42) strengthened their belief that the primary event in the mode of action of tabtoxin in tobacco tissues is the inhibition of GS. This hypothesis was further supported by the observation that tobacco leaves infiltrated with toxin or MSO along with glutamine were not subject to chlorosis. This observation was explained by postulating that glutamine protected GS against toxin inhibition in vivo although permeability effects might be involved. The question of how GS inhibition might cause chlorosis was answered by predicting that inhibition of GS, which catalyzes a principal pathway of nitrogen assimilation in plants, would lead to ammonia accumulation. Such was shown to be the case (86) in tobacco leaves infiltrated with the toxin.

Recently, however, when highly purified tabtoxins and pea GS were used, no inhibition of GS was observed, although MSO inhibited the enzyme as before (R.

D. Durbin, personal communication). Thus the thesis that tabtoxins affect plant tissues by inhibiting GS must be abandoned.

How then does tabtoxin affect plant cells? Several years ago it was reported (47) that ammonia accumulated in tobacco leaves as a response to inoculation by *P. tabaci.* The quantity of ammonia formed during the development of the disease was sufficient to account for the water-soaking and necrotic symptoms which occurred before chlorosis was evident. Goodman's (21) later work suggests that the ultrastructural degradation and chlorosis in tobacco leaf tissue inoculated with *P. tabaci* may have been also caused by ammonia.

The question which arises is: is ammonia accumulation in these cases caused by tabtoxin or some other bacterial product? Durbin (15) examined this question by studying the effects of tabtoxin on tobacco leaves. In this case, highly purified toxin was used. Four hours after treatment, degradation of chloroplast structure was already evident as was a two- to fourfold increase in concentration of ammonia. The ammonia evolved was not due to lysis of plant proteins. How ammonia affects plants is still unknown. One hypothesis (37) involves a proposal that ammonia regulates the urea and TCA cycles.

If the toxin does not inhibit GS, then what causes the ammonia buildup? Absence of proteolysis during at least the early accumulation of ammonia indicates that other enzyme reactions in the nitrogen metabolism might be involved. Before proceeding to investigate the causes of ammonia accumulation, however, definitive studies will have to be undertaken on the temporal sequence of events with regard to ammonia accumulation, on the one hand, and structural and functional changes in toxin-treated tissue, on the other. It is possible that subtle changes (e.g. chloroplast degradation) occur in treated tissues shortly after exposure to toxin which in turn induce ammonia production, thus making ammonia the effect rather than the cause of primary symptoms.

An alternative mode of action was previously proposed by Lovrekovich and associates (46) who found decreases in the concentration of RNA and soluble proteins with time and a concomitant increase in amino acids and ammonia in toxin-treated tobacco leaves. Application of kinetin counteracted the toxin effect (17, 45). Toxin action was likened to physiological changes involved in senescence in plants because both effects were reversed by kinetin. Here again, samples were taken late, no time course studies were performed, and what was observed was probably due to secondary effects.

## *Phaseotoxin*

CHEMICAL COMPOSITION *Pseudomonas phaseolicola* (Burk) Dowson, the causal organism of halo blight of bean (*Phaseolus vulgaris* L.), produces an extracellular toxin which is responsible for production of chlorotic halos in infected bean plants. I propose to call this toxin phaseotoxin rather than halo blight toxin as it has been known in the past.

Early work on this toxin is summarized elsewhere (22). Later work of Hoitink, Pelletier & Coulson (29) showed that toxin is elaborated in undefined media, produces chlorosis in bean leaves, and is thermostable and nonspecific. They also developed a viable leaf bioassay for the toxin. In subsequent work (30) using a defined medium, they attempted purification of the toxin by methanol extraction, charcoal adsorption, and thin-layer chromatography. The toxin had $R_f$ values different from that of *P. tabaci,* but identical values were observed for the toxin from *P. glycinea* Coerper, isolated from infected soybean leaves. The *P. phaseolicola* and *P. glycinea* toxins were also similar in biological activity. The molecular weight of the phaseotoxin was calculated to be less than 700.

Rudolph & Stahmann (80), used a synthetic culture medium for the first time and obtained the same amount of phaseotoxin in cultures 2 to 12 days after inoculation. They also found that the toxin is thermostable and nonspecific. Rudolph (77) later reported on its purification and chemical composition. The purified material was gel filtered on Sephadex G-15 in 0.02 *M* ammonia, and activity of fractions was determined by examining chlorosis-inducing activity on swiss chard (*Beta vulgaris* L.) leaves. The purified toxin has a molecular weight of 2100, is colorless and highly water soluble, and has no uv absorption. It is stable in alkaline conditions, very hygroscopic, and acid labile. It is composed largely of glucose (80–90%), with the remainder consisting of rhamnose and fructose. Elemental analysis showed the presence of 2% nitrogen, but no ninhydrin-positive substances are present in nitric acid hydrolysates.

Simultaneously, Patil [described in the discussion of a paper by Durbin (14, 70)] attempted the purification of phaseotoxin by using similar procedures except that the final step involved adsorption of the active aqueous extract on DEAE-cellulose, elution with 0.01 *M* NaCl and desalting on Sephadex G-10. The gelfiltration data of this material suggested a molecular weight of about 700. Further purification and characterization was accomplished with thin-layer chromatography (70) and DEAE-cellulose gradient (NaCl) chromatography (66). The latter resolves the toxin into two distinct biologically active anionic species. The two species were similar in several respects. For example, their mobilities on Sephadex G-25 were the same, their $R_f$ values on thin layers of silica gel in several solvent systems were identical, and their electrophoretic mobilities at pH 8.0 (1500 *V*) also were the same. No sugars were detected in either compound after nitric acid hydrolysis.

The more abundant of the two species was hydrolyzed (6 *N* HCl) and the resulting amino acids were converted to N-trifluoroacetyl *n*-butyl esters. The liquid chromatography of the derivatives (1.325 W/W % ethylene glycol adipate on 80–100 mesh chromsorb G) showed 2 unknown and 3 known amino acids: L-serine, L-glycine, and L-valine. When $^{14}C$ labeled L-serine was provided to a log phase culture of *P. phaseolicola,* labeled products were released in the medium. The culture filtrate was purified but it was not separated into individual components and co-chromatographed with unlabeled toxin (containing both species) on a column of Sephadex G-25. There was complete correspondence between the ratios of peaks with biological activity and radioactivity.

Thus, just as [2-serine]-tabtoxin, phaseotoxin also seems to contain L-serine. The gel filtration data, however, indicate that it has a higher molecular weight than [2-serine]-tabtoxin.

The available data on phaseotoxin had created a controversy regarding its chemical nature. However, a recent re-examination of the toxin by Rudolph (personal communication) has shown that the toxin preparation previously used for elemental analysis and gel filtration experiments was not homogenous. The claim that the toxin is a polysaccharide thus needs revision.

MODE OF ACTION It is now well accepted that phaseotoxin is the principal cause of chlorosis in tissues infected with *P. phaseolicola* or in those injected with cell-free filtrates of this bacterium. But the mechanism by which chlorosis occurs is not yet known. The serendipitous classical discovery by Patel & Walker (64, 65) that $\beta$-alanine and especially L-ornithine, both nonprotein amino acids, accumulate spectacularly in inoculated or systemically chlorotic tissues of bean leaves susceptible to the pathogen was the beginning of biochemical studies on *P. phaseolicola*-bean system. These investigators showed that methionine, asparagine-glutamine, and histidine also accumulated, but not to the same extent.

Rudolph & Stahmann (80) confirmed the accumulation of ornithine in infected tissues and found that even greater accumulation of the compound occurred in stems and petioles of infected plants. Further, sterile culture filtrates of *P. phaseolicola* which did not contain ornithine induced accumulation of ornithine in bean and swiss chard leaves. Ornithine is an important amino acid because of its involvement in the formation of citrulline and arginine (urea cycle), and inhibition of its metabolism would be expected to cause a decrease in arginine. Based on the observation that no appreciable decrease in arginine concentration of toxin-treated tissue was observed, the above researchers ruled out the possibility of inhibition of the ornithine cycle by the toxin.

In fact, closer examination of the data in one report reveals that there is almost 50% less arginine in chlorotic tissues of infected plants (64) and in another, only one fourth as much is found in toxin-treated tissues as compared to green tissue (80). Based on this, Patil, Kolattukudy & Dimond (68) examined the effect of partially purified toxin of *P. phaseolicola* on ornithine carbamoyltransferase (carbamoylphosphate: L-ornithine carbamoyltransferase, EC 2.1.3.3), the enzyme which catalyzes the carbamylation of ornithine into citrulline. The backward ornithine carbamoyl transferase (OCT) reaction was inhibited by Sephadex G-10 fractions of the crude toxin; the mobility of the halo-inducing principle was identical with that of the OCT inhibitor.

A detailed kinetic study on the inhibition of OCT (forward reaction) was done by Tam & Patil (99). Using purified bean OCT and phaseotoxin they calculated Michaelis' constants of 5.0 m*M* for ornithine and 1.7 m*M* for carbamoylphosphate (CAP). For the reversible reaction the constants were 11 m*M* for citrulline and 3.3 m*M* for arsenate. They further showed that toxin induces allosteric competitive inhibition in relation to CAP and noncompetitive mode of inhibition in relation to ornithine.

The toxin influences the CAP site of the OCT molecule; however, kinetic data show that it does so indirectly and probably is not an analog of CAP. In the backward assay, competitive inhibition was observed for both citrulline and arsenate. Toxin inhibition was shown to increase with preincubation time and the maximum inhibition at saturating concentration of toxin was 85%. Toxin binds to the enzyme protein reversibly.

The demonstration that phaseotoxin is a potent inhibitor of OCT raised three immediate questions: (*a*) what is the specificity of phaseotoxin as an enzyme inhibitor; (*b*) does it inhibit OCT in vivo; (*c*) how does the inhibition of OCT lead to chlorosis in toxin-treated leaves. To answer the question of enzyme specificity, a study was made of the effect of toxin on several enzymes related to OCT. At a hundredfold higher concentration than that needed to inhibit OCT (by 50%), the toxin did not inhibit glutamine synthetase, glutamine transferase, CAP synthetase, aspartate carbamoyltransferase, or arginase.

As regards the second question (above), the toxin does inhibit OCT in vivo (S. Patil, unpublished), although no ornithine accumulated in 48 hr after toxin treatment (69). Some ornithine may have accumulated in this short period but was metabolized by other ornithine-metabolizing enzymes. The third question cannot be answered fully at this time, but the protection from toxin-induced chlorosis by L-citrulline and its reversal by it and L-arginine-HCl, but not by several other naturally occurring amino acids, strongly suggests a cause-effect relationship between OCT inhibition and the induction of chlorosis in toxin-treated bean leaves. Rudolph (78), however, was not able to protect leaf tissue from chlorosis with L-citrulline or L-arginine-HCl. Rudolph's toxin preparations contain a contaminating polysaccharide which may interfere with protection of toxin-induced chlorosis by the amino acids. In *Euglina gracilis,* however, toxin-induced chlorosis was apparently prevented by L-citrulline even though the concentration of L-citrulline was abnormally high. From the results of reversal experiments it was surmised that induction of chlorosis does not involve irreversible structural changes in bean cell organelles. An electron microscopic study of chlorotic tissue confirmed this (69). A similar lack of ultrastructural damage (except ribosomal aggregation) in Swiss chard leaves treated with a toxin preparation apparently contaminated with a polysaccharide has been reported (44).

A complementary piece of evidence shows that *P. glycinea,* which produces a toxin with many of the same properties as phaseotoxin (30, 87), does not induce ultrastructural changes (13). A contradictory report (3) describes structural damage to chloroplasts but not to other organanelles in chlorotic halo tissues produced either by infection or by injection of culture filtrates. But the material was taken 9 days after infection or injection and observed effects cannot be compared with early stages of chlorosis. Thus, it is clear that phaseotoxin does not cause irreversible structural damage as does tabtoxin in plant cells but rather causes a specific but reversible biochemical lesion, i.e. inhibition of OCT.

How inhibition of OCT might lead to chlorosis is not known. One might speculate that the deficiency of arginine resulting from such inhibition could in turn hinder formation of chlorophyll-synthesizing enzymes and especially any of those with arginine at the active site. The time it takes to actually see significant chlorosis in toxin-treated leaves corresponds to the half-life of chlorophyll, indicating that inhibition of chlorophyll synthesis rather than acceleration of chlorophyll degradation occurs in toxin-treated tissues. Whatever the mechanism that translates OCT inhibition into chlorosis, it seems reasonably certain that the primary effect of phaseotoxin is on OCT.

Several other effects of phaseotoxin which appear to be secondary have been reported: starch accumulates in chloroplasts of toxin-treated leaf tissues (29, 44); growth of bean callus tissue is inhibited (2); and cell permeability of Swiss chard leaves treated with the toxin decreases and the osmotic value increases (106). The latter changes occur between 2–7 days after toxin treatment with the maximum change in these parameters occurring at the time of symptom expression. It was postulated that phaseotoxin inhibits the resistant reaction of plants to incompatible pathogens by decreasing rather than increasing cell permeability, which is what occurs in a resistant response. Finally, O'Brien & Wood (54) recently showed that in inoculated leaves of a variety of beans susceptible to *P. phaseolicola,* enough ammonia accumulates to account for symptoms. But as in other cases (47, 48) the contention that ammonia is the cause of symptoms is debatable.

### *Syringomycin*

Several researchers had suspected that the symptomatology of a canker disease of plum (*Prunus silicina* Lindl.) and cherry (*P. avinum* L.) caused by *Pseudomonas mors-prunorum* Wormald were due in part to a pathogenic toxin. Erikson & Montgomery (16) were the first to report the presence of a toxic factor which they considered to be a proteinacious endotoxin in culture filtrates of the pathogen. Their most important finding, however, was that sterile culture filtrates of the pathogen were able to cause severe injury only in tissues of a variety of plum susceptible to the pathogen but not in tissues of a resistant variety. Unfortunately, the toxin has never been purified and no further work on its differential effects on plum tissues has been reported.

*P. syringae,* which causes bacterial canker of peach (*P. persica*), also produces a phytotoxin. DeVay and co-workers (11) obtained a partially purified preparation of the toxin which has a wide spectrum of antibiotic activity; several genera of fungi and bacteria are affected. The toxin is heat stable, dialyzable, and labile in alkali. In peach seedlings it mimics the symptoms of the disease. The toxin was further purified by Sinden & DeVay (84) and Sinden, DeVay & Backman (85) by using ion exchange and partition chromatography. They obtained a single low-molecular-weight ninhydrin reactive compound. Acid hydrolysis of this compound yielded nine amino acids. As with the crude material, the toxin had a broad spectrum of antibiosis. The concentrations required were very low: e.g. the growth of *Geotricum candidum* was inhibited completely by 24 $\mu$g of toxin, and peach shoots immersed in a solution of toxin (600 $\mu$g/ml) for 24 hr showed phytotoxic symptoms. Enough toxin was isolated from infected shoots to account for symptoms. Although the isolation procedure used by the investigators yielded 270-fold purification of the toxin, apparently it was still only partially pure. The toxin was designated syringomycin (SR).

Although there are several differences between the toxin of *mors-prunorum* and syringomycin, they may be more apparent than real. Erikson & Montgomery's (16) conclusion that the material they studied is an endotoxin was based on the observation that only filtrates of older cultures had activity. Sinden, DeVay & Backman (85) grew their cultures on agar and harvested them after 5 days. It is not known whether syringomycin was produced extracellularly at all stages of growth or only after cell lysis. It is clear that SR has a broad spectrum of antibiotic activity and that it affects many plant species in addition to peach. No such data are available for the *mors-prunorum* toxin. Despite this there are no compelling reasons to believe that the latter toxin is not the same as SR.

After isolating the toxin, DeVay and co-workers turned to its mode of action. Based on the observations that several polypeptide antibiotics reduce surface tension and affect cell membranes, and that SR has the ability to reduce surface tension, Backman & DeVay (1) studied the sites of action of SR in vivo. Several lines of evidence showed that the primary action of SR is on cell membranes and it involves their rapid "detergent-like" lysis. They found that: (*a*) rabbit erythrocytes treated with SR rapidly leak hemoglobin and a similar leakage of (presumably) nucleic acids occurs from *G. candidum* cells treated with SR; (*b*) electron micrograph studies of similarly treated *G. candidum* cells show massive membrane disruption; (*c*) peach tissues treated with $^{14}C$-labeled SR show accumulation of label on cell and nuclear membranes. The fact that the effects of SR are counteracted by $Ca^{2+}$ and $Mg^{2+}$ further implicates the membranes. The effects of SR are also counteracted by sterols, indicating the involvement of membrane lipids in the mode of action of SR. Syringomycin also affects nucleic acid synthesis (71); however, this appears to be secondary.

### *Rhizobitoxine*

CHEMICAL COMPOSITION The bacterium *Rhizobium japonicum* (Kirch.) Buch. normally exists as a symbiont in root nodules of soybean plants [*Glycinea max.* (L.) Merrill]. But several years ago it was observed that some plants with such root nodules become chlorotic and it was thought that *R. japonicum* was somehow responsible. Involvement of *R. japonicum* was confirmed when water extracts of nodules taken from chlorotic soybean plants induced chlorosis not only in soybeans but in other plant species as well (33). Not all strains of the bacterium were capable of producing the chlorosis-inducing substances, however.

Using cation exchange and paper chromatography, Owens & Wright (62) isolated a compound from nodules of soybean plants which showed chlorosis. Preliminary studies had indicated that this compound forms an easily detected yellow product with ninhydrin; this helped in its isolation. The compound was later named rhizobitoxine (58). The nodules also contain a closely related amino compound designated "unknown Y" which forms a purple product with ninhydrin. Unknown Y is not phytotoxic. The isolated rhizobitoxine which they considered chemically homogeneous was phytotoxic to soybean as well as to sorghum seedlings in microgram quantities.

Of particular significance was their finding that soybean varieties resistant to the induction of chlorosis when infected by chlorosis-inducing strains of the bacterium required four times as much rhizobitoxine to show chlorotic symptoms as did susceptible varieties. Production of the toxin in liquid cultures of the bacterium, and the discovery that the nature of the medium affects the amount of toxin produced, demonstrated that the compound is derived from the bacterium and not from plant tissues. The compound is produced at all stages of bacterial growth; thus it is not an endotoxin. Unknown Y is also produced in culture. Based on the evidence then at hand Owens and colleagues (56) thought rhizobitoxine was a sulfur-containing amino acid with at least seven carbon atoms, and the unknown Y was tentatively identified as an ether derivative of homoserine.

Recently, Owens and co-workers (56, 58) reported the structures of rhizobitoxine and of unknown Y by using infrared, mass, and proton and $^{13}C$ NMR spectrometry. Rhizobitoxine is defined as 2-amino-4-(2-amino-3-hydroxypropoxy)-*trans*-but-3-enoic acid (Figure 1f). The unknown Y is defined as *o*-(2-amino-3-hydroxypropyl) homoserine and is designated dihydrorhizobitoxin (Figure 1g). The latter is a saturated analog of rhizobitoxine. Rhizobitoxine closely resembles the antimetabolite analog L-2-amino-4-methoxy-*trans*-but-3-enoic acid in chemical and physical properties. The latter is the first enol-ether amino acid ever reported.

Owens and associates (56, 58) at first thought rhizobitoxine was a sulfur-containing amino acid; this led them to suspect that its toxicity might be due to its role as an antimetabolite of one of the sulfur-containing amino acids. Studies on the mode of action of rhizobitoxine (discussed below) based on the above rationale show that, although sulfur is not present, the pathway of methionine synthesis in plants and other organisms is affected by rhizobitoxine.

MODE OF ACTION Owens and co-workers (60, 61) first used sorghum as a test plant, but found that toxin effects could not be reversed. For this reason they chose *Salmonella typhimurium* as a model system for study. Its growth was inhibited by low concentrations of toxin, and methionine (1.025 m$M$) added to the toxin prevented this inhibition, as did homocysteine, the methionine precursor. But earlier precursors cystathionine and homoserine did not; their failure to prevent growth inhibition could not be explained by lack of uptake. These experiments made it possible to pinpoint the reaction in the methionine biosynthetic pathway that was blocked by rhizobitoxine: the reaction involving the conversion of cystathionine to homocysteine.

The enzyme $\beta$-cystathionase which catalyzes the transformation of cystathionine to homocysteine was obtained from *S. typhimurium* and purified a hundredfold (58). The enzyme affinity for rhizobitoxine was at least three orders of magnitude greater than its substrate—cystathionine ($K_m$ = 0.36 mmole/liter). Other related enzymes were not inhibited at low concentrations of toxin.

Once the investigators knew that rhizobitoxine is a specific and potent inhibitor of $\beta$-cystathionase they turned their attention again to the effects of toxin in plants. Based on the fact that growth of *Chorella pyrenoidosa* is inhibited by the toxin (possible $\beta$-cystathionase inhibition) Giovannelli, Owens & Mudd (20) examined the effects of rhizobitoxine on $\beta$-cystathionase of spinach (*Spinacia oleracia* L.). In a study of $\beta$-cystathionase inhibition from this source they found that the toxin inhibits the enzyme in an active-site-directed, irreversible manner. The enzyme is protected from inactivation by cystathionine. The kinetics of this protection indicate that there is competition between cystathionine and rhizobitoxine for the active enzyme. The value of $K_i$ (inactivation constant) is $8.0 \times 10^{-5}$ $M$, at least two orders of magnitude less than that for the same enzyme from *S. typhimurium*. Thus, the plant enzyme is less sensitive to rhizobitoxine than the bacterial enzyme. Finally, they found that rhizobitoxine-inactivated enzymes can be reactivated by pyridoxal phosphate. This was interpreted to mean that toxin binds irreversibly to the pyridoxal phosphate prosthetic group of $\beta$-cynstathionase.

How does inhibition of $\beta$-cystathionase by rhizobitoxine result in chlorosis? This question was investigated in depth (20a) by studying the in vivo effects of rhizobitoxine on methionine biosynthesis in spinach and corn seedlings. Rhizobitoxine inhibited $\beta$-cystathionase of both plants only partially; 30 to 40% of the activity remained in the treated tissues compared to complete inhibition of spinach $\beta$-cystathionase in vitro. Several possible reasons for the partial inhibition are offered by these workers; however, chlorosis in treated seedlings cannot be fully explained because a substantial amount of enzyme activity remains.

Further evidence was sought to explain the partial inhibition of the enzyme. Rhizobitoxine-treated corn seedlings were allowed to take up $^{35}SO_4^{-2}$, and the labeling patterns of sulfur-containing amino acids were studied. Label accumulated (22-fold over control) in cystathionine extracted from toxin-treated corn tissues. But there was only a slight inhibition of accumulation of radioactivity in methionine. This indicated that another pathway (direct sulfahydration pathway) which bypasses the cystathionine $\longrightarrow$ homocysteine $\longrightarrow$ methionine channel for the label must be operative in corn seedlings. Thus it appears that direct sulfahydration pathway operates in toxin-treated corn tissues and no severe methionine deficiency can occur. But concentrations of toxin used in these experiments were more than sufficient to cause chlorosis in spinach seedlings. Thus to what extent the inhibition of $\beta$-cystathionase by rhizobitoxine in vivo is responsible for chlorosis is still an open question. Other effects of rhizobitoxine include inhibition of ethylene production (59) and herbicidal activity against weeds (57).

## *Glycopeptide and Other High Molecular Weight Toxins*

CHEMISTRY AND MODE OF ACTION Many phytopathogenic bacteria produce high molecular weight extracellular phytotoxic substances (22, 23). Several of these have been implicated in disease development (9, 27, 28, 31, 40, 43, 90, 92). Although it was generally known that the substances resembled polysaccharides, until recently no more detailed chemical characterization was available for any of them.

In a series of papers Strobel and colleagues (95, 96, 98) reported their pioneering work on the chemical characterization of a toxic substance from *Corynebacterium sepedonicum* (Spieck et Kotth.) Skapt. et Burkh., the causal organism of ring rot of potatoes. A simple isolation procedure yields a pure compound which is acidic, has a molecular weight of 2400, and is a glycopeptide. Initially it was determined (95) that the toxin was composed of glucose, mannose, L-fucose, 2-keto-3-deoxygluconic acid, and a peptide. A more sensitive detection method utilizing the alditol acetate technique for sugar analysis (98) showed the toxin to contain by weight the following: 33.4% mannose, 18.8% glucose, 1.1% L-rhamnose, 0.7% arabinose,

3.8% ribose, 4.5% galactose, 9.9% 2-keto-3-deoxygluconic acid and 4.6% peptide. L-fucose did not appear, although several new sugars were identified.

This sensitive technique disclosed that the toxin has at least 11 different sugar linkages and is a highly branched single oligosaccharide chain attached by a glucosidic linkage to a single peptide chain. Further, the oligosaccharide portion appears to be attached to the peptide portion through a mannose residue glycosidically linked to the –OH of threonine. Mannose accounts for 55% of the sugar residues in the toxin and for about half the branch points and 90% of the terminal sugars. The biologically active site or sites of the toxin exist as the carboxyl group of 2-keto-3-deoxygluconic acid (35); the peptide apparently has no part in the biological activity.

A similar toxin from *C. insidiosum* (McCull) Jens was purified by Ries & Strobel (75). Although it has a much higher molecular weight ($5 \times 10^6$) than the *C. sepedonicum* toxin, it contains the same sugar residues but in greatly differing amounts, roughly the same proportion of keto-deoxy sugar and a peptide which accounts for only 2.5% of the weight of the toxin. No information about the number of sugar linkages of this toxin is yet available. The toxin from *C. insidiosum* is bright blue and binds 75 moles of copper per mole toxin. Also, from culture filtrates of *C. michiganense* (E. Sm) Jens. three glycopeptides were isolated (73) and partially characterized (Table 1).

Many early workers in this field assumed that the wilting of plants after treatment with polysaccharide-like toxins was due to mechanical plugging. Strobel & Hess (97) clearly demonstrated that at least in the glycopeptide toxins of *C. sepedonicum,* membrane disruption is the mode of action. A similar mode of action is suspected in *C. michiganense* (74). The biological activity of *C. insidiosum* glycopeptide does not appear to be as clearly related to membrane disruption as in glycopeptides of the other two species (76). The respective glycopeptides were isolated from infected plants.

Other toxins with a possible role in pathogenesis include a lipomucopolysaccarride from *P. lachrymans,* which mimics proteolytic activity (38), and the low molecular weight toxin from *P. tolaasii* Paine (53). Sequeira & Ainslie (83) and Gardner & Kado (19) have isolated high molecular weight proteinaceous endotoxins from phytopathogenic bacteria of two different genera which appear to be involved in the induction of hypersensitive reaction in plants. L. Crosthwait and S. Patil (unpublished) have also isolated a high molecular weight endotoxin from *P. phaseolicola* that induces hypersensitive reaction in plants.

## DISCUSSION AND CONCLUDING REMARKS

From the foregoing discussion we may conclude that several bacterial phytotoxins can reproduce, wholly or in part, the symptoms of diseases their parent bacteria cause when they invade their respective host plants. In two or three cases (69, 97) even the formation of the primary lesion is reasonably well understood. Because a few of these toxins can be isolated from infected plants (30, 96) they can be described as vivotoxins (12), but none is currently considered a pathotoxin. Three of the four criteria required for a compound to qualify as a pathotoxin (105) are easily fulfilled by several of the phytotoxins described here. For example, isolates of *P. syringae* (1) and *R. japonicum* (61): (*a*) apparently produce a single toxin; (*b*) the ability of various isolates to produce toxin is correlated with their relative pathogenicity [although isolates of *P. syringae* from bean seem to be an exception (79)]; (*c*) application of the toxin at (presumably) physiological concentrations reproduces the characteristic disease symptoms in host tissues.

But none of these toxins fulfills the fourth criterion: that the pathogen and the toxin exhibit similar host specificity. Because of this it is believed that bacterial phytotoxins constitute only one of several determinants needed in the host specificity of a pathogen. Based on race differences and other considerations, Crosse (10) and Buddenhagen & Kelman (6) concluded that no single determinant can explain specificity of pathogenic bacteria for their host plants. The fourth criterion, however, is misleading because for a toxin to have a role in host specificity of the pathogen it does not itself need to be host specific.

I illustrate this point with a hypothetical host-parasite system as follows: Host *A* is resistant to pathogen *X* (limited in vivo multiplication of the pathogen) which is capable of high toxin production in culture. Host *B* is susceptible to *X* (extensive multiplication of the pathogen in the host) and shows typical disease symptoms. Isolated toxin from *X* causes symptoms in both *A* and *B*. Failure of the bacterium to cause disease in *A* could be due to its inability to grow and produce a sizable population in the host. Two possibilities could be proposed to explain this. The first and less probable of these is inadequate nutritional environment in the host; most phytobacteria are not nutritionally fastidious (18, 52). The second and more probable explanation is that the pathogen *X* cannot breach defenses of host *A*.

In animal pathology substances produced by bacterial pathogens which breach host defenses and enable the pathogen to proliferate are called *aggressins* (91). Agressins themselves may or may not be toxic to the host. I am suggesting that several of the phytotoxins described here could act as aggressins during the establishment of the pathogen and later, after reaching much higher concentrations, produce the pathological effects seen in the host plant. It has been proposed that toxins may act in more than one way (36).

The following example supports the above point. Rudolph (78) reported that resistant bean leaf tissue treated with the halo blight toxin supported much greater growth of *P. phaseolicola* than nontreated tissue. Interpreting the failure of *X* to cause disease in *A* in this light would mean that even though limited multiplication of the pathogen occurs, toxin synthesis by the pathogen is prevented by the host, thereby allowing the natural host defenses (e.g. hypersensitive reaction) to take effect and limit further pathogenic growth. Even in resistant bean tissue, *P. phaseolicola* multiplies for several generations (55). Thus, the primary event in our hypothetical host-parasite confrontation could depend on production of toxins in concentrations adequate to prevent host defenses such as hypersensitivity (24, 39).

Alternatively, looking at the susceptible combination in our hypothetical system (*B* and *X*) in light of Rudolph's finding would mean that in this system the host does not suppress toxin synthesis and that adequate quantities of toxin are produced which prevent hypersensitive host reactions.

Thus, it appears that in host tissues bacterial toxin production could be under regulatory control. At least one example supports this hypothesis: Owens & Wright (63) showed that rhizobitoxine is produced in greater concentration in nodules of susceptible plants as compared to those of resistant plants even though growth of the bacterium in nodules of both plants is the same. There are also examples of possible regulatory control of toxin production and production of other metabolites (72) in culture. Polysaccharide production by some species of pathogenic bacteria depends on the media used (22); phaseotoxin is produced only at low temperatures (29). From the above discussion it can be seen that the nonspecificity of bacterial phytotoxins does not necessarily preclude their having a role in the host specificity of bacteria which produce such toxins. The point which needs to be restressed in this connection is that the most important function of such toxins might involve prevention of hypersensitive host defences.

Another avenue of approach to investigate whether bacterial phytotoxins are the basis of specificity would involve finding out whether toxigenicity is a prerequisite to pathogenicity. The term *pathogenicity* has been variously defined (25, 82). By pathogenicity I do not mean the ability of pathogen to multiply for a few generations in host tissues but rather to cause lesions (that perpetuate) and/or systemic infection which results in sufficient damage and which in turn markedly reduces host productivity. Patil, Hayward & Emmons (67) isolated a uv-induced mutant of *P. phaseolicola* identical to the wild type except that it produces no toxin in culture and causes no chlorosis or systemic symptoms in inoculated susceptible plants. In inoculated susceptible primary leaves of beans, however, it multiplies as well as the toxin-producing parent. This behavior supports the contention that toxigenicity is a prerequisite to pathogenicity in the sense of full pathogenic capability under field conditions. The exsistence of several naturally occurring presumed nontoxigenic isolates of *P. phaseolicola, P. tabaci,* and *P. syringae* which are less effective pathogens in the field than the toxigenic isolates substantiates the above hypothesis (1, 32, 82).

Although most phytobacterial toxins are not host specific there are two or three bacterial pathogens which appear to show some degree of selective toxicity towards some plants. The toxin of *P. mors-prunorum* produces more injury in susceptible than in resistant plum tissues (16). Susceptible varieties were more sensitive to rhizobitoxine than resistant varieties (63). Recently Goodman (22a) has reported the isolation of a toxin, amylovorin, which is host specific and causes wilt symptoms in roseaceous species of plants but not in nonroseaceous species. This appears to be the first report of a bacterial toxin which conforms to the criteria for a host-specific toxin.

In conclusion, the involvement of bacterial toxins in pathogenic host specificity has intrigued pathologists for many years. Only in the past decade, however, have we learned enough about the chemistry and mode of action to begin a critical search for their role in pathogenic host specificity. It is hoped that the foregoing discussion will stimulate further interest in this promising area of research.

ACKNOWLEDGMENTS

I would like to thank Drs. Durbin, Goodman, and Rudolph for providing me with unpublished results, Dr. Buddenhagen for helpful suggestions, and Mr. Philip Youngblood for making illustrations. The work reported here is supported by a grant from NIH (AI09477). This paper constitutes Hawaii Agricultural Experiment Station, Journal Series No. 1742.

*Literature Cited*

1. Backman, P. A., DeVay, J. E. 1971. Studies on the mode of action and biogenesis of the phytotoxin syringomycin. *Physiol. Plant Pathol.* 1:215–34
2. Bajaj, Y. P. S., Saettler, A. W. 1970. Effect on halo toxin-containing filtrates of *Pseudomonas phaseolicola* on the growth of bean callus tissue. *Phytopathology* 60:1065–67
3. Bajaj, Y. P. S., Spink, G. C., Saettler, A. W. 1969. Ultrastructure of chloroplasts in bean leaves affected by *Pseudomonas phaseolicola. Phytopathology* 59:1017
4. Braun, A. C. 1950. The mechanism of action of a bacterial toxin on plant cells. *Proc. Nat. Acad. Sci. USA* 36:423–27
5. Braun, A. C. 1955. A study on the mode of action of the wildfire toxin. *Phytopathology* 45:659–64
6. Buddenhagen, I., Kelman, A. 1964. Biological and physiological aspects of bacterial wilt caused by *Pseudomonas solanacearum. Ann. Rev. Phytopathol.* 2:203–30
7. Carlson, P. S. 1973. Methionine sulfoximine-resistant mutants of tobacco. *Science* 180:1336–38
8. Clayton, E. E. 1934. Toxin produced by *Bacterium tabacum* and its relation to host range. *J. Agr. Res.* 48:411–26
9. Corey, R. R., Starr, M. P. 1957. Colony types of *Xanthomonas phaseoli. J. Bacteriol.* 74:137–40
10. Crosse, J. E. 1968. The importance and problems of determining relationships among plant-pathogenic bacteria. *Phytopathology* 58:1203–6
11. DeVay, J. E., Lukezic, F. L., Sinden, S. L., English, H., Coplin, D. L. 1968. A biocide produced by pathogenic isolates of *Pseudomonas syringae* and its possible role in the bacterial canker disease of peach trees. *Phytopathology* 58:95 101
12. Dimond, A. E., Waggoner, P. E. 1953. On the nature and role of vivotoxins in plant disease. *Phytopathology* 43: 229–35
13. Dueck, J., Cadwell, V. B., Kennedy, B. W. 1972. Physiological characteristics of systemic toxemia in soybean. *Phytopathology* 62:964–68
14. Durbin, R. D. 1971. Chlorosis-inducing *Pseudomonas* toxins: their mechanism of action and structure. In *Morphology and Biochemical Events in Plant-Parasite Interaction,* ed. S. Akai, S. Ouchi, 369–86. Tokyo: Mochizuki. 415 pp.
15. Durbin, R. D. 1972. Bacterial phytotoxins: a survey of occurrence, mode of action and composition. In *Phytotoxins in Plant Diseases,* ed. R. K. S. Wood, A. Ballio, A. Granti, 19–33. New York: Academic. 530 pp.
16. Erikson, D., Montgomery, H. B. S. 1945. The action of cell-free filtrates of *Pseudomonas mors-prunorum* Wormald and related phytopathogenic bacteria on plum trees. *Ann. Appl. Biol.* 32: 117–23
17. Farkas, G. L., Lovrekovich, L. 1963. Counteraction by kinetin of the toxic effect of *Pseudomonas tabaci. Phytopathol. Z.* 47:391–98
18. Garber, E. D. 1959. Biochemical mutants of toxigenic bacterial phytopathogens. Recent advances in botany. *Int. Bot. Congr., 9th*
19. Gardner, J. M., Kado, C. I. 1972. Induction of the hypersensitive reaction in tobacco with specific high-molecular weight substances derived from the osmotic shock fluid of *Erwinia rubrifaciens. Phytopathology* 62:759
20. Giovanelli, J., Owens, L. D., Mudd, S. H. 1971. Mechanism of inhibition of spinach $\beta$-cystathionase by rhizobitoxine. *Biochim. Biophys. Acta* 227:671–84

20a. Giovanelli, J., Owens, L. D., Mudd, S. H. 1972. $\beta$-Cystathionase in vivo inactivation by rhizobitoxine and role of the enzyme in methionine biosynthesis in corn seedlings. *Plant Physiol.* 51:492–503

21. Goodman, R. N. 1972. Phytotoxin-induced ultrastructural modifications of plant cells. See Ref. 15, pp. 311–29

22. Goodman, R. N., Kiraly, Z., Zaitlin, M. 1967. *The Biochemistry and Physiology of Infectious Plant Disease.* Princeton, NJ: Van Nostrand. 354 pp.
22a. Goodman, R. N., Huang, J. S., Huang, Pi-yu. 1974. Host-specific phytotoxic polysaccharide from apple tissue infected by *Erwinia amylovora. Science* 183:1081–82
23. Gorin, P. A. J., Spencer, J. F. T. 1961. Structural relationships of extra-cellular polysaccharides from phytopathogenic *Xanthomonas spp. Can. J. Chem.* 39:2282–89
24. Hildebrand, D. C., Riddle, B. 1971. Influence of environmental conditions on reactions induced by infiltration of bacteria into plant leaves. *Hilgardia* 41: 33–43
25. Hill, K., Coyne, D. P., Schuster, M. L. 1972. Leaf, pod, and systematic chlorosis reactions in *Phaseolus vulgaris* to halo blight controlled by different genes. *J. Am. Soc. Hort. Sci.* 97:494–98
26. Hinton, J. J. C., Moran, T. 1957. Methionine sulfoximine and the growth of the wheat embryo. *Brit. J. Nutr.* 11: 323–28
27. Hodgson, R., Peterson, W. H., Riker, A. J. 1949. The toxicity of polysaccharides and other large molecules to tomato cuttings. *Phytopathology* 39: 47–62
28. Hodgson, R., Riker, A. J., Peterson, W. H. 1945. Polysaccharide production by virulent and attenuated crown-gall bacteria. *J. Biol. Chem.* 158:89–100
29. Hoitink, H. A. J., Pelletier, R. L., Coulson, J. G. 1966. Toxemia of halo blight of beans. *Phytopathology* 56:1062–65
30. Hoitink, H. A. J., Sinden, S. L. 1970. Partial purification and properties of chlorosis-inducing toxins of *Pseudomonas phaseolicola* and *Pseudomonas glycinea. Phytopathology* 60:1236–37
31. Husain, A., Kelman, A. 1958. Relation of slime production to mechanism of wilting and pathogenicity of *Pseudomonas solanacearum. Phytopathology* 48: 155–65
32. Jensen, J. H., Livingston, J. E. 1944. Variation of symptoms produced by isolates of *Phytomonas medicaginis* var. *phaseolicola. Phytopathology* 34:471–80
33. Johnson, H. W., Means, U. M., Clark, F. E. 1959. Responses of seedlings to extracts of soybean nodules bearing strains of *Rhizobium japonicum. Nature* 183:308–9
34. Johnson, J., Murwin, H. F. 1925. Experiments on the control of wildfire of tobacco. *Wis. Agr. Exp. Sta. Res. Bull. No.* 62. 35 pp.
35. Johnson, T. B., Strobel, G. A. 1970. The active site on the phytotoxin of *Cornebacterium sepedonicum. Plant Physiol.* 45:761–64
36. Kalyanasundaram, R., Charudattan, R. 1966. Toxins in plant diseases. *J. Sci. Ind. Res. India* 25:63–73
37. Katunuma, N., Okada, M., Nishii, Y. 1966. Regulation of the urea cycle and TCA cycle by ammonia. *Advan. Enzyme Regul.* 4:317–35
38. Keen, N. T., Williams, P. H. 1971. Chemical and biological properties of a lipomucopolysaccharride from *Pseudomonas lachrymans. Physiol. Plant Pathol.* 1:247–64
39. Kiraly, Z., Barna, B., Ersek, T. 1972. Hypersensitivity as consequence, not the cause of plant resistance to infection. *Nature* 239:456–57
40. Kunz, R. 1952. Die Wirkungsweise von *Bacterium solanacearum* E.F.S. dem Erreger der tropischen Schleimkrankheit des Tabaks, auf *Solanum lycopersium* L. *Phytopathol. Z.* 20:89–112
41. Lamar, C. Jr., Sellinger, O. Z. 1965. The inhibition in vivo of cerebral glutamine synthetase and glutamine transferase by the convulsant methionine sulfoximine. *Biochem. Pharmacol.* 14:489–506
42. Lamar, C. Jr., Sinden, S. L., Durbin, R. D. 1969. The inhibition in vitro of rat cerebral glutamine synthetase by an exotoxin from *Pseudomonas tabaci. Biochem. Pharmacol.* 18:521–29
43. Leach, J. G., Lilly, V. G., Wilson, H. A., Purvis, M. R. Jr. 1957. Bacterial polysaccharides: The nature and function of the exudate produced by *Xanthomonas phaseoli. Phytopathology* 47:113–20
44. Lesemann, D., Rudolph, K. 1970. Die bildung von ribosomen-helices unter dem einfluss des toxins von *Pseudomonas phaseolicola* in blattern von mangold (*Beta vulgaris* L.). *Z. Pflanz. Physiol.* 62:108–15
45. Lovrekovich, L., Farkas, G. L. 1963. Kinetin as an antagonist of the toxic effect of *Pseudomonas tabaci. Nature* 198:170
46. Lovrekovich, L., Klement, Z., Farkas, G. L. 1964. Toxic effect of *Pseudomonas tabaci* on RNA matabolism in tobacco and its counteraction by kinetin. *Science* 145:165
47. Lovrekovich, L., Lovrekovich, H., Goodman, R. N. 1969. The role of ammonia in wildfire disease of tobacco

caused by *Pseudomonas tabaci*. *Phytopathology* 59:1713–16
48. Lovrekovich, L., Lovrekovich, H., Goodman, R. N. 1970. The relationship of ammonia to symptom expression in apple shoots inoculated with *Erwinia amylovora*. *Can. J. Bot.* 48:999–1000
49. Manning, J. M., Moore, S., Rowe, W. B., Meister, A. 1969. Identification of L-methionine S-sulfoximine as the distereoisomer of L-Methionine SR-sulfoximine that inhibits glutamine synthetase. *Biochemistry* 8:2681–85
50. Mein, F. Jr., Abrams, M. L. 1972. How methionine and glutamine prevent inhibition of growth by methionine sulfoximine. *Biochim. Biophys. Acta* 266: 307–11
51. Milner, L., Weissbach, H. 1969. Inhibition by L-methionine of the growth of *Euglena gracilis* in a glutamic acid medium. *Arch. Biochem. Biophys.* 132: 170–74
52. Misaghi, I., Grogan, R. G. 1969. Nutritional and biochemical comparisons of plant-pathogenic and saprophytic fluorescent pseudomonads. *Phytopathology* 59:1436–50
53. Nair, N. G., Fahy, P. C. 1973. Toxin production by *Pseudomonas tolaasii* Paine. *Aust. J. Biol. Sci.* 26:509–12
54. O'Brien, F., Wood, R. K. S. 1973. Role of ammonia in infection of *Phaseolus vulgaris* by *Pseudomonas* spp. *Physiol. Plant Pathol.* 3:315–25
55. Omer, M. E. H., Wood, R. K. S. 1969. Growth of *Pseudomonas phaseolicola* in susceptible and in resistant bean plants. *Ann. Appl. Biol.* 63:103–16
56. Owens, L. D. 1969. Toxins in plant disease: structure and mode of action. *Science* 165:18–25
57. Owens, L. D. 1973. Herbicidal potential of rhizobitoxine. *Weed Science* 21: 63–66
58. Owens, L. D., Guggenheim, S., Hilton, J. L. 1968. Rhizobium-synthesized phytotoxin: an inhibition of $\beta$-cystathionase in *Salmonella typhimurium*. *Biochim. Biophys. Acta* 158:219–25
59. Owens, L. D., Lieberman, M., Kunishi, A. 1971. Inhibition of ethylene production by rhizobitoxin. *Plant Physiol.* 48:1–4
60. Owens, L. D., Thompson, J. F., Fennessey, P. V. 1972. Dihydrorhizobitoxine, a new ether amino-acid from *Rhizobium japonicum*. *J. Chem. Soc. Chem. Commun.* 715
61. Owens, L. D., Thompson, J. F., Pitcher, R. G., Williams, T. 1972. Structure of rhizobitoxine, an antimetabolic enolether amino-acid from *Rhizobium japonicum*. *J. Chem. Soc. Chem. Commun.* 714
62. Owens, L. D., Wright, D. A. 1965. Production of the soybean-chlorosis toxin by *Rhizobium japonicum* in pure culture. *Plant Physiol.* 40:931–33
63. Owens, L. D., Wright, D. A. 1965. Rhizobial-induced chlorosis in soybeans: Isolation, production in nodules, and varietal specificity of the toxin. *Plant Physiol.* 40:927–30
64. Patel, P. N., Walker, J. C. 1963. Changes in free amino acids and amide content of resistant and susceptible beans after infection with the halo blight organism. *Phytopathology* 53: 522–28
65. Patel, P. N., Walker, J. C. 1963. Free amino acid and amide content of tobacco and oats infected by wildfire and halo blight bacteria. *Phytopathology* 53:855
66. Patil, S. S. 1972. Purification of the phytotoxin from *Pseudomonas phaseolicola*. *Phytopathology* 62:782
67. Patil, S. S., Hayward, A. C., Emmons, R. 1974. An ultraviolet-induced nontoxigenic mutant of *Pseudomonas phaseolicola* of altered pathogenicity. *Phytopathology*. In press
68. Patil, S. S., Kolattukudy, P. E., Dimond, A. E. 1970. Inhibition of ornithine carbamyl transferase from bean plants by the toxin of *Pseudomonas phaseolicola*. *Plant Physiol.* 46:752–53
69. Patil, S. S., Tam, L. Q., Sakai, W. S. 1972. Mode of action of the toxin from *Pseudomonas phaseolicola* I. toxin specificity, chlorosis, and ornithine accumulation. *Plant Physiol.* 49:803–7
70. Patil, S. S., Tam, L. Q., Kolattukudy, P. E. 1972. Isolation and the mode of action of the toxin from *Pseudomonas phaseolicola*. See Ref. 15, pp. 365–72
71. Penner, D., DeVay, J. E., Backman, P. A. 1969. The influence of syringomycin on ribonucleic acid synthesis. *Plant Physiol.* 44:806–8
72. Phelps, R. H., Sequeira, L. 1967. Synthesis of indoleacetic acid by cell-free systems from virulent and avirulent strains of *Pseudomonas solanacearum*. *Phytopathology* 57:1182–90
73. Rai, P. V., Strobel, G. A. 1969. Phytotoxic glycopeptides produced by *Corynebacterium michiganense* I. Methods of preparation, physical, and chemical characterization. *Phytopathology* 59:47–52
74. Rai, P. V., Strobel, G. A. 1969. Phytotoxic glycopeptides produced by

*Corynebacterium michiganense II.* Biological properties. *Phytopathology* 59: 53–57
75. Ries, S. M., Strobel, G. A. 1972. A phytoxic glycopeptide from cultures of *Corynebacterium insidiosum. Plant Physiol.* 46:676–84
76. Ries, S. M., Strobel, G. A. 1972. Biological properties and pathological role of a phytotoxic glycopeptide from *Corynebacterium insidiosum. Physiol. Plant Pathol.* 2:133–42
77. Rudolph, K. 1969. Ein Phytotoxisches polysaccharid. *Die Naturwissenschaften.* 56:569–70
78. Rudolph, K. 1972. The halo-blight toxin of *Pseudomonas phaseolicola:* influence of host-parasite relationship and counter effect of metabolites. See Ref. 15, pp. 373–75
79. Rudolph, K., Delgado, M., Baykal, N. 1973. Pathological aspects of *Pseudomonas syringae* van Hall on bush bean, *Phaseolus vulgaris* L. (Bacterial brown spot of bush bean). *Proc. I.S.P.P., 1973*
80. Rudolph, K., Stahmann, M. A. 1966. The accumulation of L-ornithine in halo-blight infected bean plants (*Phaseolus vulgaris* L.) induced by the toxin of the pathogen *Pseudomonas phaseolicola* (Burkh.) Dowson. *Phytopathology* 57:29–46
81. Scheffer, R. P., Yoder, O. C. 1972. Host-specific toxins and selective toxicity. See Ref. 15, pp. 251–72
82. Schroth, M. N., Vitanza, V. B., Hildebrand, D. C. 1971. Pathogenic and nutritional variation in the halo blight group of fluorescent pseudomonads of bean. *Phytopathology* 61:852–57
83. Sequeira, L., Ainslie, V. 1969. Bacterial cell-free preparations that induce or prevent the hypersensitive reaction in tobacco. *Int. Bot. Congr., 11th,* p. 195
84. Sinden, S. L., DeVay, J. E. 1967. The nature of the wide-spectrum antibiotic produced by pathogenic strains of *Pseudomonas syringae* and its role in the bacterial canker disease of *Prunus persicae. Phytopathology* 57:102
85. Sinden, S. L., DeVay, J. E., Backman, P. A. 1971. Properties of syringomycin, a wide spectrum antibiotic and phytotoxin produced by *Pseudomonas syringae,* and its role in the bacterial canker disease of peach trees. *Physiol. Plant Pathol.* 1:199–213
86. Sinden, S. L., Durbin, R. D. 1968. Glutamine synthetase inhibition: possible mode of action of wildfire toxin from *Pseudomonas tabaci. Nature* 219: 379–80
87. Sinden, S. L., Durbin, R. D. 1969. Some comparisons of chlorosis-inducing pseudomonad toxins. *Phytopathology* 59:249–50
88. Sinden, S. L., Durbin, R. D. 1970. A comparison of the chlorosis-inducing toxin from *Pseudomonas coronafaciens* with wildfire toxin from *Pseudomonas tabaci. Phytopathology* 60:360–64
89. Sinden, S. L., Durbin, R. D., Uchytil, T. F., Lamar, C. Jr. 1969. The production of convulsions by an exotoxin from *Pseudomonas tabaci. Toxicol. Appl. Pharmacol.* 14:82–88
90. Sloneker, J. H., Orentas, D. J., Jeanes, A. 1964. Exocellular bacterial polysaccharide from *Xanthomonas campestris* NRRL B-1459. *Can. J. Chem.* 42: 1261–69
91. Smith, H. 1968. Biochemical challenge of microbial pathogenicity. *Bacteriol. Rev.* 32:164–84
92. Spencer, J. F. T., Gorin, P. A. J. 1960. The occurrence in the host plant of physiologically active gums produced by *Cornebacterium insidiosum* and *Corynebacterium sepedonicum. Can. J. Microbiol.* 7:185–88
93. Stewart, J. M. 1961. ε-Diamino-β-hydroxypimelic acid. II. Configuration of the isomers. *J. Am. Chem. Soc.* 83:435–39
94. Stewart, W. W. 1971. Isolation and proof of structure of wildfire toxin. *Nature* 229:174–78
95. Strobel, G. A. 1970. A phytotoxic glycopeptide from potato plants infected with *Corynebacterium sepedonicum. J. Biol. Chem.* 345:32–38
96. Strobel, G. A. 1971. Comparative biochemistry of the toxic glycopeptides produced by some plant pathogenic corynebacteria. *Proc. Int. Conf. Plant Pathol. Bact., 3rd,* 357–65
97. Strobel, G. A., Hess, W. M. 1968. Biological activity of a phytotoxic glycopeptide produced by *Cornebacterium sepedonicum. Plant Physiol.* 43:1673–88
98. Strobel, G. A., Talmadge, K. W., Albersheim, P. 1972. Observations on the structure of the phytotoxic glycopeptide of *Cornebacterium sepedonicum. Biochim. Biophys. Acta* 261:365–74
99. Tam, L. Q., Patil, S. S. 1972. Mode of action of the toxin from *Pseudomonas phaseolicola.* II. Mechanism of inhibition of bean ornithine carbamoyltransferase. *Plant Physiol.* 49:808–12
100. Taylor, P. A., Durbin, R. D. 1973. The production and properties of chlorosis-

inducing toxins from a pseudomonad attacking timothy. *Physiol. Plant Pathol.* 3:9–17
101. Taylor, P. A., Maxwell, D. P., Durbin, R. D. 1971. Occurrence in Wisconsin of halo blight of timothy incited by a *Pseudomonas* species. *Plant Dis. Reptr.* 55:361–62
102. Taylor, P. A., Schnoes, H. K., Durbin, R. D. 1972. Characterization of chlorosis-inducing toxins from a plant pathogenic *Pseudomonas* sp. *Biochim. Biophys. Acta* 286:107–17
103. Wheeler, H., Luke, H. H. 1963. Microbial toxins in plant disease. *Ann. Rev. Microbiol.* 17:223–42
104. Woolley, D. W., Pringle, R. B., Braun, A. C. 1952. Isolation of the phytopathogenic toxin of *Pseudomonas tabaci,* an antagonist of methionine. *J. Biol. Chem.* 197:409–17
105. Woolley, D. W., Schaffner, G., Braun, A. C. 1955. Studies on the structure of the phytopathogenic toxin of *Pseudomonas tabaci. J. Biol. Chem.* 215:485–93
106. Zeller, W., Rudolph, K. 1972. Einfluss des toxins von *Pseudomonas phaseolicola* auf die permeabilitat und den osmotischen wert von mangoldblattern (*Beta vulgaris.* L.). *Z. Pflanzenphysiol.* 67:183–87

# RESISTANCE IN WINTER CEREALS AND GRASSES TO LOW-TEMPERATURE PARASITIC FUNGI

❖3600

*E. A. Jamalainen*
Department of Plant Pathology, Agricultural Research Center, Tikkurila, Finland

## INTRODUCTION

In countries where winters are long and snowfalls deep, snow protects hibernating plants from frost damage. Thus, little frost damage is found in central, eastern, and northern Finland, compared with Denmark and the southern regions of Sweden and Norway, where plants lack permanent snow cover and often suffer from severe frosts. In southern and western Finland, on the other hand, particularly in areas devoid of forest, the winter snow blanket may be light and frost can cause extensive damage.

Data regarding the protective effect of snow on plants has long been available. Research in Finland (30, 48, 57) showed that under a layer of snow 25–30 cm thick, the midwinter temperature did not fall below –7°C, against an ambient figure of –25°C. Even sparse winter snow usually suffices to hold soil temperatures to little below freezing point, while more substantial layers will keep them at 0°C during the severest winter (Figure 1). In snow-cover tests carried out at the Department of Plant Pathology in Tikkurila during 1950–1960 (57), timothy and winter rye were the most resistant of the graminaceous plants to lack of snow. Noticeably more susceptible were winter wheat, orchard grass, and perennial ryegrass. The effect of frost on winter growth of cereals has long been comprehensively studied. Of the studies carried out in Scandinavia, a notable example is the work done in southern Sweden (1, 1a) on the frost resistance of varieties of winter wheat under laboratory and field conditions. According to these tests, frost was the chief factor in damage arising from winter conditions. Early in this century it was widely held that frost was the principal reason for poor resistance of northern crops to winter conditions.

The significance of low-temperature parasitic fungi as biological factors causing winter damage to biennial and perennial crops later became clear. In the northern

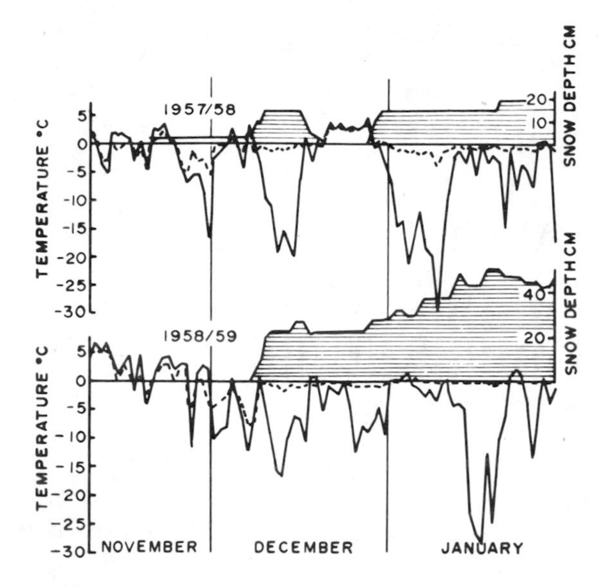


*Figure 1* The effects of snow deposit on soil surface temperature during two winters, 1957 and 1958, under differing snow conditions; solid line stands for minimum air temperature, dotted line stands for minimum soil surface temperature. (Source, A. Ylimäki).

countries the outstanding pioneer in promoting such research was Ekstrand at the Swedish Plant Protection Institute (15, 16, 17). Starting in 1930 he undertook a series of springtime journeys to examine plants that had overwintered, starting in southern Sweden and continuing with the advance of the spring season right up to the north of the country. He demonstrated that the prime cause of deterioration in winter cereals, ley grasses, and clovers in the snow-covered areas of Sweden was low-temperature parasitic fungi. Because frost was then considered the chief factor in this process, his published findings were assailed by plant breeders and plant pathologists. Critics alleged that the *Typhula* species and *Sclerotinia borealis* encountered in graminaceous plants were not parasitic fungi but were present in dormant winter crops as harmless saprophytes.

## LOW-TEMPERATURE PARASITIC FUNGI AND THEIR PREVALENCE

The most important of the low-temperature parasitic fungi injurious to graminaceous plants are *Fusarium nivale*, *Typhula* species (*T. ishikariensis* and *T. incarnata*), *Sclerotinia borealis*, and in Canada, the fungus known as LTB (low temperature basidiomycete). These fungi are able to attack dormant plants during winter at temperatures above zero. They likewise grow and can destroy plants even at subzero temperatures. Because spring temperature beneath snow is often several degrees above freezing point, these psychrophilic fungi have every opportunity to damage plants, particularly in lightly frozen soil.

### *Fusarium nivale (Ces.) Fr.*

The best-known low-temperature parasitic fungus, extensively studied in Central European countries, is *F. nivale,* pink snow mold. It is the conidial stage of the Ascomycete, *Griphosphaeria nivale* (Schaff.) Müller & von Arx (syn. *Calonectria graminicola* Berk. & Br.). The damage caused by pink snow mold and other low-temperature parasitic fungi may be ascertained shortly after the spring thaw. The leaves destroyed by *F. nivale* are pressed down against the soil, a dense layer often covering large field areas. The dead leaves are reddish in color, and when the snow has melted they become entirely dry. Pink snow mold is found throughout Finland (23, 25), but causes most damage in the central, eastern, and northern parts of the country. Field trials over many years show that in central Finland pink snow mold reduces the winter rye crop by an average of 30% and, in conjunction with *Typhula* fungi, decreases the winter wheat crop by about 20% (19).

*Fusarium nivale* is harmful to graminaceous plants in all the heavy snowfall areas of Europe, Soviet Union (39, 53), Asian Siberia (53), northern Japan (52), and the USA, including the state of Washington (4). In Canada the fungus is less important than in northern Scandinavia. Snow is certainly found over large prairie areas, but the soil freezes severely during snowless periods. However, in southern Alberta, *F. nivale* is prevalent in winter cereals and in ley grasses (32, 55). *F. nivale* is widespread and harmful in Sweden except in Skåne (15, 17); in Norway it appears in the interior where there is a continuous three month period of snow cover (2a). According to Ekstrand (17), *F. nivale* grows under laboratory conditions between –5 and 22.5°C; the optimum temperature is about 20°C, but the fungus still grows well at 0°C.

### *Typhula spp.*

Two species of *Typhula* cause winter damage in graminaceous plants: *T. ishikariensis* Lasch ex Fr. (syn. *T. idahoensis* Remsb.) and *T. incarnata* Remsb. (syn. *T. itoana* Imai and *T. graminum* Karst.). The use of the name *T. ishikariensis* has not yet become established. In the USA the fungus is usually called *T. idahoensis* (4), whereas in Europe it is referred to as *T. ishikariensis* (e.g. 1b, 2a, 24). Ekstrand has given it the designation *Typhula borealis* (15, 17). On the basis of spore characteristics, he had identified among *Typhula* material a second fungus called *T. hyperborea;* however, since these two are only distinguishable under the microscope, Ekstrand has used the name *T.* cf *borealis* in his publications. The *Typhula* fungi belong to the Basidiomycetes. Small sclerotia, dark brown (*T. ishikariensis*) or reddish brown (*T. incarnata*), and 0.5–1.5 mm in diameter appear on the dried leaves of plants in spring. In the fall the sclerotia form pale, downy, club-like fruiting bodies.

*Typhula ishikariensis* mainly attacks the Gramineae, but it has also been found as a pathogen in species of clover, alfalfa, winter turnip rape, sugar beet, and the bulbs of iris and tulip (17, 44). The fungus is found as a pathogenic agent in the same general areas as *F. nivale. T. ishikariensis* occurs throughout the whole of Sweden, though in some years its appearance is confined to Norrland in the most northerly part of the kingdom (15, 17). *T. ishikariensis* is also known as an agent of winter damage in the interior of Norway, where there is a snow cover for four months in the year (2a). The most serious effects of *T. ishikarienses* in the USA occur in winter wheat in Washington. The fungus is the most important agent of winter damage on grasses in northwest Canada, the Yukon Territory, and Alaska (32, 33, 55). The maximum temperature limit for *T. ishikariensis* is 15–20°C, the optimum being about 10°C and minimum about –5°C (4, 17).

*Typhula incarnata* appears in the same regions as *T. ishikariensis* and *F. nivale.* It causes less damage in Finland and Sweden than *T. ishikariensis* (17, 23, 28). *T. incarnata* is commoner in southern and central Finland, while in northern Finland it is very scarce. In Washington (4) and Germany (35, 36), *T. incarnata* is more widespread than *T. ishikariensis. T. incarnata* grows optimally at 8–15°C with a minimum growth temperature of –5°C and maximum about 20°C (14, 15, 17). According to Bruehl & Cunfer (8), *T. ishikariensis* is most virulent at 1.5, 0, and –1.5°C. *F. nivale* and *T. incarnata* are respectively second and third in degree of virulence, but neither caused mold at –1.5°C.

The greater aggressiveness of *T. ishikariensis* compared with *T. incarnata,* at least in Finnish conditions, may be due to the cool conditions favorable to *T. ishikariensis* in the fall and winter.

## *Sclerotinia borealis Bubák & Vleugel*

*Sclerotina borealis* (syn. *S. graminearum* Elenjev), an Ascomycete, ravages winter cereals and ley grasses (21). Patches where the plants have died appear in the fields in spring. The leaves are gray, dry, and thready. Dead leaves show numerous black sclerotia, which are globose and 2–4 mm in diameter. The following fall these sclerotia germinate and produce cup-shaped fruiting bodies (Figure 2), whose spores spread during early winter. Alternatively, the fungus spreads as the mycelium of sclerotia.

*Sclerotinia borealis,* as the name implies, is a northern species which has adapted to a cold environment. Its mycelium grows best in culture at 0 to +5°C, and is capable of invading plants in subzero conditions (Figure 3). The lowest temperature for growth of the fungus is –5°C or slightly below, and the upper limit is between

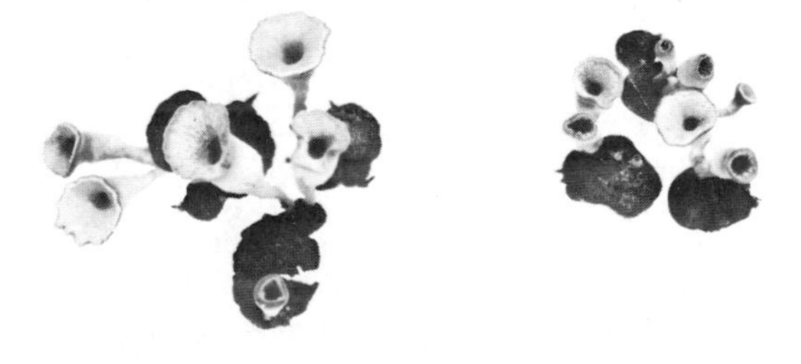

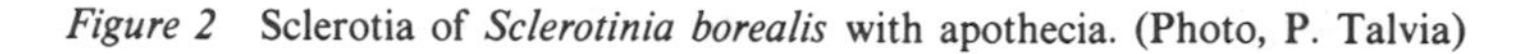

*Figure 2* Sclerotia of *Sclerotinia borealis* with apothecia. (Photo, P. Talvia)

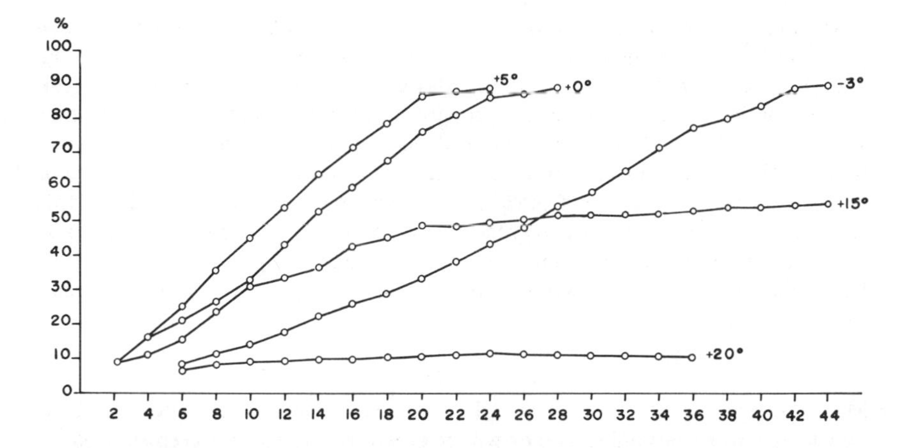


*Figure 3* *Sclerotinia borealis,* mycelial growth under low temperature conditions in petri-dish cultures. Values based on percentage area covered by mycelium (ordinate). Abscissa is time in days. Assessment made at two day intervals. (Source, P. Talvia)

15 and 20°C (17, 29, 56). Spore formation and spreading in the fall also occur at low temperatures.

The cold weather fungi have so far received relatively little attention in the literature. More interest has recently been shown in the yeasts and bacteria that thrive in cold conditions. Growth in an environment at freezing point or below indicates an enzyme system active at low temperatures. However, it is improbable that all the enzymes of an organism favored by cold conditions are specially adapted to such an environment. It is more likely that fungi favored by higher temperatures contain certain enzymic systems that cease to function at low temperatures and thus give total suspension of metabolism (56).

*Sclerotinia borealis* is usually found only in the central and northern parts of Finland. It is common during favorable winters when the average number of snow-cover days varies from 150–170 or more, between latitudes N.65° and N.62° (21, 29, 57). *S. borealis* occurred sporadically in southern Finland in the springs of 1966 and 1969.

The damage done by *S. borealis* varies greatly from one year to another. There have been winters when the fungus has been quite innocuous even in northern Finland. During other winters there has been exceptionally severe damage to ley grasses. During the winter of 1960–1961, *S. borealis* was so widely destructive that farmers had to be compensated by the state. It was equally injurious in the winters of 1965–1966 and 1970–1971; during each of these periods it completely destroyed all the plants in the ley grass variety tests at Apukka, near Rovaniemi, Finland. During the 1965–1966 winter, the damage inflicted by *S. borealis* on winter cereals was severe even near Jyväskylä (62°N). The pathogen is most destructive when the fall is damp and raw, the snow cover deep, the soil slightly frozen, and when this is followed by a slow spring thaw.

The southern limit of *S. borealis* is about the same in Finland and Sweden, where it is harmful only in the central parts and the north. *S. borealis* is found in the northern regions in Norway (2a) at altitudes where the snow cover lasts some six months. In European Russia the prevalent area of the fungus stretches much further south than in Scandinavia. It is widespread throughout northern USSR but less so in central European Russia; in western Siberia it likewise causes losses (39, 53, 54). *S. borealis* has damaged winter cereals on the Japanese island of Hokkaido. In this snow-covered region the problems concerning overwintering crops (52) seem to be the same as in northern Scandinavia. *S. borealis* occurs also on ley grasses in British Columbia and the Yukon areas of Canada (18, 34, 55), in Alaska (34), and Washington (4).

### *LTB Fungus*

Much damage to winter crops is caused in Canada by the pathogen known as LTB (low-temperature basidiomycete) or winter crown rot. It has not been possible to identify the organism, since only mycelium has been found. The disease expands radially in different directions to form large or small patches of dead plants. As far as is known, it does not occur anywhere in the Old World. Attempts to locate the disease in Finland have failed. LTB does much harm in the west-central region of Alberta and has also been found in British Columbia, Saskatchewan, Manitoba, the Yukon Territory, and Alaska. It has a wide host range, and destroys alfalfa, alsike, white clover, and ley grasses such as timothy, *Festuca rubra,* and *Bromus inermis* (10, 33). In this article no attempt is made to consider further the question of resistance to LTB.

## TESTING OF RESISTANCE TO LOW-TEMPERATURE PARASITIC FUNGI IN THE FIELD AND UNDER CONTROLLED CONDITIONS

The reasons for poor overwintering of crops have been studied at Tikkurila since 1946. An attempt has also been made to develop protective measures against low-temperature parasitic fungi. Numerous tests have been made on chemical methods of control, including seed treatment trials to control *F. nivale* on winter cereals (26) and fall treatment of stands with fungicides (27). Recently interest has been directed mainly to testing varieties offering better resistance to low-temperature parasitic fungi. The work has proceeded against a wider background, in that over the years 1966–1972 the State Scientific Committees in Denmark, Finland, Norway, and Sweden received financial assistance in carrying out tests under Finnish conditions on winter cereal and ley grass varieties and lines supplied by Scandinavian plant-breeding institutes.

The best method of testing the resistance of a variety to low-temperature parasitic fungi seems to lie in trials conducted under controlled conditions in laboratories and greenhouses. Methods for such tests have been studied mainly in Canada and North America (e.g. 4, 5, 11, 49). Bruehl et al (4) used snow mold chambers and established that some wheat lines were resistant to *F. nivale, T. ishikariensis,* and *T. incarnata* and to possible races (strains) of the pathogens. However, there was no evidence of pathogenic specialization sufficient to create a problem in breeding for resistance to these fungi (6). All the varieties tested that showed resistance in the field also displayed at least some degree of resistance in tests carried out under artificial conditions.

Tests in 1965 at Tikkurila using Cormack & Lebeau's method (11) were carried out in freezing chambers. The percentage of plants damaged by *F. nivale* for the following varieties of rye were: Dubbelstål, tetraploid, (Sweden), 10.5; Kungs II, (Sweden), 17.7; Pekka, (Finland), 23.4; Björnråg, (Sweden), 38.0; Toivo, (Finland), 44.4.

The Finnish Toivo rye variety and the north Swedish Björnråg variety proved more susceptible to *F. nivale* than the Swedish Dubbelstål and Kungs II. However, in field trials, to be described later, these results were reversed, and the two latter varieties were found to be the most susceptible to *F. nivale.*

Resistance of three Finnish and three Swedish varieties of winter wheat and winter rye was tested in 1967 at Tikkurila with isolates of *F. nivale, T. ishikariensis,* and *T. incarnata* under glasshouse conditions (3). The Finnish varieties showed a higher degree of resistance to the parasites than did the corresponding Swedish ones. This behavior agreed generally with the results of field trials. However, the Finnish Pekka showed a higher degree of resistance to *F. nivale* than Toivo and Ensi rye, although in field trials Pekka has very often proved more susceptible to the fungus. In the test the injury inflicted by *F. nivale* on winter rye was more severe than that on the winter wheat varieties. *Typhula ishikariensis* and *T. incarnata* were considerably more pathogenic to wheat than *F. nivale.* The results showed some correlation in that the hardiest varieties resisted all three fungi better than the less resistant.

Results of these trials seem to indicate that the resistance of rye varieties to low-temperature parasitic fungi under field conditions may differ from their genetic resistance revealed in glasshouse and freezing-chambers tests. The test methods devised so far for use under controlled conditions are not considered reliable in all instances.

The field conditions in an overwintering period are difficult to reproduce in detail under controlled conditions in the glasshouse or laboratory. It is often difficult to identify fungi and grade their damage in field trials. Grading does not give completely reliable results, but statistical analysis of results of trials provides a reasonable basis for determining the susceptibility of varieties to different low-temperature parasitic fungi. Possibly the best results would be obtained in the field by artificially inoculating breeding material and by controlling the depth and duration of the snow cover.

## VARIETY TESTS WITH WINTER WHEAT

Intensive efforts have been made in Washington to clarify the problems of winter wheat and its attendant low-temperature parasitic fungi and to produce varieties resistant to these pathogens (*F. nivale, T. ishikariensis, T. incarnata,* and *S. borealis*)

(4, 5, 8). The trials were carried out in chambers in the laboratory and in the field. After 1961 several thousand winter wheat strains (from the USDA World Cereal Collection) were tested (4). By 1963–1964 only 160 lines had persisted to the point of utilization as continuation trials, and of these, 9 numbered lines were considered possible material for development of resistance. In the 1965–1966 trials 15 wheat lines emerged for further testing. The resistance level of all commercial wheats was found to be too low for reliable practical value. However, some of the lines obtained by breeding displayed sufficient resistance to fungi to ensure that, apart from the rare exceptionally severe winter, their use would substantially improve shoot protection.

The most resistant of the winter wheats to *T. ishikariensis* in Sweden has been the Sammetvete native variety, the Swedish Sigyn wheat, some northern winter-wheat types, and the Finnish Olympia and Varma wheats. The more southern varieties are extremely susceptible to low-temperature parasitic fungi (17). A world-wide collection of winter wheats was tested near Leningrad in 1938 for resistance to snow mold (*F. nivale*). The 750 samples tested, grouped in 19 ecotypes, exhibited great differences in susceptibility. The ecotypes from the north overwintered well and showed little infection, whereas those from the south overwintered badly and were highly susceptible to infection (38). In tests with winter-wheat varieties at Brjansk and in the Leningrad region in 1969, the losses caused by *F. nivale* were 12%. The variety Miranovskaya 808 proved to be most resistant to the fungus (47). Several Polish winter wheat varieties have been studied in conditions prevailing in the region of Moscow. They showed some Polish varieties (Dańkowska 40, Glutenowa, Sobieszyńska 44, Wysokolitewka, Sztywnosłoma), Finnish varieties (Antti, Vakka, Varma), a Swedish variety (Banco), and others to be winter-hardy (9, 45).

## *Variety Tests With Winter Wheat in Finland*

The growing of winter wheat in Finland is restricted mainly to clay soils in the south and southwest. Even here winter wheat cannot be grown satisfactorily on other soils because of low-temperature parasitic fungi. In other parts of Finland, due to the uncertain wintering, cultivation of winter wheat is not widespread. The damage is caused by *F. nivale* and also by *Typhula* spp.

Since 1962, the Department of Plant Pathology has undertaken trials with breeding material of winter wheat and winter rye from Finland, Norway, and Sweden (28). The trials have been carried out in three localities, in Tikkurila near Helsinki (60°N), at the Häme Agricultural Experimental Station in Pälkäne (61°N), and at the Central Finland Agricultural Experiment Station in Kuusa (62°N). The test plots have been 9 meters long with three seed rows; this has proved a convenient size for studying fungal damage. There have been three replicates of each variety tested, and only nontreated seed has been used. Since 1965, a section of each plot (2 meters long) has been treated with 50% PCNB (pentachloronitrobenzene) 10 kg/ha, to show the prevalence of the pathogens (Figure 4).

The soil in these trials has been organic fine sand and fine silt containing organic matter; these soils are typical of the provinces of Häme and Central Finland, in which injury by low-temperature parasitic fungi is most common.

*Figure 4* Winter-rye variety test at Kuusa in spring 1968. Damage caused by *Fusarium nivale. Front,* susceptible varieties; *center,* resistant varieties; *back,* susceptible varieties. A section of each plot on the right (2 meters long) has been treated with PCNB in late autumn. (Photo, H. Blomqvist)

The green leaf mass in each experimental plot has been estimated by visual examination in the fall and spring. The spring assessments were made after the snow melt; often they were repeated 2–3 weeks later. A scale of 0–10 has been used, but the results have been presented as percentages, a stand of normal density being denoted as 100%. Thus, for example with *F. nivale,* 50% indicates that half of the leaf mass was destroyed by this pathogen. If the growing points and some of the leaves are uninfected, the plants may recover. A stand severely infected by *F. nivale* may also recover, particularly if weather conditions are favorable in the spring. If, on the other hand, most of the shoots are destroyed and the weather conditions in the spring are unfavorable, even partly infected plants are unable to recover. *Typhula* spp. and *S. borealis* often destroy the plants completely. The pathogens were *F. nivale, T. ishikariensis,* and occasionally *T. incarnata. Sclerotinia borealis* was frequent only at the Central Finland station in spring of 1966 and 1970.

Every year the trials included varieties under cultivation in Finland and Sweden, as well as several dozen lines of winter wheat and winter rye from plant breeding stations in Scandinavia. The damage to winter cereals caused by the fungi varies in different years and in different localities. Both *F. nivale* and *T. ishikariensis* for example did great damage in most trials during the winter seasons of 1965–1966, 1969–1970, and 1971–1972. In 1963–1964, when fungal damage was very small, there was comparatively little snow but soil freezing was deep. *F. nivale* has generally caused greater damage to winter wheat than *T. ishikariensis,* although the latter was also harmful for many years. In spite of climatic variation, the results of several years of tests in the different localities have yielded helpful data for guiding breeding work and comparing usefulness of existing varieties.

The Finnish wheat varieties have shown a higher degree of resistance to *F. nivale* and *T. ishikariensis* than Swedish varieties. The most resistant of the Finnish varieties were Elo (developed at Jokioinen) and Linna (developed at Hankkija), which wintered well both in sandy and clay soils. The latter has a lower milling value than the former. Somewhat more susceptible than the foregoing were Jyvä, Nisu, and Vakka (Jokioinen). Nisu is a good milling wheat and Vakka is the commonest winter wheat. Varma (Jokioinen) wintered somewhat less satisfactorily than any of these. All the Swedish winter-wheat varieties were very susceptible to snow molds (Table 1), viz Ertus (developed by Weibull), Norra (Weibull), Odin (Svalöf), Seba (Svalöf), Starke (Weibull), Trond (Weibull), and Virgo (Svalöf) (28). The lines of winter wheat from Finnish breeding institutes have also shown a higher degree of resistance to the pathogens than the Swedish ones in trials when fungal damage was severe. Of the Finnish lines, several had comparatively good resistance to *F. nivale* (Table 1).

Frost caused damage to winter wheat (28, 50) in the trials at Tikkurila in 1967–1968, when continued hard frosts occurred in the early winter before the soil was covered with snow. The Swedish varieties and lines were badly damaged by the frost. The Finnish wheat varieties and some of Hankkija lines, on the other hand, escaped frost damage.

### *Cultivation of Wheat in Relation to Chemical Treatment*

A total of 97 trials was carried out in Central Finland in 1960–1966 on fungicidal treatment of winter wheat stands (19). PCNB and phenyl mercury acetate (PMA) were used in the trials. The following average amounts of damage by fungi were observed in the nontreated stands: *Fusarium nivale,* 17%; *Typhula ishikariensis,* 9%; and *Sclerotinia borealis,* 5%. The figures for the treated stands were 1, 2, and 1% respectively. It was concluded that Finnish varieties of winter wheat were suitable for growing in central Finland in conjunction with chemical treatment of stands.

## VARIETY TESTS WITH WINTER RYE

From the results of resistance trials of winter-rye varieties, Ekstrand (17) divided them into two groups. Resistant were Finnish varieties Greus, Ensi, Oiva, Toivo, Pekka, the Sangaste rye of Estonian origin, the Swedish Björnråg, a group of native varieties, including Russian grown Vjatka, the Swedish Norrbottenråg, and some of the midsummer ryes. All the Finnish rye varieties orginate by selection from native strains with good wintering properties. The highest degree of resistance was shown by Greus and by the ryes from Norrbotten, particularly Norrbottenråg and midsummer ryes. Less resistant was the Finnish Onni, and still less were Pekka, Sangaste, and Björn. In a second and extremely susceptible group were Vasa II, Stål, Kungs I, Kungs II, and Malm (Swedish varieties), and Pektus ryes. The greatest damage was inflicted by *F. nivale* and *T. ishikariensis,* although in northern Sweden *S. borealis* was often an additional culprit. In Soviet Union tests with winter rye, the varieties Zima, K-10011, and K-10013 were resistant to *F. nivale.* Recovery from damage of snow mold was best in varieties from Finland and Sweden and in varieties from the nonblack soil region in the Soviet Union. Infection of seed by *F. nivale* was least on varieties Wloszanowka (Poland), Mestnaja (local Finnish variety), Karl-

**Table 1** Damage due to low-temperature fungi on winter-wheat varieties during the 1969–1970 winter of severe fungus attack at Pälkäne, Finland[a]

| Varieties and Lines | | Damage due to | | Plants surviving after overwintering |
|---|---|---|---|---|
| | | *Fusarium nivale* (%) | *Typhula ishikariensis* (%) | |
| Linna, Hankkija | F | 23 | 25 | 40 |
| Vakka, Jokioinen | F | 19 | 35 | 40 |
| Ertus, Weibull | S | 68 | 27 | 2 |
| Odin, Svalöf | S | 53 | 42 | 3 |
| Seba, Svalöf | S | 71 | 28 | 1 |
| Virgo, Svalöf | S | 77 | 21 | 1 |
| An b 495, Hankkija | F | 70 | 15 | 11 |
| An b 7800, Hankkija | F | 16 | 28 | 40 |
| An b 8536, Hankkija | F | 7 | 22 | 77 |
| An b 8735, Hankkija | F | 10 | 36 | 45 |
| An b 8837, Hankkija | F | 67 | 23 | 9 |
| An b 9143, Hankkija | F | 15 | 40 | 45 |
| An b 9491, Hankkija | F | 18 | 30 | 44 |
| An b 9549, Hankkija | F | 74 | 18 | 6 |
| Jo 0834, Jokioinen | F | 19 | 36 | 38 |
| Jo 01177, Jokioinen | F | 22 | 31 | 51 |
| Jo 01851, Jokioinen | F | 20 | 42 | 44 |
| Jo 03009, Jokioinen | F | 32 | 44 | 19 |
| Jo 03016, Jokioinen | F | 14 | 21 | 49 |
| Jo 03034, Jokioinen | F | 19 | 56 | 21 |
| Jo 03051, Jokioinen | F | 17 | 40 | 46 |
| Jo 03058, Jokioinen | F | 14 | 47 | 53 |
| 11783–68, Weibull | S | 72 | 26 | 1 |
| 14410–68, Weibull | S | 88 | 11 | 0 |
| 52–16, Holmberg | S | 33 | 57 | 11 |

[a] Soil = fine silt; F = Finnish, S = Swedish varieties.

shulder (German Democratic Republic), and Ceské (Czechoslovakia) (51). In tests with winter rye varieties in the Bryansk and Leningrad regions in 1967–1968, the most resistant varieties to *F. nivale* were Vjatka 2, Kharkovskaja 55, and Kharkovskaja 60. However, under unfavorable conditions for snow mold the yields of susceptible varieties were higher than for Vjatka variety, which is commonly cultivated in the Soviet Union. The resistance to *F. nivale* in tetraploid varieties was lower than in diploid varieties (47). Trials in the Leningrad region in 1967–1968 were arranged with 172 winter-rye varieties from 14 countries of Europe and America. Variation in susceptibility to *F. nivale* of the different varieties was 20–95%. Vjatka 2 was most resistant to *F. nivale* and gave the best grain yields. Other resistant varieties to snow mold were varieties and native strains from Finland and northern Sweden and from mountain regions in Austria and Hungary. Part of the tested varieties recovered well from damage of snow mold; such varieties were Grand de Flandre (Belgium), Petkus Kurzstroh (GDR), Malmråg (Sweden), and some varieties and strains from Finland (31). In rye variety tests in Poland in 1970 the most resistant variety to *F. nivale* was Chrobre. Dańkowskie Selekcyjne and a new variety No. 69/69 were less infected by snow mold than the majority of tested varieties (37).

### *Variety Tests with Winter Rye in Finland*

*Fusarium nivale* is very injurious to winter rye in central and eastern Finland (19). *Sclerotinia borealis* and *T. ishikariensis* may also attack winter rye, although damage by *S. borealis* only occurs in the central and northern parts of the country (20).

Trials with winter rye conducted in Tikkurila were arranged in the same manner as those with winter wheat described above. *Fusarium nivale* showed up in the test as the main pathogen, whereas *T. ishikariensis* was prevalent only in certain years. The Finnish winter-rye varieties Ensi, Toivo, Vjatka (a native variety from Russia), and Vatia (a Finnish native variety) showed the highest degree of resistance to *F. nivale.* The Finnish varieties Pekka and Voima were somewhat more susceptible. The Swedish varieties Kungs II (Svalöf), Värne (Svalöf), and Dubbelstål (Svalöf) were the most susceptible to *F. nivale.* The Norwegian lines have been tested for three years, one of which was the severe overwintering period of 1965–1966. They proved quite susceptible to fungi. On the other hand, in trials conducted in 1969–1970 one of the Norwegian lines resisted *F. nivale* well (Table 2). The Finnish lines of winter rye have also shown a higher degree of resistance to *F. nivale* than the corresponding Swedish ones (Table 2). In the Finnish winter ryes there is resistance to *F. nivale,* which may thus be used for further breeding. The Finnish varieties are long strawed and, for this reason, are susceptible to lodging. Long-strawed Finnish varieties have been crossed with short-strawed Swedish ones in an attempt to produce short and stiff-strawed varieties resistant to low-temperature parasitic fungi, and with a capacity for high yields. However, it has not proved possible to transfer the resistance to the short-strawed varieties.

From the field trials it has been possible to obtain data about the resistance to the low-temperature parasitic fungi (*F. nivale, T. ishikariensis,* and *S. borealis*) in winter-rye varieties from Finland, Norway, and Sweden. Resistance in winter-rye

**Table 2** Damage due to low-temperature fungi on winter-rye varieties in 1969–1970 at Tikkurila, Finland[a]

| Varieties and Lines | | Damage due to *Fusarium nivale* (%) | Plants surviving after overwintering |
|---|---|---|---|
| Ensi, Jokioinen | F | 27 | 97 |
| Toivo, Jokioinen | F | 22 | 94 |
| Pekka, Jokioinen | F | 51 | 63 |
| Kungs II, Svalöf | S | 80 | 20 |
| Värne, Svalöf | S | 94 | 4 |
| 1239, Hankkija | F | 28 | 80 |
| 1240, Hankkija | F | 25 | 82 |
| 1250, Hankkija | F | 30 | 80 |
| 1256, Hankkija | F | 35 | 75 |
| 1326, Hankkija | F | 47 | 55 |
| 1346, Hankkija | F | 49 | 62 |
| 1369, Hankkija | F | 37 | 72 |
| Jo 01922, Jokioinen | F | 50 | 73 |
| 67–3001, Jokioinen | F | 43 | 81 |
| 67–3057, Jokioinen | F | 35 | 84 |
| 67–3067, Jokioinen | F | 22 | 93 |
| 68–13, Jokioinen | F | 21 | 96 |
| 68–37, Jokioinen | F | 46 | 62 |
| 68–165, Jokioinen | F | 25 | 85 |
| Sv 2375, Svalöf | S | 87 | 13 |
| Sv 63570, Svalöf | S | 74 | 32 |
| Sv 63950, Svalöf | S | 93 | 4 |
| Vå 3/74 Bodø | N | 22 | 93 |

[a] Soil = fine sand; F = Finnish, S = Swedish, N = Norwegian varieties.

varieties is relative, and it is evident that under conditions favorable to the pathogens all the winter cereal varieties are more or less susceptible.

## RESISTANCE IN SCANDINAVIAN LEY GRASSES TO LOW-TEMPERATURE PARASITIC FUNGI

Low-temperature parasitic fungi cause substantial harm to ley grasses in the northern Finnish provinces of Oulu and Lapland. This damage does not occur every year. When the spring weather is favorable, diseased plants recover during the growing season, and economic losses are less than spring assessments might indicate. There is hardly any spring damage in some years, but in winters favorable to the fungi, *S. borealis, T. ishikariensis,* and to a lesser extent *F. nivale* may cause total losses of the ley. In some years it is *S. borealis,* in others it is *T. ishikariensis* that causes the greatest damage. In these circumstances, low-temperature fungi are the factors that, when they appear, make the cultivation of ley grasses uncertain under northern conditions. In certain years, water injury may occur in grasses grown on peat soils in northern Finland.

The relation of ley grass wintering to low-temperature parasitic fungi in south and central Finland differs from the situation in the far north, where the duration of snow cover is the longest in the country. In south and central Finland the Scandinavian varieties of timothy (*Phleum pratense*), meadow fescue (*Festuca pratensis*), and meadow grass (*Poa pratensis*), as in other areas with similar conditions in Scandinavia, are comparatively resistant to low-temperature parasitic fungi (29, 46). Orchard grass (*Dactylis glomerata*) is more susceptible to these fungi. *T. ishikariensis* and *F. nivale* may cause injury in certain years, even in the southern part of the country, and *S. borealis* occasionally produces losses in central Finland (29). Perennial ryegrass (*Lolium perenne*) in south Finland is very susceptible to the fungi.

The prevalence of low-temperature parasitic fungi in Swedish grassland is about the same as in Finland. This was demonstrated by Ekstrand (15, 17) who made spring observations over many years. Fungus damage is greatest in the northern parts of Sweden, whereas in the south it is scarcely apparent. According to Årsvoll (2a), damage caused in Norway by low-temperature parasitic fungi over the three-year period 1968–1971 varied between 11 and 7%, whereas the corresponding figures for damage from abiotic factors were from 10 to 7%. The most harmful fungi were *S. borealis* and *T. ishikariensis.* The latter, however, seldom occurs in coastal areas, fjord districts below 69°N, or in regions where the snow cover lasts fewer than 120 days, with a 60-day subzero maximum temperature. *Sclerotinia borealis* rarely causes obvious trouble in localities with fewer than 170 days of snow cover and subzero maximum temperatures lasting 110–120 days. This type of weather occurs in the northern Norwegian county of Finnmark, the interior of Tromsa county, and elsewhere at altitudes of 500–600 m. *Fusarium nivale* is present over the whole country and may be harmful in districts with a snow cover of fewer than about 90 days. Ice and water damage occurs principally in the Norwegian fjord and coastal areas.

## LEY GRASS TRIALS

Important information on the overwintering of ley grasses has been obtained at Apukka. Data concerning the overwintering capacity of ley grasses in northern Finland have also been obtained from studies made annually in Lapland (29). The first trials with ley grasses were carried out in 1949–1950 at Tikkurila, near Helsinki (60°N), and at Maaninka (63°N) (22). Since the beginning of 1966, field studies have been made in Finland on the resistance to low-temperature parasitic fungi of varieties and lines of ley grasses provided by plant breeding institutes in Denmark,

Finland, Norway, and Sweden. These experiments were performed at Apukka and at Kuusa (62°N). In most of the trials, the plots were 10 meters long, laid out in three replicated plots, each plot containing a strip 3 meters wide treated with 50% PCNB, 10 kg/ha, applied in late fall. This measure facilitated assessment of damage from organic causes to the plant stands in the experiments. The following is a summary of the main results yielded by various grass trials.

### *Timothy (Phleum Pratense L.)*

Timothy varieties from northern regions are substantially more resistant in Sweden (17), Norway, and Finland to *T. ishikariensis, S. borealis,* and *F. nivale* than are the more southerly varieties. In trials carried out in northern Norway (1b, 2) the Norwegian Engmo and Bodin varieties as well as the north Finnish and Swedish varieties were very resistant against *T. ishikariensis.* On the other hand, Grindstad and Forus from south Norway were most susceptible to this fungus. Engmo seed samples grown further south were less resistant to low-temperature parasitic fungi than those originating in the north (2).

Tests carried out at Apukka showed that the Finnish varieties Tammisto and Tarmo, certain local timothy strains, the Norwegian Engmo, and the Swedish Bottnia II varieties all offered good resistance to *T. ishikariensis* and *S. borealis.* The Canadian Climax and the German Lischover, in contrast, were very susceptible to these pathogens. The greatest damage was noted in first-year meadows, where *S. borealis* did more harm than *T. ishikariensis* (29). In trials conducted in 1949–1950 at Maaninka (22) *T. ishikariensis* destroyed foliage of timothy to extents varying between 10 and 80%, depending upon the variety. The lowest percentage was for Finnish Phleum 32 and Tammisto timothies. Irrespective of the intensity of disease, the proportion of completely dead plants remained low (0–10%) and most of the damaged plants recovered later in the spring. Trials at Kuusa in 1970–1971 (Table 3) showed that the weakest resistance to *T. ishikariensis* was that of the Danish Tystofte lines. The Swedish Luleå lines, on the other hand, from Svalöf in Norbotten province, showed good resistance. At the Arctic Circle station in 1968–1969 *T. ishikariensis* completely destroyed the timothy material. At this station, too, *S. borealis* was extremely harmful, destroying practically all the timothies in tests studied in 1965–1966. Fungal damage at this station was once again very extensive in 1967–1968, when *S. borealis* killed between 35 and 90% of north Finnish and north Swedish strains. Tammisto, Omnia, Grindstad, Bottnia II, Bodin, and Engmo were most resistant. The worst damage was to the Danish Tystofte lines. In the Apukka 1970–1971 trials, *S. borealis* damage was lowest (12% of destroyed foliage) in the Swedish Omnia timothy. The Swedish Luleå lines also survived well.

*Fusarium nivale* attacked timothy in Tikkurila and Maaninka, but damage in all the 13 varieties tested was small (between 2 and 10%) (22). A heavier attack by *F. nivale* at Apukka occurred only in the 1966–1967 winter season, when the fungus destroyed between 15 and 50% of foliage, depending on variety.

In north Finland, timothy is the outstanding ley grass as far as resistance to low-temperature fungi and other winter damage is concerned. Trials did not show

**Table 3** Tests with timothy and meadow fescue varieties at Kuusa, Finland in 1970–1971[a]

| Timothy (*Phleum pratense*) | | | | Meadow Fescue (*Festuca pratensis*) | | | |
|---|---|---|---|---|---|---|---|
| Varieties and Lines | | Damage due to *Typhula ishikariensis* (%) | Plants surviving after overwintering (%) | Varieties and Lines | | Damage due to *Typhula ishikariensis* (%) | Plants surviving after overwintering (%) |
| An 1160, Hankkija | F | 13 | 87 | An 156, Hankkija | F | 20 | 79 |
| Tammisto, Hankkija | F | 17 | 82 | An 2356, Hankkija | F | 39 | 60 |
| Tarmo, Jokioinen | F | 11 | 89 | Tammisto, Hankkija | F | 25 | 74 |
| Esko, Kesko | F | 12 | 88 | Paavo, Jokioinen | F | 9 | 91 |
| L-H, Kesko | F | 9 | 88 | L-H $F_1$, Kesko | F | 51 | 42 |
| Bottnia II, Svalöf | S | 8 | 92 | L-H $F_2$, Kesko | F | 56 | 42 |
| Omnia, Svalöf | S | 40 | 57 | Bottnia II, Svalöf | S | 17 | 82 |
| Luleå 0841, Svalöf | S | 4 | 93 | Sena, Svalöf | S | 27 | 72 |
| Luleå 0870, Svalöf | S | 6 | 94 | Luleå 01205, Svalöf | S | 9 | 91 |
| Luleå 0876, Svalöf | S | 7 | 93 | Luleå 01236, Svalöf | S | 13 | 86 |
| WT 48, Weibull | S | 42 | 56 | W Ås 9, Weibull | S | 71 | 25 |
| Tystofte, sample no. 1 | D | 42 | 52 | Tystofte, sample no. 1 | D | 46 | 27 |
| Tystofte, sample no. 2 | D | 40 | 58 | Tystofte, sample no. 2 | D | 43 | 28 |
| Tystofte, sample no. 3 | D | 38 | 60 | Tystofte, sample no. 3 | D | 45 | 37 |

[a]Soil = fine silt; F = Finnish, S = Swedish, D = Danish varieties.

great differences between the Scandinavian varieties in bad years, all varieties indiscriminately suffering serious harm. Varieties and strains grown in the north were nevertheless the most resistant and can be used for developing new varieties.

### *Meadow Fescue (Festuca pratensis Huds.)*

In north Swedish trials, various varieties of meadow fescue displayed great differences in resistance to *T. ishikariensis, F. nivale,* and *S. borealis.* The best were the Bottnia I and Bottnia II varieties. Damage was greater in the first-year grassland, but it could also be significant in the older fields (17). In northern Norway (1b, 2), the south Swedish and Danish varieties were susceptible to *S. borealis* and *F. nivale.* The most resistant to these fungi were the north Norwegian Vågønes and the Swedish Bottnia II. Slightly less so were the Norwegian Løken and Tjøtta.

Trials over a number of years at Apukka exhibited damage somewhat less by *S. borealis* on meadow fescue than on timothy (29). The Finnish Tammisto and Paavo were the most resistant meadow fescue varieties to *S. borealis;* the most susceptible were the Danish Hinderupgaard and German Lischover. During tests there in 1968–1969, *T. ishikariensis* destroyed all the meadow fescue varieties; neither the meadow fescue nor any of the other ley grass varieties tested in these trials showed any resistance to *T. ishikariensis.* During trials at Kuusa station in 1970–1971 (Table 3), the effects of *T. ishikariensis* on the Paavo meadow fescue and Luleå lines were small.

*Fusarium nivale* and *T. ishikariensis* appeared abundantly in all Swedish, Finnish, and Danish meadow fescue varieties tested at Tikkurila and Maaninka in 1949–1950. However the damage was ephemeral, the plants recovering later in the spring.

### *Orchard Grass* (*Dactylis glomerata L.*)

Orchard grass has proved more susceptible to low-temperature parasitic fungi than timothy and meadow fescue in Sweden (17), Norway, and Finland. In northern Norway the best resistance was shown by the northern Norwegian Hattfjelldal and Tjøtta varieties and by the Swedish Brage. Under conditions favorable to winter pathogens all orchard grass varieties suffered (1b, 2).

In trials at Maaninka during 1949–1950 orchard-grass foliage was affected by *T. ishikariensis*, and 20 to 35% of the stands were destroyed, except Finnish Tammisto II, which suffered only 5% (22). In tests in 1970–1971 at Kuusa *T. ishikariensis* caused 8% damage to the Tammisto variety; in three Danish Tystofte lines the fungal damage was between 15 and 24%.

Orchard grass at Apukka suffers more from *S. borealis* than from *T. ishikariensis*. Finnish varieties tested resisted these fungi better than Danish Daeno and Daenfeldt, the German Lischover, or the US varieties. In the winter of 1967–1968 the damage of *S. borealis* there was so severe that no variety of line of orchard grass escaped serious harm, the foliage suffering to the extent of 65–100% and the crop 25–100%. During the following winter all the orchard grass stands at this station were destroyed by *T. ishikariensis* (29).

*Fusarium nivale* damaged foliage of eight orchard-grass varieties at Tikkurila in 1949–1950 to the extent of 35–60%, but the number of dead plants nevertheless was small (22). Danish lines suffered 50–70% foliar damage from *F. nivale* at Apukka in 1971–1972. In this test the Norwegian Hattfjelldal orchard grass, with a damage factor of 10%, performed best.

Because of its susceptibility to *S. borealis* and *T. ishikariensis* the cultivation of orchard grass in north Finland is very uncertain.

### *Perennial Ryegrass* (*Lolium perenne L.*)

All tested varieties of perennial ryegrass were destroyed in Jämtland, Sweden (60°N), by *F. nivale* exept for the Swedish Valinge, which wintered comparatively well (17). Perennial ryegrass also displayed marked susceptibility to the low-temperature parasitic fungi in Finland. Thus at Tikkurila and Maaninka in 1949–1950 all varieties were seriously damaged by *F. nivale* and *T. ishikariensis*, except for the Swedish Valinge (22). In trials at Apukka no perennial ryegrass variety lasted through the winter, because of severe attack by *S. borealis* (29). In tests at Kuusa, Finland perennial ryegrass showed extensive harm by *F. nivale, T. ishikariensis*, and *S. borealis*. Here again only the Swedish Valinge variety wintered satisfactorily.

Perennial ryegrass in Finland is a highly uncertain crop because of low-temperature fungi and frost damage even in the south, and consequently it has never been widely grown in this country.

### *Other Ley Grass Species*

At Muddusniemi (69°N, in the extreme north of Finland), *S. borealis* was the chief destructive agent in ley grass trials (41). The domestic varieties wintered better than the foreign. The best survivors among the first-year grasses were *Alopecurus pratensis* L. (8.5), *Festuca pratensis* Huds. (8.7), *Poa pratensis* L. (8.0) and *Bromus inermis* Leyss (7.3). (The figures in brackets show the average winter grading for first-year stands, where 0 = complete kill and 10 = no damage). The overwintering of *Phleum pratense* (4.7) and *Dactylis glomerata* (3.4) was poor. Timothy, however, recovered well in subsequent years and gave the best average yields. In northern Norway (2) *Bromus inermis* was extensively damaged, and for this reason is unsuitable for cultivation under these conditions.

All three varieties of Italian ryegrass (*Lolium multiflorum* Lam.) in the winter of 1949–1950 at Tikkurila and Maaninka were destroyed by *F. nivale* and *T. ishikariensis* (22).

*Alopecurus pratensis, Bromus inermis, Festuca rubra* L, *Phalaris arundinacea* L., and *Poa pratensis* were tested at Apukka. *A. pratensis* showed high resistance to both *S. borealis* and *T. ishikariensis,* whereas *B. inermis* and *F. rubra* were extremely susceptible. Three lines of *Poa pratensis* obtained from Norway and tested during 1967–1968 suffered much damage from *S. borealis,* but the Norwegian Holt variety survived well (29).

The varieties of grasses cultivated in northern Finland, Norway, and Sweden are among the most resistant of all those tested for low-temperature parasitic fungi. Nevertheless, even this material is susceptible to attack when conditions favor the appearance of the fungi. The plant breeder will certainly still be called upon to do much further work in breeding varieties more resistant to these parasites for northern parts of Scandinavia.

### *Effect of PCNB on Low-Temperature Parasitic Fungi in Northern Finland*

At several trial sites in Lapland, PCNB was applied to timothy stands before the start of winter. In several of these experiments, PCNB gave good or useful control of *S. borealis* and *T. ishikariensis.* The application, however, was not sufficient to protect adequately those varieties particularly susceptible to these fungi. Application of PCNB to first-year grass leys led to yield increases over several successive years (29).

## SOME ASPECTS OF CROP RESISTANCE TO LOW-TEMPERATURE PARASITIC FUNGI

The resistance of the crops examined in this review to *F. nivale, T. ishikariensis, T. incarnata,* and *S. borealis* depends on the duration of the winter snow cover in areas where these pathogens are prevalent. These fungi tend to behave similarly on particular hosts, so that a variety resistant to one fungus is, in general, reasonably resistant also to the others. All these fungi infect graminaceous plants. However,

*T. ishikariensis* also attacks a number of other host plants. Because the hosts are many graminaceous species, the inference is that the fungi lack a high degree of pathogenic specialization, rendering more difficult the task of developing resistant varieties. In low-temperature fungi no physiologic races have been observed (4, 6, 36, 42). As stated by Cunfer (12), "the sexual recombination from a functioning basidial stage provides added potential for variation in these pathogens."

As already shown, northern Scandinavian winter-cereals and ley-grass varieties are more resistant to low-temperature fungi than those originating farther south. In areas of heavy snow, where fungal losses occur, natural selection has been and is still operating, the more susceptible material being killed, leaving the hardier plants to survive. Because systematic breeding of resistant varieties has so far been limited, the more robust types have evolved by natural selection. Good examples of this process are, (17) the Finnish winter-rye varieties, which have proved the most resistant of all the Scandinavian varieties. A number of them have been obtained by selective methods from a resistant native Finnish rye, and do not derive from hybridization. It has been found that the resistance of southern varieties increases when they are transplanted in the north; conversely, northern plant resistance to disease weakens upon cultivation further south.

*Fusarium nivale* is spread by seed and growth through soil (40). The fungus in cereal plants is distributed mainly by seed (15, 17, 43). Seeds are often so contaminated with *F. nivale* that surface chemical treatment fails to eradicate subsurface fungus hyphae in the grain. However, I am convinced that in Finland *F. nivale* infects winter cereals through soil-borne conidia and mycelium. Although the seed gives 100% germination and shows no sign of fungal growth at the germination tests, pink snow mold may develop during the ensuing winter and destroy the stands of winter cereals. In the hundreds of trials in Finland on seed treatment of winter rye and wheat, the seed has mostly shown normal germination; however, *F. nivale* has caused widespread damage to plants growing from treated seeds (26).

There are differing opinions about the manner in which the sclerotia-forming *Typhula* fungi and *S. borealis* are spread. Cunfer & Bruehl (14) and Lehmann (36) hold that vegetative hyphae of *T. ishikariensis* growing from sclerotia in the fall are the chief source of winter wheat infection, although basidiospores may also be important. Ekstrand (17) and Ylimäki (58) maintain that the chief source of infection is the basidial stage of these fungi.

The southern limit of *S. borealis* is much further north than for *Typhula* spp. and *Fusarium nivale.* Because *S. borealis* can withstand lower temperatures than the latter fungi, it tolerates extreme northern conditions better, whereas in southern Finland it has not caused any damage.

In considering the physiologic resistance of plants, Bruehl & Cunfer (8) showed that winter-wheat resistance to *T. ishikariensis* depends on three things: plant size, maintenance of a high carbohydrate supply, and unknown physiologic resistance factors. Larger plants from early sowing lose their leaves, but the crowns usually survive. Very small plants survive (7). Tests in Finland have shown that plants from later sowing, which are smaller in size during wintering, are more susceptible to *T. ishikariensis* than the larger sized plants from earlier sowing. Other trials have proved that early-sown crops, with abundant foliage growth and hence favorable conditions for the spread of *F. nivale,* are most susceptible to this pathogen. Similar conclusions were reached in Swedish tests (17). Plants with narrow leaves are more resistant to *F. nivale* than plants with broad leaves (51).

Bruehl & Cunfer (8) stated that resistant varieties show greater capacity than susceptible ones to store food reserves during long periods in darkness near freezing point. This supports the hypothesis that starvation is a factor in resistance to low-temperature parasitic fungi. Resistant varieties have a greater winter reserve of carbohydrates and utilize them more slowly than susceptible ones. The chlorophyll content of resistant wheat plants kept for 0, 4, 6, and 8 weeks in darkness at 1°C disappeared more slowly than that of susceptible wheats. Highly susceptible wheats remained greener at 1° and 0°C than resistant ones. The rate of metabolism in susceptible varieties is clearly greater than in resistant varieties. Resistance to low-temperature fungi is therefore probably due to plant food reserves and a certain minimum level of metabolism.

Breeding varieties resistant to low-temperature parasitic fungi has not yet yielded significant results either in Europe or North America. The position regarding ley grass varieties is similar; where conditions are very favorable to attack by low-temperature parasitic fungi, it is hardly possible to find a fully resistant variety. Fungus-resistant breeding material is available in the regions of major attacks. Even in the areas of the worst damage, it is possible to find plants with totally destroyed leaves but where the crown has remained alive and the plant recovers. Such plants, possessing manifestly greater food reserves than the killed plants, may form the basis for development material. Although plant resistance to low-temperature parasitic fungi exists to the extent that highly susceptible varieties are severely infected by various fungi, resistance in the same variety to different fungi can also show great variation. This is inferred from Table 1, which shows results of winter-wheat trials where the injuries caused by *F. nivale* and *T. ishikariensis* vary noticeably and independently from variety to variety. Hence the resistance of new varieties to each low-temperature parasitic fungus must be studied separately.

Cultivation of winter cereal for bread, and raising ley grass for cattle fodder in snow-covered regions are of great importance. To protect the supply of hay and cereals against the effects of extensive low-temperature parasitic fungus damage in these districts, varieties showing improvement over existing ones are essential.

*Literature Cited*

1. Åkerman, Å. 1945. Experiments on overwintering of winter cereals and hardiness of various winter wheats. *Kgl. Lantbruksakad. Tidskr.* 84:192–215

1a. Åkerman, Å. 1949. Advanced studies on the winter hardiness of wheat. *Kgl. Lantbruksakad. Tidskr.* 88:157–87

1b. Andersen, I. L. 1960. Investigations on the wintering of meadow plants in northern Norway. III. *Rept. State Exp. Sta. Holt,* Tromsφ 33:1–20

2. Andersen, I. L. 1971. Investigations on the wintering of some forage grasses. *Rept. State Exp. Sta. Holt,* Tromsø 40:121–34

2a. Årsvoll, K. 1973. Winter damage in Norwegian grasslands, 1968–1971. *Norw. Plant Prot. Inst. Div. Plant Pathol. Rept.* 56:1–21

3. Blomqvist, H., Jamalainen, E. A. 1968. Preliminary tests on winter cereal varieties of resistance to low temperature

parasitic fungi in controlled conditions. *J. Sci. Agr. Soc. Finl.* 40:88–95
4. Bruehl, G. W. et al 1966. Snow molds of winter wheat in Washington. *Wash. Agr. Exp. Sta. Bull.* 677:1–21
5. Bruehl, G. W. 1967. Correlation of resistance to *Typhula idahoensis, T. incarnata,* and *Fusarium nivale* in certain varieties of winter wheat. *Phytopathology* 57:308–10
6. Bruehl, G. W. 1967. Lack of significant pathogenic specialization within *Fusarium nivale, Typhula idahoensis,* and *T. incarnata* and correlation of resistance in winter wheat to these fungi. *Plant Dis. Reptr.* 51:810–14
7. Bruehl, G. W. 1967. Effect of plant size on resistance to snow mold of winter wheat. *Plant Dis. Reptr.* 51:815–19
8. Bruehl, G. W., Cunfer, B. 1971. Physiologic and environmental factors that affect the severity of snow mold of wheat. *Phytopathology* 61:792–99
9. Cheremisova, T. D. 1964. Summary of studies during three years of winter wheat collection in conditions prevailing in the region of Moscow. *Trudy̆ Mosk. Otdel. VIR* 1:75–106 (In Russian)
10. Cormack, M. W. 1948. Winter crown rot or snow mold of alfalfa, clover, and grasses in Alberta. I. Occurrence, parasitism, and spread of the pathogen. *Can. J. Res. Sect. C* 26:71–85
11. Cormack, M. W., Lebeau, J. B. 1959. Snow mold infection of alfalfa, grasses, and winter wheat by several fungi under artificial conditions. *Can. J. Bot.* 37: 685–93
12. Cunfer, B. M. 1971. The role of the basidial stage in the life cycle of *Typhula idahoensis. Phytopathology* 61:889
13. Cunfer, B. M., Bruehl, G. W. 1973. Role of basidiospores as propagules, and observations on sporophores of *Typhula idahoensis. Phytopathology* 63:115–20
14. Dejardin, R. A., Ward, E. W. B. 1971. Growth and respiration of psychrophilic species of the genus *Typhula. Can. J. Bot.* 49:339–47
15. Ekstrand, H. 1947. Some phytopathological views of the overwintering of winter cereals and forage grasses with special regard to experimental work in agriculture. *Stat. Växtskyddsanstalt Medd.* 49:1–48
16. Ekstrand, H. 1947. Winter cereals and the problem of winter hardiness with special regard to the resistance to certain fungi. *Stat. Växtskyddsanst. Medd.* 50:1–28
17. Ekstrand, H. 1955. Overwintering of winter cereals and forage grasses. *Stat. Växtskyddsanst. Medd.* 67:1–125
18. Groves, J. W., Bowerman, C. A. 1955. *Sclerotinia borealis* in Canada. *Can. J. Bot.* 33:591–94
19. Hänninen, P., Jamalainen, E. A. 1968. Overwintering of winter cereals in central Finland. *Ann. Agr. Fenn.* 7:194–218
20. Isotalo, A., Vogel, R. 1962. Ergebnisse von Versuchen mit Winterroggen an der Versuchsstation am Polarkreis in den Jahren 1942–1960. *Ann. Agr. Fenn.* 1:233–48
21. Jamalainen, E. A. 1949. Overwintering of Gramineae-plants and parasitic fungi. I. *Sclerotinia borealis* Bubák & Vleugel. *J. Sci. Agr. Soc. Finl.* 21: 125–42
22. Jamalainen, E. A. 1951. Occurrence of low-temperature parasitic fungi on ley grasses in Finland. *Nord. Jordbrugsforskn.* No. 2–3:529–34 (In Swedish)
23. Jamalainen, E. A. 1956. Overwintering of plants in Finland with respect to damage caused by low-temperature pathogens. *Finn. Sta. Agr. Res. Board Publ.* 148:5–30
24. Jamalainen, E. A. 1957. Overwintering of Gramineae-plants and parasitic fungi. II. On the *Typhula* sp.-fungi in Finland. *J. Sci. Agr. Soc. Finl.* 29:75–81
25. Jamalainen, E. A. 1959. Overwintering of Gramineae plants and parasitic fungi. III. Isolations of *Fusarium nivale* from gramineous plants in Finland. *J. Sci. Agr. Soc. Finl.* 31:282–84
26. Jamalainen, E. A. 1962. Trials on seed treatment of winter cereals in Finland. *Ann. Agr. Fenn.* 1:175–91
27. Jamalainen, E. A. 1964. Control of low-temperature parasitic fungi in winter cereals by fungicidal treatment of stands. *Ann. Agr. Fenn.* 3:1–54
28. Jamalainen, E. A. 1969. Resistance of Scandinavian winter cereal varieties to low temperature parasitic fungi. *Ann. Agr. Fenn.* 8:251–63
29. Jamalainen, E. A. 1970. Overwintering of ley grasses in North Finland. *J. Sci. Agr. Soc. Finl.* 42:45–58
30. Keränen, J. 1920. Über die Temperatur des Bodens und der Schneedecke in Sodankylä nach Beobachtungen mit Termoelementen. Diss. Helsingfors. 197 pp.
31. Kobylyanskiĭ, V. D., Il'ichev, G. A. 1971. World rye collection as initial material for breeding of resistance to damping-off. *Trudy̆ Prikl. Bot. Genet. Selek.* 43:101–5

32. Lebeau, J. B. 1968. Pink snow mold in Southern Alberta. *Can. Plant Dis. Surv.* 48:130–31
33. Lebeau, J. B. 1969. Diseases affecting forage production in Western Canada. *Proc. Can. Forage Crops Symp.* 1969:123–38
34. Lebeau, J. B., Logsdon, C. E. 1958. Snow mold of forage crops in Alaska and Yukon. *Phytopathology* 48:148–50
35. Lehmann, H. 1965. Untersuchungen über die Typhula-Fäule des Getreides. I. Zur Physiologie von *Typhula incarnata* Lasch ex Fr. *Phytopathol. Z.* 53:255–88
36. Lehmann, H. 1965. Untersuchungen über die Typhula-Fäule des Getreides. II. Zur Pathologie durch *Typhula incarnata* Lasch ex Fr. erkrankter Wirtspflanzen. *Phytopathol. Z.* 54:209–39
37. Maciejowska-Pokacka, Z. 1971. Preliminary studies on the resistance of rye varieties to snow mould. *Biul Inst. Ochr. Rośl.* Poznan 50:195–205
38. Markevicz, N. P. 1939. The overwintering and susceptibility to snow mold of ecotypes of winter wheat. *Bull. Plant Prot. Leningrad* 1:119–21
39. Naumov, H. A. 1958. Diseases of agricultural crops. *Sel'hozgiz.* (In Russian)
40. Neururer, H., Zwatz, B. 1965. Untersuchungen über die Bodeninfection durch *Fusarium nivale* Ces. und die Resistenz von Winterroggensorten gegen Schneeschimmel. *Pflanzenschutzberichte* 33:1–16
41. Nissinen, O., Salonen, A. 1972. Effect of *Sclerotinia borealis* on the wintering of grasses at the Muddusniemi Experimental Farm of the University of Helsinki at Inari in 1950–1965. I. The effect of weather conditions on the incidence of *S. borealis* and of the species and variety of the grass on the wintering of ley. *J. Sci. Agr. Soc. Finl.* 44:98–114
42. Peuser, H. 1931. Untersuchungen über das Vorkommen biologischer Rasen von *Fusarium nivale* Ces. *Phytopathol. Z.* 4:113–28
43. Pichler, F. 1952. Über die Prüfung von Roggensorten auf ihre Anfälligkeit für Schneeschimmel (Fusarium). *Pflanzenschutzberichte* 8:33–43
44. Procenko, E. P. 1967. *Typhula borealis* Ekstrand infesting tulips in the U.S.S.R. *Mikol. Fitopatol.* 1:107–9
45. Puhal'skiī, A. V. 1964. Polish wheats as starting-point material for the selection. *Trudȳ Mosk. Otdel. VIR* 1:8–74 (In Russian)
46. Ravantti, S. 1965. Herbage plants. *Rept. Plant Breed. Inst. Hankkija* 1965: 149–52
47. Sidorova, S. F. 1970. Role of variety in infection of winter wheat and rye by snow mould. *Byull. Vses. Nauch. Issled. Inst. Zashch. Rast.* 16:38–40
48. Simola, E. F. 1926. Untersuchungen der Landwirtschaftlichen Versuchsanstalt über das Erfrieren des Kulturlandes und das Auftauen des Bodenfrostes in Jahren 1924, 1925 und 1926. *Finn. Sta. Agr. Res. Board Publ.* 5:1–53
49. Sunderman, D. W. 1964. Modifications of the Cormack and Lebeau technique for inoculating winter wheat with snow mold-causing *Typhula* species. *Plant Dis. Reptr.* 48:394–95
50. Tapanets, M. P. 1964. Winterhardiness of winter wheat in conditions of a non-black soil area (Moscow oblast'). *Trudȳ Mosk. Otdel. VIR* 1:135–63 (In Russian)
51. Tishkov, E. N., Fedorova, L. I. 1964. Resistance of winter rye varieties to soaking, snow mould, and fusariosis of the grain. *Trudȳ Mosk. Otdel. VIR* 1:164–74 (In Russian)
52. Tomiyama, K. 1955. Studies on the snow blight disease of winter cereals. *Rept. Hokkaido Agr. Exp. Sta.* 47: 1–234
53. Tupenevich, S. M. 1965. Damping-off winter cereals. *Trudȳ Vses. Inst. Zashch. Rast.* 25:118–21
54. Tupenevich, S. M., Shirko, V. N. 1939. Measures for preventing losses of winter cereals in spring from *Sclerotinia graminearum* Elen. *Plant Prot. Leningrad* 18:85–99
55. Vaartnou, H., Elliott, C. R. 1969. Snowmolds on lawns and lawngrasses in northwest Canada. *Plant Dis. Reptr.* 53:891–94
56. Ward, E. W. B. 1966. Preliminary studies of physiology of *Sclerotinia borealis,* highly psychrophilic fungus. *Can. J. Bot.* 44:237–46
57. Ylimäki, A. 1962. The effect of snow cover on temperature conditions in the soil and overwintering of field crops. *Ann. Agr. Fenn.* 1:192–216
58. Ylimäki, A. 1969. Typhula blight of clovers. *Ann. Agr. Fenn.* 8:30–37

# FUNGUS METABOLITES TOXIC TO ANIMALS

❖3601

*Chester J. Mirocha and Clyde M. Christensen*
Department of Plant Pathology, University of Minnesota, St. Paul, Minnesota 55101

## INTRODUCTION

Mycotoxicology deals with diseases in man and animals caused by or resulting from ingestion of foods made toxic by metabolic products of fungi. Of the numerous reviews devoted to this subject over the last ten years, the following are selected: Brook & White (14), Christensen (18), Ciegler & Lillehoj (26), Goldblatt (54), Hesseltine (60), Detroy et al (35), Purchase (115), Scott (132), Wogan (177, 179), Mateles & Wogan (94), and Ciegler et al (25). This review is restricted to those mycotoxins found in feedstuffs (excluding ergot) or in the ingredients of foods and feedstuffs in the field and is little concerned with the host of toxic metabolic products of fungi produced in pure culture in the laboratory. It often is tacitly assumed that if a fungus in pure culture in the laboratory under some conditions produces a compound or compounds in some way toxic to test animals when ingested or otherwise administered (including intraperitoneal injection, which by no stretch of the imagination could occur in the feedlot or poultry house) the presence of this same fungus in food or feedstuffs is cause for alarm. This is by no means true; even when a potential toxin-producing strain of *Aspergillus flavus* is added to moist grain and allowed to grow, along with the other microflora normally present, it may produce little or no toxin.

## AFLATOXICOSES

### *The Fungus*

*Aspergillus flavus* or *A. flavus-oryzae* is a "group" species (124) that contains 11 species, including *A. flavus* Link, so that this name is used both for the group and for a separate species within the group. Another species in the group, *A. parasiticus* Speare, appears to be a more potent producer of aflatoxin than is *A. flavus* Link; the *A. flavus* that Burnside et al proved to be toxic to pigs was sent to Raper in 1954 and was identified by him as *A. parasiticus*, as was the *A. flavus* isolated in the early 1960s in England from toxic peanuts from Africa.

### *Conditions That Favor Growth of A. flavus and Production of Aflatoxin*

The ecology of aflatoxin production by *A. flavus* has been worked out rather thoroughly (33, 38, 129). The minimum moisture content for growth of *A. flavus* is that in equilibrium with a relative humidity of 85%. In the starchy cereal grains such as maize, sorghum, wheat, oats, barley, and rice, the moisture content is 18.0–18.5% wet weight basis; in peanut and sunflower seeds with a high oil content, the moisture content is 9–10%. The minimum, optimum, and maximum temperatures for aflatoxin production are 12, 27, and 42°C. Under optimum conditions for aflatoxin production in pure culture in the laboratory, some aflatoxin will be produced within 24 hr, and a maximum amount will be reached in 10 days or less, after which the amount present may decrease sharply. Later it may again increase.

The influence of mixed cultures, common in grains and seeds undergoing deterioration in nature, upon aflatoxin formation has been studied very little; according to available evidence (21), when a mixture of fungi is present, little or no aflatoxin is produced even when the moist grain is inoculated with a known aflatoxin-producing strain of *A. flavus* and held under conditions favorable for its growth. This aspect of aflatoxin production needs much more work; it is important that we know thoroughly the conditions under which aflatoxin is *not* formed.

There is abundant evidence, summarized by Christensen & Kaufmann (20), that seeds of wheat, barley, oats, rice, maize, sorghum or milo, soybeans, peas, and common beans are not invaded to any serious extent by storage fungi, including *A. flavus,* before harvest. Taubenhaus (150) in 1920 reported invasion of maize ears by *A. flavus* in the field in Texas. The ears had been damaged by earworms. A similar case of earworm damage, followed by invasion of the ears by *A. flavus* and by production of aflatoxin, occurred in southeastern Missouri in 1972. In some areas of the southern USA and in some warm and humid areas of other countries where maize is a major food crop, the hazard of aflatoxin formation before harvest may be much higher than it is in the Corn Belt of the US.

Peanut fruits and seeds also are rarely invaded by *A. flavus* before harvest (37, 53), but they may be invaded primarily or solely after the vines have been pulled and while the fruits still are attached. This is the "high-hazard" time for aflatoxin formation in peanuts.

### *High-Aflatoxin-Risk Materials*

Some materials are much more likely than others to contain appreciable amounts of aflatoxin. In part this may be due to the inherent nature of the material—some substrates are much more favorable than others for aflatoxin production, even though the fungus may grow equally well on both groups—and in part it may be due to a more favorable environment. A combination of a favorable substrate in a favorable environment naturally makes for a far higher aflatoxin risk than does an unfavorable substrate in an unfavorable environment. Among the high aflatoxin-risk materials are peanuts, brazil nuts, probably other kinds of nuts grown in warm and humid regions, cottonseeds, copra, and fishmeal.

Krogh & Hald (82) found aflatoxin in 86.5% of 52 samples of peanuts and peanut products (whole nuts, meal, and cake) imported into Denmark from ten countries for feed. One sample contained 3465 μg of aflatoxin per kg (= 3465 ppb). As a result of these findings, the Danish Ministry of Agriculture established a limit of 100 ppb of aflatoxin in imported peanut products. This is five times as high as the tolerance established by the FDA in the United States. The peanuts make up only a portion of the ration of the animals to which they are fed, probably seldom more than 25%. Assuming that the other ingredients in the ration were free of aflatoxin, this would mean that the feed consumed by the animals would contain, at most, 25 ppb of aflatoxin. Feeding tests (referred to below) have shown no injury to swine or cattle from rations containing ten times this amount of aflatoxin.

The drying, handling, storage, and inspection procedures developed for peanuts in the United States since 1965 have almost completely eliminated the possibility of peanuts with detectable amounts of aflatoxin getting into food channels here. In years of wet harvest, a considerable portion (sometimes over 50%) of the peanut harvest in a given marketing area may be judged unsuitable for food because of high invasion by *A. flavus* or high aflatoxin content. In some tropical and subtropical areas peanuts are a regular part of the diet and probably are an important source of aflatoxin as well as of protein; this most likely accounts for the fact that primary hepatocarcinomas in the human population in some tropical countries are 100 times as high as in some northern European countries.

Seeds of cereal crops and of soybeans are, in general, of low aflatoxin risk. Shotwell et al (136) stated, "Very low levels of what appeared to be aflatoxin (2 to 19 ppb) were detected by TLC in a total of 9 out of 1368 assayed samples of wheat, grain sorghum, and oats." None of these were confirmed by duckling tests. They also tested 1311 samples of corn and 866 samples of soybeans from federal grain inspection offices and found very low levels of aflatoxin $B_1$ in 30 of them, all in the poorer grades (grade 5 and sample grade).

### *High-Aflatoxin-Risk Areas*

A combination of prevalent aflatoxin-producing strains of this fungus, plus suitable substrates for aflatoxin production, plus a favorable environment (including, at times, food handling methods that favor aflatoxin production) naturally make for high aflatoxin risk. Some of the areas so identified include portions of Africa, India, and Southeast Asia, as illustrated by the following: (*a*) Aflatoxin was found in 41 of 74 samples of pulses collected in the Sudan region of Africa (57). (*b*) Of 124 samples of cottonseed from humid areas of India, 78% contained aflatoxin, and of these, 48% contained more than 500 ppb of aflatoxin (166). (*c*) Aflatoxin was found in 31% of 139 samples of rice in South Vietnam (89). (*d*) About half the samples of coconuts and coconut products tested in Ceylon had medium or high levels of aflatoxin ("medium" = 50 to 250 ppb; "high" = 250 to 1000 ppb) (3). (*e*) Aflatoxin was found in one or more specimens from 22 of 23 autopsies of children who died of acute encephalopathy and fatty degeneration of the viscera (EFDV) in Thailand (135). In some cases where they were able to identify the food contaminated with aflatoxin, it was rice that had been cooked, a portion of it eaten the same day, and the rest held for consumption on several following days. Where no refrigeration is available, this practice, in the warm and humid tropics, carries with it a very high risk of aflatoxin contamination.

### *Effects of Aflatoxins on Animals*

Animals differ greatly in sensitivity to injury by aflatoxin (178, 181, 182). Allcroft (2) lists the order of susceptibility of farm animals, from most to least susceptible, as follows: poultry: duckling > turkey poult and pheasant chick > chickens and Coturnix quail. Different breeds of chickens differ greatly in susceptibility. In mammals the order is as follows: pigs 3 to 12 weeks old, and pregnant sows > calves > fattening pigs > mature cattle > sheep. The $LD_{50}$ of a single dose for various animals is given in Table 1.

Keyl et al (76) fed graded amounts of aflatoxin to swine for 120 days and to Hereford beef steers for 4½ months. No toxic effects were detected in swine given 233 ppb or less, or in beef cattle given 300 ppb or less of aflatoxin in the ration. A small proportion of the aflatoxin consumed by dairy cattle is excreted in the milk in a slightly changed but still toxic form called $M_1$. Brewington et al (11) tested many hundreds of samples of fluid milk from different marketing areas in the United States and from farms, and many samples of dried milk, and found no aflatoxin in any of them.

### *Aflatoxin Derivatives*[1]

The most commonly encountered aflatoxins are $B_1$, $G_1$, $B_2$, and $G_2$. These toxins may occur together, although not necessarily, and their concentration and occurrence may vary. For example, Hesseltine et al (61) found that $B_1$ need not occur with $G_1$ but not vice versa and that some isolates produce all four principal aflatoxins, but the ratio of one to the other may vary. An unusual isolate of *A. flavus* from black pepper produced only $B_2$ (131) but generally $B_1$ is more frequently encoun-

**Table 1** Single $LD_{50}$ values of aflatoxin for various animal species[a]

| Species | $LD_{50}$ mg/kg body weight | Species | $LD_{50}$ mg/kg body weight |
|---|---|---|---|
| Rabbit | 0.3–0.5 | Guinea pig | 1.4–2.0 |
| Duckling | 0.34–0.56 | Sheep | 2.0 |
| Cat | 0.55 | Monkey | 2.2 |
| Pig | 0.62 | Chick | 6.5–16.5 |
| Rainbow trout | 0.81 | Mouse | 9.0 |
| Dog | 1.0 | Hamster | 10.2 |
| | | Rat | 5.5–17.9 |

[a] Toxin administered orally for all species except trout (110).

[1] See Figure 1 for abbreviations used in text.

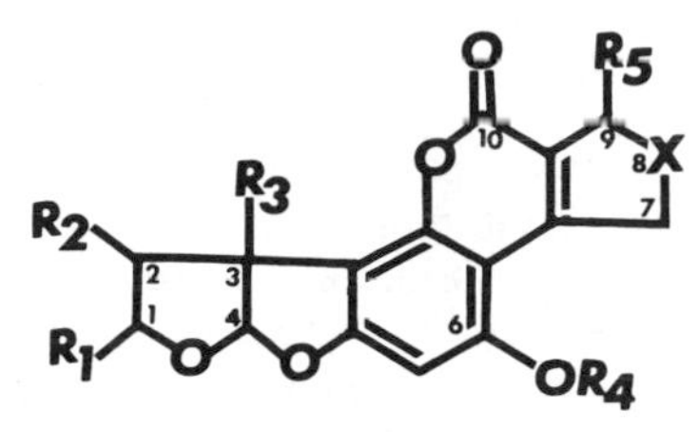


| Derivative | $R_1$ | $R_2$ | $R_3$ | $R_4$ | $R_5$ | X |
|---|---|---|---|---|---|---|
| $B_1$ | H | H | H | $CH_3$ | O | $H_2$ |
| $G_1$ | H | H | H | $CH_3$ | O | $-OCH_2$(lactone) |
| $B_2$ | $H_2$ | $H_2$ | H | $CH_3$ | O | $H_2$ |
| $G_2$ | $H_2$ | $H_2$ | H | $CH_3$ | O | $-OCH_2$(lactone) |
| $B_{2a}$ | OH | $H_2$ | H | $CH_3$ | O | $H_2$ |
| $G_{2a}$ | OH | $H_2$ | H | $CH_3$ | O | $-OCH_2$(lactone) |
| 1-methoxy-$B_1$(2 isomers) | $O-CH_3$ | $H_2$ | H | $CH_3$ | O | $H_2$ |
| 2-methoxy-$B_1$ | $H_2$ | $O-CH_3$ | H | $CH_3$ | O | $H_2$ |
| 2-ethoxy-$B_1$ | $H_2$ | $O-CH_2CH_3$ | H | $CH_3$ | O | $H_2$ |
| 2-ethoxy $G_1$ | $H_2$ | $O-CH_2CH_3$ | H | $CH_3$ | O | $-OCH_2$(lactone) |
| Aflatoxicol ($R_o$) | H | H | H | $CH_3$ | OH | $H_2$ |
| $P_1$ | H | H | H | H or glucuronide | O | $H_2$ |
| $M_1$ | H | H | OH | $CH_3$ | O | $H_2$ |
| $M_2$ | $H_2$ | $H_2$ | OH | $CH_3$ | O | $H_2$ |
| $GM_1$ | H | H | OH | $CH_3$ | O | $-OCH_2$(lactone) |
| Dihydroaflatoxicol | $H_2$ | $H_2$ | H | $CH_3$ | OH | $H_2$ |

| | $R_1$ | $R_2$ | $R_3$ |
|---|---|---|---|
| O methyl sterigmatocystin | H | H | $CH_3$ |
| Aspertoxin | OH | H | $CH_3$ |
| 5-methoxysterigmatocystin | H | $O-CH_3$ | H |
| 5-hydroxysterigmatocystin | H | OH | H |
| Sterigmatocystin | H | H | H |

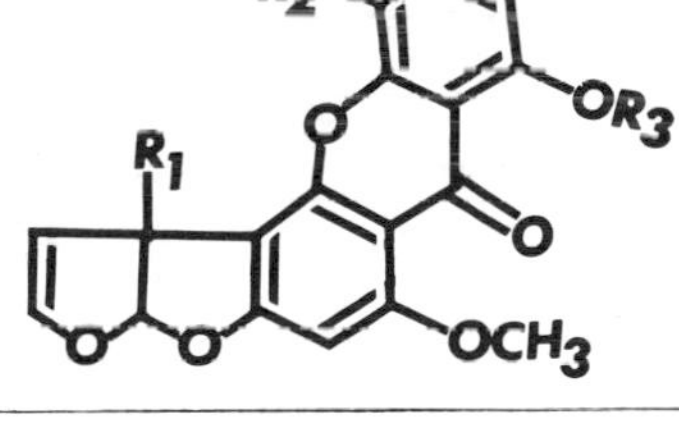


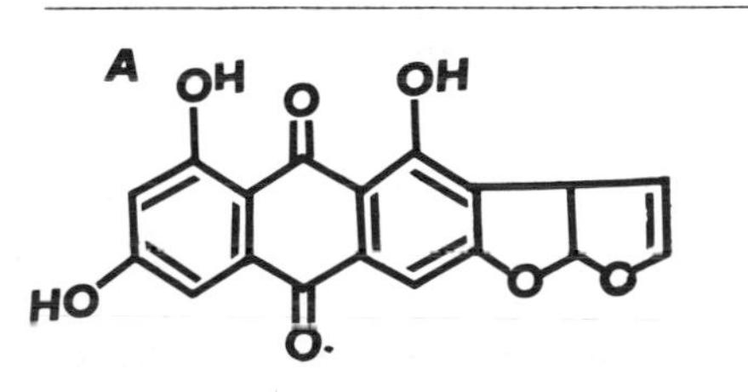


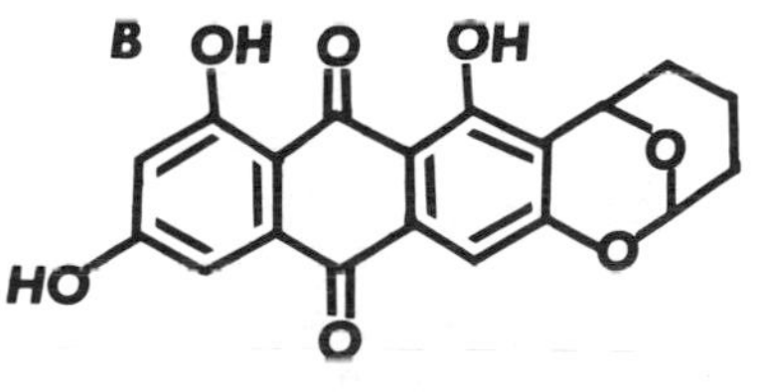


*Figure 1* Various derivatives of aflatoxin, sterigmatocystin, and (A)Versicolorin A and (B) Averufin.

tered. $M_1$ is a product of metabolism of the animal and usually occurs in milk and urine, but is also produced in cultures of *A. parasiticus* and *A. flavus* (61). Its activity is almost equal to that of $B_1$; it occurs in milk of lactating humans and animals who have ingested $B_1$ and is carcinogenic. According to Wogan (177), the incidence of $B_1$ in meat is rare, whereas the incidence of $M_1$ in milk is not. $M_2$ is similar to $M_1$ and $M_2$ varies only with unsaturation in the terminal furane ring. The chemical structures of $M_1$ and $M_2$ were determined by Holzapfel et al (66), Masri et al (93), and Buchi & Weinreb (15). The intensity of fluorescence of $M_1$ and $M_2$ is approximately 3 times that of $B_1$ (66).

Two other closely related derivatives are $B_{2a}$ and $G_{2a}$ which have a hydroxyl group in the 2 position of the furane ring (42, 43). Hydroxylation in this position is important because such derivatives are virtually nontoxic and might be a detoxification mechanism of the host. These two derivatives have been found in cultures of *A. flavus.* $B_{2a}$ is 1/200th as effective as $B_1$ in initiating bile duct hyperplasia in the duckling (86). Further, Pons et al (114) have carried out a study describing the acid catalyzed conversion of $B_1$ and $G_1$ to $B_{2a}$ and $G_{2a}$ as a means of detoxifying soapstocks used in the recovery of fatty acids.

Irradiation of $B_1$ in methanol results in the production of two isomers of 1-methoxy aflatoxin $B_1$ (169). Dutton & Heathcote (41) described three of four derivatives of $B_1$ that may arise during the isolation of aflatoxin. The compounds were the 2-methoxy-3-hydro addition product of $B_1$, the 2-ethoxy-3-hydro addition compound of $B_1$, and the 2-ethoxy-3-hydro addition compound of $G_1$. The fourth compound was unrelated to the aflatoxins and appeared to be a long chain dihydric alcohol. This study explains the presence of artifacts in extracts of *A. flavus* that could be attributed to the method of isolation. As an example, the methoxy or ethoxy derivatives occur as a result of methanol or ethanol in the medium under slightly acid conditions. The ethyl alcohol present in chloroform as a stabilizer or that due to fermentation is sufficient to explain the presence of these addition compounds. Simple purification of the chloroform (free of ethyl alcohol) was sufficient to prevent formation of these derivatives under acid conditions. Hence, O-alkyl derivatives of $B_{2a}$ and $G_{2a}$ can arise from extraction of the culture medium with impure chloroform. To overcome these problems the material should be neutralized prior to extraction and extracted with freshly prepared chloroform free from ethyl alcohol. The 2-methoxy-3-hydro addition compound of $B_1$ is identical to that reported by Waiss & Wiley (169).

Detroy & Hesseltine (34) reported reduction, by various molds, of the keto group on the terminal cyclopentene ring of $B_1$ to a hydroxyl group. They called this derivative $R_0$ but suggest "aflatoxicol" as a more appropriate name.

Another toxic metabolite produced by *A. parasiticus,* which may be a precursor of $B_1$, is called parasiticol (147). It resembles $B_1$ except for an ethanol group in place of the terminal cyclopentene group. Parasiticol was also found in a culture of *A. flavus* and was called $B_3$ (58).

Pong & Wogan (112, 113) compared the toxicity of synthetic $M_1$ and $B_1$ (racemic) versus the naturally occurring $M_1$ and $B_1$ (nonracemic). The latter were lethal to rats at a dose of 0.6 mg/kg and higher, while the former racemic aflatoxins were lethal at 1.5 mg/kg. Both synthetic and naturally occurring aflatoxins suppressed precursor incorporation into liver RNA and caused similar fine structure changes in liver cells, but again a higher dose of the synthetic toxins was necessary to incite a response equal to that of the natural toxins. This study is important because it indicates that only one isomer of each racemic mixture of $B_1$ and $M_1$ is biologically active.

The various derivatives described vary in their toxicity as one might expect. A study of the structure-activity relationships indicates that the furofuran ring of aflatoxin is essential for toxic and carcinogenic activity (28, 180). The double bond in the terminal furan ring determines potency, particularly for acute and chronic effects in rats. The rat liver can detoxify $B_1$ by opening the furane ring and forming a branched aliphatic side chain with terminal aldehydic groups (1). The lactone portion of the molecule is also important as illustrated by greater activity of $B_1$ than $G_1$. Reduction of the ketone of the terminal cyclopentene ring of $B_1$ to form aflatoxicol also causes a loss in potency (34). Reduction of $B_1$ and $G_1$ to $B_{2a}$ and $G_{2a}$ also results in a diminution of toxicity (86).

## *Biosynthesis of Aflatoxin*

An excellent detailed account of the synthesis of aflatoxin is presented by Detroy et al (35) and is recommended reading.

Biollaz et al (8, 9), using $^{14}C$-labeled acetate and *A. parasiticus*, reported on the distribution of $^{14}C$-label in $B_1$ based on chemical degradation. The labeling pattern showed that the aflatoxin molecule is totally derived from acetate and that methionine supplies the methoxymethyl group.

In the study of possible precursors of $B_1$, Hsieh et al (71) have shown the conversion of $^{14}C$-sterigmatocystin into $B_1$. The $^{14}C$-labeled sterigmatocystin was isolated from cultures of *Aspergillus versicolor* supplemented with 1 $^{14}C$ acetate and then converted to $B_1$ by resting mycelium of *A. parasiticus*. This pathway is supported by earlier work (64, 149) which has shown the similarity in label distribution in sterigmatocystin and $B_1$ when 1-$^{14}C$-acetate or 1-$^{13}C$-acetate is used as precursor. Further, O-methyl sterigmatocystin (16) and aspertoxin (hydroxylated derivative of O-methyl sterigmatocystin) (127) have been found in aflatoxin-producing strains of *A. flavus*. Moreover, $^{14}C$-averufin (Figure 1), a quinone fungal pigment produced by *A. parasiticus*, is also incorporated into $B_1$ and hence is postulated as an intermediate in the biosynthesis of $B_1$ from acetate (88).

As a hypothesis for synthesis of $B_1$, Hsieh (69) proposes that averufin, by oxidation and subsequent recyclization, results in the formation of versicolorin A (Figure 1), which is then converted to 5-hydroxysterigmatocystin and then to sterigmatocystin (63, 64). The terminal ring of sterigmatocystin is oxidized next, resulting in cleavage and subsequent cyclization to form a cyclopentenone ring. After cyclization, the terminal carboxyl group is eliminated as $CO_2$, resulting in $B_1$ (9, 88).

In the production of aflatoxin on solid, natural substrate, the insecticide dichlorvos (dimethyl-2,2-dichlorovinyl phosphate) at 20 ppm inhibited the biosynthesis of aflatoxin in wheat, corn, rice, and peanuts inoculated with *A. flavus* (56). Hsieh (70) found that dichlorvos at 10 ppm inhibited the incorporation of 1-$^{14}C$-acetate into

$B_1$ synthesized by *A. flavus.* The inhibitory step occurred early in the acetate pathway of metabolism. In both studies, there was no apparent inhibition of mycelial growth.

There is considerable practical importance in limiting aflatoxin synthesis by chemicals or varietal resistance. Nagarajan & Bhat (101) found a considerable difference in production of aflatoxin on seven hybrid varieties of maize. Opaque-2 as substrate resulted in low (0.05–0.25 ppm) aflatoxin production apparently due to a low molecular weight protein, present in Opaque-2, which acted as an inhibitor to biosynthesis but not to mycelial growth. Similar varietal differences in aflatoxin production have been reported in peanuts (40, 123).

### *Sterigmatocystin*

The toxin sterigmatocystin (ST) is produced primarily by *A. versicolor* (130), but may also be produced by some strains of the *A. flavus* group, and by *Penicillium luteum* (65). Administered orally, it has an $LD_{50}$ of 166 mg/kg in male rats and 120 mg/kg in female rats (119), and in excess of 800 mg/kg in mice (87). Purchase & Pretorius (117) found 1143 ppb of ST in a sample of coffee beans and considered it to be a potentially dangerous food contaminant. Scott et al (134) reported 0.3 ppm of ST in a sample of moldy wheat, but otherwise it has rarely been found in foodstuffs. Kurata & Tanabe (84) tested 494 samples of nonmoldy and moldy foodstuffs in Japan and found no ST; similar results have been reported from other parts of the world, and it seems probable that this toxin does not pose much of a health problem for man or for domestic animals.

ST, O-methyl-ST, aspertoxin, 5-methoxy-ST (Figure 1), and demethyl-ST occur naturally. Of these, ST (121, 122) and demethyl-ST [Engelbrecht & Altenkirk (45), quoting Dr. J. S. E. Holker, University of Liverpool] have been shown to be carcinogenic. Engelbrecht & Altenkirk (45) found that, similar to the aflatoxins (Wogan et al, 180), the unsaturated $^{1,2}$furobenzofuran ring of ST and its derivatives were more toxic than the unsaturated compounds.

## TOXINS PRODUCED BY FUSARIUM

### *Zearalenone*

SIGNS OF THE DISEASE Zearalenone or F-2 is a natural metabolite produced by *Fusarium roseum, F. tricinctum, F. oxysporum,* and *F. moniliforme;* consumed by swine, it can result in the estrogenic syndrome. [See reviews by Mirocha et al (95, 96).]

The estrogenic syndrome in swine involves primarily the genital system: in the prepuberal gilt, the vulva becomes swollen and edematous, and in severe cases it may progress to rectal and vaginal prolapse; the uterus is enlarged, edematous, and tortuous; there is a general atrophy of the ovaries; and pregnant sows may abort. The abortion part of this syndrome is not due to zearalenone, but perhaps to some other metabolic product of *Fusarium.* Infertility and reduced litter size as well as weak piglets are part of the syndrome, and results in production losses. The young males may undergo a feminizing effect with atrophy of the testes and enlargement of the mammary glands. Besides the estrogenic signs, other signs associated with ingestion of Fusarium-infected corn are diarrhea, emesis, refusal of feed, loss of weight gain, and hemorrhagia. Not all of these signs are caused by zearalenone, but rather by other unidentified toxins present in the feedstuff and presumably produced by *Fusarium.*

*Fusarium roseum* may invade corn in the field or after it is stored in cribs. A period of low temperature, or of alternating moderate and low temperature, is necessary for the production of the toxin. Other animals experimentally affected by zearalenone are rats, mice, guinea pigs, rabbits, turkey poults, chicks, and lambs. In nature, it is most commonly reported to affect swine.

CHEMISTRY AND NATURALLY OCCURRING DERIVATIVES[2] The chemical structure of zearalenone (I) is 6-(10-hydroxy-6-oxo-*trans*-1-undecenyl) $\beta$-resorcylic acid lactone (163). When the ketone in the 6' position is reduced to a hydroxyl group and the double bonds at carbons 1' and 2' are reduced, two diasterioisomers called zearalanol (II) result. The isomer with the higher melting point (178–180°C) is about four times more active than the parent molecule (zearalenone), and the isomer with the lower melting point (146–148°C) is only slightly more active. The latter thus far have not been found in nature. Zearalanol (178–180°C) is used as an anabolic agent (RALGRO®) in promoting growth in beef cattle and also as a chemotherapeutant for the alleviation of postmenopausal discomfort. Mirocha et al (96) reported at least seven other naturally occurring derivatives of zearalenone, the better known being F-5-3 and F-5-4, which are sterioisomers differing in melting point and identified as 8'-hydroxyzearalenone (III; 73). The latter (III) were also described by Bolliger & Tamm (10) as well as 7'-dehydrozearalenone (IV) and 5-formylzearalenone (V). Steele (142) found 6',8'-dihydroxyzearalenone in cultures of *F. roseum* growing on shredded wheat. The structure of the other derivatives described by Mirocha et al (96) are not known as yet.

BIOSYNTHESIS The enzymes responsible for biosynthesis of zearalenone are either induced or activated at low temperatures (12–14°C), a temperature not optimum for growth, but rather one of physiological stress. Although trace amounts of zearalenone are produced at room temperature (22–27°C), copious amounts (10,000–30,000 ppm) are produced on corn or rice substrate at low temperatures.

The metabolic pathway of biosynthesis of zearalenone by *F. roseum* was studied by Steele et al (143). Acetate-1-$^{14}$C and diethylmalonate-2-$^{14}$C were rapidly incorporated. Chemical degradation of the $^{14}$C-labeled zearalenone yielded $CO_2$, oxalic acid, succinic acid, glutaric acid, and the aromatic ring. The relative molar activity of the degradation fragments agreed with the postulate of synthesis via the malonyl-Co-A-acetate pathway.

The biosynthesis of zearalenone can be inhibited by Vapona (dimethyl-2,2-dichlorovinyl phosphate) with no apparent effect on mycelial growth (175). Treat-

[2]See Figure 2 for chemical structures.

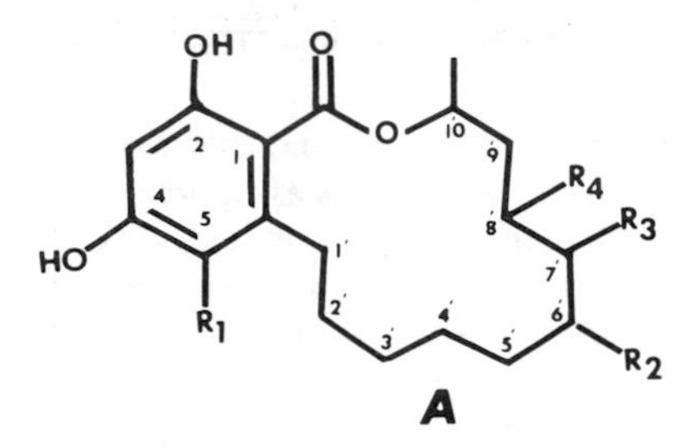


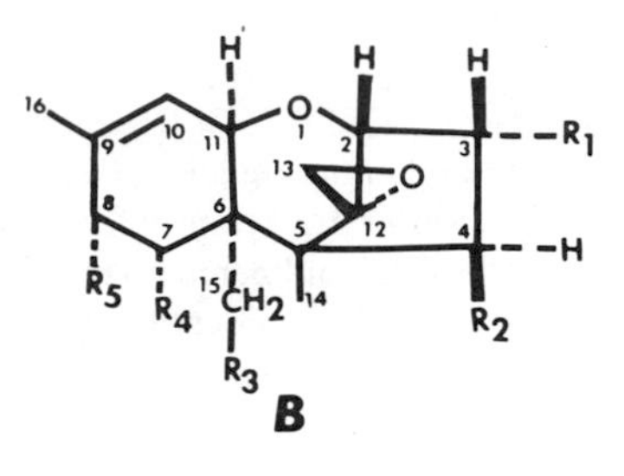


ZEARALENONES

| No. | Name | $R_1$ | $R_2$ | $R_3$ | $R_4$ | $C_{1'}$ and $C_{2'}$ |
|---|---|---|---|---|---|---|
| I | zearalenone | H | =O | $H_2$ | $H_2$ | CH=CH (trans) |
| II | zearalanol | H | OH | $H_2$ | $H_2$ | $CH_2$-$CH_2$ |
| III | 8'-hydroxyzearalenone | H | =O | $H_2$ | OH | CH=CH (trans) |
| IV | 7'-dehydrozearalenone | H | =O | H | H | CH=CH (trans) |
| V | 5-formylzearalenone | HC=O | =O | $H_2$ | $H_2$ | CH=CH (trans) |
| VI | 6',8'-dihydroxyzearalenol | H | OH | $H_2$ | OH | CH=CH (trans) |

TRICHOTHECENES

| No. | Name | $R_1$ | $R_2$ | $R_3$ | $R_4$ | $R_5$ |
|---|---|---|---|---|---|---|
| VII | T-2 toxin | OH | OAc | OAc | H | isovaleroxy |
| VIII | HT-2 toxin | OH | OH | OAc | H | isovaleroxy |
| IX | Neosolaniol | OH | OAc | OAc | H | OH |
| X | T-2 tetraol | OH | OH | OH | H | OH |
| XI | Diacetoxyscirpenol | OH | OAc | OAc | H | H |
| XII | Fusarenone-X | OH | OAc | OH | OH | =O |
| XIII | Nivalenol | OH | OH | OH | OH | =O |
| XIV | Monoacetoxyscirpenol | OH | OH | OAc | H | H |
| XV | Trichothecin | H | isocrotonyloxy | H | H | =O |
| XVI | Verrucarol | H | OH | OH | H | H |
| XVII | Dihydronivalenone(vomitoxin) | OH | H | OH | OH | =O |
| XVIII | Roridin E | C-15-O-C(=O)-CH=CMe-$CH_2$-$CH_2$-O-CH(MeCHOH)-CH=CH-CH=CH-C(=O)-O-C-4 | | | | |

*Figure 2* Derivatives of (A) zearalenone and (B) trichothecene toxins.

ment of stored grain with Vapona has potential for controlling production of zearalenone. Vapona also inhibits synthesis of aflatoxin as described earlier.

SEX HORMONE ACTIVITY IN F. ROSEUM Although zearalenone was first studied because of its estrogenic activity in animals, it is also a sex-regulating hormone in *F. roseum* (176). When amounts of zearalenone from 0.1–10.0 ng are applied to a 1.0 cm diameter disc of Coons medium in agar on which *F. roseum* is growing, perithecial formation is enhanced by as much as 100%. Amounts in excess of 10.0 μg inhibit its growth. There is also periodicity (accessibility to regulating mechanism) in its action insofar that greatest enhancement occurs when applied 1–4 days of culture age and greatest inhibition (with inhibitory concentrations) at 3–4 days of culture age. Zearalenone regulates the sexual stage of various other fungi (103), but the latter do not synthesize zearalenone. The sex-regulating activity of various derivatives of zearalenone in *F. roseum* closely parallels that in female mice and rats (176).

NATURAL OCCURRENCE AND SIGNIFICANCE Although a simple and rapid method of analysis for zearalenone has not yet been devised, the incidence of zearalenone in nature has been found to be high. Of 65 samples ranging from corn-on-the-cob to commercial feed preparations, 45% of the samples contained zearalenone in concentrations ranging from 0.1 to 2909 ppm based on the dry weight of the feedstuff (95, 99). In a survey conducted by the FDA in 1973, 17% of 223 samples of corn contained 0.1–5.0 ppm zearalenone (48). The samples were from food processing plants in the midwest USA and were intended to be used for food, feed, starch, or export. Concentrations of zearalenone between 1 and 5 ppm consumed by swine will cause vulvovaginitis.

### *Trichothecenes*

The 12,13-epoxytrichothecenes (Figure 2) are a group of chemically related and biologically active secondary metabolites produced by *Fusarium tricinctum, F. roseum, F. oxysporum, F. solani, F. nivale, F. lateritium, F. rigidiusculum,* and *F. episphaeria* [named according to the system of Snyder & Hanson (151)]. These various trichothecene derivatives are also produced by species of *Cephalosporium, Myrothecium, Trichoderma,* and *Stachybotrys* (140). Presently, there are about 27 naturally occurring derivatives, six of which are thought to be somehow involved in Fusarium toxicosis. Comprehensive reviews on this subject have been written by Bamburg & Strong (6) and Smalley & Strong (140).

BIOLOGICAL ACTIVITY The trichothecenes, when applied to the shaved skin of the rat or rabbit, produce a strong dermatitic reaction; this is a crude but sensitive bioassay for metabolites having trichothecene-like activity (172). The reaction is characterized by severe local irritation, inflammation, desquamation, and subepidermal hemorrhaging and general necrosis. When given orally or by intraperitoneal injection (ip), the trichothecenes are acutely toxic at low concentrations. The T-2 toxin in rats, for example, has an $LD_{50}$ of 3.04 mg/kg body weight when administered by ip injection (185) and 3.8 when administered orally (78). Bamburg & Strong (6) list the toxicity of the various derivatives to different animals and plant species.

When the trichothecenes are administered orally, the usual method of ingestion by animals, the test animals become listless or inactive and develop diarrhea and rectal hemorrhaging. Necrotic lesions may develop in the mouth parts. The mucosal epithelium of the stomach and small intestines erodes, accompanied by a hemorrhage which may develop into a severe case of gastroenteritis, followed by death. The cells of the bone marrow, lymph nodes, and intestines undergo a pathological degeneration. In larger animals, massive hemorrhages develop in the lumen of the small intestines (140).

Small amounts of T-2 toxin (1–16 $\mu$g/g) in a diet fed to chickens cause abnormal positioning of the wings, hysteroid seizures, and an impaired righting reflex (183).

Various trichothecenes affect protein and DNA synthesis in rabbit reticulocytes, Ehrlich ascite tumor cells, rat liver, and bacteria (155). The activity is not restricted to rabbit reticulocytes, but was also demonstrated in guinea pigs and hens. Although both DNA and protein synthesis were depressed in tumor cells, there was no noticeable effect on RNA synthesis. Trichothecenes studied were more potent protein inhibitors than the antibiotics puromycin and cycloheximide (155).

Besides inhibition of protein synthesis, the trichothecenes also affected cell membranes, as demonstrated by the reversible alteration of polyribosomes by Fusarenon-X (XII). Trichothecenes such as T-2 toxin, HT-2 toxin, diacetoxyscirpenol, and neosolaniol, which have substituents other than a ketone at the C-8 position, may have a higher affinity for bonding and altering membranes than the C-8 ketone type (155).

Fusarenon-X, T-2 toxin, and related trichothecenes caused in mice, rats, and cats a severe cellular destruction and karyorrhexis in the actively dividing cells of the bone marrow, in mucosal epithelium of the small intestine, and in testes (154–156). The activity on bone marrow cells is particularly important because it is one of the signs of alimentary toxic aleukia (ATA) described by Soviet scientists (7, 74, 104).

Although the trichothecenes were not found to be carcinogenic (90), they do inhibit DNA and protein synthesis as do the aflatoxins, but by a different mechanism.

Moldy corn toxicosis as described by Hsu et al (72), Smalley et al (139), and Marasas et al (91) in northern climates is associated with many different fungi, predominantly *F. tricinctum.* The signs of this disease are similar to those described by Forgacs (51), namely general digestive disorders, bloody diarrhea, and hemorrhagic lesions in the stomach, heart, intestines, lungs, bladder, and kidneys. Hsu et al (72) provided the first cause and effect relationship between the T-2 toxin found in the feedstuff and a case of moldy corn poisoning, in which 7 of 35 lactating Holstein cows died over a period of 5 months. The T-2 concentration found was 2 ppm.

Alimentary toxic aleukia (ATA) in man was described by Olifson (7, 104) and Joffe (74) in the USSR. This disease was associated with moldy millet primarily, but also with wheat, rye, oats, and buckwheat. The signs described were "typical spots on the skin, leukopenia, agranulocytosis, necrotic angina, hemorrhagic diathesis, sepsis and exhaustion of the bone marrow" (74). Of the fungi involved, *F. sporotrichiodes* (syn. *F. tricinctum*) and *F. poae* were thought to be involved, and the toxins were described as poaefusarin and sporofusarin (Figure 3, see page 320).

The symptoms and signs of this disease resembled toxicity of the trichothecenes as reported in the USA. An authentic sample of poaefusarin was obtained and analyzed for the $C_{24}H_{28}O_5$ steroid, but no trace of a steroid could be found. However, 2.5% of the sample was made up of the T-2 toxin, an amount sufficient to explain the toxicity found in the rat and rabbit skin tests. In addition, neosolaniol (0.14%), T-2 tetraol (0.6%), and zearalenone (0.43%) were present as constituents of the sample (98).

CHEMISTRY AND NATURAL OCCURRENCE Bamburg & Strong (6) and Smalley & Strong (140) summarized the information on the chemistry, derivatives, and physical properties of the trichothecenes. All the naturally occurring trichothecenes have a basic tetracyclic sequiterpene structure as shown in Figure 2. The verrucarins and roridins, exemplified by roridin E (Figure 2), are classified as a special subgroup of the trichothecenes because they have a macrocyclic ring linking carbons 4 and 15 with either an ester or ester-ether bridge. This large cyclic carbon bridge makes isolation difficult, and hydrolysis of the extract is usually necessary to detect these compounds.

The 12,13-epoxide is characteristic of the trichothecenes, and its removal results in loss of toxicity to animals (6). Grove & Mortimer (55) found similar results when they removed the epoxy oxygen from diacetoxyscirpenol. Hydrolysis of the naturally occurring derivatives to their parent alcohols also results in a diminution of toxicity.

A new naturally occurring trichothecene is monoacetoxyscirpenol (XIV). It is produced by an isolate of *F. roseum* and was the sole toxic constituent present in cultures of this fungus that accounted for the toxicity noted in rat and turkey poult tests (108). This derivative is similar in structure and toxicity to diacetoxyscirpenol (XI), originally discovered by Brian et al (12).

Of 173 samples of corn tested by the FDA (48), 94 (54%) contained a "skin-irritating factor" identified as the T-2 toxin in concentrations of 0.05 to 1.0 $\mu g/g$ of corn.

## *Toxins Responsible for Refusal of Corn by Swine and Emesis*

During certain years in the Midwest USA, corn infected with *F. roseum (Gibberella zeae)* as well as other species of *Fusarium* can cause multiple problems when fed to swine: (*a*) hyperestrogenism, (*b*) emesis, and (*c*) refusal of feed. Such corn is usually associated with wet weather that precedes and delays harvest, as in 1965 and 1972. Three separate toxins appear to be involved, which may occur singly or together. Hyperestrogenism has already been described and is caused by zearalenone; emesis is caused by low concentrations of T-2 toxin (44) or Fusarenone-X or neosolaniol (157), or vomitoxin (167), and refusal of corn is caused by a metabolite as yet unidentified. Curtin & Tuite (31) described the condition in swine and list pertinent references of the early work.

Emesis in swine is now thought to be caused by various trichothecenes in the moldy corn. Ueno et al (157) found that Fusarenone-X, T-2, and neosolaniol cause emesis in cats and ducklings and that emesis is a common response to the trichothecenes. This symptom is associated with scabby grains in Japan (154) and the USA (31, 68). Ellison et al (44) report that the T-2 toxin is responsible for much of the emesis noted with moldy corn, but did not confirm this in swine, while Vesonder et al (167) claim vomitoxin (XVII) to be the causal agent and did test it on swine. Vesonder et al (167) further speculate that vomitoxin may be the factor responsible for refusal of corn by swine.

The refusal factor is so named because corn containing it or rations containing such corn are refused by swine. Five percent infected kernels in the ration is sufficient to cause refusal. Both metabolites (emetic and refusal) are produced in the field. The only reliable method for detecting the presence of the "refusal factor" is to offer the suspected corn to swine. Results from our laboratory as well as from that of Curtin & Tuite (31) indicate that the metabolite is water soluble and can be separated by column chromatography using Sephadex G-10 and on TLC with Silica Gel-G. Once the toxin is formed, it is persistent and cannot be disguised with feed additives, nor does it disappear with age. Corn refused by swine may be accepted by cattle and poultry.

## TOXINS PRODUCED BY STACHYBOTRYS

### Introduction and Biological Signs

*Stachybotrys* is a saprophyte frequently encountered colonizing straw or other cellulose products used as bedding for animals, although Ashworth & George (4) reported *S. atra* as a pathogen causing a root disease of cotton.

Stachybotryotoxicosis was first reported in the USSR as a mycotoxicosis affecting horses after ingestion of fodder colonized by *S. atra* as well as by other fungi. Forgacs (50, 51) has an excellent review of the early Russian literature and complete description of the clinical signs in animals. The disease also affects man.

In horses, Forgacs (50) describes two forms of the disease: the typical form due to chronic exposure to toxin and the atypical or acute form. In both, the pathological findings are characterized by a profuse hemorrhage and necrosis in many tissues including the stomach, small and large intestines, liver, kidney, and heart. Other domestic animals affected are sheep, cattle, swine, poultry, and dogs. Aerosols formed after combustion of contaminated hay have also been reported to affect man. Ozegovic et al (105) in Yugoslavia described a toxic dermatitis, conjunctivitis, rhinitis, pharyngitis, and laryngitis in cattle and farm workers handling straw colonized by *S. atra.* The authors interpreted these symptoms as due to a direct toxic action rather than to allergic reaction. Toxicity was noted on the scrotum and mucous membranes of the nose, eye, and throat. Szabo & Szeky (148) described similar signs in swine held on straw colonized by *S. atra.* Palyusik et al (106, 107) reported that the stachybotryotoxins cause pathological signs in chickens almost indistinguishable from those caused by fowl pox virus. Voluntir et al (168) in Romania and Chernov (17) in the USSR, described effects of these toxins in swine.

Walhberg et al (170) reported death of European Elk foraging on hay infected with *Stachybotrys alternans.* The case was diagnosed as stachybotryotoxicoses characterized by a severe gastrointestinal disorder complicated by hemorrhagic diathesis.

Korpinen (77) reported finding toxic strains of *S. alternans* from animal feeds, fodder, bedding, commercial feed, and wheat intended for human consumption. She suggested that *S. alternans* is often missed in routine isolations because of overgrowth by other fungi. She recommends exposing culture plates to ultraviolet radiation. Under these conditions, *Stachybotrys* survives and grows out.

### *Identification of the Toxins*

Initial attempts at isolation and characterization of the stachybotryotoxins were done by Fialkov & Serebriany (49) in the years preceding 1949. The ether extracts were the most potent when tested on rabbit's skin test, when injected into a horse with production of typical signs of the toxicosis, and when tested on an isolated frog heart. The toxins were described as "cardiac poisons which stop the heart in a systolic state and whose effect is similar to the glycosides of the digitalis group, and which besides produce haemolysis." Stachyobotryotoxin A (SB-A) was the name given to an ether extractable, colorless or light yellow amorphous powder, sparingly soluble in water, soluble in organic solvents, with an empirical formula of $C_{25}H_{38}O_6$ (430.3) or $C_{26}H_{38}O_6$ (446.3).

Yuskiv (186) confirmed the work of Fialkov & Serebriany, reporting that the acutely toxic fraction of stachybotryotoxin (2 components) gives a colored reaction with resorcinol and developed a colorimetric test for its detection.

If one constructs a model based on the information supplied in (49), using both the chemical and physiological data, one finds a steroid structure almost identical to that described by Olifson (104) and Bilai (7); i.e. poaefusarin and sporofusarin were reported to be produced by *Fusarium sporotrichiodes* (Figure 3). Mirocha et al (97) and Pathre et al (109) found two similar components, one of which has cardiac activity, and two others that incite a necrotic dermatosis in the rabbit skin test. These compounds are the following: SB-3($C_{23}H_{30}O_5$, MW 386; SB-4($C_{27}H_{34}O_8$, MW 486); SBF-5 ($C_{23}H_{32}O_5$, MW 388); and SB-A($C_{23}H_{29}O_5N$, MW 399). SB-3 and SB-4 are not active in the rabbit skin test, whereas SBF-5 and SB-A are. It is difficult to determine which of these toxins corresponds to the original stachybotroyotoxin A.

On the other hand, Bamburg & Strong (6) and Smalley & Strong (140) mention that the symptoms as described for stachybotryotoxicosis are almost identical to those produced by the trichothecene toxins, and on that basis they suggested that the latter are the actual agents of the disease. In support of this hypothesis, Eppley & Bailey (47) reported that *S. atra* when grown on oats produced five compounds toxic to brine shrimp. Three of these are 12,13-epoxy- $\Delta^9$-trichothecenes, one is roridin E (Figure 2, XVIII), and the other two were hydrolyzed to verrucarol, the product of roridin and verrucarin hydrolysis.

In summary, it appears that stachybotryotoxicosis may be caused by a number of toxins, some of which are steroid-like as originally described, and the others may be the trichothecenes produced by *Fusarium* and other fungi.

## TOXINS PRODUCED BY ASPERGILLUS AND PENICILLIUM

### *Ochratoxins*

Ochratoxins (Figure 3) are produced by some members of the *A. ochraceus* group (62) and by *Penicillium viridicatum* (165). *A. ochraceus* is seldom isolated from more than a small percentage of surface disinfected kernels of stored grains, even when they are undergoing deterioration by fungi, presumably because it cannot compete effectively with other storage fungi (19). *P. viridicatum,* on the other hand, is sometimes common in stored corn (100); it is able to grow at temperatures from just below zero to about 35°C and at moisture contents in equilibrium with relative humidities of 83% and above, and so might be a more common producer of ochratoxin than is *A. ochraceus.*

Ochratoxin A (OA) was first isolated in South Africa from *A. ochraceus* (164) growing on maize meal and was toxic to day-old Pekin ducklings, mice, and rats. Its $LD_{50}$ in ducklings is 150 $\mu$g per duckling (116); 20–22 mg/kg *per os* in rats (118); 4.67 mg/kg in rainbow trout (39); and 3.3–3.9 mg/kg in day-old chicks (111).

OA causes tubular necrosis of the kidney, mild degeneration of the liver, and enteritis of the small intestine, the main target organs being the liver and kidney. OA was found to be noncarcinogenic in the rat (120).

Still et al (146) found that with a single dose of 12.5 mg/kg, given on the tenth day of gestation, of a total of 88 implantations in seven rats, 81.8% were either dead or resorbed (rats usually resorb the fetus rather than abort). They suggest that *A. ochraceus* may account for the large number of bovine and porcine abortions associated with the ingestion of moldy feedstuff. However, Ribelin et al (126) reported that 8 g of OA given to a pregnant cow produced anorexia, diarrhea, milk cessation, but not abortion. They concluded that in low dosages the rumen is a good detoxifier.

Krogh & Hasselager (83) in Denmark described porcine nephropathy, which occurs with high frequency after harvest of barley and rye preceded by extended wet periods. The disease is characterized by degeneration of the kidney, consisting of epithelial destruction of the proximal tubules followed by diffuse formation of connective tissue and cysts. This disease is thought to be caused by toxins produced by *A. ochraceus* and *P. viridicatum,* both of which have been found in moldy feedstuff fed to the animals. *P. viridicatum* produces ochratoxin and citrinin simultaneously (79, 133). Citrinin produces nephropathy in animals and *Penicillium* may be more important than *A. ochraceus* as a causal organism. Both OA and citrinin were found in moldy barley and rye and were associated with swine nephropathy (81). Krogh (80) studied organs and meat from pigs suffering from this disease and detected OA in all cases studied, the maximum amount being 67 ppb. He concluded that OA can be transmitted through animals to the human food chain and may pose a public health problem.

Choudhury et al (22) found that 1 ppm of OA in pullets delayed sexual maturity and decreased egg production. Morbidity and mortality was severe at 2 and 4 ppm with damage to the kidney and liver. Hatchability of fertile eggs was reduced, and those chicks that hatched had a reduced growth rate for 2 weeks and recovered after

4 weeks. Tucker & Hamilton (153) stated that ochratoxin is more toxic to broilers than are any other mycotoxins previously investigated.

CHEMICAL DERIVATIVES AND BIOLOGICAL ACTIVITY Steyn (144) reviewed the chemistry of the ochratoxins recently. Other derivatives of OA are the following: ochratoxin B (OB), the dechlorinated product of OA which is virtually nontoxic, and ochratoxins C (OC), the methyl and ethyl esters of OA which have similar toxicity to OA. Steyn & Holzapfel (145) reported the isolation of the methyl and ethyl esters of both OA and OB. The esters of OB, unlike OA, are relatively nontoxic, and toxicity appears to be related with the presence of the Cl atom in the aromatic ring.

Doster et al (39) suggest that in trout, OA is metabolized to nontoxic, water-soluble OC which is readily excreted. The phenolic hydroxyl in the dissociated form (phenoxide ion) is necessary for toxicity (52). They reasoned that the chlorine atom on OA may have a direct effect on the dissociation of the phenolic hydroxyl, but we believe that the lactone carbonyl exerts a greater effect on its ionization.

Natori et al (102) examined 33 strains of *A. ochraceus* isolated from foods in Japan and found that 2 isolates produced OA and 28 produced penicillic acid. They associated the toxicity of these isolates primarily with penicillic acid (carcinogenic) and thought this metabolite more important than OA as a causal agent of mycotoxicoses.

OCCURRENCE OF OCHRATOXINS IN NATURE In a survey of 283 corn samples of all grades from commercial markets, Shotwell et al (137) found only one sample with OA (110–150 ppb). In another survey, 293 corn samples of all grades destined for export were collected from 10 different ports; 3 of these contained OA from 83–166 ppb (138). Scott et al (133) detected 18 ochratoxin positive samples out of 29 samples of heated grain (primarily wheat) ranging in concentration from 0.03 to 27 ppm. Citrinin was also present (0.07–80 ppm) in 13 of these samples. They also reported finding OA in peanuts, dried white beans, and mixed animal feeds. Levi et al (85) found OA in 6 samples of molded green coffee beans (20–230 ppb).

It appears that OA occurs at least occasionally in feedstuff and may at times pose a threat to animal health, but its significance in public health is not known.

## *Yellowed-Rice Toxins*

Stored rice is invaded by fungi similar to those found on other cereals, depending on the moisture content, temperature, and length of storage. Most notable of the toxicoses associated with rice are those referred to as "yellowed rice toxicoses" (128) and reported to be associated with various diseases, such as cardiac beriberi, nervous and circulatory disorders, degeneration of the kidneys, liver cirrhosis, and hepatoma. The fungi most likely responsible are *Penicillium citreo-viride, P. citrinum,* and *P. islandicum.*

CITREOVIRIDIN *P. citreo-viride* Biourge is highly toxic and causes paralysis of an ascending type with respiratory and circulatory disturbances. The toxin thought to

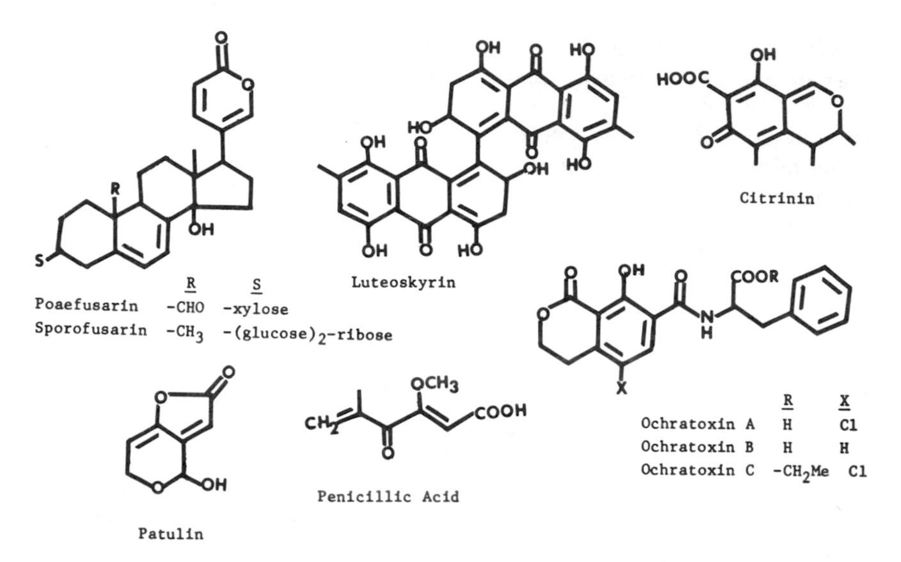


*Figure 3* Chemical structures of various mycotoxins.

be responsible is citreoviridin, a neurotoxic yellow pigment which in experimental animals causes signs similar to those of beriberi in man, i.e. multiple neuritis, general weakness, paralysis, mental deterioration, and heart failure. Beriberi has been described as a thiamine deficiency, but Uraguchi (128, 161) cites evidence that Vitamins $B_1$ and C play only a subordinate role in this disease and suggests that it is quite possibly a mycotoxicosis. Beriberi is prominent in rice-eating parts of the world and correlates well with ingestion of moldy rice. Ueno & Ueno (156) further support this view by demonstrating a symptomatic and toxicological correlation of citreoviridin and cardiac beriberi in animals.

CITRININ Citrinin (Figure 3) was first isolated from *P. citrinum,* but it can be produced by various species of *Penicillium,* most important of which are *P. citrinum,* occurring on rice, barley, and dried fish, and *P. viridicatum,* commonly found on stored maize. It is also produced by various species of *Aspergillus* and by a flowering plant, *Crotalaria crispata* F. Muell. It gained practical importance because it causes renal damage in animals and has been implicated in yellowed rice toxicoses. It is not certain whether citrinin is the actual toxic principle of *P. citrinum* growing on rice, because there are some symptomatic differences between the administration of authentic citrinin and cultures of *P. citrinum* to experimental animals. However, it is considered along with luteoskyrin, islanditoxin, citreoviridin, and cyclochlorotine as one of the toxins of yellowed rice.

*Penicillium citrinum* usually grows on polished rice in storage and causes a characteristic yellowish color on its surface, which fluoresces under uv irradiation. As a diagnostic sign, citrinin causes an enlarged turbid kidney in rats with degeneration and dilatation of the lower nephrons and renal lesions resembling glomerulonephrosis. When given to rats, it causes an increase in urination, whereas pure cultures of *P. citrinum* cause renal damage but no increase in urination. It is not a carcinogen and is considered a slow-acting toxin. Its implication in nephropathy of swine has been discussed earlier.

Citrinin has been found naturally occurring in wheat, rye, and barley (134), but there are no documented reports of its natural occurrence in yellowed rice implicated in mycotoxicoses.

LUTEOSKYRIN AND CYCLOCHLOROTINE Rats fed moldy rice overgrown with *P. islandicum* develop severe cirrhosis of the liver and hepatoma. One of the toxins responsible is luteoskyrin (LS) (Figure 3), one of a number of quinoid pigments produced by *P. islandicum,* as well as the chlorine-containing peptide called cyclochlorotine (CC). Both LS and CC are carcinogenic and primarily affect the liver (46). The administration of LS to animals leads to a preferential development of liver necrosis, whereas CC causes cirrhosis and fibrosis. Cyclochlorotine is water soluble and quick acting, whereas LS is water insoluble and acts more slowly. Luteoskyrin affects the function of mitochondria (160), binds with DNA and inhibits synthesis of nuclear RNA in Ehrlich ascite tumor cells (158), and has some antitumor activity (159).

Islanditoxin, as described by Marumo (92), has also been implicated in toxicosis of moldy rice. It is similar in structure to CC (162) but not identical (46).

Although ample evidence exists to implicate *P. citrinum, P. islandicum,* and *P. citreo-viride* with mycotoxicoses (152), except for citrinin, the toxins produced by these fungi have not been found in nature.

### *Tremorgenic Toxins*

As the name implies, these mycotoxins cause marked body tremors and diuresis, followed by convulsive seizures which in many cases are fatal. Wilson (173) first described this syndrome in rats caused by extracts from *Aspergillus flavus* and indicated that metabolites capable of causing sustained trembling are rare. Trembling was not inhibited by administration of drugs.

Wilson (173) isolated another tremorgen from an isolate identified as either *P. cyclopium* or *P. crustosum,* and Ciegler (23) found the identical toxin in *P. palitans* and called it tremortin A (67). Tremortin A has an empirical formula of $C_{37}H_{44}NO_6Cl$ (173). Hou et al (67) characterized another tremorgen from *P. palitans* called tremortin B ($C_{37}H_{45}NO_5$) and tremortin C, unidentified as yet. Yamazaki et al (184) added to this list by isolating fumitremorgin A and B from *A. fumigatus* with empirical formulas of $C_{33}H_{45}O_6N_3$ and $C_{26}H_{29}O_6N_3$ respectively. Cole et al (30) reported still another tremorgen (verruculogen) from *P. verruculosum* and partially characterized it as a 6-O-methylindole (29), which is similar to those reported by Yamazaki (184).

The isolates of *Penicillium,* particularly those of Wilson (173), all came from feedstuff suspected of being involved in disease of sheep, horses, and cows. Ciegler & Pitt (27) surveyed over 200 strains of *Penicillium* and found that some of the species like *P. crustosum* (tremorgen producer) are commonly found on refrigerated foods, grains, and cereal products. Although tremorgenic toxins have not been found in foodstuff on the farm, signs in afflicted animals suggest that some of the toxicoses may be due to this group of mycotoxins. As an example, Cysewski & Pier (32) report that calves fed ground mycelium of *P. puberulum* (known tremortin A producer) exhibit signs of tremor, ataxia, muscular rigidity, and convulsions. Such signs are often associated with intoxication of farm animals.

### *Penicillic Acid*

Penicillic acid (PA) (24) (Figure 3) is a carcinogen and was first described as a product of *Penicillium puberulum* and subsequently found as a product of various species of *Penicillium* and *Aspergillus* including *A. ochraceus.* It was shown to be carcinogenic (36) in rats and mice, causing local tumors when administered by subcutaneous injection. It is produced readily in laboratory culture but has rarely been detected in nature. Snow et al (141) detected between 0.11 and 0.22 ppm of PA in two samples of visibly molded tobacco with trace amounts detectable in tobacco smoke after combustion. Scott et al (134) could not find PA in visibly molded samples of cereal grains and mixed feeds although known PA-producing fungi were isolated from the feedstuff.

*Aspergillus ochraceus* produces both ochratoxin A (OA) and PA on mixed poultry feed (5). A combination of low temperature (15–22°C) and low moisture (unfavorable conditions for optimum growth of the fungus) favored PA production, whereas high temperature (30°C) and moisture content favored OA production. This suggests that feed not visibly moldy may still harbor PA in amounts sufficient to be harmful to poultry. The question still remains as to what effect the natural competitive microflora of the feed would have on growth of *A. ochraceus* and PA production.

### *Patulin*

Patulin (Figure 3) is a carcinogenic lactone and a metabolite of several species of *Penicillium* and *Aspergillus* (24). It was given various names such as clavacin, claviformin, expansin, mycoin c, and penicidin in the past. It is an antibiotic, toxic to bacteria and to some fungi and to higher plants and animals. Patulin is an important mycotoxin because it is carcinogenic, affects the balance of constituents of the blood, increases vascular permeability, causes edema, and has been found naturally occurring in foodstuff.

Wilson & Nuovo (174) inoculated apples with *P. expansum* (a common cause of decay in apples) and found between 9 and 146 mg patulin/liter of expressed apple juice. Five of 100 samples of fresh, commercially produced apple juice, contained patulin ranging in concentrations up to 45 ppm, and 4 of 91 samples contained up to 25 ppm (174). Patulin was also found in spontaneously molded bread and bakery products (125). Ware et al (171) found between 49 and 309 ppb in 8 of 13 commercial samples of apple juice, while Brian et al (13) found it in decaying apples. Patulin is stable in apple products, and it may constitute a health hazard.

## CONCLUSIONS

Many fungi are capable of producing toxic substances when grown under special conditions in the laboratory, but relatively few mycotoxins have been found to occur naturally. One requirement for production of toxins by fungi appears to be the occurrence of a toxin-producing strain in practically pure culture in a suitable substrate. Starchy cereal seeds and most legumes are of relatively low aflatoxin risk, whereas peanuts, brazil nuts, cottonseed, fishmeal, and copra, especially in warm humid climates, may be of high aflatoxin risk. Acute aflatoxicosis is rare, but chronic aflatoxin poisoning may be common both in man and in domestic animals in high-aflatoxin-risk countries of the tropics. Chronic exposure to aflatoxin may predispose animals, and presumably man, to other infectious agents.

In technologically advanced countries, high-aflatoxin-risk crops are well monitored for aflatoxin, and the incidence is found to be low. Moreover, when found, it is usually less than 50 ppb, a concentration that will not cause acute signs of toxicity. However, in countries such as India, South Africa, Uganda, Thailand, and the Phillipines, the concentration of aflatoxin $B_1$ has been found to range as high as 1000 ppb, with an average of about 225 ppb, and may be detrimental to public health.

Of the toxins produced by *Fusarium,* the zearalenones are most commonly encountered in nature and a clear cause-effect relationship exists in hyperestrogenism in swine. The relationship between T-2 toxin (trichothecene derivative) and field cases of mycotoxicity is less clear. Although signs of trichothecene toxicity (hemorrhaging) are often encountered in animals in nature, the T-2 toxin is rarely found in feed samples implicated in such apparent mycotoxicoses. A likely explanation exists in the fact that more often, *Fusarium roseum* rather than *F. tricinctum* is encountered in feedstuff corn, and the former is not normally a T-2 producer, i.e. in the same manner as *F. tricinctum. F. roseum* is also toxic to animals, however, by production of trichothecene derivatives such as monoacetoxyscerpenol for which chemical analyses normally are not made. We feel that the key to toxicity of *Fusarium* to animals lies in *F. roseum* and its characteristic array of trichothecene derivatives. Our data over the last year lead us to conclude that toxicity of natural products from *Fusarium* is of greater incidence and perhaps of more importance than aflatoxins.

Ochratoxin and citrinin may also be significant in causing diseases of swine, especially porcine nephropathy. Citrinin is perhaps the only toxin of the yellowed-rice toxins found in nature and appears to be more directly involved in porcine nephropathy than ochratoxin. The other toxins produced by species of *Penicillium* have not as yet been firmly authenticated as causing mycotoxicoses other than in laboratory tests.

The following mycotoxins have been found in nature: aflatoxins, zearalenone, T-2 toxin, citrinin, ergot alkaloids, ochratoxin, slaframine (causes excessive salivation), sporodesmins, penicillic acid, patulin, and sterigmatocystin. Of these, the first eight have been implicated in natural cases of mycotoxicity.

Whether a given sample of food or feed contains a fungus toxin can be determined only by extraction and chemical identification of the toxin, and whether a given sample is toxic can be determined only by feeding tests; presence or absence of a toxin-producing fungus is not diagnostic.

*Literature Cited*

1. Abdel Kader, M. M., El-Aaser, A. B. A., El-Merzabani, M. M., King, L. J. 1971. The metabolism of aflatoxins in the rat. *Acta. Physiol. Acad. Sci. Hung.* 39: 375–81
2. Allcroft, R. 1969. Aflatoxicosis in farm animals. See Ref. 54, p. 237
3. Arseculeratne, S. N., DeSilva, L. M., Wijesundera, S., Bandunatha, C. H. S. R. 1969. Coconut as a medium for the experimental production of aflatoxin. *Appl. Microbiol.* 18:88–94
4. Ashworth, L. J. Jr., George, A. G. 1971. *Stachybotrys atra:* a cause of root disease of cotton. *Phytopathology* 61: 1320 (Abstr.)
5. Bacon, C. W., Sweeney, J. G., Robbins, J. D., Burdick, D. 1973. Production of penicillic acid and ochratoxin A on poultry feed by *Aspergillus ochraceus:* temperature and moisture requirements. *Appl. Microbiol.* 26:155–60
6. Bamburg, J. R., Strong, F. M. 1971. 12, 13-epoxytrichothecenes. See Ref. 75, Vol. VII: 207
7. Bilai, V. I., Pidoplichko, N. M. 1970. *Toxic-Producing Microscopic Fungi and Diseases of Man and Animals.* Naukova Dumka, Kiev: Ukr. Acad. Sci., 289 pp.
8. Biollaz, M., Buchi, G., Milne, G. 1968. Biosynthesis of aflatoxins. *J. Am. Chem. Soc.* 90:5017
9. Biollaz, M., Buchi, G., Milne, G. 1968. The biogensis of bisfuranoids in the genus *Aspergillus. J. Am. Chem. Soc.* 90:5019
10. Bolliger, G., Tamm, C. 1972. Vier neue Metabolite von *Gibberella zeae:* 5-formyl-zearalenon, 7'-dehydrozearalenon, 8'-hydroxy und 8'-epihydroxyzearalenon. *Helv. Chim. Acta* 55: 3030–48
11. Brewington, C. R., Weihrauch, J. L., Ogg, C. L. 1970. Survey of commercial milk samples for aflatoxin M. *J. Dairy Sci.* 53:1509–10
12. Brian, P. W., Dawkins, A. W., Grove, J. F., Hemming, H. G., Lowe, D., Norris, G. L. F. 1961. Phytotoxic compounds produced by *Fusarium equiseti. J. Exp. Bot.* 12:1–12
13. Brian, P. W., Elson, G. W., Lowe, D. 1956. Production of patulin in apple fruits by *Penicillium expansum. Nature* 178:263
14. Brook, P. J., White, E. P. 1966. Fungus toxins affecting mammals. *Ann. Rev. Phytopathol.* 4:171–94
15. Buchi, G., Weinreb, S. M. 1969. The total synthesis of racemic aflatoxin $M_1$ (milk toxin). *J. Am. Chem. Soc.* 91:5408
16. Burkhardt, H. J., Forgacs, J. 1968. O-methylsterigmatocystin, a new metabolite from *Aspergillus flavus,* Link ex Fries. *Tetrahedron* 24:717–20
17. Chernov, K. S. 1970. Mycotoxicoses of swine and their differential diagnosis. *Vet. Microbiol.* 16:803 (In Russian)
18. Christensen, C. M. 1971. Mycotoxins. *CRC Crit. Rev. Environ. Contr.* 2:57–80
19. Christensen, C. M. 1962. Invasion of stored wheat by *Aspergillus ochraceus. Cereal Chem.* 39:100–6
20. Christensen, C. M., Kaufmann, H. H. 1969. *Grain Storage: the Role of Fungi in Quality Loss.* Minneapolis: Univ. Minnesota Press. 153 pp.
21. Christensen, C. M., Nelson, G. H., Speers, G. M., Mirocha, C. J. 1973. Results of feeding tests with rations containing grain invaded by a mixture of naturally present fungi plus *Aspergillus flavus* NRRL 2999. *Feedstuffs* 45: 20–21
22. Choudhury, H., Carlson, C. W., Semeniuk, G. 1971. A study of ochratoxin toxicity in hens. *Poultry Sci.* 50: 1853
23. Ciegler, A. 1969. Tremorgenic toxin from *Penicillium palitans. Appl. Microbiol.* 18:128–29
24. Ciegler, A., Detroy, R. W., Lillehoj, E. B. 1971. Patulin, penicillic acid and other carcinogenic lactones. See Ref. 25, p. 409
25. Ciegler, A., Kadis, S., Ajl, S. J. 1971. *Microbial Toxins,* Vol. VI *Fungal Toxins.* New York: Academic. 563 pp.
26. Ciegler, A., Lillehoj, E. B. 1968. Mycotoxins. *Advan. Appl. Microbiol.* 10:155–219
27. Ciegler, A., Pitt, J. I. 1970. Survey of the genus *Penicillium* for tremorgenic toxin production. *Mycopathol. Mycol. Appl.* 42:119

28. Clifford, J. I., Rees, K. R. 1966. Aflatoxin: a site of action in the rat liver cell. *Nature* 209:312
29. Cole, R. J., Kirksey, J. W. 1973. The mycotoxin verruculogen: a 6-o-methylindole. *J. Agr. Food Chem.* 21:927
30. Cole, R. J. et al 1972. Tremorgenic toxin from *Penicillium verruculosum. Appl. Microbiol.* 24:248–56
31. Curtin, T. M., Tuite, J. 1966. Emesis and refusal feed in swine associated with *Gibberella zeae*-infected corn. *Life Sci.* 5:1937
32. Cysewski, S. J., Pier, A. C. 1973. A mycotoxicosis produced by a tremorgen from *P. puberulum:* clinical pathology. *2nd Int. Congr. Plant Pathol., Minneapolis, Minn.,* St. Paul, Minn.: Am. Phytopathol. Soc. Pap. No. 0412 (Abstr.)
33. Davis, N. D., Diener, U. L. 1970. Environmental factors affecting the production of aflatoxin. See Ref. 59, p. 43
34. Detroy, R. W., Hesseltine, C. W. 1970. Aflatoxicol: structure of a new transformation product of aflatoxin $B_1$. *Can. J. Biochem.* 48:830–32
35. Detroy, R. W., Lillehoj, E. B., Ciegler, A. 1971. Aflatoxin and related compounds. See Ref. 25, p. 3
36. Dickens, F., Jones, H. E. H. 1965. Carcinogenic action of certain lactones and related substances. *Brit. J. Cancer* 19:392
37. Diener, U. L., Davis, N. D. 1969. Aflatoxin formation by *Aspergillus flavus.* See Ref. 54, p. 13
38. Diener, U. L., Davis, N. D. 1969. Production of aflatoxin on peanuts under controlled environments. *J. Stored Prod. Res.* 5:251
39. Doster, R. C., Sinnhuber, R. O., Wales, J. H. 1972. Acute intraperitoneal toxicity of ochratoxins A and B in rainbow trout (Salmo gairdneri). *Food Cosmet. Toxicol.* 10:85
40. Doupnik, B. Jr. 1969. Aflatoxins produced on peanut varieties previously reported to inhibit production. *Phytopathology* 59:1554
41. Dutton, M. F., Heathcote, J. G. 1969. O-alkyl derivatives of aflatoxins $B_{2a}$ and $G_{2a}$. *Chem. Ind. London,* p. 983–86
42. Dutton, M. F., Heathcote, J. G. 1968. Metabolites of *Aspergillus flavus. Chem. Ind. London,* p. 418
43. Dutton, M. F., Heathcote, J. G. 1967. Two new hydroxyaflatoxins. *Biochem. J.* 101:22–23
44. Ellison, R. A., Kotsonis, F. N. 1973. T-2 toxin as an emetic factor in moldy corn. *Appl. Microbiol.* 26:540–43
45. Engelbrecht, J. C., Altenkirk, B. 1972. Comparison of some biological effects of sterigmatocystin and aflatoxin analogues on primary cell cultures. *J. Nat. Cancer Inst.* 48:1647–55
46. Enomoto, M., Saito, M. 1972. Carcinogens produced by fungi. *Ann. Rev. Microbiol.* 26:279–312
47. Eppley, R. M., Bailey, W. J. 1973. 12-13-epoxy-$\Delta^9$-trichothecenes as the probable mycotoxins responsible for Stachybotryotoxicosis. *Science* 181:758
48. Eppley, R. M., Stoloff, L., Chung, C. W. 1973. Survey of corn for *Fusarium* toxins. *Abstr. 87th Ann. Meet. AOAC.* p. 40
49. Fialkov, T. A., Serebriany, J. B. 1949. Production of toxic substances by *Stachybotrys alternans* in culture and research of their chemical nature. Kiev: Acad. Sci., Ukr. S.S.R., Inst. Microbiol. D. K. Zabolotny
50. Forgacs, J. 1972. Stachybotryotoxicosis. See Ref. 75, Vol. VIII: 95
51. Forgacs, J. 1965. Stachybotryotoxicosis and moldy corn toxicosis. See Ref. 179, p. 87
52. Fun-Sun-Chu, N. I., Chang, C. C. 1972. Structural requirements for ochratoxin intoxication. *Life Sci.* 11:503
53. Garren, K. H., Christensen, C. M., Porter, D. M. 1969. The mycotoxin potential of peanuts (groundnuts): the USA viewpoint. *J. Stored Prod. Res.* 5:265
54. Goldblatt, L. A., Ed. 1969. *Aflatoxin, Scientific Background, Control, and Implications.* NY: Academic 472 pp.
55. Grove, J. F., Mortimer, P. H. 1969. Cytotoxicity of some transformation products of diacetoxyscirpenol. *Biochem. Pharmacol.* 18:1473
56. Gundu Rao, H. R., Harein, P. K. 1972. Dichlorvos as an inhibitor of aflatoxin production on wheat, corn, rice and peanuts. *J. Econ. Entomol.* 65:988–89
57. Habish, H. A., 1972. Aflatoxin in Haricot bean and other pulses. *Exp. Agr.* 8:135
58. Heathcote, J. G., Dutton, M. F. 1969. New metabolites of *Aspergillus flavus. Tetrahedron* 25:1497–1500
59. Herzberg, M., Ed. 1970. *Toxic Micro-organisms: Mycotoxins-Botulism.* UJNR Joint Panels Toxic Micro-organisms and the US Dept. Interior. Washington DC: GPO
60. Hesseltine, C. W. 1969. Mycotoxins. *Mycopathol. Mycol. Appl.* 39:371–83
61. Hesseltine, C. W. et al 1970. Production of various aflatoxins by strains of the *Aspergillus flavus* series. See Ref. 59, p. 202

62. Hesseltine, C. W., Vandegraft, E. E., Fennell, D. I., Smith, M. L., Shotwell, O. L. 1972. *Aspergilli* as ochratoxin producers. *Mycologia* 64:539–50
63. Holker, J. S. E., Kagal, S. A. 1968. 5-methoxysterigmatocystin, a new metabolite from a mutant strain of *Aspergillus versicolor. Chem. Commun.* 1968:1574–75
64. Holker, J. S. E., Mulheirn, L. J. 1968. The biosynthesis of sterigmatocystin. *Chem. Commun.* 1968:1576
65. Holzapfel, C. W., Purchase, I. F. H., Steyn, P.S., Gouws, L. 1966. The toxicity and chemical assay of sterigmatocystin, a carcinogenic mycotoxin, and its isolation from two new fungal sources. *S. Afr. Med. J.* 40:1100
66. Holzapfel, C. W., Steyn, P. S., Purchase, I. F. H. 1966. Isolation and structure of aflatoxins $M_1$ and $M_2$. *Tetrahedron Lett.* 2799–2803
67. Hou, C. T., Ciegler, A., Hesseltine, C. W. 1971. Tremorgenic toxins from penicillia. II. A new tremorgenic toxin, tremortin $B_1$ from *Penicillium palitans. Can. J. Microbiol.* 17:599. III. Tremortin production by *Penicillium* species on various agricultural commodities. *Appl. Microbiol.* 21:1101–3
68. Hoyman, W. G. 1941. Concentration and characterization of the emetic principle present in barley infested with *Gibberella saubinetti. Phytopathology* 31: 871–85
69. Hsieh, D. P. H. 1973. Biosynthetic pathways in the natural production of aflatoxins. See Ref. 32, Pap. No. 1101
70. Hsieh, D. P. H. 1973. Inhibition of aflatoxin biosynthesis by Dichlorvos. *Agr. Food Chem.* 21 (3):468–70
71. Hsieh, D. P. H., Lin, M. T., Yao, R. C. 1973. Conversion of sterigmatocystin to aflatoxin $B_1$ by *Aspergillus parasiticus. Biochem. Biophys. Res. Commun.* 52: (3):992–97
72. Hsu, I. C., Smalley, E. B., Strong, F. M., Ribelin, W. E. 1972. Identification of T-2 toxin in moldy corn associated with a lethal toxicosis in dairy cattle. *Appl. Microbiol.* 24:684–90
73. Jackson, R. A. 1973. *The chemistry of some derivatives of the macrolide, zearalenone.* PhD thesis. Univ. Minnesota, Minneapolis. 185 pp.
74. Joffe, A. Z. 1971. Alimentary toxic aleukia. See Ref. 75, Vol. VIII: 139
75. Kadis, S., Ciegler, A., Ajl, S. J., Eds. 1971. *Microbial Toxins,* Vol. VII, *Algal and Fungal Toxins,* 401 pp; Vol. VIII, *Fungal Toxins,* 1972, 400 pp. NY: Academic
76. Keyl, A. C., Booth, A. N., Masri, M.S., Gumbmann, M.R., Gagne, W. E. 1970. Chronic effects of aflatoxin in farm animal feeding studies. See Ref. 59, p. 72
77. Korpinen, E. L. 1973. Studies on *Stachybotrys alternans* I. Isolation of toxicogenic strains from Finnish grains and feeds. *Acta Pathol. Microbiol. Scand.,* Sect. B., 81:191–97
78. Kosuri, N. R., Smalley, E. B., Nichols, R. E. 1971. Toxicologic studies of *Fusarium tricinctum* (Corda) Snyder et Hansen from moldy corn. *Am. J. Vet. Res.* 32:1843
79. Krogh, P. 1973. Nephropathy caused by mycotoxins from *Penicillium* and *Aspergillus. J. Gen. Microbiol.* 73:R34–35
80. Krogh, P. 1973. Natural occurrence of ochratoxin A and citrinin in cereals associated with swine nephropathy. See Ref. 32, Pap. No. 0360
81. Krogh, P. 1972. Natural occurrence of ochratoxin A and citrinin in cereals associated with field outbreaks of swine nephropathy. Abstracts, IUPAC-Sponsored Symposium: *Control of Mycotoxins,* Goteburg, Sweden, p. 19
82. Krogh, P., Hald, B. 1969. ForeKomst of alfatoksin i importerede jordnprodukter. *Nord. Veterinaer Med.* 21:398 (In Danish, with English summary)
83. Krogh, P., Hasselager, E. 1968. Studies on fungal nephrotoxicity. *Roy. Vet. Agr. Coll. Yearb. Copenhagen,* pp. 198–214
84. Kurata, H., Tanabe, H. 1973. Natural occurrence of mycotoxins in foodstuffs of Japan. See Ref. 32, Pap. No. 0361
85. Levi, C. P., Trenk, H. L., Mohr, H. K. 1973. A study of the occurrence of ochratoxin A in green coffee beans. See Ref. 100. (not abstracted but presented)
86. Lillehoj, E. B., Ciegler, A. 1969. Biological activity of aflatoxin $B_{2a}$. *Appl. Microbiol.* 17:516–19
87. Lillehoj, E. B., Ciegler, A. 1968. Biological activity of sterigmatocystin. *Mycopathol. Mycol. Appl.* 35:373–76
88. Lin, M. T., Hsieh, D. P. H. 1973. Averufin in the biosynthesis of aflatoxin $B_1$. *J. Am. Chem. Soc.* 95:1668
89. Lucas, F. V. Jr., Monroe, P., Pham-Van, N., Townsend, J. F., Lucas, F. V., 1970. Mycotoxin contamination of South Vietnamese rice. *J. Trop. Med. Hyg.* 73:182
90. Marasas, W. F. O. et al 1969. Toxic effects on trout, rats, and mice of T-2 toxin produced by the fungus *Fusarium tricinctum. Toxicol. Appl. Pharmacol.* 15:471

91. Marasas, W. F. O., Smalley, E. B., Bamburg, J. R., Strong, F. M. 1971. Mycotoxicoses associated with moldy corn. *Phytopathology* 61:1488–91
92. Marumo, S. 1959. Islanditoxin, a toxic metabolite produced by *Penicillium islandicum* Sopp. III Structure of islanditoxin. *Bull. Agr. Chem. Soc. Jap.* 23: 248
93. Masri, M. S., Lundin, R. E., Page, J. R., Garcia, V. C. 1967. Analysis for aflatoxin M in milk. *Nature* 215:753
94. Mateles, R. I., Wogan, G. N. 1967. *Biochemistry of Some Foodborne Microbiol Toxins.* Cambridge, Mass.: MIT 171 pp.
95. Mirocha, C. J., Christensen, C. M. 1974. Oestrogenic mycotoxins synthesized by *Fusarium.* In *Mycotoxins,* ed. I. F. H. Purchase. Amsterdam: Elsevier. In press
96. Mirocha, C. J., Christensen, C. M., Nelson, G. H. 1971. F-2 (zearalenone) estrogenic mycotoxin from *Fusarium.* See Ref. 75, VII: 107
97. Mirocha, C. J., Palyusik, M., Pathre, S. V., Schauerhamer, B. 1972. Mycotoxins from *Stachybotrys alternans* grown on oats. *Phytopathology* 62:778 (Abstr.)
98. Mirocha, C. J., Pathre, S. 1973. Identification of the toxic principle in a sample of poaefusarin. *Appl. Microbiol.* 26: 719–24
99. Mirocha, C. J., Schauerhamer, B., Pathre, S. V. 1973. Natural occurrence of Zearalenone (F-2) in maize. See Ref. 32, Pap. No. 0350
100. Mislevic, P. P., Tuite, J. 1970. Temperature and relative humidity requirements of species of *Penicillium* isolated from yellow dent corn kernels. *Mycologia* 62:75–88
101. Nagarajan, V., Bhat, R. V. 1972. Factor responsible for varietal differences in aflatoxin production in maize. *J. Agr. Food Chem.* 20 (4):911–14
102. Natori, S. et al 1970. Chemical and cytotoxicity survey on the production of ochratoxins and penicillic acid by *Aspergillus ochraceus* Wilhelm. *Chem. Pharm. Bull.* 18:2259
103. Nelson, R. R. 1971. Hormonal involvement in sexual reproduction in the fungi, with special reference to F-2, a fungal estrogen. In *Morphological and Biochemical Events in Plant-Parasite Interaction,* ed. S. Akai, S. Ouchi. Tokyo: Phytopathological Soc. Jap.
104. Olifson, L. E. 1965. *Chemical and biological characteristics of poisonous materials of cereals (grain crop) affected by the fungus Fusarium sporotrichiella.* PhD thesis. Tech. Inst. Food Ind., Moscow (In Russian)
105. Ozegovic, L., Pavlovic, R., Milosev, B. 1971. Toxic dermatitis, conjunctivitis, rhinitis, pharyngitis, and laryngitis in fattening cattle and farm workers caused by molds from contaminated straw (Stachybotryotoxicosis?). *Veterinaria* 20:263–67 (In Yugoslavian with English summary)
106. Palyusik, M., Bamberger, K., Nagy, Z. A. 1970. Comparative study of oral lesions associated with stachybotryotoxicosis and fowl pox. *Acta Vet. Acad. Sci. Hung.* 20:165–70
107. Palyusik, M., Bamberger, K., Zoltan, N. 1971. Experimental stachybotryotoxicosis of day-old and growing chickens with special regards to its diagnostic differentiation from fowl pox. *Magy. Allatorv. Lapja* 6:304–6
108. Pathre, S. V., Behrens, G. C., Mirocha, C. J., Christensen, C. M. Unpublished data
109. Pathre, S. V., Mirocha, C. J., Palyusik, M. 1973. Toxic metabolites from *Stachybotrys atra.* See Ref. 32, Pap. No. 0978
110. Patterson, D. S. P. 1973. Metabolism as a factor in determining the toxic action of the aflatoxins in different animal species. *Food Cosmet. Toxicol.* 11:287–94
111. Peckham, J. C., Doupnik, B., Jones, O. H. 1971. Acute toxicity of ochratoxins A and B in chicks. *Appl. Microbiol.* 21:492–94
112. Pong, R. S., Wogan, G. N. 1971. Toxicity and biochemical and fine structural effects of synthetic aflatoxins $M_1$ and $B_1$ in rat liver. *J. Nat. Cancer Inst.* 47:585–92
113. Pong, R. S., Wogan, G. N. 1970. Time course and dose-response characteristics of aflatoxin $B_1$ effects on rat liver RNA polymerase and ultrastructure. *Cancer Res.* 30:294–304
114. Pons, W. A. Jr., Cucullu, A. F., Lee, L. S., Janssen, H. J., Goldblatt, L. A. 1972. Kinetic study of acid-catalysed conversion of aflatoxins $B_1$ and $G_1$ to $B_{2a}$ and $G_{2a}$. *J. Am. Oil Chem. Soc.* 49:124–28
115. Purchase, I. F. H., Ed. 1971. *Symposium on Mycotoxins in Human Health.* London: Macmillan 306 pp.
116. Purchase, I. F. H., Nel, W. 1967. Recent advances in research on ochratoxin. Part I. Toxicological aspects. See Ref. 94, p. 153
117. Purchase, I. F. H., Pretorius, M. E. 1973. Sterigmatocystin in coffee beans.

*J. Assoc. Offic. Anal. Chem.* 56:225–26
118. Purchase, I. F. H., Theron, J. J. 1968. The acute toxicity of ochratoxin A to rats. *Food Cosmet. Toxicol.* 6:479
119. Purchase, I. F. H., Van Der Watt, J. J. 1971. The acute and chronic toxicity of sterigmatocystin. See Ref. 115, p. 209
120. Purchase, I. F. H., Van Der Watt, J. J. 1971. Acute toxicity of the mycotoxin cyclopiazonic acid to rats. *Food Cosmet. Toxicol.* 9:681
121. Purchase, I. F. H., Van Der Watt, J. J. 1970. Carcinogenicity of sterigmatocystin. *Food Cosmet. Toxicol.* 8:289–95
122. Purchase, I. F. H., Van Der Watt, J. J. 1968. Carcinogenicity of sterigmatocystin. *Food Cosmet. Toxicol.* 6:555–56
123. Rao, K. S., Tulpule, P. G. 1967. Varietal differences in groundnut in the production of aflatoxin. *Nature* 214:738
124. Raper, K. B., Fennell, D. I. 1965. *The Genus Aspergillus.* Baltimore: Williams & Wilkins
125. Reiss, J. 1972. Nachweis von Patulin in spontan verschimmelten Brot und Gebäck. *Naturwissenschaften* 1:1
126. Ribelin, W. E., Smalley, E. B., Strong, F. M. 1973. Effect of ochratoxin and other mycotoxins on cattle. See Ref. 32, Pap. No. 0865
127. Rodricks, J. V., Lustig, E., Campbell, A. D., Stoloff, L. 1968. Aspertoxin, a hydroxy derivative of O-methylsterigmatocystin from aflatoxin-producing cultures of *Aspergillus flavus. Tetrahedron Lett.,* p. 2975–78
128. Saito, M., Enomoto, M., Tatsuno, T. 1971. Yellowed rice toxins, luteoskyrin, and related compounds, chlorine-containing compounds, and citrinin. See Ref. 25, p. 299
129. Sanders, T. H., Davis, N. D., Diener, U. L. 1968. Effect of carbon dioxide, temperature and relative humidity on production of aflatoxin in peanuts. *J. Am. Oil Chem. Soc.* 45:683–85
130. Schindler, A. F., Abadie, A. N. 1973. Sterigmatocystin production by *Aspergillus versicolor* on semi-synthetic and natural substrates: A comparison of 28 isolates and studies on the effect of time. See Ref. 32, Pap. No. 0492
131. Schroeder, H. W., Carlton, W. W. 1973. Accumulation of only aflatoxin $B_2$ by a strain of *Aspergillus flavus. Appl. Microbiol.* 25:146–48
132. Scott, P. M. 1973. Mycotoxins in stored grain, feeds, and other cereal products. In *Grain Storage: Part of a System,* ed. R. N. Sinha, W. E. Muir. Westport, Conn: 481 pp.
133. Scott, P. M., Van Walbeek, W., Harwig, J., Fennell, D. I. 1970. Occurrence of a mycotoxin, ochratoxin A, in wheat and isolation of ochratoxin A and citrinin producing strains of *Penicillium viridicatum. Can. J. Plant Sci.* 50:583
134. Scott, P. M., Van Walbeek, W., Kennedy, B., Anyeti, D. 1972. Mycotoxins (Ochratoxin A, Citrinin and Sterigmatocystin) and toxigenic fungi in grains and other agricultural products. *J. Agr. Food Chem.* 20:1103–9
135. Shank, R. C., Bourgeois, C. H., Keschamras, N., Chandavimol, P. 1971. Aflatoxins in autopsy specimens from Thai children with an acute disease of unknown aetiology. *Food Cosmet. Toxicol.* 9:501
136. Shotwell, O. L. et al 1969. Survey of cereal grains and soybeans for the presence of aflatoxin: I. Wheat, grain, sorghum and oats. II. Corn and soybeans. *Cereal Chem.* 46:446, 454
137. Shotwell, O. L., Hesseltine, C. W., Goulden, M. L., Vandegraft, E. E. 1970. Survey of corn for aflatoxin, zearalenone, and ochratoxin. *Cereal Chem.* 47:700–7
138. Shotwell, O. L., Hesseltine, C. W., Vandegraft, E. E., Goulden, M. L. 1971. Survey of corn from different regions for aflatoxin, ochratoxin and zearalenone. *Cereal Sci. Today* 16:266
139. Smalley, E. B. et al 1970. Mycotoxicoses associated with moldy corn. See Ref. 59, p. 163
140. Smalley, E. B., Strong, F. M. 1974. Toxic Trichothecenes. See Ref. 95
141. Snow, J. P., Lucas, G. B., Harvan, D., Pero, R. W., Owens, R. G. 1972. Analysis of tobacco and smoke condensate for penicillic acid. *Appl. Microbiol.* 24: 34–36
142. Steele, J. A. 1974. Biogenesis and metabolism of zearalenone by *Fusarium roseum.* PhD thesis. Univ. Minnesota, St. Paul
143. Steele, J. A., Lieberman, J. R., Mirocha, C. J. 1974. Biogenesis of zearalenone (F-2) by *Fusarium roseum* 'Graminearum.' *Can. J. Microbiol.* In press
144. Steyn, P. S. 1971. Ochratoxin and other dihydroisocoumarins. See Ref. 25, p. 179
145. Steyn, P. S., Holzapfel, C. W. 1967. The isolation of the methyl and ethyl esters of ochratoxins A and B, metabolites of *Aspergillus ochraceus* Wilh. *J. S. Afr. Chem. Inst.* 20:186
146. Still, P. E., Macklin, A. W., Ribelin, W. E., Smalley, E. B. 1971. Relationship of ochratoxin A to foetal death in

laboratory and domestic animals. *Nature* 234:563
147. Stubblefield, R. D., Shotwell, O. L., Shannon, G. M., Weisleder, D., Rohwedder, W. K. 1970. Parasiticol: A New Melatiolite from *Aspergillus parasiticus*. *J. Agr. Food Chem.* 18:391–93
148. Szabo, I., Szeky, A. 1970. Scaly and scabious skin affection and rhinitis (Stachybotryotoxicosis) in pig stocks. II. Histological investigations *Magy. Allatorv. Lapja* 92:633 (In Hungarian with English summary)
149. Tanabe, M., Hamasaki, T., Sato, H. 1970. Biosynthetic studies with carbon-13:[13]C nuclear magnetic resonance spectra of the metabolite sterigmatocystin. *Chem. Commun.* 1970:1539–40
150. Taubenhaus, J. J. 1920. A study of the black and the yellow molds of ear corn. *Texas Agr. Exp. Sta. Bull.* No. 270
151. Toussoun, T. A., Nelson, P. E. 1968. *Identification of Fusarium species*. Univ. Park, Pa.: Pennsylvania State Univ. 51 pp.
152. Tsunoda, H. 1970. Microorganisms which deteriorate stored cereals and grains. See Ref. 59, p. 143
153. Tucker, T. L., Hamilton, P. B. 1971. The effect of ochratoxin in broilers. *Poultry Sci.* 50:1637
154. Ueno, Y. et al 1971. Toxicological approaches to the metabolites of Fusaria. I. Screening of toxic strains. *Jap. J. Exp. Med.* 41:257–72
155. Ueno, Y. et al 1973. Comparative toxicology of trichothecene mycotoxins: Inhibition of protein synthesis in animal cells. *J. Biochem.* 74:285–96
156. Ueno, Y., Ueno, I. 1972. Isolation and acute toxicity of citreoviridin, a neurotoxic mycotoxin of *Penicillium citreoviride* Biourge. *Jap. J. Exp. Med.* 42:91–105
157. Ueno, Y. et al 1971. Toxicological approaches to the metabolites of Fusaria. III. Acute toxicity of Fusarenon-X. *Jap. J. Exp. Med.* 41:521–39
158. Ueno, Y., Ueno, I., Mizumoto, K. 1968. The mode of binding of Luteoskyrin, a hepatotoxic pigment of *Penicillium islandicum* Sopp. to deoxyribonucleic acid. *Jap. J. Exp. Med.* 38:47–64
159. Ueno, Y. et al 1971. Toxicological approach to (+) rugulosin, and anthraquinoid mycotoxin of *Penicillium rugulosum* Thom. *Jap. J. Exp. Med.* 41:177–88
160. Ueno, I., Ueno, Y., Tatsuno, T., Uraguchi, K. 1964. Mitochondrial respiratory impairment of luteoskyrin, a hepatotoxic pigment of *Penicillium islandicum*. *Jap. J. Exp. Med.* 34:135–52
161. Uraguchi, K. 1969. Mycotoxin origin of cardiac beriberi. *J. Stored Prod. Res.* 5:227–36
162. Uraguchi, K. et al 1972. Chronic toxicity and carcinogenicity in mice of the purified mycotoxins, luteoskyrin and cyclochlorotine. *Food Cosmet. Toxicol.* 10:193–207
163. Urry, W. H., Wehrmeister, H. L., Hodge, E. B., Hidy, P. H. 1966. The structure of zearalenone. *Tetrahedron Lett.* No. 27:3109–14
164. Van Der Merwe, K. H., Steyn, P. S., Fourie, L. 1965. Mycotoxins. II. The constitution of ochratoxins A, B, and C, metabolites of *Aspergillus ochraceus* Wilh. *J. Chem. Soc. London,* pp. 7083–88
165. Van Walbeek, W., Scott, P. M., Harwig, J., Lawrence, J. W. 1972. *Penicillium viridicatum* Westling: A new source of ochratoxin A. *Can. J. Microbiol.* 15: 1281–85
166. Vedanayagam, H. S., Indulkar, A. S., Raghavendar-Rao, S. 1971. Aflatoxins and *Aspergillus flavus* Link in Indian cottonseed. *Indian J. Exp. Biol.* 9:410
167. Vesonder, R. S., Ciegler, A., Jensen, A. H. 1973. Isolation of the emetic principle from Fusarium graminearum-infected corn. *Appl. Microbiol.* 26: 1008–10
168. Voluntir, V. 1971. Aspects of Stachybotryotoxicosis and Fusariotoxicosis in swine. *Rev. Zooteh. Med. Vet.* 21:(1)68
169. Waiss, A. C., Wiley, M. 1969. Anomalous photochemical addition of methanol to 6-methoxydifurocoumarone. *Chem. Commun.*, pp. 512–13
170. Wahlberg, C., Korpinen, E. L., Henriksson, K. 1973. A suspected outbreak of Stachybotryotoxicosis among European Elk (*Alces alces*) in the Helsinki Zoo. XV *Int. Symp. Erkrankungen Zootiere.* Kolmarden. p. 81.
171. Ware, G. M., Thorpe, C. W., Pohland, A. E. 1973. Liquid chromatographic method for determination of patulin in apple juice. See Ref. 48, p. 50
172. Wei, R. D., Smalley, E. B., Strong, F. M. 1972. Improved skin test for detection of T-2 toxin. *Appl. Microbiol.* 23:1029–30
173. Wilson, B. J. 1971. Miscellaneous *Aspergillus* toxins. See Ref. 25, 207. Miscellaneous *Penicillium* toxins. See Ref. 25, 459
174. Wilson, D. M., Nuovo, G. J. 1973. Patulin production in apples decayed by

*Penicillium expansum*. *Appl. Microbiol.* 26:124–25

175. Wolf, J. C., Lieberman, J. R., Mirocha, C. J. 1972. Inhibition of F-2 (zearalenone) biosynthesis and perithecium production in *Fusarium roseum* 'Graminearum.' *Phytopathology* 62: 937–39
176. Wolf, J. C., Mirocha, C. J. 1973. Regulations of sexual reproduction in *Gibberella zeae (Fusarium roseum* 'Graminearum') by F-2 (Zearalenone). *Can. J. Microbiol.* 19:725–34
177. Wogan, G. N. 1969. Alimentary Mycotoxicoses. In *Food Borne Infections and Intoxications,* ed. H. Riemann, p. 395. New York: Academic
178. Wogan, G. N. 1968. Biochemical responses to aflatoxins. *Cancer Res.* 28:2282
179. Wogan, G. N., Ed. 1965. *Mycotoxins In Foodstuffs.* Cambridge, Mass.: MIT. 291 pp.
180. Wogan, G. N., Edwards, G. S., Newberne, P. M. 1971. Structure-activity relationships in toxicity and carcinogenicity of aflatoxins and analogs. *Cancer Res.* 31:1936–42
181. Wogan, G. N., Newberne, P. M. 1967. Dose-response characteristics of aflatoxin $B_1$ carcinogenesis in the rat. *Cancer Res.* 27:2370–76
182. Wogan, G. N., Shank, R. C. 1971. Toxicity and carcinogenicity of aflatoxins. *Advan. Environ. Sci. Technol.* 2:321–50
183. Wyatt, R. D., Colwell, W. M., Hamilton, P. B., Burmeister, H. R. 1973. Neural disturbances in chickens caused by dietary T-2 toxin. *Appl. Microbiol.* 26:757–61
184. Yamazaki, M., Suzuki, S., Miyaki, K. 1971. Tremorgenic toxins from *Aspergillus fumigatus* Fres. *Chem. Pharm. Bull.* 19:1739
185. Yates, S. G., Tookey, H. L., Ellis, J. J., Berkhardt, H. J. 1968. Mycotoxins produced by *Fusarium nivale* isolated from tall fescue. *Phytochemistry* 7:139
186. Yuskiv, R. V. 1966. Quantitative method of determining stachybotryotoxin. *Mikrobiol. J.* 28:68 (In Russian)

# RECENT ADVANCES IN THE GENETICS OF PLANT PATHOGENIC FUNGI

❖3602

*R. K. Webster*
Department of Plant Pathology, University of California, Davis, California 95616

## SCOPE

An attempt to review recent advances of any endeavor immediately imposes difficult decisions in deciding what constitutes recent. The breadth of the present topic has intensified the need for decision-making during the preparation of this article. At the onset I would hasten to point out that the subject matter chosen for consideration is meant to be representative and it is not implied to be complete, as it is drawn from a rapidly growing literature of research reports, books, review, and symposium articles. I have attempted to avoid cataloging numerous reports under related topic areas but rather state my impressions of current status where possible. Articles providing the basis for conclusions offered are cited where appropriate.

The key to understanding variability in the fungi clearly lies in their reproductive systems. Mechanisms outside the sexual cycle which contribute and provide variability in these organisms must be understood. The relationship of reproductive systems to life cycles, their genetic control and function, and the involvement of extrasexual mechanisms need extensive evaluation. The role of life cycles and mating systems in aiding or impeding genetic investigations of pathogenicity requires careful consideration.

In many fungi self-fertilization is minimized or precluded by a variety of mechanisms which insure outbreeding and genetic reassortment in the diploid and dikaryotic forms and delay the elimination of certain genes, which may reduce fitness when homozygous. In contrast, there are many fungi which utilize a variety of inbreeding systems or for which no sexual mechanism is known. The need for genetic diversity in these organisms is either left to chance through mutation, occasional out-crossing, or other nonsexual mechanisms.

Classically, it is accepted that mutation, heterokaryosis, parasexuality, physiological adaptation, and cytoplasmic determinants provide variation aside from a standard sexual cycle. Their occurrence in the fungi has been well documented in recent years and the general presumption that they also operate in nature appears to prevail. The literature on variation and genetics of plant pathogens is proliferating as a result of the demonstration that a known phenomenon occurs in an ever-growing number of genera and species. This trend has both favorable and unfavorable aspects. Similarities denoting relatedness or minor deviations from the known provide insight into the evolution of these fungi. The less desirable aspect seems to be the near monotonous generation of series of papers describing the same phenomenon for different species without relating the biological significance of the observed similarities or deviations. The line between is thin and their ultimate value is difficult to gauge. Nevertheless, a major challenge cited by previous reviewers has been an assessment of the occurrence of these phenomena in nature. Consequently, I have attempted to emphasize this aspect in the present review.

## SOURCES AND MECHANISMS FOR VARIATION

### *Mutation*

The assumption that mutation provides the fundamental basis underlying most of the observed variation in the fungi is valid regardless of the rate with which mutations occur or of the characters that are affected. Single mutations affecting characters such as host range, virulence, fitness, tolerance to chemicals, and blocks in sexual or asexual reproduction have been demonstrated (16, 27, 38, 41–43, 57, 107).

It is believed that most mutations adversely affect basic processes in the fungi and would thus be eliminated from wild populations. On the other hand, mutation providing for an extension of host range or virulence, increased fitness, and factors encouraging out-crossing should be viewed as an advantage. When considering the extensive reproductive potential possessed by the fungi through spore production, one might assume that mutation alone would provide adequate variation to allow for adaptation and survival of advantageous genotypes. This possibility is suggested for some asexual fungi that function in the wild as haploids (28), although conclusive evidence for such is not easily obtained. In this case mutants are immediately exposed and selected. Conversely, it is well known that fungal elements may be multinucleate or heterokaryons. The role and necessity of the dikaryon in many of the basidiomycetes, particularly the rusts and smuts, is recognized. Recently considerable interest has been generated regarding the potential occurrence of diploids in many fungi previously believed to be haploid (13, 45, 79, 81, 127, 128, 140). Thus, mechanisms involving heterokaryosis or diploidy account for the major portion of conservation of variation provided by mutation in thalli of fungi. Without such, many mutations which appear to be detrimental at the time would be eliminated rapidly from populations restricted to homokaryotic or haploid vegetative growth.

The genetics of the higher fungi and the systems involved in conserving variation have been thoroughly reviewed by Raper (109). Mechanisms for conserving variation in the ascomycetes are considerably more unstable, as many of these fungi produce uninucleate cells. In cases where multinucleate cells occur, associations of different nuclear types are frequently precluded during conidial formation. Numerous studies have shown, however, that species characteristically producing single nucleate conidia also produce a few conidia with two or more nuclei, either by accident or by some unknown mechanism. Nevertheless, conidial and sexual spore formation appear to be a primary means of dissociation of nuclear types in many of the fungi.

The majority of studies on the genetics of plant pathogenic fungi are concerned with the mechanisms which allow for the sorting out and expression of variation as nuclei carrying mutant characters are dissociated from the complementing characters in the wild-type nuclei or in heterozygous diploid nuclei. As our knowledge of these mechanisms has evolved, a number of reports have also considered variation to occur by adaptation, typified as a mechanism which does not involve genetic change. Adaptation in fungi was reviewed by Buxton (16) in 1960 and more recently by Person (120). The concept of adaptation is controversial because of the difficulties in distinguishing between nongenetical changes and mutation followed by selection. Induced adaptive changes involving cytoplasmic elements (84) as opposed to nuclear elements must also be considered. There is considerable need for extensive study in this area in regard to reports of fungicide tolerance or resistance in plant pathogens.

## *Nuclear Cytology and Cytogenetics*

The mitotic cycle is undoubtedly the most important mechanism known to occur with regularity in living organisms. It insures that information coded in DNA is passed from one cell generation to the next. Although there are chemical and physical agents known which can upset or rearrange the mitotic cycle, it occurs with remarkable regularity. The replication and separation of the genetic information into daughter nuclei is the main function of the mitotic cycle in fungi, as nuclear replication and division is not always accompanied by the production of new cells. The manner of nuclear division in the fungi has invoked a major controversy during the last 25 years. Conflicting reports on mode of duplication and division of nuclei exist within all classes. Olive (103) concluded after an extensive review of the literature that the fungal nucleus divides by classical mitosis and that contrary interpretations resulted mainly from inadequate techniques and the inability to discern nuclear divisions in the traditionally small nuclei of fungi. Various schools of thought have since emerged (110). Pontecorvo (122) described the parasexual cycle in fungi which provides for exchange of whole chromosomes or chromosome parts during somatic division of diploid nuclei. Concurrently, the literature describing various mechanisms for nuclear division in the fungi is enhanced while the parasexual cycle seemingly dictates classical mitosis.

Most treatments of fungal genetics (38, 41–43, 122) apparently have assumed nuclear division in the fungi to be analogous to that known for higher organisms. This is understandable inasmuch as nuclei participating in fusion nuclei arise by somatic nuclear division and behave as would be expected in inheritance studies and in maintaining genetically pure clones.

Olive (106) reviewed meiosis in fungi and concluded that the mechanism in the fungi has more in common than in variance with other organisms. Differences noted in the fungi include the onset of synapsis by contracted chromosomes, intranuclear spindles, and asynchronous disjunction of chromosomes at anaphase. The fundamental nature and the products of fungal meiosis are similar to that in higher forms as attested by extensive genetics studies, particularly in the Ascomycetes.

The behavior of fungal nuclei during somatic division must now be viewed as somewhat further removed from classical mitosis than was concluded by Olive (103). Known differences include: 1. division occurs entirely within the nuclear membrane; 2. the absence or failure of homologs to align on a metaphase plate; and 3. the possible occurrence of chromosomes attached end-to-end by a threadlike structure, although the nature of the thread and its function are not known. In addition, the nature and function of a spindle apparatus and the nature and role of the nucleolus are poorly understood (24, 110).

Although much of the literature dealing with nuclear division in the fungi is concerned with saprophytes, a number of plant pathogens emerge in the midst of the controversy. Notable examples are *Fusarium oxysporum* (1), *Ceratocystis fagacearum* (1, 2), *Alternaria tenuis* (65), *Rhizoctonia solani* (46, 134), *Macrophomina phaseoli* (89, 90), *Verticillium albo-atrum* (71, 115), *Penicillium expansum* (44, 92), *Thielaviopsis basicola* (80), *Piricularia oryzae* (58), *Puccinia graminis* (96), *Phymatotrichum omnivorum* (78), and others. Tinline & MacNeill (139) cited the need for reconciliation of cytologic and genetic data in the fungi in their review on parasexuality. Considerable difficulty exists in interpreting parasexuality, particularly the steps involving mitotic crossing-over and recombination in the absence of a division process that does not allow for exchange of segments of homologous arms of chromosome pairs during division of heterozygous diploid nuclei. Alan Day (24) has ably reviewed areas where clarification is needed to accomplish this reconciliation. After an extensive study and comparison of published figures, he described models of division which, in his opinion, are widespread in most fungi other than Phycomycetes. The two-tract type of division as suggested by Day allows for nonalignment at metaphase and the possible interconnection of the chromosome complement by a nuclear thread. It differs slightly from models proposed by Robinow & Caten (111) and by Duncan & Macdonald (35). At this time it is not possible to determine which of the three is most likely. However, all models with appropriate assumptions account for most of the nuclear configurations reported by many authors and also allow a reconciliation with genetic data. Paramount to this point is the fact that there appears to be no clear basis for assuming that nuclear division in all fungi is similar. Whatever mechanisms finally prevail, it is evident that a level of genetic continuity similar to that provided by classical mitosis is achieved. Evidence is conclusive that the genome maintains its integrity during somatic growth and production of the nuclei that participate in the fusion nucleus during sexuality, regardless of the limits of our understanding of the processes involved. Nevertheless, the fact that many workers have resisted the possibility that nuclear division in somatic cells of at least some fungi may replicate and divide by a process different from conventional mitosis may have retarded progress on understanding basic phenomena in the fungi.

PHYTOPHTHORA CYTOLOGY AND GENETICS Cytology and genetics of species in the genus *Phytophthora* have received considerable attention in recent years. Prior to 1962, it was generally believed that meiosis in the oomycetes occurred at germination of the zygote (oospore). However, as a result of cytological studies on species of *Achlya, Pythium,* and *Phytophtora,* Sansome (125–128) reported that meiosis occurs in the sex organs, which implies that the vegetative thallus of these fungi is diploid. Barksdale (5) recently provided cytological evidence that *Achlya ambisexualis* is diploid and Bryant & Howard (13) suggested that *Saprolegnia terrestris* is diploid on the basis of a microspectrophotometric analysis of nuclear DNA. Subsequent studies with *Phytophthora* spp. (135) have both supported and refuted this view. Analysis of a few of these studies provides a striking example of the need for understanding life cycles and sexuality in attempting to interpret variation and genetic studies in plant pathogens.

Savage et al (131) categorized various classes of sexuality in *Phytophthora* as 1. homothallic with predominantly paragynous antheridia, 2. homothallic with amphigynous antheridia, or 3. heterothallic with compatibility types A1 and A2 as determined by mating with testor strains of *Phytophthora infestans.* Gallegly (53) suggested that the heterothallic species may be haploid and the homothallic species diploid. Support for this conclusion is not available. To the contrary, Galindo & Zentmyer (52) reported cytological observations on the heterothallic *Phytophthora drechsleri* which were similar to those of Sansome, while Marks (95) published observations on the heterothallic *Phytophthora infestans* which do not support the conclusions of Sansome. Analysis and interpretation of available genetic data for this group of fungi is frustrating due to the uncertainty of their ploidy. Often data are discussed from both viewpoints (33, 53, 55, 121, 130, 138).

Galindo & Gallegly (51) assumed that *P. infestans* was haploid in its vegetative stage and therefore interpreted compatibility types A1 and A2 to be allelic. The occurrence of these two types in Mexico in a one-to-one ratio was cited as evidence. Reports by Romero & Erwin (112, 113) for this species reveal the predominance of A1 types in the progeny of an A1 X A2 cross. A one-to-one mating type ratio in progeny in crosses with strains of *P. drechsleri* was observed by Galindo & Zentmyer (52). On the other hand, Satour & Butler (130) obtained a two-to-one ratio in progeny of *P. capsici,* while Polach & Webster (121) reported a one-to-one ratio in this species. Timmer (138) suggested that mating type may be controlled by three independently segregating alleles.

Genetic studies should provide evidence revealing the ploidy of *Phytophthora* species, but thus far none have done so. The ratios of progeny types obtained by Galindo & Zentmyer (*P. drechsleri*), Satour & Butler (*P. capsici*), Romero & Erwin (*P. infestans*), Timmer (*P. capsici*), and Polach & Webster (*P. capsici*) are subject to either haploid or diploid interpretation. Sansome (129) interpreted the data of Galindo and Zentmyer as being indicative of a diploid somatic state and credited the atypical segregation ratios of characters to the possibility that selfing had occurred in one of the parent isolates. A report by Haasis & Nelson supports this view (61). This reviewer believes that the significance of progeny ratios as a means to resolve the question of ploidy has been overemphasized. This is based on the following: 1. genotypes of strains used and genetic control of many characters employed such as colony morphology and mating type are poorly defined; and 2. in all studies known to me, germination percentages of oospores are very low, which suggests the possibility that progeny obtained do not represent random samples of the populations under study. In addition, there is no evidence that fusion of nuclei occurs in the possible selfed oospores suspected to influence ratios.

Mating type of isolates has been studied as a segregating character in most reports on the genetics of *Phytophthora.* Whether the vegetative stages of these fungi prove to be diploid or haploid, a reevaluation of genetic control of compatibility may be necessary. Gallegly (54) suggests dominance and recessiveness might be involved if the nucleus is diploid and the character is under monogenic control and cites the possibility that one type could be *a a* and the other *A a.* With appropriate assumptions, the published results with other possible combinations of alleles that could occur at the mating type locus in matings can be reconciled. An additional interpretation of mating type control that appears attractive to me invokes rationale from a number of sources. It is generally agreed that single strains of *Phytophthora* species considered to be heterothallic produce a few oospores in culture under proper conditions. However, appropriate matings between different isolates yield oospores profusely at the point of contact, demonstrating an indisputable interaction between them. Brasier indicates that a similar induction of oospore formation in isolates of heterothallic species can be brought about by pairings with an isolate of *Trichoderma viride* (10). Thus the established practice of utilizing a designated A1 or A2 mating type of *Phytophthora infestans* to determine the mating type of isolates of numerous other species of the genus merits examination. In this case, the inability to germinate these oospores has precluded genetic analysis of the presumed products of interspecific matings and thus evidence that they represent hybridization is lacking. Early papers on sexuality in fungi are of interest at this point. Raper (107, 108) separated genetic factors controlling sexuality into two categories: 1. essential genetic determinants for culmination of sexual reproduction, and 2. factors that affect the quantitative expression of sex, such as nutrition. Olive (104, 105) emphasized the importance of distinguishing between hybridization and induced selfing in a basically self-fertile organism that possesses the basic genetic determinants for sexuality as described in category 1 by Raper.

The production of some oospores in single cultures of most isolates of *Phytophthora,* the induction of oospores by *T. viride* in single isolates, and the use of A1 and A2 mating types of *P. infestans* to determine mating reactions in other species of *Phytophthora* suggest that a portion of the oospores formed between mated cultures in this genus are the result of nutritional or other type of induction. Whether they represent selfing is dependent upon the occurrence of a fusion nucleus, a fact thus far not adequately demonstrated. As stated earlier, Sansome has explored this possibility in reinterpreting available genetic data in the genus *Phytophthora.* Thus it appears logical to assume the mating reaction attributed to compatibility between isolates of *Phytophthora* could in effect be complementation between a number of genes, in most cases closely linked. Support for this conclusion can be found in an example with *Phytophthora capsici.* Polach & Webster (121), in a cross between two isolates of this species each apparently self-sterile and cross-fertile, obtained progeny from 391 individual oospores. Of these, 389 proved to be either of two mating groups in back-crosses to the parent isolates. 180 mated only with one parent isolate, while 209 mated only with the other parent isolate, which suggests that mating type is controlled by a single locus with two alleles. Oospores were produced abundantly in the back-cross matings within 10 days. The remaining two progeny apparently were homothallic recombinants, as they produced oospores in equal abundance in single zoospore cultures when compared with the back-cross matings of 387 other progeny. On the other hand, the occurrence of the two apparent recombinant isolates shown to be homokaryons and homothallic may indicate that recombination between factors for oospore production occurred in the cross of these two isolates. Unfortunately oospores produced by the self-fertile recombinants failed to germinate. Whether the homothallic recombinants represent intra-allelic recombination at a compatibility locus or between factors controlling ability to produce oospores cannot be concluded. In either event, these data suggest "compatibility loci" in heterothallic isolates of *Phytophthora* may not be identical to those known for Ascomycetes (40).

It seems probable that the sexual process (with or without a major locus for compatibility) is controlled by a segment of a chromosome containing several loci with each locus responsible for a particular step in oospore development. If all genes are present in functional form, oospores will be produced and the isolate is homothallic. The maleness or femaleness of isolates as described by Galindo & Gallegly (51) could best be explained under this condition assuming that only segments controlling the male properties or female properties occurred in functional form of a particular isolate in question. In this case oospores might be formed through cross-feeding as described by Olive (104, 105) if the two isolates were in close proximity as in a mating. It could account for results reported by Brasier (10) and data utilizing the mating types of *P. infestans* (131) in determining mating capacity of other isolates. In addition, the cross-feeding possibility could explain the predominance of parental types in many genetic studies of *Phytophthora* spp. Nonetheless, at least a portion of the oospores produced in pairings between genetically related strains represent hybridization. Future studies involving larger numbers of progeny should provide information regarding possible numbers of genes and their relationship.

In some respects, it might be advantageous if we could begin anew with studies on life cycles, nuclear cytology, and sexuality in *Phytophthora.* The accumulated literature appears to be interwoven by conventional precepts that have prevailed with the result that answers to basic questions on the biology of these organisms have eluded us.

### *Heterokaryosis*

*Heterokaryosis* is generally defined as the coexistence of genetically different nuclei in cytoplasmic continuity. Various authors have differed as to whether definitions of a heterokaryotic condition should be restricted to a single cell or a single thallus.

These differences have been discussed by Parmeter et al (117). The cells in actively growing mycelium, whether it be multinucleate or uninucleate, are continuous with one another due to the occurrence of septal pores (8, 148). Cytoplasmic streaming and translocation (37, 72) in the fungi through the cytoplasm have been well documented. The validity of considering the individual cell as a separate unit of organization is a matter of some question in regard to the filamentous fungi. Davis (23) considers the mycelium as an integrated organism with a population of nuclei.

Considerable general discussion of the potential advantages of heterokaryosis exists in the literature. Authors generally agree that heterokaryosis provides a haploid organism with many of the advantages of heterozygosity enjoyed by diploids. An additional advantage may result from the lack of control mechanisms guaranteeing maintenance of the heterokaryon in some fungi. Simple changes in nuclear proportions could provide additional somatic flexibility to heterokaryons when confronted with changing environments (23). These appear to be immediate advantages. The main farther-reaching advantage of heterokaryosis appears to be the fact that it is a requisite step for genetic recombination either by interchanges of whole chromosomes or through mitotic crossing over. The significance of heterokaryosis in this regard would appear to favor the homothallic and the imperfect fungi. This subject has been extensively reviewed by Caten & Jinks (20).

The primary theme of Parmeter et al (117) was the need for conclusive evidence that heterokaryosis is, in fact, occurring. They described appropriate methodology for such demonstrations. Davis (23) has also stressed the need for careful methodology and considered measurement of nuclear ratios, dominance in heterokaryons, the genetics of heterokaryon formation, and its occurrence in the fungi in general.

THE ORIGIN OF HETEROKARYONS Heterokaryons may arise by a number of methods including mutation in a multinucleate thallus, anastomosis and nuclear exchange between strains of closely related fungi, or formation of multinucleate spores. The role of mutation in this regard is obvious. Parmeter (116) discussed the need for additional information on anastomosis and its relationship to heterokaryosis. Examples of heterokaryon incompatibility are well known (23, 75, 85) and progress has been made in determining the genetic control of such systems. Many workers who were unable to demonstrate heterokaryosis have attributed their lack of success to the failure of strains to anastomose or to various types of post-anastomosis incompatibility. Little is known of the mechanisms involved. Considerably more information on anastomosis control and incompatibility will be required before conclusive assessment of the genetic potential of this phenomenon in nature can be drawn. In this regard, the review of Caten & Jinks (20) is of interest. Briefly, these authors divide heterokaryons into three types: forced, neutral, and naturally occurring. Forced heterokaryons are characterized by the use of auxotrophic mutants. In this case, a demonstration of heterokaryosis depends on the complementation in the heterokaryon formed between two strains carrying nonallelic auxotrophic mutants. The extreme selection favoring the formation of the forced heterokaryon is obvious. The second category includes what are termed *neutral* heterokaryons. These include examples of heterokaryosis where the heterokaryon is not deliberately or artificially placed at a selective advantage over the homokaryon components. The third category cited is naturally occurring heterokaryons that are identified through the examination of wild-type strains under nonselective conditions. To identify this class requires considerably more labor and essentially consists of isolating samples of the fungus from the wild and determining directly the proportion of heterokaryons in the sample. As pointed out by Caten & Jinks (20), this latter approach has been attempted only on a few occasions. These authors have convincingly argued that forced heterokaryons may not reveal valid information regarding the true situation in nature. Ample evidence exists to suggest that forced heterokaryons are considerably more unstable than the other categories when removed from extreme selection pressure.

HETEROKARYOSIS AND PATHOGENICITY Heterokaryosis has been demonstrated in a number of pathogenic fungi. Unfortunately, the majority of these demonstrations utilized techniques similar to those resulting in forced heterokaryons. By strict definition, the dikaryon of the rusts and smuts represent true heterokaryons, although some authors justifiably prefer to consider them a special type. The heterokaryotic nature and role of dikaryons in rusts and smuts has been well established. On the other hand, little is known regarding the basic differences between the pathogenic dikaryon vs the saprophytic monokaryon in the smuts. This question and the situation of nuclear alteration on alternate hosts in the rusts has intrigued physiologists and pathologists for years. Hopefully, recent advances in methodology allowing the culture of certain of these organisms will provide a means for increased understanding in this area (29, 133).

Parmeter et al (117) and Caten & Jinks (20) conclude that many of the claims for heterokaryosis in plant pathogens have not been substantiated to occur in nature, and evidence for the significance of the phenomenon in fungi other than the rusts and smuts is in general lacking. A recent report by Spector & Phinney (136) has shown that two genes control gibberellin synthesis in *Gibberella fujikuroi.* This finding, combined with a report by Ming, Lin & Yu (98), provides a demonstration of the importance of heterokaryosis in relation to natural variation in virulence in this fungus. Ming, Lin & Yu (98) described three distinctly colored strains of *Fusarium fujikuroi:* purple, red and white. Upon isolation of single uninucleate microconidia, these strains propagated true to type. The three differ significantly in quantitative production of gibberellin in culture. The white form produced very little gibberellin, the purple form at least 500 times that of the white form, and the red form was intermediate. Gibberellin production in this species is closely correlated with virulence. For all practical purposes the white form was avirulent, or nonpathogenic. The relative significance of these observations is that these three genotypes occur regularly as components of wild heterokaryotic isolates in China and may be demonstrated through the method described by Caten & Jinks (20) for naturally occurring heterokaryons. Most heterokaryons found contained only two of the three genotypes in a single hyphal tip. Obviously, this system provides the opportunity for further study and comparison of the frequency and significance of heterokaryons as they occur in wild populations.

Heterokaryosis has received considerable attention in studies relating to *Rhizoctonia solani* (47). The synthesis of heterokaryons in *R. solani* was reported by Whitney & Parmeter in 1963 (147). This observation was supported by later findings of Garza-Chapa & Anderson (56), Vest & Anderson (143), and Stretton & Flentje (137). Collectively, these works show that the occurrence of heterokaryosis in *R. solani* is limited by two groups of factors. One category is described as *anastomosis groups* by Parmeter, Sherwood & Platt (118). Isolates belonging to different anastomosis groups are unable to anastomose with each other and thus the formation of heterokaryons is precluded between groups. Recently, the second group of factors, termed *H factors* by Anderson et al (4), have been clarified. These authors clearly demonstrate that within each anastomosis group the formation of heterokaryons is controlled by heterokaryon incompatibility factors. These may occur as alleles at at least two loci and allelic series have been identified. They demonstrated that only isolates belonging to the same anastomosis group that carry different H factors are able to fuse and form heterokaryons, and they described at least 17 such H factors. Isolates with common H factors do not fuse.

The majority of reports of heterokaryosis in *R. solani* are based on observed variation and mating reaction between colonies derived from single basidiospores obtained from self-fertile isolates from nature. The fact that many isolates induced to fruit have been considered to be heterokaryotic has promoted speculation that heterokaryosis plays a significant role in the pathogenicity and virulence of this species. Reports are available to support this view. For example, Garza-Chapa & Anderson (56) and Vest & Anderson (143) have demonstrated that virulence in *R. solani* on flax may be increased as a result of heterokaryosis. Unfortunately, conclusive proof on the formation of heterokaryons used in the foregoing work was not presented. McKenzie et al (97) paired nonpathogenic mutants and obtained heterokaryons which showed virulence equal to the wild type on radish seedlings. In the above examples the component strains of the heterokaryons showing increases in virulence were derived from single basidiospore cultures and therefore involved closely related genomes. That these authors have demonstrated the role of heterokaryosis in maintaining virulence in this organism is not disputed. Their results, however, do not support claims that heterokaryosis in this organism may be the basis for its ability to attack a wide range of host plants. Bolkan & Butler (6, 15), prompted by the discovery of Day (29) that charcoal incorporated into the media enhanced the possibility of growing heterokaryons in culture, have been able to demonstrate that heterokaryons can be formed between different heterokaryons of *R. solani* provided the conditions imposed by the anastomosis group and H factors are met. Consequently, additional genomes may be introduced into an already heterokaryotic isolate. The results obtained by these authors in regard to the impact of newly formed heterokaryons on virulence to a number of hosts differ significantly from those obtained by the previous authors (6). For example, when a heterokaryon between two highly virulent isolates was formed (already heterokaryotic), the newly synthesized heterokaryon showed a near complete loss of virulence to both hosts upon which the component isolates were highly virulent. This suggests that additional heterokaryon formation is disadvantageous to *R. solani* in regard to virulence. A genetic interpretation of these results awaits further study. The authors (6) suggest that the newly synthesized avirulent heterokaryon has a survival advantage. Thus, upon reassortment of the nuclear types, virulence would again be available in the basidiospore progeny.

The major drawback in future studies with *R. solani* appears to be an inability to produce a sexual stage with regularity. There is no evidence at this time that either the anastomosis group or H factor genes has an effect on basidiospore production in this organism. The cytological studies by Flentje et al (46) describing near conjugate division of the nuclei in the vegetative mycelium of *R. solani* indicate that the primary method for disassociation among heterokaryons is basidiospore formation. Therefore, further study on this aspect of the biology of *Rhizoctonia* is necessary to parallel progress on studies of heterokaryosis.

The recent reports or implications that diploidy may occur more frequently than previously assumed in Fungi Imperfecti, particularly *Verticillium albo-atrum* (81, 140), and in the Oomycetes diminish the essentiality of heterokaryon formation prior to mitotic recombination. Further discussion of heterokaryons in *Verticillium* is included in the next section. On the other hand, studies such as those mentioned on *Verticillium* (70, 140) and *Rhizoctonia* have contributed essential new knowledge in relationship to the role and significance of heterokaryosis in maintaining virulence and survival of these pathogens in nature. The demonstration that particular substrates may in fact influence nuclear ratios in heterokaryons (23) suggests the need for more studies on tolerance to fungicides in plant pathogens and the effects or impact of heterokaryosis on its occurrence and stability (22, 145, 146) in wild populations.

## *Parasexuality*

Pontecorvo (122) pioneered the way for genetic analysis of asexual fungi with his discovery and descriptions of parasexuality. Numerous reviews have dealt with the subject in recent years (9, 18, 114, 139).

Tinline & MacNeill (139) considered the parasexual cycle in *Aspergillus nidulans* as a model for comparisons of the phenomenon as it occurs in other fungi. In *A. nidulans*, a demonstration of heterokaryosis, formation of heterozygous diploids, and recovery of diploid and haploid segregant genotypes constitute proof for the occurrence of parasexuality. The availability and employment of extensive marker genes and established methodology for detecting diploids facilitate demonstrations of the phenomenon in this species. Where unequivocal demonstrations of parasexuality in plant pathogens exist, similar methodology has been employed. Tinline & MacNeill (139) have recently reviewed parasexuality in plant pathogens. Therefore, discussion here is limited to those cases where there has been progress in demonstrating the phenomenon in nature or near natural situations.

With some obligate plant pathogens, such as the rust fungi, it is not possible to use certain established culture techniques utilized in studies on parasexuality. In this case, the investigator may employ strong selection forces inherent in the host to screen for certain genotypes in the parasite, permitting the phenomenon to be studied under conditions more similar to those encountered in nature. Consequently, new strains have been identified on hosts inoculated with mixtures of urediospores of two races of *Melampsora lini, Puccinia graminis* f. sp. *tritici, P. recondita* f. sp. *tritici,* and between *P. graminis* f. sp. *trici* and *P. graminis* f. sp. *secalis.* Explanations for the origin of the new races include parasexuality (11, 14, 36, 144), asexual variation (64), somatic segregations (60), somatic variations (49), asexual crossing, unstable heterokaryosis (102), and mutation and reassortment of whole heterokaryotic nuclei (95).

Whether such a variety of mechanisms actually exists awaits further study. The hesitancy of authors to term their observations a result of parasexuality is understandable, as all essential steps of parasexuality cited earlier were not observed. The reassortment of whole nuclei (94) is a valid explanation in some cases, although it should not be invoked when the multiplicity of new strains exceeds the number that can be accounted for by this mechanism. The main complication appears to be alternative explanations to parasexuality introduced by the dikaryotic stage characteristic of these organisms. In vitro studies with certain of the rusts (133) should expedite elucidation of the mechanisms involved.

Recent examples of parasexuality in the smut fungi are noteworthy. Kozar (91) inoculated barley (*Hordeum vulgare*) with compatible mixtures of sporidia of *Ustilago hordei* carrying biochemical markers and obtained evidence for nuclear fusion in the infective dikaryon which produced diploid sporidia. Chemical and cytological tests revealed them to be uninucleate and to contain approximately twice the amount of DNA when compared to a known haploid of *U. hordei.* Infection tests showed their ability to attack host plants, indicating the diploids contained both compatibility types. Ruling out normal sexuality, solopathogenicity, and mutation, the author concluded that recombinants recovered from the host plant arose from an operative parasexual cycle.

*Ustilago violacea,* the anther smut fungus, has been utilized by A. Day & Jones (25) in demonstrating the production of vegetative diploids and mitotic haploidization. They noted no discrepancies in linkage data for 33 markers between meiotic segregation and mapping by mitotic haploidization. Thus in this pathogen all essential steps of the parasexual cycle as outlined by Roper (114) have been demonstrated.

P. Day & Anagnostakis (30) analyzed meiotic products from natural infections on corn (*Zea mays*) by *U. maydis.* In this case whole tetrads from single teliospores produced colonies that could be distinguished on the basis of compatibility reactions between the meiotic products of the teliospore. They identified incomplete tetrads, solopathogens (unreduced diploids), and a group containing only one cell type for compatibility, none of which were solopathogens but proved to be diploid based on DNA content. The latter were considered homozygous for the *b* factor. The origin of this class may be accounted for by assuming mitotic crossing-over prior to teliospore formation and the failure of teliospores homozygous for *b* to undergo meiosis upon germination. The significance of these observations lies for the most part in the following: 1. mitotic recombination is well known in *U. maydis* (73, 74) but prior to this its demonstration has been dependent on the use of auxotrophic markers, and 2. the methods developed and employed by these workers (30) allow its demonstration in field material from natural infections, negating the need for continued assumptions that mitotic recombination may be but a laboratory phenomenon in this pathogen.

The parasexual cycle has been described in outline in *Verticillium albo-atrum* by Hastie (66–68). These and more recent studies reveal differences in detail from the *A. nidulans* model. For example, heterokaryons and heterozygous diploids of *V. albo-atrum* are less stable than are those in *A. nidulans.* Hastie (70) suggests the easiest method of demonstrating heterokaryosis in *Verticillium* is to search for heterozygous diploids, implying that their occurrence indicates a transient heterokaryotic phase. He further states that all heterozygotes in *Verticillium* frequently form homozygous diploid segregants by mitotic recombination. Estimates of 0.2 mitotic recombination/nuclear division are suggested for *V. albo-atrum.* Hastie observed frequencies of mitotic nondisjunction higher than known in *A. nidulans* and suggested this accounts for a low recovery of diploids. This contrasts with observations by Ingram (81) who observed a very low frequency of mitotic nondisjunction but a high frequency of diploids in *V. dahliae longisporum.* It will be interesting to see how the recent report on diploidization and heritable gene repression depression on variability in *V. albo-atrum* by Tolmsoff (140) is reconciled with earlier studies. The results of Roth & Brandt (115) also suggest that the occurrence of multinucleate conidia is more frequent than previously assumed. Whether or not these observations invalidate conclusions drawn from individual phialide (69) analyses remains to be seen. Nevertheless, the demonstration of hyphal fusions (132) between germinating conidia in contact with soil and the extensive occurrence of diploid nuclei in *V. dahliae longisporum* (81) and *V. albo-atrum* (140) indicate the opportunity for culmination of the parasexual cycle exists in this genus and that for all practical purposes it may occur in nature.

Asexual variation in the genus *Phytophthora* has intrigued many investigators in the past and the subject has been extensively reviewed (21, 39, 55, 135). It has been generally concluded by these authors that since anastomosis occurs at best only rarely, if heterokaryons occur they arise most likely from mutation in the multinucleate thalli. If, however, the asexual stages of these fungi are diploid, as suggested by Sansome, the occurrence of heterozygous diploid nuclei followed by mitotic crossing-over could account for at least a portion of the variation observed in these organisms. In this case haploidization may or may not be involved since recombinant homozygous diploid nuclei have a high probability of being established as separate strains during the formation of sporangia and zoospores. On the other hand, the assumption by proponents of the hypothesis that meiosis occurs in sex organs does not preclude the possibility that somatic hyphae are haploid and that fusion occurs in the sex organs just prior to meiosis.

A few recent reports have suggested the occurrence of parasexuality in *P. infestans.* Denward (32) studied variable reactions of eight different races on two sets of differential hosts. Somatic recombination on an intra-allelic basis resulting in the creation of a new allele for pathogenicity on corresponding R genotypes was suggested in another paper (33) to account for the apparent recombinants. In this study, however, the possibility of a temporarily self-replicating extranuclear structure in sectors of the mycelium was not ruled out. Leach & Rich (93) noted three genotypes from a single sporangium differing from the original combination employed, and suggested heterokaryosis and parasexuality as mechanisms to explain the occurrence of the apparent recombinants. These workers observed infection of potato plants with R genes by certain combinations of races in which the complementary p gene was absent (in all of the original races) and attributed it to cytoplasmic inheritance. The fact that both of the above reports invoke a cytoplasmic mechanism to explain a portion of their results further stresses the need for a clearer understanding of the somatic nuclear cycle in the genus. Conclusive evidence either for or against the existence of a plasmatic genetical system and parasexuality awaits its resolution.

## *Cytoplasmic Factors*

Earlier reviews have generally concluded that few, if any, reports of extranuclear inheritance of factors conditioning pathogenicity in fungi are available. This generalization does not appear to apply when considering factors affecting virulence (aggressiveness) and fitness in strains. Reports by Caten (19), Leach & Rich (93), Caten & Jinks (21), and Denward (33) support this conclusion in the case of *Phytophthora,* and by d'Oliviera (34), Johnson (86), and Green & McKenzie (59) in *Puccinia.*

The suitability for study and widespread occurrence of extranuclear inheritance in the fungi has been reviewed by Jinks (84) and Fincham & Day (43). A wide spectrum of morphological, physiological, and biochemical variants have been identified in the fungi in which their genetic control can be attributed to extrachromosomal mutation. Thus the assumption that a similar mechanism for variation can exist in plant pathogens seems valid. It may be that the genetic mechanisms pertaining to the host-parasite relationship in general reside in the nuclei, but the accounts cited above indicate this is not always the case. Unfortunately much of the information implying extranuclear inheritance has been carried out with strains in which the genetic background is poorly defined. This can also be said for many studies of genetic analysis utilizing sexual processes. Until better-defined strains are available and a more complete knowledge of the nuclear cytology of many pathogens is at hand, it is likely that difficulties in interpreting results on a Mendelian basis will continue to give rise to suggestions of extrachromosomal determinants. By the same token, further study will also provide more clearly defined examples of non-Mendelian inheritance. The seeming absence of clear-cut demonstrations of extrachromosomal control of pathogenicity does not justify an assumption that this source of variation plays no role in the genetics of plant disease. The fact that extrachromosomal mutations lead to losses of particular developmental pathways such as conidiophore and conidial production serve as examples that a normal extrachromosomal complement is essential (82–84). Further, extensive investigations on control of perithecial production in sexual Aspergilli have revealed influence of an extrachromosomal system (84). On the other hand, as extrachromosomal variants such as reduced sporulation can be considered as defects, these types would not likely play a significant role in variation and adaptation in nature and may be eliminated. Such would be expected in the modified races of *Puccinia graminis* and *P. anomala [P. hordei]* reported by Johnson (86) and d'Oliveira (34). The extrachromosomal variants that showed extended pathogenic capabilities in *Phytophthora* would result in an adaptive advantage.

An interesting conclusion from intensive investigations on various plant and animal genera is that extrachromosomal systems produce barriers to out-crossing, thus enhancing the emergence of species (17). Similar conclusions for fungi are not possible at this time, but the extrachromosomal-chromosomal interactions between strains of *U. maydis* as described by Puhalla (123) present an interesting possibility. In this case, interstrain inhibition occurs due to the excretion of a heat-labile protein substance (63) by certain strains (P1) which inhibit growth of P2 strains. P2 strains (P2s) produce no inhibitor and are either sensitive or insensitive. The difference between the sensitive and insensitive P2 strains is controlled by a nuclear gene with sensitivity dominant to insensitivity. Another type, P3, produces no inhibitor and does not express the sensitivity nuclear gene. All strains are stable in culture. Appropriate analysis has shown that production of inhibitor is cytoplasmically determined and that loss of the cytoplasmic element is irreversible. A similar system has been described by Anagnostakis (3).

Day & Anagnostakis (31) have shown that "cytoplasmic determinants [I] or production of the 'killer' protein and the factor for resistance [S] are transmitted at dikaryon formation and [I] is lost at low frequencies during vegetative growth." They were not successful in attempts to transfer the [S] factor to P3 cells. Evidence that the [S] factor is a virus-like particle is presented by Wood & Bozarth (149). Superficially it would seem that the existence of a system such as that emerging would constitute an extravagance on the part of *U. maydis*, as no one would argue against the effectiveness of the known system enforcing outbreeding. On the other hand, its suitability (73) as a genetic tool combined with recent advances in methodology (29, 124) have provided further insight into mechanisms for variation in fungi. The demonstration of the virus-like nature of the [I] and [S] factors in *U. maydis* should stimulate further study in fungi along a different line of thinking.

## GENETICS OF THE HOST–PARASITE INTERACTION

Since the rediscovery of Mendel's papers, studies on the inheritance of the disease reaction between host and parasite have progressed through three more or less distinct phases. In the earliest studies, emphasis was concerned with determining the number of resistance-conditioning genes in the host. This resulted in identification of gene pairs segregating in the many individual host hybrids studied. Little information on distinctiveness or duplication of disease-resistant genes in various resistant varieties was revealed. During the next phase, identification of individual disease-resistance conditioning genes came to the forefront. Host plant hybrids were tested against single physiologic races of pathogens in experiments designed to identify individual genes for host resistance. This was followed by testing the disease reaction of these individual host genes to various known races of the pathogen. Perhaps the third phase could be identified as the genetics of the host-pathogen interaction. It is characterized by studies revealing specific genes for pathogenicity or nonpathogenicity as identified in the pathogen and relating them to specific genes for resistance or susceptibility in the host. This phase is characterized by the gene-for-gene concept as proposed by Flor (48).

Advancement of our knowledge in these phases has differed widely for various host-pathogen relationships, mainly due to the disparity among the various hosts and pathogens in their suitability for precise genetic study. Unfortunately, our inability to perform the necessary genetic analysis has not tempered assumptions of the universality of the gene-for-gene relationship. A notable example is the *Phytophthora infestans-Solanum tuberosum* system. Nevertheless, some generalizations can be stated. In most instances when segregation studies to determine pathogenicity vs nonpathogenicity have been carried out, the allele for nonpathogenicity is dominant.[1] Only a few reports of linkage between pathogenicity genes in a parasite have been offered. Therefore, pathogenicity genes usually segregate independently from other genes for pathogenicity in the same strain of the parasite. A somewhat parallel situation exists in the host wherein studies to determine the genetic nature of resistance have revealed the resistant allele of these gene pairs is generally dominant. A major difference between inheritance studies of genes for resistance in the host and genes for pathogenicity in the parasite appears to be that alleleism and frequently close linkage between factors in the host that condition resistance have been discovered. Extensive studies on the variability observed in plant pathogenic fungi allow some further conclusions. Pathogenicity is generally inherited as a Mendelian character. Cases where pathogenicity has been reported or suggested to be controlled by cytoplasmic factors have been cited earlier. Natural species of pathogenic fungi are usually made up of many physiologic races with different pathological capabilities. The extent to which we have identified the physiological specialization is determined primarily through the availability of suitable host material or host genotypes. Physiologic races appear to be quite stable, but upon hybridization and sexual reproduction single races may produce numerous other races as the result of segregation of genes governing pathogenicity or nonpathogenicity to specific hosts or to a variety of hosts.

Two general areas emerge with respect to variation in pathogenic behavior. On the one hand, pathogenicity, considered herein as the ability of an organism to attack a given host, appears to be controlled by a relatively few genes in any given host-parasite interaction. On the other hand, changes in virulence, considered herein as the relative amount of disease caused by a pathogenic organism, are usually the result of the action of a large number of genes. Little information is available on the precise inheritance of factors conditioning virulence or fitness. The genetic data supporting these conclusions are too numerous to review here. A number of recent reviews are cited for convenience of the reader (7, 26, 27, 53, 62, 76, 77, 87, 88, 99, 120).

[1]The use of the terms *dominant* and *recessive* is clearly justifiable in considering inheritance studies in dikaryotic (functionally diploid) rusts and smuts. Whether it is equally justified in describing gene action in haploids may be questionable although complementation studies in haploid heterokaryons in general support the validity of their use. In haploids pathogenicity and nonpathogenicity alleles of a single locus segregate 1:1.

The significance of the variation provided pathogenic fungi by the mechanisms discussed in the foregoing sections lies in the production of novel genotypes in the pathogen and their ultimate behavior in wild populations. The adaptative or selective value of these variants is usually recognized by plant pathologists as increased levels of disease. Epidemic levels are often attributed to the occurrence of new strains. The attributes of the pathogen affected by or resulting from the release of variability by these mechanisms have been categorized by Nelson (101) as: 1. ability to incite diseases; 2. relative ability to cause a greater amount of disease (virulence); 3. greater efficiency in causing disease; 4. ability to incite disease over a broader environmental range; and 5. increased ability to persist.

Strictly speaking, this could be termed the ecology of plant disease. Plant pathologists have not always considered population genetics of the pathogen as an integral part of epidemiological studies. The strategy has been to choose clear-cut genetic markers, such as physiologic races, in determining patterns of population variation and perhaps ignoring traits of obvious adaptative significance, such as rate of disease increase, sporulation, and ability to survive. Considerable progress toward an integration of population genetics of the pathogen and ecology of plant disease into a coherent evolutionary biology of coexisting populations of host and pathogen has been made in recent years.

The pioneering work of Flor essentially led the way. The biological implication and the current status of the gene-for-gene concept have been reviewed recently by Day (27), Person (119), and Flor (50) and will not be repeated here.

Some of van der Plank's (141) concepts have been a catalyst to the expanding studies of population behavior of the host and parasite. His mathematical treatments of the means by which resistance genes affect both disease onset and the amount and rate of disease increase speak directly to population behavior. His concept of "vertical" resistance genes which react differentially to races of a pathogen has among other things prompted Flor to conclude that the gene-for-gene concept applies only in instances in which resistance is conditioned by major vertical genes. Genetic evidence for such a conclusion has been available for many years.

Van der Plank's views on the role of vertical resistance genes in directing shifts in racial frequencies are embodied in his controversial concept of stabilizing selection which equates pathogenic simplicity to superior fitness (142). The concept revolves about genetically simple and complex members of a host and its parasite. Complex host cultivars possess several VR genes while simple cultivars have only a few or none. Complex races of the parasite can attack both simple and complex host cultivars. Simple races can attack only simple cultivars. The unnecessary genes for virulence possessed by the complex race are assumed to place it at a disadvantage to a simple race when a genetically simple host provides the only substrate. The rationale explaining differences in fitness in simple races of obligate parasites differs from that employed to explain the assumption in nonobligate parasites. The stabilizing factor for obligate parasites would be the simple host. In nonobligate parasites, the stabilization is imposed by the saprophytic phase of the parasite.

The concept, and data supporting or refuting it, has been extensively reviewed by Nelson (100, 101) and Watson (144), and alluded to by others (12). It is obvious from these works that the impact of van der Plank's proposals, regardless of their validity, has stimulated evaluation of old and development of new philosophies in approaching the problems of breeding for disease resistance.

The future role of studies designed to evaluate the factors influencing population shifts in plant pathogens is clearly delineated. The integration of population genetics with studies on the ecology and epidemiology of plant diseases is essential to the judicious use of measures designed to prevent plant diseases from reaching levels of economic consequence.

## *Concluding Remarks*

TOLERANCE TO FUNGICIDES It is well known that fungi may acquire tolerance to fungicides and other chemicals in laboratory culture. Even so, reports on their occurrence and persistence in nature or where disease control has failed due to tolerance are few in number and inconclusive. Nevertheless, the trend toward development of more specific chemicals suggests the possibility that tolerance to fungicides could become of significant importance in the future. Study on mode of action, fate, and persistence of both primary chemicals and their breakdown products in nature and effects on nontarget organisms should be pursued.

Most reports on fungicide tolerance allude to, but do not resolve, the nature of the mechanisms involved in acquisition of tolerance in organisms. Genetic analysis has shown that in certain instances tolerance is heritable and controlled by nuclear genes, but most reports simply identify tolerant strains. Adaptation via activation of enzyme systems and mutation followed by selection are offered as possible explanations. A primary question largely unresolved is whether the presence of the chemicals influences mutation rates, in other words, are fungicides mutagenic? There is limited evidence that certain of them are. Failure to provide answers for these questions has surely influenced the intensity of recent public scrutiny of the overall impact of pesticides.

PHYSIOLOGY OF THE HOST-PARASITE INTERACTION In the past two decades a significant proportion of resources expended for research on plant disease has been devoted to physiological studies, resulting in volumes of literature expanding our knowledge in this area. A close scrutiny of this literature, however, quickly reveals that a meaningful marriage between physiology and genetics of plant disease has not been fully consummated. Studies on inheritance and action of toxins such as victorin have been the most rewarding although not yet fully exploited. Hopefully, future endeavors will resolve difficulties encountered in responding to this challenge.

### Acknowledgments

The author wishes to express appreciation to Dr. P. R. Day, Dr. E. E. Butler, and Dr. H. Bolkan for providing manuscripts in press. Special thanks to Dr. R. R. Nelson and Dr. R. G. Grogan for stimulating discussion, helpful advice, and reviewing the manuscript, and to Mrs. Barbara Overson for clerical assistance.

*Literature Cited*


1. Aist, J. R. 1969. The mitotic apparatus in fungi, *Ceratocystis fagacearum* and *Fusarium oxysporum*. *J. Cell Biol.* 40:120–35
2. Aist, J. R., Wilson, C. L. 1967. Nuclear behavior in the vegetative hyphae of *Ceratocystis fagacearum*. *Am. J. Bot.* 54:99–104
3. Anagnostakis, S. L. 1971. Cytoplasmic and nuclear control of an interstrain interaction in *Ustilago maydis*. *Mycologia* 63:94–7
4. Anderson, N. A., Stretton, H. M., Groth, J. V., Flentje, N. T. 1972. Genetics of heterokaryosis in *Thanatephorus cucumeris*. *Phytopathology* 62:1057–65
5. Barksdale, A. W. 1968. Meiosis in the antheridium of *Achlya ambisexualis* E 87. *J. Elisha Sci. Soc.* 84:187–94
6. Bolkan, H. A., Butler, E. E. Studies on heterokaryosis and virulence of *Rhizoctonia solani*. In press
7. Boone, D. M. 1971. Genetics of *Venturia inaequalis*. *Ann. Rev. Phytopathol.* 9:297–318
8. Bracker, C. E. 1967. Ultrastructure of fungi. *Ann. Rev. Phytopathol.* 5: 343–74
9. Bradley, S. G. 1962. Parasexual phenomena in microorganisms. *Ann. Rev. Microbiol.* 16:35–52
10. Brasier, C. M. 1972. Observations on the sexual mechanism in *Phytophthora palmivora* and related species. *Trans. Brit. Mycol. Soc.* 58:237–51
11. Bridgman, G. H., Wilcoxson, R. D. 1959. New races from mixtures of urediospores of varieties of *Puccinia graminis*. *Phytopathology* 49:428–29
12. Browning, J. A., Frey, K. J. 1969. Multiline cultivars as a means of disease control. *Ann. Rev. Phytopathol.* 7: 355–82
13. Bryant, T. R., Howard, K. L. 1969. Meiosis in the Oomycetes: I. A microspectrophotometric analysis of DNA in *Saprolegnia terrestris*. *Am. J. Bot.* 56: 1075–83
14. Bugbee, W. M., Line, R. F., Kernkamp, M. F. 1968. Pathogenicity of progenies from selfing race 15B and somatic and sexual crosses of races 15B and 56 of *Puccinia graminis* f. sp. *tritici*. *Phytopathology* 58:1291–93
15. Butler, E. E., Bolkan, H. 1973. A medium for heterokaryon formation in *Rhizoctonia solani*. *Phytopathology* 63: 542–43
16. Buxton, E. W. 1960. Heterokaryosis, saltation and adaptation. In *Plant Pathology*, ed. J. G. Horsfall, A. E. Dimond 2: Chap. 10:359–405. New York: Academic. 715 pp.
17. Caspari, E. 1948. Cytoplasmic inheritance. *Advan. Genet.* 2:1–66
18. Casselton, L. A. 1965. Somatic recombination in fungi. *Sci. Progr. Oxford.* 53:107–15
19. Caten, C. E. 1970. Spontaneous variability of single isolates of *Phytophthora infestans*. II. Pathogenic variation. *Can. J. Bot.* 48:897–905
20. Caten, C. E., Jinks, J. L. 1966. Heterokaryosis: Its significance in wild homothallic Ascomycetes and Fungi Imperfecti. *Trans. Brit. Mycol. Soc.* 49:81–93
21. Caten, C. E., Jinks, J. L. 1968. Spontaneous variability of single isolates of *Phytophthora infestans*. I. Culture variation. *Can. J. Bot.* 46:329–48
22. Da Costa, E. W. B., Kerruish, R. M. 1965. The comparative wood-destroying ability and preservative tolerance of monocaryotic and dicaryotic mycelia of *Lenzites trabea* (Pers.) Fr. and *Poria vaillantii* (DC ex Fr.) Cke. *Ann. Bot. N. S.* 29:241–52
23. Davis, R. H. 1966. Mechanisms of inheritance. 2. Heterokaryosis, *The Fungi*, ed. G. C. Ainsworth, A. S. Sussman, 2:567–88. New York: Academic. 805 pp.
24. Day, A. W. 1972. Genetic implications of current models of somatic nuclear division in fungi. *Can. J. Bot.* 50: 1337–47
25. Day, A. W., Jones, J. K. 1969. Sexual and parasexual analysis of *Ustilago violacea*. *Genet. Res. Cambridge* 14: 195–221
26. Day, P. R. 1960. Variation in phytopathogenic fungi. *Ann. Rev. Microbiol.* 14:1–16
27. Day, P. R. 1966. Recent developments in the genetics of the host-parasite system. *Ann. Rev. Phytopathol.* 4:245–68
28. Day, P. R. 1970. The significance of genetic mechanisms in soil fungi. In *Root Diseases and Soil-borne Pathogens*, ed. T. A. Toussoun, R. V. Bega, P. E. Nelson, 69–74. Berkeley: Univ. Calif. Press. 252 pp.
29. Day, P. R., Anagnostakis, S. L. 1971. Corn smut dikaryon in culture. *Nat. New Biol.* 231:19–20
30. Day, P. R., Anagnostakis, S. L. 1971. Meiotic products from natural infections of *Ustilago maydis*. *Phytopathology* 61:1020–21

31. Day, P. R., Anagnostakis, S. L. 1973. The killer system in *Ustilago maydis:* Heterokaryon transfer and loss of determinants. *Phytopathology* 63:1017–18
32. Denward, T. 1967. Differentiation in *Phytophthora infestans.* I. A comparative study of eight different biotypes. *Hereditas* 58:191–220
33. Denward, T. 1970. Differentiation in *Phytophthora infestans.* II. Somatic recombination in vegetative mycelium. *Hereditas* 66:35–48
34. d'Oliveira, B. 1939. Studies on *Puccinia anomala.* I. Physiological races on cultivated barley. *Ann. Appl. Biol.* 26: 56–82
35. Duncan, E. J., Macdonald, J. A. 1965. Nuclear phenomena in *Marasmius androsaceus* (L. ex Fr.) and *M. rotula* (Scop. ex Fr.). *Trans. Roy. Soc. Edinburgh* 66(7)
36. Ellingboe, A. H. 1961. Somatic recombination in *Puccinia graminis* var. *tritici. Phytopathology* 51:13–15
37. Ellingboe, A. H. 1964. Nuclear migration in dikaryotic-homokaryotic matings in *Schizophylum commune. Am. J. Bot.* 51:133–39
38. Emerson, S. 1967. Fungal genetics. *Ann. Rev. Genet.* 1:201–17
39. Erwin, D. C., Zentmyer, G. A., Galindo, J., Niederhauser, J. S. 1963. Variation in the genus *Phytophthora. Ann. Rev. Phytopathol.* 1:375–96
40. Esser, K. 1966. Incompatibility. See Ref. 23, 2: Chap. 20: pp. 661–76
41. Esser, K., Kuenen, R. 1967. *Genetics of Fungi.* Transl. E. Steiner. New York: Springer. 500 pp.
42. Fincham, J. R. S. 1970. Fungal genetics. *Ann. Rev. Genet.* 4:347–72
43. Fincham, J. R. S., Day, P. R. 1965. *Fungal Genetics.* Oxford: Blackwell. 326 pp.
44. Fjeld, A., Laane, M. M. 1970. The nuclear division and parasexual cycle in *Penicillium. Genetica* 41:517–24
45. Flanagan, P. W. 1970. Meiosis and mitosis in *Saprolegniaceae. Can. J. Bot.* 48:2069–76
46. Flentje, N. T., Stretton, H. M., Hawn, E. J. 1963. Nuclear distribution and behavior throughout the life cycles of *Thanatephorus, Waitea,* and *Ceratobasidium* species. *Aust. J. Biol. Sci.* 16:450–67
47. Flentje, N. T., Stretton, H. M., McKenzie, A. R. 1970. Mechanism of variation in *Rhizoctonia solani.* In *Rhizoctonia solani: Biology and Pathology,* ed. J. R. Parmeter Jr., 52–65. Berkeley: Univ. Calif. Press. 255 pp.
48. Flor, H. H. 1956. The complementary genic systems in flax and flax rust. *Advan. Genet.* 8:29–54
49. Flor, H. H. 1964. Genetics of somatic variation for pathogenicity in *Melampsora lini. Phytopathology* 54:823–26
50. Flor, H. H. 1971. Current status of the gene-for-gene concept. *Ann. Rev. Phytopathol.* 9:275–96
51. Galindo, J., Gallegly, M. E. 1960. The nature of sexuality in *Phytophthora infestans. Phytopathology* 50:123–28
52. Galindo, A. J., Zentmyer, G. A. 1967. Genetical and cytological studies of *Phytophthora* strains pathogenic to pepper plants. *Phytopathology* 57:1300–5
53. Gallegly, M. E. 1968. Genetics of pathogenicity of *Phytophthora infestans. Ann. Rev. Phytopathol.* 6:375–96
54. Gallegly, M. E. 1970. Genetical aspects of pathogenic and saprophytic behavior of the *Phycomycetes* with special reference to *Phytophthora.* See Ref. 28, pp. 50–54
55. Gallegly, M. E. 1970. Genetics of *Phytophthora. Phytopathology* 60:1135–41
56. Garza-Chapa, R., Anderson, N. A. 1966. Behavior of single-basidiospore isolates and heterokaryons of *Rhizoctonia solani* from flax. *Phytopathology* 56:1260–68
57. Georgopoulos, S. G., Zaracovitis, C. 1967. Tolerance of fungi to organic fungicides. *Ann. Rev. Phytopathol.* 5: 109–29
58. Giatong, P., Frederiksen, R. A. 1969. Pathogenic variability and cytology of monoconidial subcultures of *Piricularia oryzae. Phytopathology* 59:1152–57
59. Green, G. J., McKenzie, R. I. H. 1967. Mendelian and extrachromosomal inheritance of virulence in *Puccinia graminis* f. sp. *avenae. Can. J. Genet. Cytol.* 9:785–93
60. Green, G. J., Kirmani, M. A. S. 1969. Somatic segregation in *Puccinia graminis* f. sp. *avenae. Phytopathology* 59: 1106–8
61. Haasis, F. A., Nelson, R. R. 1963. Studies on the biological relationship of species of *Phytophthora* as measured by oospore formation in intra- and interspecific crosses. *Plant. Dis. Reptr.* 47: 705–9
62. Halisky, P. M. 1965. Physiologic specialization and genetics of the smut fungi. III. *Botan. Rev.* 31:114–50
63. Hankin, L., Puhalla, J. E. 1971. Nature of a factor causing interstrain lethality in *Ustilago maydis. Phytopathology* 61: 50–53

64. Hartley, M. J., Williams, P. G. 1971. Genotypic variation within a phenotype as a possible basis for somatic hybridization in rust fungi. *Can. J. Bot.* 49: 1085–87
65. Hartman, G. C. 1964. Nuclear division in *Alternaria tenuis. Am. J. Bot.* 51: 209–12
66. Hastie, A. C. 1962. Genetic recombination in the hop-wilt fungus *Verticillium albo-atrum. J. Gen. Microbiol.* 27: 373–82
67. Hastie, A. C. 1964. The parasexual cycle in *Verticillium albo-atrum. Genet. Res.* 5:305–15
68. Hastie, A. C. 1967. Mitotic recombination in conidiophores of *Verticillium albo-atrum. Nature* 214:249–52
69. Hastie, A. C. 1968. Phialide analysis of mitotic recombination in *Verticillium. Mol. Gen. Genet.* 102:232–40
70. Hastie, A. C. 1970. The genetics of asexual phytopathogenic fungi with special reference to *Verticillium.* See Ref. 28, pp. 55–62
71. Heale, J. B., Gafoor, A., Rajasingham, K. C. 1968. Nuclear division in conidia and hyphae of *Verticillium albo-atrum. Can. J. Genet. Cytol.* 10:321–40
72. Hill, E. P. 1965. Uptake and translocation. 2. Translocation. See Ref. 23, 1: 457–64. 748 pp.
73. Holliday, R. 1961. The genetics of *Ustilago maydis. Genet. Res.* 2:204–30
74. Holliday, R. 1961. Induced mitotic crossing-over in *Ustilago maydis. Genet. Res.* 2:231–48
75. Holloway, B. W. 1955. Genetic control of heterokaryosis in *Neurospora crassa. Genetics* 40:117–29
76. Holton, C. S., Hoffman, J. A., Duran, R. 1968. Variation in the smut fungi. *Ann. Rev. Phytopathol.* 6:213–42
77. Hooker, A. L. 1967. The genetics and expression of resistance in plants to rusts of the genus *Puccinia. Ann. Rev. Phytopathol.* 5:163–82
78. Hosford, R. M. Jr., Gries, G. A. 1966. The nuclei and parasexuality in *Phymatotrichum omnivorum. Am. J. Bot.* 53:570–79
79. Howard, K. L., Bryant, T. R. 1971. Meiosis in the Oomycetes. II. A microspectrophotometric analysis of DNA in *Apodachlya brachynema. Mycologia* 63: 58–68
80. Huang, H. C., Patrick, Z. A. 1972. Nuclear division and behavior in *Thielaviopsis basicola. Can. J. Bot.* 50:2423–29
81. Ingram, R. 1968. *Verticillium dahliae* var. *longisporum,* a stable diploid. *Trans. Brit. Mycol. Soc.* 51:339–41
82. Jinks, J. L. 1957. Selection for cytoplasmic differences. *Proc. Roy. Soc. B* 146: 527–49
83. Jinks, J. L. 1958. Cytoplasmic differentiation in fungi. *Proc. Roy. Soc. B* 148: 314–21
84. Jinks, J. L. 1966. Mechanisms of inheritance. 4. Extranuclear inheritance. See Ref. 23, 2:619–57
85. Jinks, J. L., Grindle, M. 1963. The genetical basis of heterokaryon incompatibility in *Aspergillus nidulans. Heredity* 18:407–11
86. Johnson, T. 1946. Variation and the inheritance of certain characters in rust fungi. *Cold Spring Harbor Symp. Quant. Biol.* 11:85–93
87. Johnson, T. 1960. Genetics of pathogenicity. See Ref. 16, 2: Chap. 11: 407–59
88. Johnson, T. 1968. Host specialization as a taxonomic criterion. See Ref. 23, 3:543–56. 738 pp.
89. Knox-Davis, P. S. 1966. Nuclear division in the developing pycnospores of *Macrophomina phaseoli. Am. J. Bot.* 53:220–24
90. Knox-Davis, P. S. 1967. Mitosis and aneuploidy in the vegetative hyphae of *Macrophomina phaseoli. Am. J. Bot.* 54:1290–95
91. Kozar, F. 1969. Mitotic recombination in biochemical mutants of *Ustilago hordei. Can. J. Genet. Cytol.* 11:961–66
92. Laane, M. M. 1967. The nuclear division in *Penicillium expansum. Can. J. Genet. Cytol.* 9:342–51
93. Leach, S. S., Rich, A. E. 1969. The possible role of parasexuality and cytoplasmic variation in race differentiation in *Phytophthora infestans. Phytopathology* 59:1360–65
94. Little, R., Manners, J. G. 1969. Somatic recombination in yellow rust of wheat (*Puccinia striiformis*). I. The production and possible origin of two new physiologic races. *Trans. Brit. Mycol. Soc.* 53:251–58
95. Marks, G. E. 1965. The cytology of *Phytophthora infestans. Chromosoma* 16:681–92
96. McGinnis, R. C. 1953. Cytological studies of chromosomes of rust fungi. I. The mitotic chromosomes of *Puccinia graminis. Can. J. Bot.* 31:522–26
97. McKenzie, A. R., Flentje, N. T., Stretton, H. M., Mays, M. J. 1960. Heterokaryon formation and genetic recombination within one isolate of *Thanatephorus cucumeris. Aust. J. Biol. Sci.* 22:895–904

98. Ming, Y. N., Lin, P. C., Yu, T. F. 1966. Heterokaryosis and *Fusarium fujikuroi* (Sacc.). *Wr. Scientia Sinica* 15:371–78
99. Moseman, J. G. Genetics of powdery mildews. *Ann. Rev. Phytopathol.* 4: 269–90
100. Nelson, R. R. 1972. Stabilizing racial populations of plant pathogens by use of the resistance genes. *J. Environ. Qual.* 1:220–27
101. Nelson, R. R. 1973. *Breeding Plants for Disease Resistance—Concepts and Applications.* University Park, Pa./London: Penn. State Univ. Press. 401 pp.
102. Nelson, R. R., Wilcoxson, R. D., Christiansen, J. J. 1955. Heterocaryosis as a basis for variation in *Puccinia graminis* var. *tritici. Phytopathology* 45:639–43
103. Olive, L. S. 1953. The structure and behavior of fungus nuclei. *Bot. Rev.* 19: 439–586
104. Olive, L. S. 1958. On the evolution of heterothallism in fungi. *Am. Nat.* 42: 233–51
105. Olive, L. S. 1963. Genetics of homothallic fungi. *Mycologia* 55:93–103
106. Olive, L. S. 1965. Nuclear behavior during meiosis. See Ref. 23, 1:143–61
107. Raper, J. R. 1960. The control of sex in fungi. *Am. J. Bot.* 47:794–808
108. Raper, J. R. 1963. Patterns of sexuality in fungi. *Mycologia* 55:79–92
109. Raper, J. R. 1966. *Genetics of Sexuality in Higher Fungi.* New York: Ronald. 283 pp.
110. Robinow, C. F., Bakerspigel, A. 1965. Somatic nuclei and forms of mitosis in fungi. See Ref. 23, 1:119–42
111. Robinow, C. F., Caten, C. E. 1969. Mitosis in *Aspergillus nidulans. J. Cell Sci.* 5:403–31
112. Romero, S., Erwin, D. C. 1967. Genetic recombination in germinated oospores of *Phytophthora infestans. Nature* 215: 1393–94
113. Romero, S., Erwin, D. C. 1969. Variation in pathogenicity among single-oospore cultures of *Phytophthora infestans. Phytopathology* 59:1310–17
114. Roper, J. A. 1966. Mechanisms of inheritance. 3. The parasexual cycle. See Ref. 23, 2:589–617
115. Roth, J. N., Brandt, W. H. 1964. Nuclei in spores and mycelium of *Verticillium. Phytopathology* 54:363–64
116. Parmeter, J. R. Jr. 1970. Mechanisms of variation in culture on soil. See Ref. 28, pp. 63–68
117. Parmeter, J. R. Jr., Snyder, W. C., Reichle, R. E. 1963. Heterokaryosis and variability in plant-pathogenic fungi. *Ann. Rev. Phytopathol.* 1:51–69
118. Parmeter, J. R. Jr., Sherwood, R. T., Platt, W. D. 1969. Anastomosis grouping among isolates of *Thanatephorus cucumeris. Phytopathology* 59:1270–78
119. Person, C. 1959. Gene-for-gene relationships in host: parasite systems. *Can. J. Bot.* 37:1101–30
120. Person, C. 1968. Genetical adjustment of fungi to their environment. See Ref. 23, 3:395–415
121. Polach, F. J., Webster, R. K. 1972. Identification of strains and inheritance of pathogenicity in *Phytophthora capsici. Phytopathology* 62:20–26
122. Pontecorvo, G. 1958. *Trends in genetic analysis.* New York: Columbia Univ. Press. 145 pp.
123. Puhalla, J. E. 1968. Compatibility reactions on solid medium and interstrain inhibition in *Ustilago maydis. Genetics* 60:416–74
124. Puhalla, J. E. 1969. The formation of diploids of *Ustilago maydis* on agar medium. *Phytopathology* 59:1771–72
125. Sansome, E. 1961. Meiosis in the oogonium and antheridium of *Pythium debaryanum* Hesse. *Nature* 191:827–28
126. Sansome, E. 1963. Meiosis in *Pythium debaryanum* Hesse and its significance in the life cycle of the biflagellate. *Trans. Brit. Mycol. Soc.* 46:63–72
127. Sansome, E. 1965. Meiosis in diploid and polyploid sex organs of *Phytophthora* and *Achlya. Cytologia Tokyo* 30:103–17
128. Sansome, E. 1966. Meiosis in the sex organs of the oomycetes. *Chromosomes Today,* ed. C. D. Darlington, K. R. Lewis, 1:77–83. Edinburgh: Oliver & Boyd
129. Sansome, E. 1970. Selfing as a possible cause of disturbed ratios in *Phytophthora* crosses. *Trans. Brit. Mycol. Soc.* 54:101–7
130. Satour, M. M., Butler, E. E. 1968. Comparative morphological and physiological studies of the progenies from intraspecific matings of *Phytophthora capsici. Phytopathology* 58:183–92
131. Savage, E. J. et al 1968. Homothallism, heterothallism, and interspecific hybridization in the genus *Phytophthora. Phytopathology* 58:1004–21
132. Schreiber, L. R., Green, R. J. Jr. 1966. Anastomosis in *Verticillium albo-atrum* in soil. *Phytopathology* 56:1110–11
133. Scott, K. J., Maclean, D. J. 1969. Culturing of rust fungi. *Ann. Rev. Phytopathol.* 7:123–46
134. Shatla, M. N., Sinclair, J. B. 1966. *Rhizoctonia solani:* mitotic division in

vegetative hyphae. *Am. J. Bot.* 53: 119–23
135. Shaw, D. S., Khaki, I. A. 1971. Genetical evidence for diploidy in *Phytophthora*. *Genet. Res. Cambridge* 17: 165–67
136. Spector, C., Phinney, B. O. 1968. Gibberellin biosynthesis: Genetic studies in *Gibberella fujikuroi*. *Physiol. Plant.* 21:127–36
137. Stretton, H. M., Flentje, N. T. 1972. Inter-isolate heterokaryosis in *Thanatephorus cucumeris*. I. Between isolates of similar pathogenicity. *Aust. J. Biol. Sci.* 25:293–303
138. Timmer, L. W., Castro, J., Erwin, D. C., Belser, W. L., Zentmyer, G. A. 1970. Genetic evidence for zygotic meiosis in *Phytophthora capsici*. *Am. J. Bot.* 57:1211–18
139. Tinline, R. D., MacNeill, B. H. 1969. Parasexuality in plant pathogenic fungi. *Ann. Rev. Phytopathol.* 7:147–70
140. Tolmsoff, W. J. 1972. Diploidization and heritable gene repression-derepression as major sources for variability in morphology, metabolism and pathogenicity of *Verticillium* species. *Phytopathology* 62:407–13
141. van der Plank, J. E. 1963. *Plant Diseases: Epidemics and Control*. New York: Academic. 349 pp.
142. van der Plank, J. E. 1968. *Disease Resistance in Plants*. New York: Academic. 206 pp.
143. Vest, G., Anderson, N. A. 1968. Studies on heterokaryosis and virulence of *Rhizoctonia solani* isolates from flax. *Phytopathology* 58:802–7
144. Watson, I. A. 1970. Changes in virulence and population shifts in plant pathogens. *Ann. Rev. Phytopathol.* 8: 209–30
145. Webster, R. K., Ogawa, J. M., Moore, C. J. 1968. The occurrence and behavior of variants of *Rhizopus stolonifer* tolerant to 2,6-dichloro-4-nitroaniline. *Phytopathology* 58:997–1003
146. Webster, R. K., Ogawa, J. M., Bose, E. 1970. Tolerance of *Botrytis cinerea* to 2,6-dichloro-4-nitroaniline. *Phytopathology* 60:1489–92
147. Whitney, H. S., Parmeter, J. R. Jr. 1963. Synthesis of heterokaryons in *Rhizoctonia solani* Kuhn. *Can. J. Bot.* 41:879–86
148. Wilson, C. L., Aist, J. R. 1967. Motility of fungal nuclei. *Phytopathology* 57: 769–71
149. Wood, H. A., Bozarth, R. F. 1973. Heterokaryon transfer of viruslike particles associated with a cytoplasmically inherited determinant in *Ustilago maydis*. *Phytopathology* 63:1019–20

# COMPARISON OF EPIDEMICS ❖3603

*Jürgen Kranz*
Tropische Phytopathologie, Justus Liebig-Universität, 63 Giessen, W. Germany

Epidemiology is the study of populations of pathogens in populations of hosts and the resulting disease under the influence of the environment and human interference. *Epidemic* is used here as a synonym of disease progress, which can, but need not, conform to the classical definition of an epidemic, i.e. the steep rise, and decline of disease within a limited period of time (18). Any increase or decrease of disease $y$ between $0 < y \leq 1$, which is equivalent to disease ratings from near 0–100%, is an epidemic. This leaves no place for *endemics*, although there are endemic diseases in the sense of having been indigenous in an area. Their epidemics may differ in behavior from those of *pandemic* diseases, which are still spreading throughout the world or parts of it.

Comparison of epidemics is one facet of the newly developing comparative epidemiology, which essentially should be quantitative and experimental. Comparison is the basis for comprehension and is a major scientific tool. Apart from defining what is different, similar, or identical among diseases, comparative epidemiology derives models and principles of more general application from the multitude of singular events in plant diseases and their epidemics. Comparative epidemiology can be approached "horizontally" or "vertically."

A horizontal comparison is that of leaf vs root disease, diseases in open vs closed environment, or diseases of annual vs perennial crops. Comparisons of this kind are often useful and may help to point out existing differences or similarities. Although it also allows for inferences, one can envisage problems with this method in the implementation of experiments for verification. More suitable for experimental studies is the vertical approach. Here the elements or phases of disease (e.g. primary inoculum, spore formation, spore release, spore dispersal, landing, infection, incubation, latent period, infectious period), their structures, and individual requirements are the objectives.

Simultaneous measurements of all these elements and structures of disease progress for as many diseases as possible under exactly the same conditions would be ideal. But this is beyond feasibility and beyond the competence of a single research worker. Therefore, the study of one of the above elements in several diseases seems to be a solution. One can, however, also compare the integral aspect of all these various elements and structures, i.e. the epidemic as a whole. This will be our subject.

## THE GRAPH OF AN EPIDEMIC: THE DISEASE PROGRESS CURVE

The disease progress curve is the graph of an epidemic. Mathematically it is the dependent variable. Each disease progress curve expresses the various effects of pathogen, host, environment, and human interferences within the disease quadrangular and time effects. Disease progress curves result from measuring disease $y$ (Van der Plank's $x$) in various ways (1, 30, 40, 41, 59, 62, 70, 72, and others) at various dates from a starting point of the epidemic. Mostly it is disease severity (i.e. the proportion of the susceptible host tissue affected) or disease frequency (i.e. the proportion of host plants or leaves diseased) which are estimated. But other criteria have also been suggested, such as new infections (1, 41, 59). All these criteria are acceptable as long as they ensure substantially correct and quantitative measurements.

### *Kinds of Disease Progress Curves*

Disease progress may be plotted as cumulative or frequency curves of growth rates. Both kinds of curves can be deduced from each other (Figure 1). Cumulative curves sum up the progress of a disease and are commonly used. Such a curve aptly describes the additive amount of disease $y$ at all dates $t_i$ and consequently the course of an epidemic in time. Ideally this curve is sigmoid (62, 65). Baker (5) divides it into four phases: (*a*) true logarithmic scale, (*b*) synergistic (or more commonly, exponential) slope, (*c*) transitional slope, and (*d*) plateau.

A frequency curve relates the increments of disease (over the last measurement) more directly to changes in resistance, environment, or control. They also show more clearly the underlying distribution, and thus indicate which transformations are required. Rates of changes can be any difference in disease between two dates, $t_{i-1}-t_i$, or infection rates, as for example (from ref. 62).

$$r = \frac{1}{t_2-t_1} \ln x_2\,(1-x_1) - \ln x_1\,(1-x_2) \text{ or (from ref 1)}$$

$$BS' = \ln\,(1/1 - y_{i+1} / y_{max} - \ln 1/1 - y_i/y_{max}$$

Most disease progress curves are still depicted as cumulative curves. Van der Plank (62) has based his models and his whole book essentially on cumulative curves. Few examples of frequency curves can be noted in the literature so far (1, 42, 59).

As plant pathologists are nearly exclusively concerned with crops that are harvested, they either do not know or are not interested in the complete course of an epidemic. This is why phytopathological literature is full of unilateral rather than

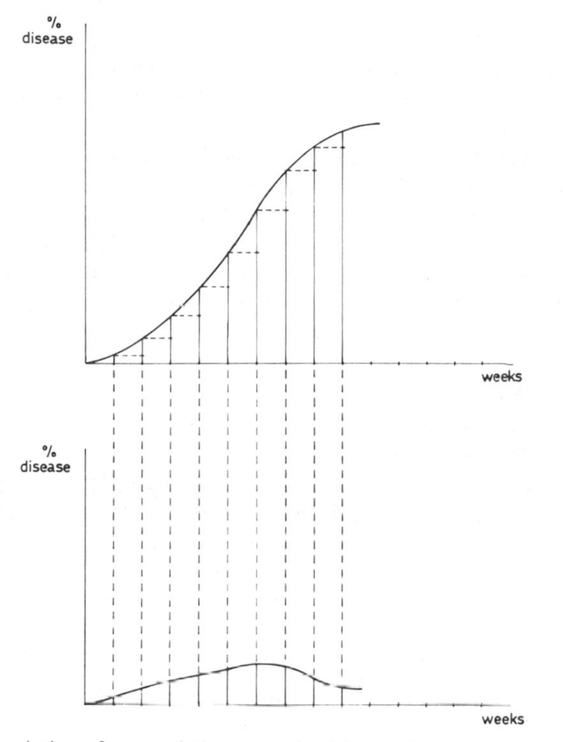


*Figure 1* Transcription of a cumulative curve (*top*) into a frequency or growth-rate curve (*bottom*) showing disease progress (time in weeks on abscissas, growth rate in percent disease on ordinates).

bilateral curves. Harvest mostly cuts short the degressive leg of the curve, which nevertheless occurs if an epidemic is left alone (18, 32, 37).

For any comparison of epidemics we have to choose a basic time unit. Generally the vegetation period of the host in days is the most suitable time unit. A vegetation period is distinct for annuals, although it may be argued whether to start with the date of sowing or emergence. For pathogens that attack only aerial parts of their hosts, the date of emergence seems to be appropriate. The vegetation period for perennial hosts may be defined either on the basis of growth rhythm (refoliation and date of flowering) or favorable climatic periods (rainy season or summer), or both. The vegetation period ends by the natural disappearance or harvest of the host or of the parts relevant to the disease. There are, however, cases where a sequence of some years may be better time units.

This review is limited to disease progress curves and thus only to epidemics in time. Spatial aspects, however, may also be a subject of a comparison of epidemics. The reader is referred to Gregory (19, 20) and Kiyosawa & Shiyomi (29) for a comparison of disease gradients.

*Elements of Disease Progress Curves*

In disease progress curves, whether complete or incomplete cumulative and frequency curves, one may discern and propose various curve elements (32), which can be quantified (Figure 2):

1. Amount $BD = X_0$ measured in days on the abscissa from $B$ = beginning of vegetation period or refoliation, until the first discernible symptoms at $D$.

2. Amount $DP_1$ or $DT = X_p$ in days of the progressive leg of the curve measured on the abscissa; $P_1$ is the beginning of the asymptote, $P_2$ the end of the plateau.

3. Amount $P_1P_2 = X_a$ in days of asymptote (or plateau); $X_a = 0$ = no plateau.

4. Amount $P_2D = X_d$ in days of degressive leg of the curve.

5. Amount $DF = X_t$ in days, the total duration of a naturally completed epidemic.

6. Amount $FB = X_i$ in days the interval between disappearance of last and reappearance of new epidemics; this may be of interest in recurrent epidemics in perennial or natural vegetation.

7. $tg\beta = b$, the regression coefficient. Angle between progressive leg of the curve and abscissa at $D$ through $Y_0$, which may be a parallel to the abscissa (shown somewhat elevated in Figure 2.)

8. $Y_0$ at $D$, amount of disease ($0 < Y_0 \leq 1$) in primary foci etc.

9. $Y_{max}$ at $T$ ($0 < Y_{max} \leq 1$), maximum disease incidence (may be identical with asymptote).

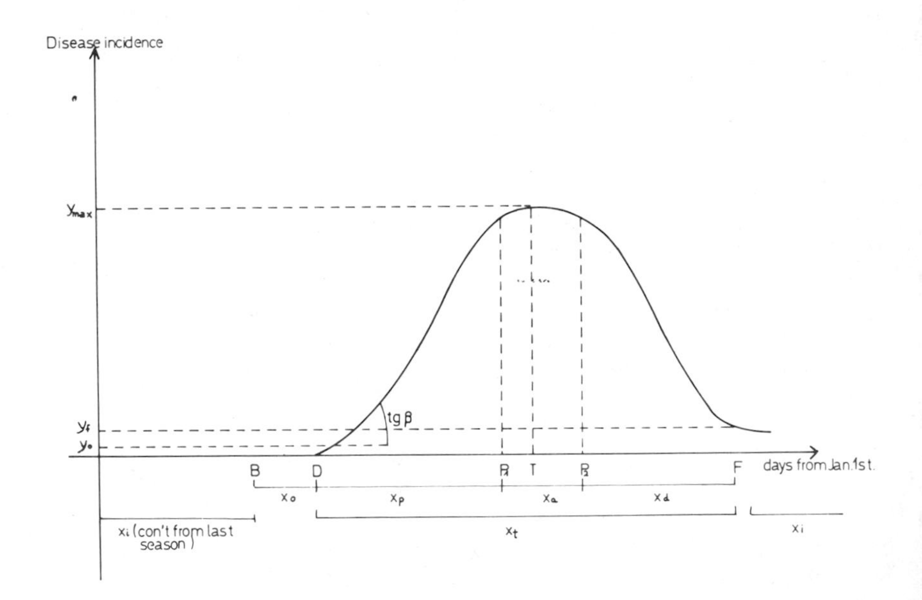


*Figure 2* The elements of a disease progress curve as graph of an epidemic. For explanations, see text.

10. $Y_f$ at $F$, amount of disease ($0 < Y_f \leq 1$) at the final stage of the epidemic (end of epidemic), $Y_f = Y_{max}$ and $Y_f = 0$ are possible.

11. Area under disease progress curve. (62).

It is obvious that values of these parameters should be in classes rather than absolute figures.

Curve elements are thought to represent certain biological facets that are integrated in the dependent variable disease progress curve. Curve elements may then be regarded as gauges for the effects of independent variables and as a basis for quantitative comparisons of epidemics.

Some of these proposed curve elements have already been studied (32) or are under study (35). Little can be said yet as to how far these elements will carry for a quantitative immediate comparison and classification. This will largely depend on the variability of each single element for each disease, which still has to be investigated more extensively.

All disease progress curves found in the literature either show or permit derivation of some of the above elements. Elements 4, 5, 6, and 10 are frequently lacking because the crop is harvested before a disease can complete its full cycle or because the plant pathologist loses interest.

## WHY COMPARISON OF EPIDEMICS?

Comparison of epidemics (or more precisely, that of disease progress curves) is a common practice in epidemiology. There actually is no need to justify comparison of epidemics since it has proved useful for the interpretation of experimental results. This certainly applies to comparison of effects of experimental manipulation of a given disease, as well as to studies of basic differences and/or similarities in the behavior or reaction of epidemics of various diseases.

However, one may ask how it is done, what can be achieved by these methods, what are their limits, and how they can possibly be improved. This is the subject of later sections. Methods for more or less automatic comparisons of epidemics are also discussed. These methods will help us to cope with the vast and often confounded data confronting one in epidemiology, and with data banks or information otherwise stored.

Any comparison can achieve its ends by both, or either, two main methods: (*a*) immediate comparison or (*b*) classification (not in the sense of taxonomy). For both approaches it must be ensured, usually by common experimental precautions, that the epidemics are comparable. Criteria and methods of comparison must be quantitative and equally applicable to all epidemics included in the comparison. For classifications, delimitation of classes may prove difficult, as is shown later. But this might still be due to lack of specific research efforts.

An immediate approach has its place in the evaluation of effects of different treatments or resistance, on the course of the epidemic. To an extent such a comparison always implies data analysis. Classification provides for the assignment of epidemics to known models or events, for automatic retrieval, and similar operations. A class (or model) necessarily implies a higher degree of abstraction than is required for an immediate comparison. However, such an abstraction may be advantageous as it can do away with much of the confusing though statistically insignificant variations in disease progress curves, or eliminate the undue weight of certain elements and their structures. Thus, disease progress curves that look different when plotted may belong to the same class when stripped of their chance variation.

One might object to a comparison of disease progress curves as graphs of epidemics on grounds of their variability from year to year and place to place. Variability cannot be denied, nor can complexity. However, there are indications (1, 13, 17, 42, 59) that only few, though possibly changing, factors influence disease progress significantly. One might therefore suspect that the progress curves of a given disease can have only limited scope of variation. These variations, no doubt, may overlap with other disease as conditioned by environment, varieties, different pathogen population, or control measures. But it is up to comparative epidemiology to establish this and reduce the variations of effects, as well as elements and structure, to the significant ones. Further treatment of the subject emphazises quantitative approaches to a comparison of epidemics, some of which are still rather theoretical.

## QUALITATIVE NOTATIONS IN COMPARISON OF EPIDEMICS

Before we discuss quantitative means of comparison of epidemics we should refer to some qualitative notation in use to classify diseases and their epidemics. Although descriptive codewords, they may also be the framework for quantitative comparisons. In the first comprehensive, though descriptive, treatment of plant epidemics, Gäumann (18) made a distinction between annual and secular (or perennial) epidemics with various intermediate forms as well as between endemy and pandemy. Van der Plank distinguished between primary and secondary epidemics (61) and between epidemics caused by monocyclic and those caused by polycyclic diseases ("simple" and "compound interest" diseases respectively) for annual epidemics with the year or the vegetation cycle as time unit (62). Focal and general epidemics (18, 62) refer to the distance covered by an epidemic.

Various qualitative types of disease progress have been described (31) according to their form (type), number of peaks, and whether they are continuous or discontinuous from one season to the next. This already approaches quantitative treatment, for these types represent various forms of Gaussian normal curves and skewed forms. Though there appeared to be some variability among 59 host-parasite combinations studied (30), the form of a disease progress curve seems often to be determined by the ontogenetic change in susceptibility of the host and by rapidity of pathogen buildup. Weather can have a modifying effect. Annual epidemics are rarely continuous, even in the humid tropics. They are normally discontinuous, i.e. their visible course ends within the season they began. This is due either to the effect of the disease that kills the host or causes the diseased parts of it to disappear, or to the life cycle of the hosts. This criterion, as other qualitative ones, leaves only limited scope for comparison. A frequent combination of these three criteria is the more or less bell-shaped, one-peaked disease progress curve, which is discontinuous by virtue of the pathogen's action (31).

Epidemics of diseases may also be qualitatively classified according to the kind and source of their primary inoculum (seed-, soil-, or airborne). Another criterion may be the effect of an epidemic on the host (31, 49). For example, in relation to yield loss, short and long epidemics with medium or severe attack (50) and early and late predictive disease (25) have been distinguished.

## QUANTITATIVE COMPARISON OF EPIDEMICS

Epidemiology and comparative epidemiology are basically quantitative empirical sciences. We therefore also need adequate quantitative methods for vertical comparison of epidemics. These methods can range from simple plotting, mapping, or tabulation of quantitative data, to elaborate statistical, mathematical, and computer methods.

### *Plotting of Disease Progress Curves*

Plotting disease curves as cumulative curves (5, 10, 15, 18, 39, 51, 52, 62, 65) or curves of growth rates (1, 42, 59) is a common base for comparison of (mostly unilateral) epidemics. The majority of epidemics in the literature are compared within a single disease, rather than between diseases. The former are almost exclusively meant to analyze the effect of certain experimental treatments. A typical and extensive example is the compilation of Cox & Large (15) of smoothed disease progress curves of late blight of potatoes (in %) plotted over time. The effects investigated are varieties, years, sprayed vs unsprayed, number of sprays, and soil types. One can, *inter alia,* compare the influence of various doses of chlorcholinchloride (CCC) and nitrogen on two spring barley cultivars on epidemics of *Septoria nodorum* (11), the fungicidal effects on the coffee berry disease, *Colletotrichum coffeanum* (21, 46, 59), or that of differing altitude (24, 43, 45), or microclimatic effects on *Polyspora lini* (55) in relation to sowing dates, varieties, and years.

Mapping severity of plant disease is an alternative to plotting when a comparison of the date of appearance and $Y_{max}$ of a disease in relation to geographic features is attempted (53, 58, 69).

Although versatile, the plotting of curves (and mapping) is essentially data reduction, which yields indications and sometimes permits inferences. If data are large and comprehensive enough, in some cases even models may be deduced. Plotting, however, can make only limited contributions to parameter estimation and to predictions, for it has the serious handicap of not allowing for reliable tests of hypotheses and their significance. Thus two apparently different curves may represent chance variation.

Consequently, plotted curves are of limited use as a method of comparison, unless subjected to the following treatments: (*a*) smoothing curves by moving averages to minimize accidental variations mostly due to experimental errors; (*b*) transforming curves into linearity or curvilinearity by appropriate methods; (*c*) verifying whether differences among disease progress curves are statistically significant.

After these procedures, disease progress will allow for comparisons on more objective criteria. But before this can be achieved, we must refer to transformation.

### *Transformation of Disease Progress Curves*

Transformation is based on actual distribution of events (in our case disease incidence) and not on assumption, as Baker (5) feels. These distributions have a biological background and can be described by mathematical models.

Transformation serves a double purpose in the context of our subject. First it is a statistical prerequisite. Merit and scope of various methods of transformation to obtain linearity have been discussed by Analytis (1), Baker (4, 5), Dimond & Horsfall (16), and Van der Plank (62), and distributions have been discussed by Bald (6).

A second, and here more interesting aspect of transformation is for classification of epidemics. For it has become obvious that there is nothing like a unique method for transformation of disease progress curves (1, 5, 16, 28). If a goodness-of-fit test proves that the disease progress curves of a given disease can be better straightened by means of a monomolecular growth function than by the logistic function (1), this is due to different distributions. As already noted, such a distribution is conditioned by ontogenetic changes in host susceptibility, changing density of inoculum, or the weather. Thus a variety of transformation models are needed to cater to possible classes of epidemics. It appears reasonable that epidemics may be classified according to the transformation model that ensures the best fit of a transformed curve to observed disease measurements. In the case of apple scab, for example, the epidemic follows a skewed progress curve and not a symmetric normal distribution, as required for the probit and logistic transformation (1, 2).

Van der Plank (61, 62) uses the multiple infection and the logit transformations, $\ln (1/1-x)$ and $\ln (x/1-x)$, respectively, for the disease progress curves of two different kinds of diseases. These are the ones that do or do not multiply in the host population (compound and simple interest diseases). Here transformation as criterion for the comparison of epidemics may lead to a classification into two groups of epidemics. Baker (5) found that the three transformations he used on published results [the semilogarithmic multiple infection transformation $\ln 1/(1-x)$, probit–log, and $\ln 1/(1-x)$, log transformations] of soil pathogens performed differently in slope and position with the various diseases in relation to his four models.

All three transformations suffer from the shortcoming of being suitable only for curves without a distinct asymptote. Here Analytis' (1) transformation equations derived from growth functions and Kiyosawa's (28) equation $dy/dt = ry(1-t/T)$, where $T$ is the time where the increase in lesions stops, take account of the asymptote. These and similar transformations will certainly provide a wider scope for transformation as a criterion for the comparison of epidemics. We return to this when dealing with curvilinear disease progress curves.

### *Linear Disease Progress Curves*

In the past, plotted curves have increasingly been replaced by lines to make comparison of epidemics more meaningful. These lines, which mostly represent unilateral disease progress curves, in principle allow for the following comparions: (*a*) starting points of curves ($D$ and $x_0$ of Figure 2), (*b*) growth rates, (*c*) slopes (partly $b$ in Figure 2), (*d*) existence and, if so, duration ($P_1P_2, X_a$) of an asymptote, and (*e*) maximum disease incidence, $Y_{max}$ as in Figure 2.

Probits and probit lines are widely used to compare the effect of fungicides or spray regimes on the course of epidemics (39). Such a linear curve is characterized mainly by slope and position. Experimental results and implications with probit lines relevant in this context have been reviewed by Baker (5) and Dimond & Horsfall (16). The latter authors are confident that the use of inoculum density/disease curves may help "in discriminating qualitatively different mechanisms of resistance as between resistant varieties of a host or of pathogenesis between races of a pathogen." This may well apply to other factors, for example the effect of inoculum density in soil (5).

Other criteria for a comparison of disease progress lines are the logarithmic and apparent infection rates (62), $r_l$ and $r$, provided they are computed for the whole unilateral epidemic. They have been extensively used by Van der Plank (62–65) to compare a wide range of effects on epidemic models. Merrill (41) used $r_l$ to compare the changes in the spread of oak wilt (*Ceratocystis fagacearum*) in Pennsylvania and West Virginia from 1956–1965 and to predict their future course. Browning & Frey (12) compared $r$ with the effect of resistance in multiline cultivars in control of crown rust in oats. Kranz (33) employed $r_l$ for two epidemics of Sigatoka disease of bananas (*Mycosphaerella musicola*) in a humid and a dry site.

These apparent infection rates seem to be very sensitive (31, 62) and therefore suitable for immediate comparisons of effects on the epidemic behavior. "For a combination of host and pathogen, $r$ . . . is a function of the environment . . . (and) . . . need change little to affect the course of disease profoundly" (66). But just this consequence severely limits the suitability of $r_l$ and $r$ for comparison by classification (31) unless fairly natural classes of diseases can be delimited according to the $r$ values they most likely attain under defined conditions. However, a serious shortcoming of these infection rates are that they do not provide for tests of significance.

Various types of regression analyses have been the most frequent method by which disease progress lines are compared in literature. The degree of similarity is expressed in $b$, the regression coefficient, and $a$, the $y$ intercept. Cooke & Jones (14) compared the disease progress curves of *Septoria tritici* and *S. nodorum* on two cultivars. In one variety the $b$ values were nearly identical, whereas in the other one the $b$ values for the same regression were distinct for each disease. If disease progress curves, for example, of 96 epidemics of potato late blight (25) or from 55 locations of wheat leaf rust (13, 17) are subjected to linear regression analysis and do not differ significantly, they may be considered to form one class. An equation can then be derived as a model that may generate predictions of general application, as with these two diseases. But this is by no means valid for all diseases.

The comparison of linear disease progress curves is complicated by another aspect: a linear disease progress curve may be described more exactly by several instead of only one $b$ as derived from multivariate regression analyses. Each variable that eventually appears in such a regression equation may be regarded as an element, though in this case borne by independent effects, not as parts of the dependent variable, e.g. disease progress curve (Figure 2). Classifications and immediate comparisons based on elements that describe reactions to specific environmental factors could be envisaged, provided there is agreement on which variables are to be measured during the experiments.

All three methods reviewed here have the advantage of quick and straightforward comparison of the *y* intercepts (or position) and slopes on grid papers. But it is only the regression coefficient *b* that can be tested for statistical significance within and between *b* values. Both are essential to avoid—as often is done—conclusions from differences in disease progress curves merely due to chance variation. Infection rates permit only immediate comparisons of effects on an epidemic. With regression equations and probit lines a classification of disease progress curves can be attempted.

Not all epidemics, however, have linear curves. Therefore, we also need objective methods that allow for the comparison of nonlinear disease progress curves.

### *Curvilinear Disease Progress Curves*

Curvilinear models often describe incomplete disease progress curves more naturally than do linear models with their rigid requirements. However, curvilinear models call for tedious and complicated mathematics. This is why these models have little appeal to plant pathologists, who still prefer to put their epidemics into the "bed of Procrustes" of linearity. Little work with curvilinear models has yet been published, and knowledge of their elements is thus scanty (1, 3, 26). Nearly all the models used in these papers, whether growth or growth rate functions (1), have only a handful of constants or variables. These could be used as criteria, or elements, for the comparison of epidemics.

There is always the upper asymptote with the notation $Y_{max}$ as in Figure 2, which often is set as 1. If one accepts that maximum incidence of diseases is not in every case equal to 100% host surface affected, but say 36% as with apple scab (1), then the value of $Y_{max}$ for two diseases may be distinct within their ranges. Also growth rates $k$ may be suitable, as they are determined to a certain extent by the length of latent periods which, despite large variations, can differ markedly between diseases. However, as growth rates primarily seem to express effects of the environment, variability may be too large and the growth rate may not be a criterion for some comparisons. A position factor $b$ of the Bertalanffy and Gompertz function (1) may be a fairly stable criterion for a given disease if this is related to the beginning of host vegetation, or similar stages ($D$ and $X_0$ in Figure 2). Skewness $n$ of the disease progress curve in the growth functions of Bertalanffy & Mitscherlich (1) may be a good criterion for comparison, as it describes the natural coincidence between susceptible stages of the host, the amount of inoculum present in the course of an epidemic, or both.

Curvilinear epidemics may be compared and subsequently classified according to one or more of the above elements or according to which growth functions describe best the observed course of epidemics commonly produced by a disease. For apple scab these are the Bertalanffy ($n$=2 and 3) and the Gompertz functions (1, 2).

Disease progress curves may, however, be curvilinear for both disease frequency (prevalence) and disease severity, linear, or alternative combinations of both. It may be worthwhile to investigate the bearing of this criterion on comparison of epidemics, as did Analytis & Kranz (3) for *Venturia inaequalis* and *Cronartium ribicola.* Also the lag phase between the same degree of intensity in the severity and frequency curves may be of interest.

### *Entire Disease Progress Curves*

Curvilinear models approach entire curve models and become identical whenever they describe the naturally completed course of an epidemic. These curves are essential for the epidemiological treatment of the full life cycles of pathogen and host populations. For them, all elements in Figure 2 apply.

In this context the following questions may be raised: (*a*) what kind or classes of disease progress curves do exist, (*b*) how do they vary, (*c*) what kind of curves can a particular disease attain and with which other diseases does it tend to form clusters, (*d*) what are the relevant elements of disease progress curves, their individual range of variation, and intercorrelations, and (perhaps some time later) (*e*) which curve results from which combination of the more essential independent variables?

A comparison of entire epidemics as bilateral cumulative curves was attempted by Kranz (30–32, 35) and Kranz & Lörincz (37). Forty such curves for each of two years have been derived from a concurrent study of 59 host-parasite combinations. The curves studied by Kranz showed high diversity, and each seemed to have its own morphology. But after having characterized the curves of 13 elements (32, 27, Figure 2), a certain grouping of epidemics in a few curve models could be achieved (Figure 3).

Robinson (48) presented similar epidemic patterns for *Pseudomonas solanacearum* on 16 potato clones. Their shape in conjunction with changes in the clone populations' size indicates whether resistance was vertical, horizontal, or both.

THE RELATIVE IMPORTANCE OF CURVE ELEMENTS The question of which elements in Figure 2 are most important to characterize a disease progress curve is largely determined by the degree of variability and intercorrelation (30–32). A Principal Axis Factor Analysis (PAFA) revealed some intercorrelation between 13 elements of 40 different host-pathogen combinations investigated for two years (Table 1). From the rotated correlation matrix, 6 factors were extracted. Underlined elements in Table 1 show the highest intercorrelation. The more interesting correlations here are in the 1, 5, and 6 factors. The first proves a high and positive intercorrelation between initial disease incidence and maximum disease incidence, which in turn is tied in with "area under disease progress curve" (62) and "consequences of the epidemic" (defined in ref. 31). Therefore, a high primary infection very often results in a high final level of disease, a large area under disease progress curve, and a high rate of plants killed. This all appears biologically sound.

Factor 5 indicates a negative correlation between the lag between "date of emergence to first visible disease" and "consequences of the epidemic." Early appearance

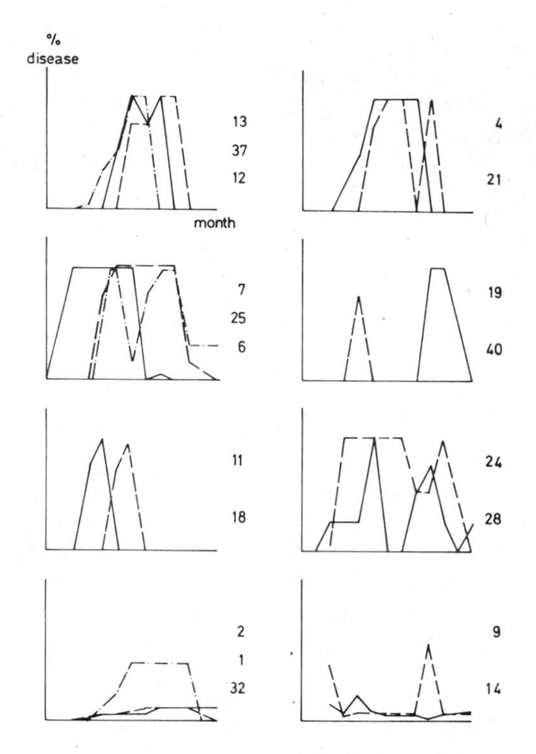


*Figure 3* Clusters of disease progress curves obtained by means of the computer program CLUS. (Adapted from refs. 32 and 37) Numbers refer to pathogen-host combinations (30) (time in months on abscissas, disease incidence in percent on ordinates).

of disease is thus more likely to result in severe disease, and vice versa. Hamilton & Stakman (22) related considerable differences in $Y_{max}$ of wheat stem rust observed for 42 years in the Mississippi Basin with the date of the first rust appearance: The earlier the first incidence the higher $Y_{max}$ tends to be. The delay $X_0$ has proved to be of prognostic value, for so-called "negative prognosis" as for *Phytophthora infestans* on potato (56, 59, 60) and *Leveillula taurica* on tomatoes and pepper (47). In spite of its fairly high consistency, $X_0$ remains somewhat variable in the field (32) and in the phytotron (10). This $X_0$ also determines the position of dosage-response curves.

The last factor, 6, conforms with Van der Plank's (62) finding of a close correlation between the apparent infections rate (not included in above elements) and his "area under disease progress curve." Apart from $X_0$, $X_p$ and $X_a$ appeared most consistent, whereas $r$ proved most variable (31, 32). The $r$ in this study ranged from 0.02 to 0.30 among the 59 host-parasite combinations. The infection rate here behaved as a function of environment rather than of the disease. In only five of the diseases studied $r$ did not differ between these two years. We have therefore dropped it as a curve element for Figure 2.

Table 1 Factor analysis of curve elements

| Element[a] | Factor number and factor load | | | | | |
|---|---|---|---|---|---|---|
| | 1 | 2 | 3 | 4 | 5 | 6 |
| $x_0$ | .03 | −.24 | −.29 | −.17 | <u>.81</u> | −.12 |
| $x_t$ | −.06 | <u>.48</u> | <u>.53</u> | <u>.65</u> | −.19 | .01 |
| $x_p$ | −.05 | −.18 | <u>.94</u> | .04 | −.10 | −.21 |
| $x_a$ | −.01 | <u>.97</u> | −.14 | .05 | −.10 | .10 |
| $x_d$ | .00 | −.01 | −.01 | <u>.99</u> | −.06 | .07 |
| FB | <u>.73</u> | .12 | .33 | .23 | −.08 | <u>.38</u> |
| $y_0$ | <u>.76</u> | .12 | −.14 | −.21 | .07 | .24 |
| $y_{max}$ | <u>.87</u> | −.18 | −.05 | .05 | −.13 | −.19 |
| $r$ | .11 | .09 | −.19 | .06 | −.11 | <u>.93</u> |
| BF | <u>.57</u> | −.16 | −.26 | −.05 | <u>−.64</u> | .05 |

[a] See Figure 2; $r$ = apparent infection rate (62); FB = area under disease progress curve (62); BF = consequences of attack (31, 32). (Adapted from ref. 32.)

When studying 59 host-pathogen combinations in two years (31) the diseases have been grouped according to whether their hosts occurred in one of the four ecological situations: sunny site–dense stand; sunny site–open stand; shady site–dense stand; and shady site–open stand. Among the curve elements of the respective diseases only $X_t$ was practically not affected. All the other ones were more or less distinctly elongated. Thus $X_a$ and $X_d$ became shorter under "open," the area under the disease progress curve, and smaller under "shady"; both $r$ and $X_i$ increased under "sunny–dense," whereas $Y_0$ and $Y_{max}$ decreased under "shady–dense." Lower values due to both factors showed $X_0$ at dense and shady and $X_p$ at dense and sunny. There were also differences among the values of curve elements when *Cercospora, Phyllachora,* and *Puccinia* spp. were grouped. For Cercosporas the $X_0$ was markedly shorter, and $X_d$ longer than with other genera. All this information is only indicative, although it proves variations of curve elements. But this seems to be connected with factors inherent in a disease, as ecological requirements of host and/or pathogen.

For comparison of entire epidemics, elements that exhibit the least variation in the disease progress curves of a disease are the most useful and should rank high in any weighting system. But no clues so far have been found to a realistic and appropriate weighting in this context.

THE VARIABILITY OF CURVE MODELS With a program for cluster analysis (CLUS) 40 host-pathogen combinations on wild hosts could be grouped into 11 and 16 clusters in the humid and in the dry year, respectively, with 2 to 9 host-pathogen combinations each (32). Similar clusters emerged with the program MCEF (37), which computes and arranges matrices of similarity coefficients. However, the same curve model does not always comprise the same host-pathogen combinations in each of the two years. On the other hand, out of 18 of them, which make up the first 5 clusters in 1962, 13 appear again in the first 5 clusters of 1963. Ten more pathogens show up again in at least adjacent clusters. This variation is due to the variability of the elements as just outlined. Further research—with some emphasis on the weighting of elements—are certain to help reduce this variability.

Disease progress curves that appear in the same cluster (Figure 3) of course differ somewhat in their course, because they have not been grouped according to their shape alone. Rather, the relative variability of each element has been weighted by the cluster analyses. Selected test parameters (for details, see ref. 32) thereafter performed the clustering. For example, with a test parameter of 2.00 CLUS produces only groups of two epidemics with the closest resemblance. There clusters are useful when we are looking for the most similar epidemic among the stored ones. It is obvious that components of clusters are determined to a large extent by kind and number of epidemics tested and available for comparison in the computer (32, 37).

It becomes clear in these studies that any disease can have various disease progress curves, depending on environment, host, pathogenicity etc. It is therefore unlikely that the same disease always appears in the same cluster. However, with improved knowledge of measurement and weighting of elements, clustering methods, and cluster delimitation, it will become more probable that chance fluctuation of a disease between classes of epidemics can be reduced. However, we should not aim at fixed assignment of a disease to one and only one epidemic taxon. We should rather expect a given disease to have an epidemic of class *a* in a wet year and that of class *b* in a dry year.

It would therefore be worthwhile to know which different shapes disease progress curves of a given disease may attain, and with which other diseases it tends to be associated in this respect. Such models of known disease progress curves then could be yardsticks stored in the computer against which incoming information is tested.

Knowledge of this kind will enhance automatic comparison. We can already check for the nearest similar epidemics for any given issue. For instance, in combination with a sorting program one may ask for potato late blight epidemics of a particular region and year and can continue this *ad libitum* (other sites, years, varieties, diseases, or planting times). However, if we wish to go beyond this we would need still much more and specified research which also will lead to an agreement among plant pathologists as to which elements are to be measured and weighted, and how it is to be done. It certainly does not require much imagination to see the potential of such a perfected method.

## MATHEMATICAL EXPRESSIONS FOR ENTIRE DISEASE PROGRESS CURVES

Disease progress curves may also be regarded as periodic or serial functions, or as parts of it. This applies not only to diseases of perennials, though it is more evident with them as they are often recorded for successive years, as with coffee leaf rust, *Hemileia vastatrix* (8, 9). For such an approach it will be required that for diseases of annual or deciduous plants also the negative (and invisible) disease progress curve, i.e. the fate of the overwintering inoculum, is measured. And "any continuous or reasonably discontinuous periodic function can be represented by a trigonometric polynomial with any degree of accuracy" (7). The fitting of periodic data by a trigonometric polynomial is achieved by approximation techniques known as Fourier analyses. But the simpler empirical functions, such as exponential functions and their inverses, can also serve this purpose. Also "frequency curves . . . are very important because of the empirical fact, that it is but rarely that we find in practice an empirical distribution, which could not be satisfactorily fitted by any such curves" (44). The interested reader may consult a good textbook on mathematics. He then will discover that curve models in Figure 3, which resulted from a few discrete measurements, could be described by mathematical approximation. We may find an underlying function that describes the general behavior of the model continuously, not necessarily utilizing the curve elements of Figure 2.

Such models of the epidemics graph can be useful for prediction, systems analyses, and simulation on analog and digital computers. When parameterized for a particular situation the model also may permit extrapolations. This rather formal approach eventually could also reveal causal relationships in terms of differential equation(s). This procedure is well known in physiology and in epidemiology from models adopted from meteorology (20, 54).

Among the functions that fit an observed disease progress curve the simplest will be the best as long as its accuracy suffices. It certainly would be of interest to know which other diseases obey a function estabished by approximation or which different functions belong to a disease. In many cases approximation will always remain arbitrary, but even then would mean a gain in comparison of epidemics. The mathematical methods that deal with oscillations of curves would provide a useful tool to describe and compare, for instance, the course of introduced diseases. Also the disease vulnerability of crops and the temporal behavior of diseases in vertically and horizontally resistant crops could be more exactly and succinctly compared by means of oscillations.

## COMPUTER METHODS FOR THE COMPARISON OF EPIDEMICS

Computers are not a *conditio sine qua non* in comparison of epidemics. Very often grid papers will suffice, particularly if we are interested only in positions and slopes. When we become more exacting and wish to compare the epidemic in its entirety or at least essential parts of it, data can become too massive and, with many elements involved, too confusing. As already indicated in the previous section the computer is useful in such instances and serves in research, retrieval, and subsequent identification.

Epidemics recorded according to some appropriate rules and procedures or even data not yet evaluated to these ends can be stored on punchcards, magnetic tape etc for further use. This data bank information system may be of interest in the future. Here models of disease progress curves may already be of advantage for "grading" of data before storage. Identification according to the type of epidemic of stored or incoming information is easiest and most quickly done by digital computers.

As far as storage and retrieval are concerned, general computer routine applies. We should, however, briefly review computer types and programs as to their relevance in comparison of epidemics.

### *Analog Computer*

Analog computers have a potential in comparing epidemics as soon as an algebraic or differential equation as a model is at hand. Once available, this instrument permits a rather quick operation and rapid comparison. The limitation lies in the fact that it remains a "black box," with only few integrated variables as position factor $b$ and growth rate $k$ (1). The reaction to environmental factors and the suitability of transformations can nevertheless be compared. No other information has obviously been published so far for plant diseases.

### *Digital Computer*

STATISTICAL PROGRAMS We already have referred to statistical programs available in computer centers, particularly for regression analyses of various types, which can be used for comparison as well. Multivariate analysis (34) is usually followed by arithmetic simulation of disease progress curves. Their potential for comparison has been proved by Analytis (1). Arithmetic simulations of apple scab epidemics gave the same good fit to the observed disease progress curves with virtually the same two variables in two years in a row.

A Principal Axis Factor Analysis (23) with subsequent rotation according to the varimax criterion (27) was used to test multivariate intercorrelation among the characteristics in Table 1 with the programs PAFA and BIMD[1] (32). This secures the selection of the more relevant characteristics to be used for further work. An attempt then was made to group the epidemics of host-pathogen combinations into classes or clusters according to their similarity in three qualitative and ten quantitative characteristics with the programs CLUS and MCEF[1] (32, 37). Both methods have been measured against a dendrogram, constructed on the basis of the number of characteristics that two given epidemics have in common. The clusters obtained by MCEF and CLUS agreed quite well. In addition, MCEF gave a slightly better fit of groupings and permitted a fully automatic comparison. CLUS secures a more objective grouping, which can only be influenced by the choice of the test parameters. Weighting of elements in these experiments did not always improve the results. Also promising for some aspects of the comparison of epidemics seem to be the multiple and stepwise discriminant analyses which have been used by Hindorf (24) and Kranz & Knapp (36) in studies of pathogen populations. Factor, discriminant, and cluster analyses are tools borrowed from applied mathematics, which can help the epidemiologist to file, sort, and reduce the overwhelming number of seemingly different epidemics and make them handy. Using them seems likely to disprove Van der Plank's (62) apprehension that there are as many kinds of epidemics as there are diseases.

[1]All these programs belong to the program library of the Deutsche Rechenzentrum, Darmstadt, W. Germany.

SYSTEMS ANALYSIS AND SIMULATION Programs for systems analysis and simulation of real epidemics are available (34, 38, 67, 68). These are quick and inexpensive programs that permit rapid comparison of numerous epidemics of apple scab, early blight of tomatoes, and southern leaf blight from various sites, years, and treatments, as long as weather data are available. This applies to factorial analysis of curves, relative weight of phases in the life cycles of the pathogen and host, as well as to the entire course of the epidemic. It is likely that there will be programs that allow different diseases to be compared at the same site, year, and treatment. Less comprehensive programs (71) are valuable in testing the effects of elements or single parameters on resultant epidemics.

Systems analysis is likely to be a most versatile, flexible, and comprehensive tool for comparison. This, nevertheless, will require prerequisites outlined in previous sections.

## CONCLUSION

We have discussed methods of rendering comparison of epidemics objectively and experimentally accessible. Our reference has been the graph of an epidemic, the disease progress curve, either with its various elements or as an unilateral or bilateral entity. As such, it may be used for immediate comparisons of (*a*) the effects of experimentally manipulated factors or (*b*) the summation of factors during the natural course of an epidemic. Also classification of epidemics has been suggested. Various criteria obviously can be adopted for the determination of classes. These may be transformation models, classes of slopes (e.g. slow-medium-fast rates of increase) or $Y_{max}$, degree of skewness, and/or entire curve models with or without a sophisticated mathematical description, to mention a few. All have a biological reasoning that gives each class a particular epidemiological meaning.

A class thus becomes a model. Any disease whose epidemics can be assigned to one or more such model(s) exhibits the characteristics of each model. In light of this, the variation we have noted in both elements and whole disease progress curve cannot be disturbing. It shows rather that comparison of epidemics could be handled flexibly and in an object-orientated way. This fits well with the possibilities of automatic comparison, particularly with large bodies of data. Perhaps it may not be too soon to anticipate a sort of data sheet issued for plant diseases, as with chemical compounds, giving various epidemiological characteristics of the kind discussed here.

Our approach to comparison of epidemics is still in its infancy; other approaches are possible. Our purpose has been to stimulate thoughts and experiments rather than to report on achievements.

*Literature Cited*


1. Analytis, S. 1973. Zur Methodik der Analyse von Epidemien dargestellt am Apfelschorf (*Venturia inaequalis* (Cooke) Aderh.). *Acta Phytomed.* 1:1–79
2. Analytis, S. 1974. Der Einsatz von Wachstumsfunktionen zur Analyse der Befallskurven von Plfanzenkrankheiten. *Phytopathol. Z.* In press
3. Analytis, S., Kranz, J. 1972. Über die Korrelation zwischen Befallshäufigkeit und Befallsstärke von Pflanzenkrankheiten. *Phytopathol. Z.* 73: 201–7
4. Baker, R. 1965. The dynamics of inoculum. In *Ecology of Soil-borne Plant Pathogens,* ed. K. F. Baker, W. C. Snyder, 395–403. Berkeley: Univ. Calif. Press. 571 pp.
5. Baker, R. 1971. Analyses involving inoculum density of soil-borne plant pathogens in epidemiology. *Phytopathology* 61:1280–92
6. Bald, J. G. 1970. Measurement of host reactions to soilborne inoculum. See Reference 4, pp. 37–41
7. Batschelet, E. 1971. *Introduction to Mathematics for Life Scientists.* Berlin: Springer. 495 pp.
8. Bock, K. R. 1962. Seasonal periodicity of coffee leaf rust, and factors affecting the severity of outbreaks in Kenya colony. *Trans. Brit. Mycol. Soc.* 45:289–30
9. Bock, K. R. 1962. Control of coffee leaf rust in Kenya colony. *Trans. Brit. Mycol. Soc.* 45:301–13
10. Brettschneider-Herrmann, B., Langerfeld, E. 1971. Untersuchungen über Beziehungen zwischen CCC Behandlung und *Septoria*-Befall bei Sommerweizen unter klimatisch-kontrollierten Bedingungen. *Acker- und Pflanzenbau* 133:137–56
11. Brönnimann, A. 1969. Einfluβ von Chlorcholinchlorid (CCC) und verschiedener Stickstoffdüngung auf den Befall von zwei Sommerweizensorten durch *Septoria nodorum* Berk. *Mitt. Schweiz. Landwirt.* 17:29–36
12. Browning, J. A., Frey, K. J. 1969. Multiline cultivars as a means of disease control. *Ann. Rev. Phytopathol.* 7: 355–82
13. Burleigh, J. R., Roelfs, A. P., Eversmeyer, M. G. 1972. Estimating damage to wheat caused by *Puccinia recondita tritici. Phytopathology* 62:944–46
14. Cooke, B. M., Jones, D. G. 1970. The epidemiology of *Septoria tritici* and *S. nodorum* II. Comparative studies of head infection by *Septoria tritici* and *S. nodorum* on spring wheat. *Trans. Brit. Mycol. Soc.* 54:395–404
15. Cox, A. E., Large, E. C. 1960. Potato blight epidemics throughout the world. *US Dep. Agr. Agr. Handb.* 174:1–230
16. Dimond, A. E., Horsfall, J. G. 1965. The theory of inoculum. See Ref. 4, pp. 404–15
17. Eversmeyer, M. G., Burleigh, J. R. 1970. A method of predicting epidemic development of wheat leaf rust. *Phytopathology* 60:805–11
18. Gäumann, E. 1951. *Pflanzliche Infektionslehre.* Basel: Birkhäuser. 2nd ed. 681 pp.
19. Gregory, P. H. 1961. *The Microbiology of the Atmosphere.* London: Leonard Hill. 251 pp.
20. Gregory, P. H. 1968. Interpreting plant disease dispersal gradients. *Ann. Rev. Phytopathol.* 6:189–212
21. Griffiths, E., Gibbs, J. N., Waller, J. M. 1971. Control of coffee berry disease. *Ann. Appl. Biol.* 67:45–74
22. Hamilton, L. M., Stakman, E. C. 1967. Time of stem rust appearance on wheat in the Western Mississippi Basin in relation to the development of epidemics. *Phytopathology* 57:609–14
23. Harman, H. H. 1967. *Modern Factor Analysis.* Chicago: Univ. Chicago Press. 2nd ed. 400 pp.
24. Hindorf, H. 1972. *Qualitative und quantitative Unterschiede in der Colletotrichum-Population auf Coffea arabica L. in Kenia.* PhD thesis. Univ. Giessen, Giessen. 146 pp.
25. James, W. C., Shih, C. S., Hodgson, W. A., Callbeck, L. C. 1972. The quantitative relationship between late blight of potato and loss of tuber yield. *Phytopathology* 62:92–96
26. Jowett, D., Haning, B. C., Browning, J. A. 1973. Non-linear disease progress curves. *2nd Int. Congr. Plant Pathol., Minneapolis, Minn.* (Abstr. 0244)
27. Kaiser, H. F. 1958. The Varimax criterion for analytic rotation in factor analysis. *Psychometrika* 23:187–200
28. Kiyosawa, S. 1972. Mathematical studies on the curve of disease increase. *Ann. Phytopathol. Soc. Japan* 38:30–40
29. Kiyosawa, S., Shiyomi, M. 1972. A theoretical evaluation of the effect of mixing resistant variety with susceptible variety for controlling plant diseases. *Ann. Phytopathol. Soc. Japan,* 38:41–51
30. Kranz, J. 1968. Eine Analyse von annuellen Epidemien pilzlicher Parasiten. I. Die Befallskurven und ihre Abhäng-

igkeit von einigen Umweltfaktoren. *Phytopathol. Z.* 61:59–86

31. Kranz, J. 1968. Eine Analyse von annuellen Epidemien pilzlicher Parasiten. II. Qualitative und quantitative Merkmale der Befallskurven. *Phytopathol. Z.* 61:171–90
32. Kranz, J. 1968. Eine Analyse von annuellen Epidemien pilzlicher Parasiten. III. Über Korrelationen zwischen quantitativen Merkmalen von Befallskurven und Ähnlichkeit von Epidemien. *Phytopathol. Z.* 61:205–17
33. Kranz, J. 1968. Zur Infektion und Erkrankung der Banane durch *Mycosphaerella musicola* Leach. *Z. Pflanzenkr.* 75:518–27
34. Kranz, J. 1973. Bemerkungen zur Simulation von Epidemien. *EDV Med. Biol.* 4:41–45
35. Kranz, J. Unpublished data
36. Kranz, J., Knapp, R. 1973. Qualitative und quantitative Unterschiede der Vergesellschaftung parasitischer Pilzarten in verschiedenen Vegetationseinheiten. *Phytopathol. Z.* 77:235–51
37. Kranz, J., Lörincz, D. 1970. Methoden zum automatischen Vergleich epidemischer Abläufe bei Pflanzenkrankheiten. *Phytopathol. Z.* 67:225–33
38. Kranz, J., Mogk, M., Stumpf, A. 1973. EPIVEN - ein Simulator für Apfelschorf. Vorläufige Mitteilung. *Z. Pflanzenkr.* 80:181–87
39. Large, E. C. 1945. Field trials of copper fungicides for the control of potato blight. I. Foliage protection and yield *Ann. Appl. Biol.* 32:319–29
40. Large, E. C. 1966. Measuring plant disease. *Ann. Rev. Phytopathol.* 4:9–28
41. Merrill, W. 1967. The oak wilt epidemics in Pennsylvania and West Virginia: An analysis. *Phytopathology* 57: 1206–10
42. Mogk, M. 1973. *Untersuchungen zur Epidemiologie von Colletotrichum coffeanum Noack sensu Hindorf in Kenia.* PhD thesis. Univ. Giessen, Giessen. 163 pp.
43. Mulinge, S. K. 1971. Effect of altitude on the distribution of the fungus causing coffee berry disease in Kenya. *Ann. Appl. Biol.* 67:93–98
44. Neyman, J. 1939. On a new class of contagious distributions applicable in entomology and bacteriology. *Ann. Math. Statist.* 10:35–37
45. Nutman, F. J., Roberts, F. M. 1969. The effect of fungicidal treatments on sporulation capacity in relation to the control of coffee berry disease. *Ann. Appl. Biol.* 64:101–12
46. Nutman, F. J., Roberts, F. M. 1969. Seasonal variation in the sporulation capacity of the fungus causing coffee berry disease. *Ann. Appl. Biol.* 64:85–99
47. Palti, J. 1971. Biological characteristics, distribution and control of *Leveillula taurica* (Lév.) Arn. *Phytopathol. Mediter.* 10:139–53
48. Robinson, R. A. 1968. *The concept of vertical and horizontal resistance as illustrated by bacterial wilt of potatoes. Commonw. Mycol. Inst. Phytopathol. Pap.* 10:1–37.
49. Röder, W. 1967. Über Beziehungen zwischen Witterungsverlauf, Befall mit Mehltau (*Erysiphe graminis* DC f. sp. *hordei* Marchal) und Ertrag der Sommergerste bei den langjährigen Sortenversuchen in Bad Lauchstädt. *Arch. Pflanzenschutz* 3:121–30
50. Romig, R. W., Calpouzos, L. 1970. The relation between stem rust and loss in yield of spring wheat. *Phytopathology* 60:1801–5
51. Rotem, J., Palti, J., Lomas, J. 1970. Effects of sprinkler irrigation at various times of the day on development of potato late blight. *Phytopathology* 60: 839–43
52. Rotem, J., Palti, J., Rawitz, E. 1962. Effect of irrigation method and frequency on development of *Phytophthora infestans* on potatoes under arid conditions. *Plant Dis. Reptr.* 46: 145–49
53. Schnathorst, W. C. 1962. Comparative ecology of downy and powdery mildews of lettuce. *Phytopathology* 52:41–46
54. Schrödter, H. 1960. Dispersal by air and water—the flight and landing. *Plant Pathology—an Advanced Treatise,* ed. J. G. Horsfall, A. E. Dimond, 3:170–227. New York: Academic. 675 pp.
55. Schrödter, H., Hoffmann, G. M. 1961. Der Einfluß der Witterung und des Mikroklimas auf den Befall des Leins durch *Polyspora lini* Laff. *Zentralbl. Bakteriol.* II, 114:15–44
56. Schrödter, H., Ullrich, J. 1965. Untersuchungen zur Biometeorologie und Epidemiologie von *Phytophthora infestans* (Mont.) de By. auf mathematisch statistischer Grundlage. *Phytopathol. Z.* 54:87–103
57. Schrödter, H., Ullrich, J. 1966. Weitere Untersuchungen zur Biometeorologie und Epidemiologie von *Phytophthora infestans* (Mont.) de By. Ein neues Konzept zur Lösung des Problems der epidemiologischen Prognose. *Phytopathol. Z.* 56:265–78

58. Stephan, S. 1968. Zur Epidemie des Getreidegelbrostes im Jahre 1967. *Nachrichtenbl. Deut. Pflanzenschutzdienst Berlin* 22:197–200
59. Steiner, K. G. 1973. *Wechselwirkung zwischen Witterung, Wirt, Parasit und Fungiziden bei der Kaffeekirschen-Krankheit (Colletotrichum coffeanum Noack).* PhD thesis. Univ. Giessen, Giessen. 154 pp.
60. Ullrich, J., Schrödter, H. 1966. Das Problem der Vorhersage des Auftretens der Kartoffelkrautfäule (*Phytophthora infestans*) und die Möglichkeit seiner Lösung durch eine "Negativ-Prognose." *Nachrichtenbl. Deut. Pflanzenschutzdienstes* 18:33–40
61. Van der Plank, J. E. 1960. Analysis of epidemics. See Ref. 54, pp. 229–89
62. Van der Plank, J. E. 1963. *Plant disease: Epidemics and Control.* New York: Academic. 349 pp.
63. Van der Plank, J. E. 1965. Dynamics of plant disease. *Science* 147:120–24
64. Van der Plank, J. E. 1965. Epidemiology of fungicidal action. *Fungicides—an Advanced Treatise*, ed. D.C. Torgeson, 1:63–92. New York: Academic. 697 pp.
65. Van der Plank, J. E. 1968. *Disease Resistance in Plants.* New York: Academic. 206 pp.
66. Waggoner, P. E. 1965. Microclimate and plant disease. *Ann. Rev. Phytopathol.* 3:103–26
67. Waggoner, P. E., Horsfall, J. G. 1969. EPIDEM. A simulator of plant disease written for a computer. *Conn. Agr. Exp. Sta. Bull.* 698: 1–80
68. Waggoner, P. E., Horsfall, J. G., Lukens, R. J. 1972. EPIMAY. A simulator of southern corn leaf blight. *Conn. Agr. Exp. Sta. Bull.* 729:1–84
69. Weltzien, H. C. 1972. Geophytopathology. *Ann. Rev. Phytopathol.* 10: 277–98
70. Zadoks, J. C. 1961. Yellow rust on wheat, studies in epidemiology and physiological specialization. *Neth. J. Plant Pathol.* 67:69–256
71. Zadoks, J. C. 1971. Systems analysis and the dynamics of epidemics. *Phytopathology* 61:600–10
72. Zakoks, J. C. 1972. Methodology of epidemiological research. *Ann. Rev. Phytopathol.* 10:253–76

# HEAVY METALS: IMPLICATIONS FOR AGRICULTURE[1]

❖3604

*Donald Huisingh*
Department of Plant Pathology, North Carolina State University,
Raleigh, North Carolina 27607

## INTRODUCTION

Heavy metals, and in particular mercury, have been used since the early 1900s in the control of a wide variety of plant diseases. During the last twenty years, inorganic forms were replaced by the vastly more effective aryl, alkoxy, and alkyl organomercurials which were used extensively, especially as seed dressing in the control of a number of seedling diseases. Although the toxicity of Hg has been known since ancient times, until comparatively recently little attention was paid to possible environmental or human health consequences of the prescribed use of the mercurials.

In the 1950s populations of seed-eating and predatory bird species began to decline drastically in Sweden. By 1965, when it was established that these bird-population changes were due to organomercurials, primarily the alkyls, a ban on their continued use in Sweden was imposed. In 1970 the United States also banned the use of alkyl mercurials and in 1972 banned the use of all mercurials in plant disease control. Similar restrictions or complete prohibitions of the use of mercurials in plant disease control have been imposed or are being considered in other countries.

In part, these actions occurred as a result of a growing concern about the effects of man upon his environment. Attention also has been given to the biological effects of cadmium and lead. Within the last few years reviews on Cd (15, 16, 18, 39, 41, 61, 66, 67); Pb (1, 15, 25, 39, 41, 61, 64, 66, 67); and Hg (13, 15, 19, 24, 39, 41, 45, 47, 61, 66, 67) have provided comprehensive analyses of the present understanding of the effects of heavy metals upon pathogens, plants, animals and humans.

The purposes of this paper are (*a*) to provide a perspective on heavy metals in the environment, (*b*) to describe the avenues by which heavy metals enter the

[1]Journal Series Paper No. 4210 of the North Carolina State University Agricultural Experiment Station, Raleigh, North Carolina 27607.

biosphere, (*c*) to discuss factors that influence the toxicity of heavy metals, and (*d*) to point out some unanswered questions about the effects of heavy metals in the agri-ecosystem.

## PERSPECTIVE

Both Pb and Hg have been utilized for centuries. Extensive use of Pb dates back to the pre-Christian era (25). Although classical lead poisoning in man has been known for centuries, Pb has continued to be one of man's most useful heavy metals.

While Hg was first mentioned in the writings of Aristotle in the fourth century BC, it was mined as cinnabar in China as early as 1100 BC. In 200 AD Galen declared that all forms of Hg were so toxic that they should never be used as medicines (24). Throughout the subsequent 1700 years, however, Hg has played a very important role in the control of human, animal, and plant diseases. Thus, both Pb and Hg have been used by man for centuries, but Cd is a comparative newcomer to the list of man's useful elements.

Cadmium was discovered in 1817 AD by Strohmeyer as an impurity in zinc carbonate (41). The geochemical kinship of Zn and Cd is important in terms of cadmium's origin and is a clue to its tendency to accumulate in mammals. Cadmium tends to replace Zn in certain enzymes, thereby altering their catalytic capacities (39). Technological uses of these metals have provided and continue to provide many benefits to mankind. But within the last 10 years, concern has developed about the effects of the widespread redistribution of increasing amounts of these elements in various ecosystems. The following is a brief recounting of a few of the findings that have contributed to this concern.

The best chronological record of the role of man in the redistribution of Pb is found in the various strata of the Greenland ice sheet. Lead concentrations increased from 0.005 μg/kg to more than 0.2 μg/kg in ice from 800 BC to 1965 AD (25). Most of this 400-fold increase has occurred since the industrial revolution, with the sharpest rise occurring after 1940. The question raised by these data is: what are the long-term environmental effects of such extensive redistribution of Pb?

Experimental evidence in rats (53, 55, 56) and analyses of human kidney tissue from normotensive and hypertensive patients suggests that cadmium may be an important contributor to cardiac hypertension and thus a factor in one of the leading causes of death in humans. However, other workers have not confirmed these results, and uncertainty currently exists about the role of Cd in cardiac hypertension (19).

In a recent review (45) the following statement is made:

> Wildlife served its traditional role as an early warning system for man when problems with Hg arose in Sweden. The first alarm sounded in the middle 1950's when populations of the yellow bunting dropped catastrophically. Deaths, reproductive failures and population declines of both seed-eating and raptorial birds were soon noted. Toxicity from Hg, used in seed dressing to prevent cereal diseases, was suspected. Experimental, chemical and pathological work demonstrated that the suspicions were correct.

Broader and deeper studies were initiated as a result of the Japanese problem, the so-called "Minamata disease," with human symptoms that include progressive blindness, deafness, incoordination, and intellectual deterioration. Methylmercury was responsible for the more than 150 serious cases and 52 deaths in Japan; it was ingested in fish that had grown in rivers and bays heavily polluted by Hg used as a catalyst in the manufacture of vinyl chloride and acetaldehyde. Shellfish in this area of Japan were found to have 38–102 ppm, and the people who died from Minamata disease had 13–144 ppm Hg in their kidneys (45).

Evidence that American rivers were polluted with Hg was presented by the Sports Fishing Institute in August 1969 (41). In March 1970, a Norwegian scientist working in Canada announced that he had found high concentrations of Hg in fish taken from Lake St. Clair. The Canadian and US governments instituted immediate restrictions on fishing and the sale of fish products and initiated extensive analyses of fish for Hg. Many of the fish contained 0.3 to 0.9 ppm Hg and thus were near to or exceeded the "action level" of 0.5 ppm Hg recommended in 1971 by the US Food and Drug Administration (6).

Outbreaks of Hg poisoning involving hundreds of persons have taken place in Iraq, Pakistan, and Guatemala, where Hg-treated seed was consumed directly instead of being planted (4, 23). In 1969 a farmer and five of his neighbors in the area of Alamogordo, New Mexico obtained waste seed wheat treated with methylmercury guanidine. The grain, containing 32 ppm Hg, was fed to swine which in turn were slaughtered and eaten by the family over a 3.5 month period (11). The report continues (11):

> In early December, one child of Mr. H's family became ill. By late December two other members of the family developed the same illness. Mercury poisoning caused by the ingestion of contaminated pork was suspected. Analyses of the muscle of the hog in question revealed a concentration of 29.4 ppm Hg. At the time, the mother of the children was pregnant. Three months later, the baby was born. The baby is blind and suffers from serious mental retardation.

This episode of indirect Hg poisoning, combined with others in which humans suffered because of direct ingestion of Hg-treated seed, have resulted in the prohibition of the use of agricultural mercurials in the US. Banning the use of a material is one approach to reducing the amount of the material that comes in contact with the food chain. To what concentration should humans restrict their intake?

As our analytical techniques have improved and the lower limits of detection extended to the nanogram or picogram range, it has become evident that all soil, water, and biological materials contain detectable quantities of Pb, Cd, and Hg. Such ubiquity makes it obvious that the concept of "zero concentration" or "zero tolerance" is not only meaningless and unattainable but may actually be deleterious (24). The concept of "tolerable concentration" would be more useful and would be in keeping with the fact that all elements, essential and nonessential alike, are potentially toxic, depending upon their form, concentration, ratio with other elements, and duration and route of exposure (4) (see Figure 1).

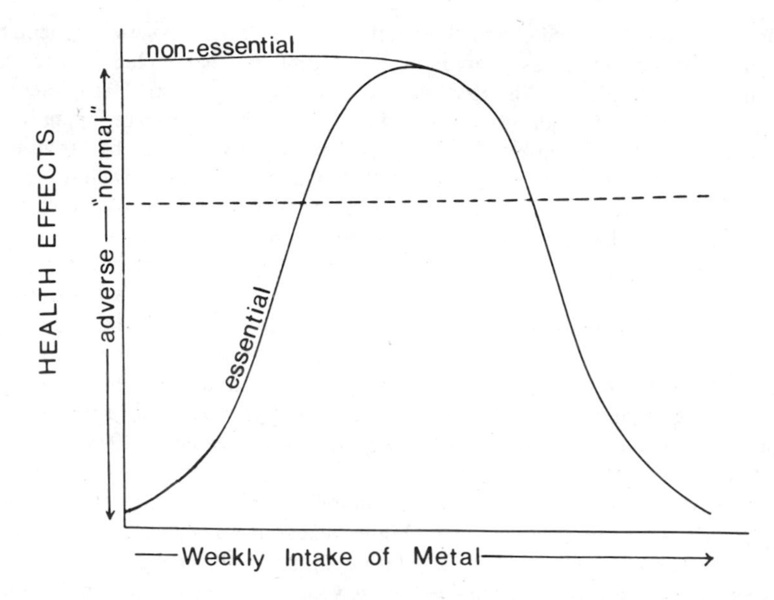


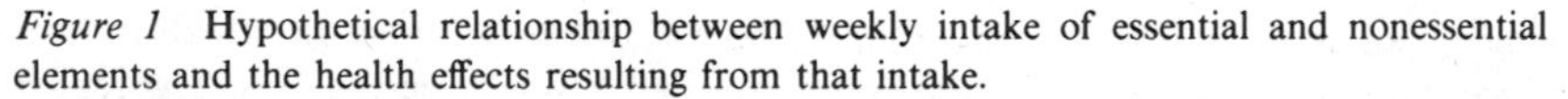


*Figure 1* Hypothetical relationship between weekly intake of essential and nonessential elements and the health effects resulting from that intake.

## AVENUES OF HEAVY METAL INTRODUCTION INTO THE BIOSPHERE

### *Geological*

Because of its volatility, mercury is released to the atmosphere naturally at a rate influenced by such land disturbances as earthquakes and volcanoes (68). Such natural sources account for the finding that the Hg content in museum specimens of tuna and swordfish caught 90 years ago are in the same concentration range as those caught recently (5, 44). Lead and Cd are present in all soils and waters. Except near areas of geological concentrations, natural sources do not account for any appreciable environmental burden.

### *Anthropogenic*

BIOCIDES AND ADDITIVES These include heavy metals purposely added to our environment for their beneficial effects. In a list of 280 different pesticides, heavy metals were present in 112 of the formulations [Hg, 45; Cu, 28; Zn, 24; Cd, 13; Mn, 12; Cr, 7; Sn, 5; Ni, 4; Fe, 3; and Pb, 1 (38)]. Annual agricultural use of fungicides in the US amounts to about 90,000,000 kg (14). In 1970, mercurial fungicides accounted for more than 100,000 kg or slightly over 16% of the total yearly US Hg consumption for all purposes (14). On a worldwide basis, 10,000,000 kg of Hg are produced annually, and agricultural uses account for 500,000 kg (23). What are the benefits and risks of the continued use of mercurial fungicides?

For more than twenty years organomercurials have been vital for the control of a wide range of diseases in barley, wheat, oats, sweet potatoes, Irish potatoes, fruits, and turf grasses (59). It is estimated that the agricultural uses of organomercurials in the US in 1967 resulted in a monetary gain of $222, $110, and $79 millions on wheat, cotton, and barley respectively (36). Thus, their use for seed treatment on three major crops returned $411 million to the US farmers in one year. Similar economic gains from their use were obtained in other parts of the world.

How much of the Hg applied as a fungicide and used according to recommendations appears in the plant parts utilized by man and his animals? When hydroxymercurichlorophenols are applied to sweet potato mother roots as a fungicidal dip, the Hg is translocated to all the new plant parts during the early part of the growing season, but little if any is translocated to the new fleshy roots (30). In addition it was found that the Hg content of soil from fields in which Hg-treated sweet potatoes had been grown for 20–30 years was not elevated above the Hg content of soil from adjacent nontreated areas (30). Soil Hg content varies broadly; one worker, after analyzing 273 Swedish soils, reported a range of 4–922 $\mu$g Hg/g soil, with a mean of 60 $\mu$g/g. Fourteen African topsoils had a mean of 23 $\mu$g Hg/g. The cultivated Swedish soils averaged 7.8 $\mu$g Hg/g greater than those not cultivated, and part of this was attributed to the use of mercurial fungicides (2). Residues in the soil were limited almost entirely to the uppermost 5 cm. However, both the maximum and minimum concentrations were found in uncultivated soils.

Crops grown from Hg-treated seeds of barley and corn have approximately 0.03 ppm of Hg, or 2 times the amount in seed produced on plants from untreated seeds (37). Similar results have been obtained from a number of other crops. When foliar sprays of phenylmercuric acetate are used, concentrations of 0.1–1.0 ppm Hg have been found in the new rice seed (13). In studies to determine translocation patterns of Hg applied as a foliar spray on rice plants, it was found that about 50% of the Hg was absorbed by the plants and nearly 5% of the amount absorbed was translocated to the seed, resulting in Hg concentrations of 0.04–0.12 ppm (13).

Scandinavian workers have shown that the Hg concentrations in Swedish pork and poultry products decreased from 0.03–0.06 ppm before the ban to 0.008–0.02 ppm after the alkylmercurials were banned in 1967. In Denmark, where alkylmercurials were never used extensively, the Hg concentrations remained at the lower values of 0.003–0.009 ppm found in the Swedish products after the ban (13). The alkoxy and aryl forms of Hg, when used as seed treatment materials, are not translocated into the seed in as large amounts as is the case of the alkyls. In summary, although Hg used in prescribed ways in seed treatment is translocated to the new plant parts, the concentrations are quite small. Substantially larger amounts of Hg are translocated to new plant parts following foliar sprays.

Experiments with corn, soybean, and peanut were done in 1972 to determine the translocation and accumulation patterns of foliarly applied $CdCl_2$ and Cd succinate. The results (D. Huisingh, unpublished observations) reveal two- to three-fold increases in Cd content of the seed of all three crops as a result of early season sprays. The mean Cd concentrations of seeds for control and treated plants were respectively: (*a*) corn 0.49 and 1.23 ppm; (*b*) soybean 0.73 and 2.60 ppm; and (*c*) peanut

0.45 and 4.57 ppm. The species difference in accumulation is noteworthy and is being studied further.

Cadmium also may be translocated from apple leaves to fruits after a single early season foliar spray of $CdCl_2$. The movement and accumulation patterns were similar to those of foliarly applied mercurials (51). Little information is available on the amount of Pb that may enter the food chain as a result of its use as the pesticide Pb arsenate. The use of Pb arsenate decreased from $16 \times 10^6$ kg in 1950 to $1.5 \times 10^6$ kg in 1968, but since the ban on the use of DDT its use has increased especially on certain fruit trees, berry species, and Irish potatoes (39).

CONTAMINANTS AND WASTE PRODUCTS Large amounts of heavy metals are used in alloys, instruments, electrical apparatus, and industrial processes from which they may be lost to the air, water, or soil and subsequently enter man's food chain. In the US in 1965, for example, 450,000 kg of Hg were lost primarily to water from the chloralkali industry (62). Approximately 1,365,000 kg of Hg are released each year by the burning of coal in the US (7, 33). Cadmium may also enter man's food chain as a contaminant. Like Hg, it is present in fossil fuels and is released to the atmosphere during combustion. As much as 20 metric tons are emitted annually in the US in the manufacture of plastics, batteries, pigments, alloys, and fertilizers (39). Cadmium is present in most soils at less than 0.2 ppm, but soils near highways may contain 10 times that much (40). Sources of this Cd are lubricating oils (0.20–0.26 ppm), tires (20–90 ppm), and diesel oils (0.42–0.54 ppm).

Widespread contamination of agricultural soils with Cd results from the heavy use of superphosphate fertilizers. Species differences in uptake have been found, with parsnips, accumulating very high concentrations of Cd (54). Tobacco leaves accumulate Cd to such an extent that cigarettes contain 1.5 $\mu$g Cd per cigarette.

This Cd may be an important contributor to man's body burden of Cd for the following reasons: (*a*) 38–50% of the Cd is volatilized during the combustion of tobacco products thereby releasing 13–15 $\mu$g per pack (43); (*b*) the average daily US diet contains 23–92 $\mu$g Cd (39); (*c*) only about 5% of ingested Cd is absorbed by the human, but 30% of inhaled Cd is absorbed (39); and (*d*) the FAO/WHO provisional tolerable dietary intake of Cd is 55–70 $\mu$g/day.

Lead is released to the atmosphere by metal smelters, battery- and pigment-producing plants, city refuse incinerators, and by coal and oil burners. The combustion of coal and oil are believed to release annually 3500 and 50 metric tons of Pb, respectively. More than 80% of the total Pb released is due to combustion of leaded gasoline. In 1968 US traffic released 141,000 metric tons of Pb into the atmosphere. This Pb may be inhaled and absorbed or may be deposited on plants and soil and subsequently enter the food chain (39). The amount of Pb deposited on plants along highways is a function of distance from the road, prevailing wind direction, and traffic density. The decrease in Pb deposition is approximately 30% for each 30 m of distance from the roadway. Tomato fruit produced at 23 m from roadways contained 6.02 ppm Pb compared to 0.95 ppm Pb at 183 m (39); 70% of this Pb was removable by washing. While corn seed and lima bean seed produced under the same conditions contained only small amounts of Pb, alfalfa contained 30 ppm Pb

(20). In other experiments, vegetation grown in the vicinity of a smelter had 20–200 ppm Pb, with concentrations in dicotyledons significantly higher than in monocotyledons. Concentrations of 120–150 ppm Pb in forages are considered toxic for cattle and horses (10). Lead uptake in tobacco can induce symptoms resembling the early phases of "frenching," a disease characterized by the development of narrow, elongated leaves (12).

In addition to the sources of heavy metals discussed in the foregoing paragraphs, toxic concentrations and forms of metals may be present in municipal, industrial, and agricultural wastes. The elements of most concern in these wastes are Cd, Co, Cr, Cu, Hg, Ni, Pb, and Zn. It is anticipated that in the coming years agricultural land spreading will be the method employed for the use or disposal of the elements present in these wastes. However, before this practice becomes fully accepted, many questions about the possible effects of the metals contained in them on soil, plants, animals, and humans must be answered.

The current average weekly dietary intake of these metals is estimated to be 490 μg Cd, 175 μg Hg, and 2100 μg Pb (13, 18, 25). The FAO/WHO recently established provisional tolerable limits or weekly dietary intakes for Cd at 400–500 μg, Hg at 200–300 μg, and Pb at 3000 μg (15). Thus, the average dietary intake of these metals is currently near the provisional tolerable limits.

In summary, heavy metals enter the agri-ecosystem from natural and man-generated sources. How significant any one may be and what approaches should be used to reduce the deleterious effects of heavy metals from that source are dependent upon a number of factors. Some of those factors are discussed in the following section.

## FACTORS INFLUENCING TOXICITY OF HEAVY METALS

### *Biotransformation*

Prior to 1967 it was thought that once Hg was discharged into surface waters it would soon be incorporated in bottom sediments, where it would be unavailable to living organisms. The discovery that inorganic Hg can be methylated to the mono and dimethyl forms by anaerobic microorganisms proved the hazard of this untested assumption (32). The significance of this finding is summarized by the statement, "The biological conversion process undoubtedly is the key to the biological concentration of Hg in the aquatic ecosystem" (13). But more recently it has been demonstrated that there are also microorganisms present in lake sediments capable of degrading methyl Hg (63). Similar types of organisms are present in soils. In conjunction with a number of physical-chemical binding mechanisms, they are important in their effect upon the circulation of Hg (2).

The alkyl forms of Hg, particularly methyl and ethyl, are both significantly more toxic and biologically more mobile than the other forms (19). Dietary methylmercury is absorbed almost completely (90–100%), is stable in the human body, and has a longer biological half-life (0.2 years) compared to the other organic forms, the alkoxys and aryls, that have half-lives of 0.08 years. From 2–20% of these are absorbed, transformed in the body of inorganic Hg, and excreted rapidly. The inorganic form is even more poorly absorbed from the diet than are the alkoxys and aryls. Therefore, toxicity assessment must be done with full knowledge of the form(s) and concentrations of the mercurials present.

Much less is known about the biological transformations of Cd and Pb or the relative toxicity of different forms of these metals. Both Cd and Pb have extremely long biological half-lives in man (16–33 and 4 years, respectively) and therefore tend to accumulate in the body. Because of the long half-lives, attention should not be addressed solely to acute toxicity but also to the long-term effects of chronic exposure to low concentrations.

### *Organismal Tolerance Mechanisms*

Because all organisms, including man, have been exposed to heavy metals throughout evolution, organisms may have developed mechanisms such as selective absorption and excretion to deal with these metals safely. Metallothionein appears to be an example of a protective protein produced by an animal to protect it against Cd. This protein contains unusually large amounts of both Zn and Cd and sometimes Hg (42, 34). Although it first was thought to be a Cd enzyme necessary to the cell, more recently metallothionein has been considered to play a protective role by sequestering Cd. Small initial doses of Cd have been shown to render an animal immune to subsequent doses of Cd that would have been toxic, presumably by inducing the synthesis of a Cd-binding protein similar to metallothionein which can then bind the large dose of Cd (18).

Heavy metal tolerance mechanisms are known to be present in fungi (3) and higher plants (65). Although responsible mechanisms have not been elucidated thoroughly, they are known to be under genetic control. Tolerance for one element does not usually confer tolerance to other elements. In the future, it may be desirable or necessary to select agricultural crop genotypes based upon their metal tolerance just as selections are now made for disease, insect, drought, and frost tolerance.

### *Interactions with Other Elements*

Another factor modifying the toxicity of the heavy metals is their metabolic interaction with other elements. Methylmercury added to a tuna-corn-soya diet for Japanese quail was found to be considerably less toxic than an equivalent amount added to a corn-soya diet (21). This suggests that other factors present in the diet may have modified the toxicity of the methylmercury. The protective factor in tuna was found to be 0.5 ppm Se. Selenium added to corn-soya diets (0.5 ppm) fed either to quail or rats was found to decrease methylmercury toxicity (22). Other species of fish have been shown to contain similarly high levels of Se. This work indicates that the accumulation of Se is likely to accompany the accumulation of Hg and that the presence of Se reduces the toxicity of the methylmercury. Selenite increases the retention of Hg in the tissues of rats while decreasing the toxicity of Hg (49), suggesting that data on the ratio of the concentrations of these two elements are likely to provide information of more value in predicting toxicity than is information on the concentration of either element alone.

The role of Se in protecting animals against the toxic effects of certain heavy metals has been investigated in a number of toxicological and physiological studies over the past 17 years in Parizek's laboratory in Prague (46). He found that small amounts of injected selenite will protect animals against the effects of Cd-related damage to the reproductive organs as well as against acute toxicity of larger doses of Cd. Although nutritional studies similar to those described above with Hg and Se have not been done for dietary concentrations of Cd and Se, it is likely that a similar protective effect will be found.

The toxicity of Hg, Cd, and Pb not only may be modified by other elements such as Se but may also interfere with normal metabolism of essential trace elements. Over the past decade, certain chemical properties or parameters of trace elements and heavy metals useful for predicting interactions have been identified and have led to the development of the "chemical parameter concept" (27).

In animal dietary studies, it was found that $Cd^{2+}$ interferes with both copper and zinc metabolism, apparently because $Cd^{2+}$ has the same chemical parameters as $Zn^{2+}$ and $Cu^{+}$. But $Hg^{2+}$ does not interfere with either Cu or Zn metabolism because it differs in the chemical parameters of geometric configuration and coordination number (8, 28). This interference of Cd with normal Zn metabolism is of particular concern in light of the important role of Zn in wound healing (48). Currently, nutritionists are concerned about marginal trace element deficiencies, especially of Fe and Zn. These marginal deficiencies should be evaluated with knowledge of the prevalent concentrations and effects of dietary intake of heavy metals. That the chemical parameter concept also may be useful in studies with heavy metals in plants is suggested by the observations that plants may absorb Cd preferentially by as much as an 18 to 1 ratio over Zn and by a 22 to 1 ratio over Pb and Cu. Many factors, such as soil type, soil pH, plant species, and forms of the element all influence the amount of the element incorporated by the plant (38, 40). Cadmium uptake can be reduced or kept to a minimum if the Cd:Zn ratio in the soil is maintained at 1:1000 and the pH is kept near neutrality.

Phenylmercuric acetate foliar sprays used to control the coffee berry disease have been shown to induce Zn-deficiency symptoms in treated coffee plants. Treated plants contained one fourth as much Zn as nontreated controls. This suggests that some modifications may be necessary in the application of the chemical parameter concept to plants.

Foliar sprays of Hg or Cd, containing fungicidal formulations applied during the first 4–8 weeks to corn, soybean, and peanut resulted in 40–55% reductions in the Fe content of the new corn seed, but had little or no effect upon the Fe content of the soybean and peanut (31). While most interest in this review has focused upon Hg, Cd, and Pb, a comment about Cu toxicity is relevant here. Copper accumulations have been reported in light, low organic matter soils subject to heavy applications of Cu in fertilizers and fungicidal sprays. Accumulated Cu was found to be the main cause of an imbalance between soil Cu and Fe and to be responsible for Fe chlorosis in the citrus (50).

In addition to having pronounced effects upon the metabolism of trace elements, heavy metals also affect the metabolism of some major elements. For example, dietary calcium and phosphorus concentrations influence the deposition of Pb in soft tissue and bone of animals (60). Recently it has been reported that Ca and P affect the intestinal absorption of Pb (35) and that lowering the dietary Ca increases the toxicity of Pb in rats (58).

The amount of Pb soybean plants take up from the soil and incorporate into the seed has been shown to be influenced by applications of lime (19). The Pb content decreased from 60 ppm Pb at 0 lime to 27 ppm Pb at 1000 kg of lime/hectare. The increased pH and $Ca^{2+}$ activity may diminish the physiological capacity of the plant roots to take up Pb. In general, with all toxic metals addressed in this review, less uptake will result if the soil pH is maintained at 6.5–7.0 than at 5.5.

Interactions of heavy metals with nonmineral dietary ingredients also have been found. For example, Cd toxicity is decreased with dietary ascorbic acid supplements (17). It is particularly important to keep such interactions in mind in setting and evaluating standards for safe levels of exposure to heavy metals. The determination of "normal" on the diagram in Figure 1 cannot be determined for one element or compound alone but should be established in the light of its interrelationships with other factors. While some countries have already set "safe" or "tolerable" concentrations of heavy metals in foods, reevaluation of present standards or future establishment of such safe concentrations should take into consideration the interactions of various factors that influence the toxicity of Hg, Cd, and Pb. These concentration limits, when set, should not become so rigid that they can not be changed in the light of new evidence, particularly modifying factors that may increase or decrease the toxicity of a particular heavy metal. In my estimation, safe exposure standards should be based more upon ratios of the concentrations of Hg:Se, Pb:Ca, and P and Cd:Se, Cd:Zn, or Cd:Cu, and upon the form of the element than upon the quantity of a particular element present. More thorough elucidation of these types of interactions will be of value in predicting and avoiding heavy metal toxicities and in developing remedial procedures where toxicity problems already exist.

## INTERACTIONS TO ALTER DISEASE SUSCEPTIBILITY

Much of the concern about heavy metal toxicity has been due to the acute toxicity of these elements. Comparatively less attention has been given to chronic effects. An excellent review by Adamson (1) suggests that chronic low levels of Pb to which urban man is subjected may be predisposing him to enhanced susceptibility to infection by various pathogens. This conclusion is based on animal experiments (26, 53, 57) that indicate enhancement by Pb of susceptibility to infection in two ways: by increased morbidity and mortality from bacterial infections after Pb exposure and by increased sensitivity to directly administered bacterial endotoxins.

In one series of experiments, Pb was supplied at 5 ppm in the drinking water of the test groups of 50 rats of each sex during their entire lifetime. The rats receiving Pb showed a 16% increase in infant mortality, a 10% increase in the number that died from an epidemic of pneumonia, and a 26% reduction in longevity (53). Lead analyses performed on various organs of these rats at death, on wild rats, and on humans showed that in most cases the Pb content was lower in the treated rats than in wild rats or in adult humans, indicating that the wild rats and humans had been exposed to greater than 5 ppm Pb.

In 30 day experiments, mice were preexposed to ~3 mg Pb/kg per day ip as Pb acetate. No signs of Pb toxicity were observed during this period. The treated and control mice were then exposed to a standard dose of *Salmonella typhimurium*. Within 7 days 54% of the Pb-treated animals but only 13% of the controls died of the Salmonella infection. In terms of $LD_{50}$ of *Salmonella,* the control animals were 10 times as resistant as the Pb-treated ones (26).

A single dose of Pb acetate caused a 100,000-fold increase in sensitivity of rats to endotoxins of several types of Gram-negative bacteria. The largest dose of Pb used (27 mg/kg body weight as a single dose) was well tolerated in the absence of endotoxin administration with no mortality being observed with Pb or purified endotoxin alone. If both Pb and endotoxin were administered, however, 80–100% mortality resulted within 24 hr. Even at doses as low as 5.5 mg Pb/kg plus endotoxin 30% mortality resulted (57). There is a report of similar dramatic increases in sensitivity of baboons to endotoxin, after exposure to Pb (29).

What is the relevance of these data for humans? Are humans currently being predisposed to greater susceptibility to various types of infections because of their body burden of Pb? What are the differences in response to exposure of experimental animals to Pb for short periods and exposure of humans to low levels for a lifetime? No one has approached these questions directly, but there is evidence that some human populations already contain body burdens at which such effects can be expected, even though they do not have high enough concentrations to produce symptoms of Pb poisoning.

What are the physiological and molecular bases for the observed changes in disease susceptibility? What are the effects of chronic exposure to several heavy metals simultaneously? Are these metals having a comparable effect upon plant health? A limited amount of work with plant parasitic nematodes revealed that various inorganic ions commonly contained in fungicidal formulations stimulate increased hatch of nematodes and support more rapid population increases with the consequence of more severe disease losses (9).

Normal symbiotic nitrogen fixation patterns in bean have been reported to be altered by low dosages of various insecticides and fungicides (52). In research with soybean and peanut on the effects of foliar applications of a number of metal-containing fungicides, it was found that nodulation was severely depressed in both crops when sprays containing phenyl mercuric acetate, mercuric chloride, cadmium chloride, or cadmium succinate were used at nonphytotoxic levels. Incidentally, foliar applications with triphenyl tin hydroxide stimulated significant increases in nodulation in soybean but not in peanut (D. Huisingh, unpublished observations).

Heavy metals, whether applied advertently as components of pesticides or inadvertently as contaminants with other materials, may alter the plant's trace element metabolism and alter its capacity for normal healthy development. An assessment of these types of interactions should become an integral part of all pesticide evaluation programs so that benefits from the control of one pathogen on one host are not nullified by concomitant accentuation of other diseases or undue inhibition of normal metabolic functions. Care should also be taken to determine the effects of each year's cultural practices upon subsequent crops in a rotation sequence.

## UNANSWERED QUESTIONS AND RECOMMENDATIONS FOR RESEARCH

Heavy metals have always been present naturally in the biological world; thus, all organisms have developed a degree of tolerance, and some organisms have developed special, genetically controlled, mechanisms for tolerating very high concentrations of the heavy metals. Man has increased the concentrations of these elements in the environment to the point that some species and isolated individual organisms have suffered from acute heavy metal toxicity. Hazards can be lessened by diligent enforcement of reasonable regulations designed to reduce man's emission of metal wastes into the environment.

However, at present while billions of dollars are being spent upon construction of facilities to reduce the flow of pollutants into the environment, very little is being spent on research devoted to understanding the interactions of pollutants, such as the heavy metals, with ecosystems we seek to save. With the objective of minimizing the future negative impact (risks) of heavy metals upon us and our environment, research should be undertaken to do the following:

(*a*) Increase our understanding of the mechanisms involved in the movement and transformation of the heavy metals through the soil and into agricultural plants.

(*b*) Develop feasible mechanisms for slowing or blocking the flow of heavy metals from the soil into the food chain.

(*c*) Develop an understanding of the heavy metal tolerance mechanisms present in some plants and microorganisms.

(*d*) Develop a breeding and selection program to select species and varieties of crops that will tolerate heavy metals but will transmit less of them into the food chain.

(*e*) Develop soil management practices that will reduce the availability of heavy metals to the plants.

(*f*) Develop a more complete understanding of the long-term effects of low concentrations of heavy metals upon man and upon select organisms.

(*g*) Establish a standardized international food monitoring network to monitor continuously the concentrations and forms of the heavy metals in our total diets. Concentrations of elements that modify toxicity of the heavy metal should also be determined.

(*h*) Develop and monitor model ecosystems designed to provide an "early warning" of adverse biological effects.

*Literature Cited*

1. Adamson, L. F. 1973. The effect of lead on susceptibility to infection. A review. Washington DC: The Environmental Defense Fund
2. Andersson, A. 1967. Mercury in soil. *Grundförbättring* 95:3–4
3. Ashida, J. 1965. Adaptation of fungi to metal toxicants. *Ann. Rev. Phytopathol.* 3:153–74
4. Bakir, F. et al 1973. Methylmercury poisoning in Iraq. *Science* 181:230–41
5. Barber, R. T., Vijayakumar, A., Cross, F. A. 1972. Mercury concentrations in recent and ninety-year-old benthopelagi fish. *Science* 178:636–39
6. Beasley, T. M. 1971. Mercury in selected fish protein concentrates. *Environ. Sci. Technol.* 5:634–35

7. Dillings, C. E., Watson, W. R. 1972. Mercury emissions from coal combustion. *Science* 176:1232–33
8. Bunn, C. R., Matrone, G. 1966. In vivo interactions of Cd, Cu, Zn, and Fe in the mouse and rat. *J. Nutr.* 90:395–99
9. Clarke, A. J., Shepherd, A. M. 1966. Inorganic ions and the hatching of *Heterodera* spp. *Ann. Appl. Biol.* 58: 497–508
10. Cox, W. J., Rains, D. W. 1972. Effect of lime on lead uptake by five plant species. *J. Environ. Qual.* 1:167–69
11. Curley, A. et al 1971. Organic mercury identified as the cause of poisoning in humans and hogs. *Science* 172:65–67
12. David, D. J., Wark, D. C., Mandryk, M. 1955. Lead toxicity in tobacco resembles an early symptom of frenching. *J. Aust. Inst. Agr. Sci.* 21:182–85
13. D'Itri, F. M. 1972. *The Environmental Mercury Problem.* Cleveland, Ohio: CRC Press. 124 pp.
14. Easton, B. M. 1971. *The Pesticide Review-1970.* Washington DC: USDA/ ASCS
15. FAO-WHO Joint Expert Committee on Food Additives: Sixteenth Report. 1972. Geneva. WHO Tech. Rept. Ser. No. 505
16. Flick, D. F., Kraybill, H. F., Dimitroff, J. M. 1971. Toxic effects of cadmium: a review. *Environ. Res.* 4:71–85
17. Fox, M. R. S., Fry, B. E. 1970. Cadmium toxicity decreased by dietary ascorbic acid supplements. *Science* 169: 989–91
18. Friberg, L., Piscator, M., Nordberg, G. 1971. *Cadmium in the Environment.* Cleveland, Ohio: CRC Press, 166 pp.
19. Friberg, L., Vostal, J. 1972. *Mercury in the Environment, An Epidemiological and Toxicological Appraisal.* Cleveland, Ohio: CRC Press, 214 pp.
20. Ganje, T. J., Page, A. L. April 1972. Lead concentrations of plants, soil and air near highways. *Calif. Agr.* 26:7–9
21. Ganther, H. E. et al 1972. Selenium: relation to decreased toxicity of methylmercury added to diets containing tuna. *Science* 175:1122
22. Ganther, H. E. et al 1973. *Trace Substances in Environmental Health—VI,* ed. D. D. Hemphill. Columbia, Mo.: Univ. Missouri Press. 247 pp.
23. Goldwater, L. J. 1971. Mercury in the environment. *Sci. Am.* 224:16–21
24. Goldwater, L. J. 1972. *Mercury: A History of Quicksilver.* Baltimore: York. 318 pp.
25. Hammond, P. B. 1972. *Airborne Lead in Perspective.* Washington DC: Nat. Acad. Sci. 330 pp.
26. Hemphill, F. E., Kaeberle, M. L., Buck, W. B. 1971. Lead suppression of mouse resistance to *Salmonella typhimurium. Science* 172:1031–32
27. Hill, C. H., Matrone, G. 1970. Chemical parameters in the study of *in vivo* and *in vitro* interactions of transition elements. *Fed. Proc.* 29:1474–81
28. Hill, C. H. et al 1963. *In vivo* interactions of cadmium with Cu, Zn and Fe. *J. Nutr.* 80:227–35
29. Holper, K., Trejo, R. A., Brettschneider, L. 1973. Enhancement of endotoxin shock in the lead sensitized subhuman primate. *Surg. Gynecol. Obstet.* 136:593–601
30. Huisingh, D., Nielsen, L. W. 1972. Mercury content of sprouts and harvested roots from treated sweet potato mother roots. *Phytopathology* 62:804
31. Huisingh, D. 1973. Agri-ecology of heavy metals in plant disease relationships. Presented at 2nd Int. Congr. Plant Pathol., Minneapolis, Minn., Sept. 1973. Paper No. 1017 (Abstr.)
32. Jensen, S., Jernelov, A. 1969. Biological methylation of mercury in aquatic organisms. *Nature* 223:753–54
33. Joensuu, O. I. 1971. Fossil fuels as a source of mercury pollution. *Science* 172:1028–29
34. Kägi, J. H. R., Vallee, B. L. 1961. Metallothionein: A cadmium and zinc-containing protein from equine renal cortex. *J. Biol. Chem.* 236:2435–42
35. Kostial, K., Simonovic, I., Pisonic, M. 1971. Reduction of lead adsorption from the intestine in newborn rats. *Environ. Res.* 4:360–63
36. Krauscue, K. 1968. *Nat. Agr. Chem. Assoc.* Cited in Ref. 13
37. Lagervall, M., Westöö, G. 1969. Mercury content in eggs from hens fed with grain from corn treated with mercury-containing pesticides and from untreated corn. *Var Foeda* 21:9
38. Lagerwerff, J. V. 1967. Heavy-metal contamination of soils. In *Agriculture and Quality of Our Environment,* ed. N. C. Brady, No. 85:353–64. Washington DC:AAAS
39. Lagerwerff, J. V. 1972. Lead, mercury and cadmium as environmental contaminants. In *Micronutrients in Agriculture,* 593–636. Madison, Wis.: Soil Sci. Soc. Am.
40. Lagerwerff, J. V., Specht, A. W. 1970. Contamination of roadside soil and vegetation with cadmium, nickel, lead and zinc. *Environ. Sci. Technol.* 4:583–86

41. Lee, D. H. K., Ed., *Metallic Contaminants and Human Health.* New York: Academic. 241 pp.
42. Margoshes, M., Vallee, B. L. 1957. A cadmium protein from equine kidney cortex. *J. Am. Chem. Soc.* 79:4813
43. Menden, E. D. et al 1972. Distribution of cadmium and nickel of tobacco during cigarette smoking. *Environ. Sci. Technol.* 6:830–32
44. Miller, G. E. et al 1972. Mercury concentrations in museum specimens of tuna and swordfish. *Science* 175: 1121–22
45. Nelson, N. 1971. Hazards of mercury. *Environ. Res.* 4:1–69
46. Parizek, J. et al 1971. The detoxifying effects of selenium interrelations between compounds of selenium and certain metals. In *New Trace Elements in Nutrition,* ed. W. Mertz, W. E. Cornatzer, 85–122. New York: Dekker
47. Pecora, W. T. 1970. Mercury in the environment. *Geol. Surv. Prof. Pap.* No. 713
48. Pories, W. J., Strain, W. H., Rob, C. G. 1971. Zinc deficiency in delayed healing and chronic disease. *Geol. Soc. Am. Mem.* No. 123:73–95
49. Potter, S. D., Matrone, G. 1973. Effect of selenite on the toxicity and retention of dietary methylmercury and mercuric chloride. *Fed. Proc.* 32:929
50. Reuther, W., Smith, P. F. 1952. Iron chlorosis in Florida citrus groves in relation to certain soil constituents. *Proc. Fla. State Hort. Soc.* 65:62–69
51. Ross, R. G., Stewart, D. K. R. 1969. Cadmium residues in apple fruit and foliage following a cover spray of cadmium chloride. *Can. J. Plant Sci.* 49:49–52
52. Ruschel, A. P., Costa, W. F. 1966. Symbiotic nitrogen fixation in beans (*Phaseolus vulgaris* L.). III. Effect of some insecticides and fungicides. *Pesqui. Agropecuar. Brasil.* 1:147–49
53. Schroeder, H. A. 1964. Cadmium hypertension in rats. *Am. J. Physiol.* 207:62–66
54. Schroeder, H. A. Balassa, J. J. 1963. Cadmium: uptake by vegetables from superphosphate in soil. *Science* 140:819
55. Schroeder, H. A., Balassa, J. J., Vinton, W. H. 1965. Chromium, cadmium and lead in rats: effects on life span, tumors, and tissue levels. *J. Nutr.* 86:51–66
56. Schroeder, H. A. et al 1967. Essential trace metals in man: zinc relation to environmental cadmium. *J. Chronic Dis.* 20:179–210
57. Selye, H., Tuchweber, B., Bertok, L. 1966. Effect of lead acetate on the susceptibility of rats to bacterial endotoxins. *J. Bacteriol.* 91:884–90
58. Six, K. M., Goyer, R. A. 1970. Experimental enhancement of lead toxicity by low dietary caldium. *J. Lab. Clin. Med.* 76:933–40
59. Smart, N. A. 1968. Use and residues of mercury compounds in agriculture. *Residue Rev.* 23:1–36
60. Sobel, A. E. et al 1940. The biochemical behaviour of lead. I. Influence of calcium, phosphorus and vitamin D on lead in blood and bone. *J. Biol. Chem.* 132:239–65
61. Somers, E., Smith, D. M. 1971. Source and occurrence of environmental contaminants. *Food Cosmet. Toxicol.* 9: 185 93
62. Sommers, H. A. 1965. The chlor-alkali industry. *Chem. Eng. Progr.* 61:99–109
63. Spangler, W. J. et al 1973. Methylmercury: bacterial degradation in lake sediments. *Science* 180:192–93
64. Tepper, L., Levin, L. 1972. A survey of air and population lead levels in selected American Communities. Report submitted to EPA, June 1972
65. Turner, R. G. 1969. Heavy metal tolerance in plants. In *Ecological Aspects of Mineral Nutrition of Plants,* ed. I. H. Rorison. Oxford, England: Blackwell. 484 pp.
66. Vallee, B. L., Ulmer, D. D. 1972. Biochemical effects of mercury, cadmium, and lead. *Ann. Rev. Biochem.* 41:91–128
67. Walker, C. 1971. Heavy metals. In *Environmental Pollution by Chemicals.* London, England: Hutchinson. 198 pp.
68. Weiss, H. V., Koide, M., Goldberg, E. D. 1971. Mercury in the Greenland Ice Sheet: Evidence of recent input by man. *Science* 174:692–94

# EPIDEMIOLOGY AND CONTROL OF FIRE BLIGHT

❖3605

*M. N. Schroth, S. V. Thomson, and D. C. Hildebrand*
Department of Plant Pathology, University of California, Berkeley, California 94720

*W. J. Moller*
Department of Plant Pathology, University of California, Davis, California 95616

## INTRODUCTION

Fire blight of pome fruits continues to be one of the most intensively studied bacterial diseases of plants. It has generated two important concepts for science (10): that bacteria may cause diseases of plants, and that insects provide an efficient means of dissemination of bacterial pathogens. The amount of literature concerning the disease is awesome, commencing with a 1794 record of fire blight by Denning (28) in New York and followed by a steady flow of research reports beginning with Burrill (22) in 1877. In spite of this effort, the disease still is not satisfactorily controlled; it continues to spread throughout continental Europe and remains a major concern in most countries where pome fruits are grown.

A critical evaluation of past and current research is long overdue since the literature contains numerous unsubstantiated claims, contradictory reports, and working hypotheses which, with time and succeeding citations, became accepted as facts. The reports provide an interesting chronicle of how scientific discoveries usually result from cumulative efforts of many scientists and also how attractive, provocative ideas or research contributions are periodically rediscovered, slightly embellished and presented anew, using updated terminology. Nevertheless, the literature embodies an impressive amount of detailed information concerning the disease cycle of the pathogen. A thorough understanding of the epidemiology of the disease now seems close at hand and should provide an opportunity to markedly improve control methods. This perhaps could be accomplished best by assembling an epidemimetric simulator, such as EPIMAY and EPIDEM (150, 151) and could be effected by a judicious winnowing of accumulated data, accompanied by a cooperative effort by the various laboratories now studying fire blight epidemiology.

With this in mind, we have critically examined past methodologies and data in reference to speculations, claims, and conclusions. Little significance was attached to abstracts or reports where techniques and data could not be evaluated. We have emphasized the first reports of research findings and concepts, rather than fully listing subsequent confirmations. Our aim, therefore, is to present a conservative assessment of what is factually known about the epidemiology and control of fire blight, to discuss this in terms of current, ongoing research and to identify aspects that need further investigation. An excellent historical account of fire blight of pome fruits was written by Baker (10).

## THE BACTERIUM

Recent descriptions of *Erwinia amylovora* reveal the usual inconsistencies inherent in taxonomic studies. For example, Billing et al (14) reported that considerable physiological homogeneity exists among English isolates, whereas Martinec & Kocur (88), using an assortment of strains, found that only 57 of 71 biochemical tests were relatively constant, and some of these did not agree with Dye's findings (31). The species, however, may be characterized as being nutritionally fastidious, utilizing relatively few compounds for growth. Strains vary in virulence (5, 123, 140), morphology (5, 65, 149), serology (120, 121), and phage typing (39, 40). The strains are somewhat unstable when cultured continuously (5, 60), producing colony types which vary in virulence. Strains also differ in growth rates with generation times of 45–100 min at 30°C (60, 140) and temperatures for growth (5, 14, 71, 88) with minimum at 3–8°C, optimum at 23–30°C, and maximum at 35–37°C. These findings vary among authors and in part may be the result of using different strains and test media, but also may reflect the use of strains that were repeatedly subcultured; this can lead to the development of laboratory curiosities.

Mutants resistant to streptomycin occur readily in culture (140), and have developed in nature, since steptomycin-resistant strains are now present in California pear orchards (95, 97). All naturally occurring strains that exhibited streptomycin resistance were resistant at levels over 800 ppm. But the generation times decreased with each incremental addition of streptomycin over 400 ppm.

Little is known concerning factors contributing to the pathogenicity of *E. amylovora.* It has been suggested that cell-wall degradation is not an important factor in disease development because there was a general absence of hydrolytic enzyme activity (103, 122). Pierstorff (103) reported that fire blight ooze contained a toxin that caused wilting of young apple shoots. The shoots usually recovered if they were removed from the toxin and placed in fresh water after the ends had been excised. Hildebrand (61) confirmed these results using pear shoots and indicated that the toxic principle was thermostable and caused cell plasmolysis. It recently has been characterized as a high molecular weight heteropolysaccharide with galactose, glucose, mannose, and glucuronic acid residues (34). Whether or not this toxin is causally related to disease initiation rather than the result of infection is not known. The latter may be the case because the bacterium is frequently found in symptomless tissues far in advance of necrotic diseased areas.

Another hypothesis was that fire blight symptoms were caused by the production of ammonia in tissues infected by *E. amylovora* (84). Ammonia was also considered the cause of the hypersensitive reaction in tobacco (85), but Goodman (49) subsequently reevaluated its role and found that it was not causally related. We suspect that this is also true with fire blight symptoms and that ammonia production is the result of tissue degradation. In a similar system, Stall et al (129) found that ammonia apparently formed after multiplication of the pathogen and injury and thus found it difficult to accept the position that symptoms were caused by ammonia. Other work by Hurst et al (67) showed that ammonia production followed infection.

It is apparent from the aforementioned literature that surprisingly little information exists concerning the characterization of *E. amylovora,* especially when considering the widespread seriousness of the disease and years of research. A thorough understanding of epidemiology is dependent upon this kind of information. For example, what is known concerning the genetic variability of the pathogen in nature, and how does such diversity relate to epidemiology? Are there substantial differences among strains in their capacity to survive and multiply at temperature extremes? Do some strains survive better than others because of ooze production or because of their tolerance to host-produced metabolites, and do they differ in their potential for dissemination because of strand formation, ooze production, and generation time? Are there differences in the potential of strains to attack different plant species and cultivars, and do they vary in capacity to infect flowers, leaves, or twigs? These and other suggestions have been asked by growers and researchers who have observed the varying patterns of the disease over the years.

We believe it is timely to initiate a detailed, comprehensive study of the physiological, biochemical, and nutritional characters of *E. amylovora,* using a heterogeneous selection of freshly isolated or promptly lyophilized isolates from different hosts and geographical areas. Ideally, the isolates should be single celled, since there are examples where a research finding such as repeated reversions in virulence, may have been the result of using cultures composed of several different strains of the pathogen. This is supported by Hildebrand's observation (62) that single-celled isolates of *E. amylovora* were more stable for virulence than single-colonied isolates. Also, relying on standard single-colony isolation techniques for complete insurance against contamination is hazardous because some bacteria are difficult to purify (29, 111).

Aside from taxonomic purposes, knowledge gained from a thorough characterization of *E. amylovora* would assist in developing techniques for rapid strain identification. At present the number of traditional biochemical differences that distinguish *E. amylovora* from related species are "small and few" (31). Thus, the identification of *E. amylovora,* as with the majority of plant pathogenic bacteria, has relied to a large extent on the production of symptoms in a host plant. Therefore, a degree of skepticism is appropriate in reviewing data developed by the use of so-called avirulent isolates, because such strains do not produce symptoms. The isolates must be comprehensively characterized as to their similarity to *E. amylovora* by use of newer tests; production of 2-ketogluconate and $\alpha$-ketoglutarate (132, 133), the formation of characteristic nucleotides by nucleoside phosphotransferase (78), serology (120, 121), and absence of polypectate gel pitting (57) etc may help to authenticate the identity of avirulent strains. We note in some cases that the so-called avirulent strains differ substantially from virulent strains in physiology, growth rate, colony color, and morphology, thus suggesting that the "avirulent" strains were selected from mixed cultures and may be contaminants. It is possible, however, that an avirulent strain could result from the loss of an episome or be a multiple auxotroph, thus differing from the virulent strains in physiology and generation time. Klement & Goodman's (77) use of the hypersensitive reaction (HR) as a test for identification of avirulent strains of *E. amylovora* is a convincing step towards authentication because it is unlikely that a saprophyte would produce such a reaction, although many other members of the plant pathogenic Enterobacteriaceae are capable of this (73).

It is not practical to identify isolates by an array of biochemical tests when studying the epidemiology of *E. amylovora.* For example, the monitoring of *E. amylovora* populations to assist California growers in timing of sprays (142) necessitates the rapid identification of thousands of colonies. The development of a selective, differential medium (95) has enabled the identification and enumeration of *E. amylovora* within 60 hr. Although identification by use of the medium is virtually 100% certain when used by a trained technician, supplementary tests may be used for verification if desired. These tests include those for HR, pathogenicity, presence of peritrichous flagella, and the Hugh-Liefson fermentation test (66). The HR was the most useful supplementary test because its elicitation by *E. amylovora* in 8–12 hr distinguished it from occasional *E. herbicola*-like strains which approach *E. amylovora* in appearance. Other bacteria that elicit the HR either do not grow on the medium or differ substantially from *E. amylovora* in colony appearance. The use of a specific antiserum in agglutination tests may also be used for rapid identification of *E. amylovora.*

## SURVIVAL AND OVERWINTERING

*Erwinia amylovora* survives and overwinters in pear and apple trees (Figure 1) as a result of infections occurring the previous year. The primary source of inoculum in spring appears to be determinate and indeterminate cankers plus latent infections. The number of active holdover cankers are reported as varying from 2–46% (94, 143). Some investigators (45, 146), however, questioned the role of cankers as a primary source of inoculum on the basis that they found only a small percentage that were active, e.g. 2–11%. We do not agree with this for several reasons: 1. It is based primarily on the assumption that cankers are easily discernible. In fact, it is impossible to locate many of the existing cankers, because, as Parker (101) and others (20, 94, 103, 113, 135, 143) have reported, holdover cankers occur on small twigs, some as small as 6 mm in diameter. Rosen (113) noted that blighted twigs on apples and pears served as sources for overwintering bacteria and aptly stated that "the importance of large limb and body cankers acting as sources of overwintering has been exaggerated in the past." Tullis (143) further observed that some cankers develop without definite margins, thus complicating the detection of an infection. 2. The common assumption (113, 146) that visible ooze is a necessary indicator of the presence of inoculum is erroneous. Recently, fire blight bacteria

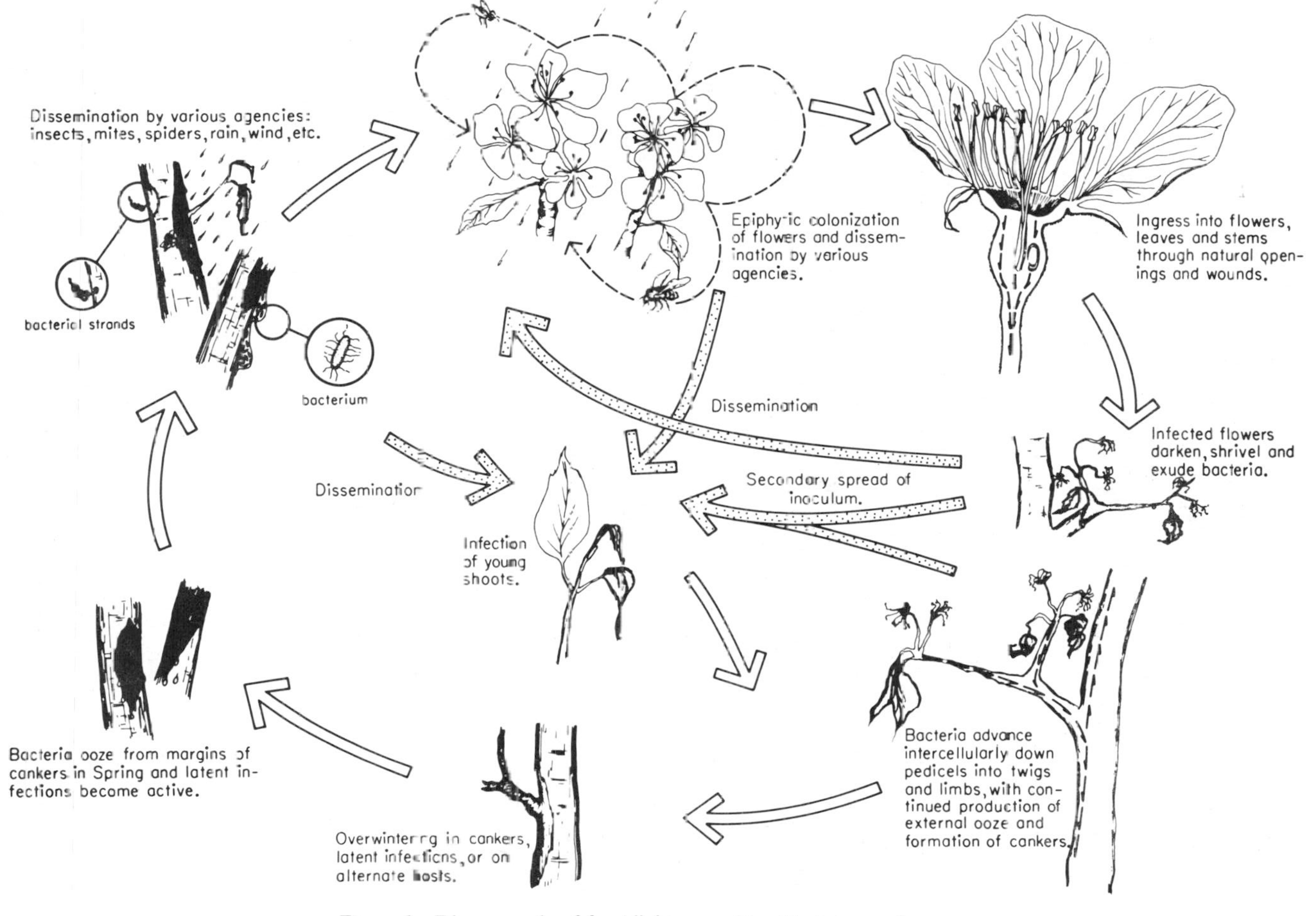


*Figure 1* Disease cycle of fire blight caused by *Erwinia amylovora.*

were isolated from the surface of 33% of the detected cankers prior to any visible evidence of ooze in spring (142). 3. Furthermore, the percentage of surviving cankers may not be as important as the absolute number of cankers present in an orchard.

Bacteria from a low percentage of lesions or holdover cankers have been considered more than ample to initiate an epidemic under weather conditions ideal for disease development (94). This is not surprising when considering the epiphytic nature of the bacterium (95), its rapid generation time, and the various efficient means for its dissemination to susceptible host parts (Figure 1). Studies monitoring the population of *E. amylovora* have shown that multiplication and dissemination occur at remarkable rates after the bacterium is first detected in an orchard (142).

Latent infections in various plant parts also appear to be an important means of overwintering. The observation that bacteria occur in apparently healthy tissues ahead of an advancing margin of the disease was among the major discoveries concerning the disease made by Burrill, according to Smith (125). By 1904, this led to the standard recommendation (55) to remove all diseased wood from trees by making cuts 30 cm or more distal from diseased parts. Many investigators have subsequently reaffirmed that bacteria occur in symptomless tissues (8, 20, 53, 74).

R. A. Lelliott recently stated (personal communication) that fire blight bacteria persist in localized areas of the xylem for 500 days after inoculation without signs of disease on the exterior of the stem. Eden-Green (34) further noted symptomless infections in which bacteria remained confined to the medullary xylem below infected leaves or leaf scars, frequently following inoculation of leaf parts of more resistant apple cultivars. These findings support the observation of Rosen (113) who reported that *E. amylovora* was present in "xylem tubes without calling forth any disease symptoms." Rosen found it remarkable that even though xylem invasion had been noted by Burrill and others, little attention had been given to it. The pathogen apparently readily enters the xylem through injuries in leaves (25). The role of these latent xylem infections to disease development in spring still remains obscure.

Keil & van der Zwet (74) stated that *E. amylovora* seemed to be a common resident in apparently healthy apple and pear tissues during the growing season and suggested that this was an important mechanism for survival. However, this intriguing theory needs to be supported with field data showing that the bacterium is still present in the tissues during winter and early spring. Our experience has been that *E. amylovora,* as with many bacteria, invades the plant through natural openings during moist conditions and persists for extended periods of time, but later dies without producing symptoms of disease. In a related study, Gowda & Goodman (53) reported that within two weeks after inoculation of apple shoots, bacteria were found in apparently healthy tissues 30 cm in advance of the visible infection. However, the population of bacteria subsequently declined markedly, and they concluded, thus confirming past work, that "survival of *E. amylovora* is best in or near infected tissues." This experiment supports our observations that rosaceous hosts seem to have the capacity through host defense mechanisms to purge the majority of pathogenic bacteria from their tissues during the season and especially during dormancy. This is why fire blight bacteria are not isolated from the majority of cankers after the winter season. The climatic conditions during dormancy appear to be a major factor influencing the survival of *E. amylovora* in host tissues.

Buds have been suggested as overwintering sites for *E. amylovora,* but this is not supported by substantive published data. Rosen (114) first thought that buds infected the previous year could serve as a possible vehicle for survival of fire blight bacteria. This was largely based on finding minute infections in the base of what appeared to be a "perfectly sound" bud. However, he did not pursue his isolations throughout the dormancy period. Later, *E. amylovora* was reported (30) to remain viable on fruit spurs following blossom infection until bud movement the following spring. Baldwin & Goodman (11) subsequently claimed to have isolated *E. amylovora* from apple buds on diseased trees and suggested that the buds were infected during the previous growing season. The identification of the isolates as *E. amylovora* is doubtful, because none were virulent and many were yellowish-white. It was later claimed (46) that the yellowish-white mucoid colonies from buds were a mixture of a virulent *E. amylovora* and another bacterium, but this claim was based on meager evidence. Morrill (98), however, indicated that authentic, virulent fire blight bacteria were prevalent in pear buds in early spring. J. Dueck (personal communication) also found large numbers of *E. amylovora* cells in buds taken from apparently healthy shoots of pear in February and March. Future data will presumably indicate whether or not the presence of bacteria in buds is the result of internal colonization of healthy tissues or of latent infections from the previous year. Positive results also might reflect extrusions of bacteria in xylem from previous leaf and stem infections, if any stem tissue is included during bud sampling procedures. Over the past three years, thousands of buds have been examined in California for the presence of *E. amylovora* with negative results. However, bud infection may be precluded by California climatic conditions which are not conducive for twig blighting.

*Erwinia amylovora* has a large host range (79, 118, 135, 139, 147) and may overwinter on alternate hosts. *Erwinia amylovora* commonly overwinters on *Pyracantha* spp. in California, and populations of the bacterium often are as great as $10^5$ per flower (142) in spring. Dissemination of bacteria from these flowers may not affect primary flowers of pears since flowering of *Pyracantha* is generally later than pear. However, this later flowering period of *Pyracantha* is particularly important as a source of inoculum for the infection of secondary pear flowers (commonly called rattails), which is a major problem in California. In England, the hawthorn (*Crataegus* spp.) appears to be a major overwintering host (43).

Some caution should be exercised before accepting the validity of the various reported hosts, since early workers did not understand the nature of hypersensitivity and the fact that large dosages of *E. amylovora* inocula elicit this reaction in most plants. Thus, a necrotic, hypersensitive reaction may have been accepted as proof of pathogenicity.

*Erwinia amylovora* may survive in nonsterile soil for short periods of time (4) and possibly overwinter in diseased fruits (3, 45). However, these aspects are of questionable importance in the disease cycle. Although still viable in dried ooze after two or more years in the laboratory (61, 103), bacteria in dried exudate survived only a few weeks during "dry weather" conditions in the field (61). Presumably dew and other sources of water may have moistened the exudate in the field, causing rapid cell death. Nevertheless, the survival of bacteria in ooze or exudate for even short periods indicates that infected plant parts should not be left in the field after pruning, because the bacteria might persist long enough for insects or other vectors to transmit the pathogen to susceptible plant parts.

The capacity of fire blight bacteria to survive on air-borne particles in open air for 2 hr or more (128) carries significant implications for field dissemination. Southey (128) comprehensively discussed the difficulty of accurately assessing the survival of fire blight bacteria in air using laboratory techniques and indicated that little is known concerning the occurrence and sizes of bacterial clumps or strands in the air (128). Thus, he considered the survival time of 2 hr or more as an underestimation.

The importance of bee hives in the disease cycle of *E. amylovora* has attracted considerable attention and generated some controversy (50–52, 64, 104, 114, 136). In reviewing the literature, we found no evidence to suggest that bacteria survive in hives for longer than 20 days when introduced under natural conditions. Although blossom blight developed when bees were fed freshly contaminated sucrose solutions (64), there were no conclusive data showing that bacteria were returned to flowers after natural deposition in a hive (104, 135).

The capacity of *E. amylovora* to survive in different osmotic concentrations of sugars was studied (64, 69) because it was a commonly held hypothesis that flower infection was favored by low-nectar concentrations. This theory remained surprisingly popular, even though the findings indicated that *E. amylovora* was relatively resistant to high concentrations of sugars: growth occurred in 30% glucose, 35% artificial nectar, 58% sucrose, and it could survive in 40% sucrose at 21°C for five weeks (64). Hildebrand (59) accordingly reported that "blossom infection may readily occur with nectar at the secretion concentration," and further stated that Thomas & Ark "probably have exaggerated the importance of nectar dilution by rain in bringing about blossom infection, because openings other than those found in the nectary (and particularly the stigmas) were left out of consideration by them, as infection courts." Although the nectar concentration theory for infection has been an attractive, well-accepted concept for years, we found no evidence that low concentrations of nectar sugars per se are responsible for increased infection. Low nectar concentrations are probably little more than an incidental occurrence during periods of high moisture, which favor the invasion of natural openings by *E. amylovora.*

## DISSEMINATION AND TRANSMISSION

### *Short-Distance Spread*

It seems reasonable to assume that any moving object, such as birds, insects, mites, spiders, man, wind, water, or mechanical equipment can disseminate fire blight bacteria (Figure 1). Some investigators (20, 52, 94), considered water to be the most important agent for fire blight dissemination, whereas others considered insects to be of equal or greater importance (101, 103, 114, 135). No doubt climatic differences between regions where investigations have been made influence this consideration. Water, for example, would be particularly important in areas such as Wisconsin or England where wind-driven rains are frequent (43, 94). In contrast, the incidence and timing of fire blight infections in Arkansas and California has not always been correlated with rain; here, insects were considered of greater importance (114, 135). Either agent, however, could conceivably disseminate the bacteria effectively from cankers, lesions, and latent infections to susceptible sites. Hence the argument as to which agent is most important seems futile.

From the time that Waite (152) first demonstrated that honeybees carry fire blight bacteria, reports have continued to emerge listing additional insects as vectors (20, 21, 36, 37, 52, 72, 95, 101, 131, 153, 154). Presumably any insect that visits infested or infected plant material is capable of disseminating bacteria. Insects are a very efficient inoculating agent, carrying up to $10^5$ cells per insect (95). This is particularly important, because, as Hildebrand (58) and R. A. Lelliott (personal communication) have shown, inoculum concentration is critical for disease development; the deposition of a few cells in a pear flower usually did not result in multiplication. Thus, contrary to many other plant diseases, fire blight may be an example of a disease where inoculation often is accomplished in nature by massive rather than low dosages of inoculum.

The fire blight disease cycle in California begins in spring by the attraction of various insects to bacteria and bacterial exudates on twigs and limbs which then transmit the bacteria to flowers. Subsequent dissemination, following new infections or epiphytic multiplication of the bacteria in flowers is effected by bees and other insects and rain (Figure 1). The possibility exists that some insects might injure the flowers, thus providing additional modes of entry. For example, bees have sharp tarsal claws and stiff bristles which could injure or abrade some of the flower parts while scraping and gathering pollen (H. V. Daly, personal communication). This idea needs investigation and might complement other studies that have related infections to injuries caused by chewing, biting, and sucking insects (21, 70, 92, 131, 143, 157, 159). The bacterium apparently persists in some of these insects up to 6 days (7, 105).

Wind is important in dissemination primarily as a transport medium for insects, contaminated pollen (59), clumps of bacteria, water-borne bacteria (13), and aerial strands. The long hair-like strands of *E. amylovora* found on pear plants parts by Ivanoff & Keitt in 1937 (68) were commonly considered relevant to dissemination, but were not studied until Bauske (12) discussed their nature and importance. The strands vary in morphological structure and readily disintegrate in water (75) and under conditions of high relative humidity (35). Their importance in dissemination, however, is still only conjectural. Dissemination presumably would be a function of wind, moisture, and the capacity of the cells to survive in air and on plant surfaces. In California, strands are commonly seen on plant parts during an epidemic of fire blight. However, their importance as a primary mode of dissemination is probably minimal when compared with other agents, especially because flowers are usually infested with *E. amylovora* before infections or strands are observed in the field.

*Long-Distance Spread*

The Atlantic Ocean long served as a barrier to the spread of fire blight from North America to Europe. However, sometime prior to 1957, the barrier was breached and the resultant spread of fire blight bacteria across northern Europe serves as an example of long-distance spread of the pathogen.

Fire blight in Europe was first observed in pear orchards in Kent, England in 1957 (24). Serious efforts were made to eradicate the disease through a program of inspection and burning. Although 9620 infected fruit trees were destroyed between 1958–1960, the disease gained a foothold in London and its suburbs (79). Because of this, the eradication policy was changed in 1961 to one of containment and delay. This was fairly successful until 1966 when unusually large numbers of secondary blossoms became infected. Within a year, the disease had spread throughout two-thirds of the pear-producing acreage in England (79).

The next outbreak was discovered in the Netherlands in 1966 (91). Infections apparently had started in an old hawthorn hedge as early as 1965; by the next year when the disease was discovered, several secondary infections had become established nearby. Eradication measures were thought to be successful in this case because the area was purely agricultural and rather isolated. However, some disease foci were found again in 1971. Fire blight was also discovered in Melobadz, Poland in 1966 (18), and as in the Netherlands, it was apparently established in 1965 (19). This locus was presumably eradicated (147), but small outbreaks occurred at Stierniewice in 1968 and in subsequent years.

Two years later, fire blight was reported in an orchard at northern Falster, Denmark. Although prompt eradication measures were undertaken, new infections were discovered in August 1969 at Falster, Lolland, and several small islands north of Lolland. In 1970 it was located in Langland and Seeland, and in 1971 it was found in the southwestern area of the peninsula of Jutland, near the German border. In 1971, fire blight was observed in West Germany, first on the island of Sylt which faces the Danish areas of infection of southwest Jutland, and later on other islands off the west coast. It was also detected in some continental orchards in the Schleswig-Holstein region. Most recent reports indicate that it is now present in France and Belgium (148) near the Franco-Belgian border (100).

The means of long-range dissemination to England and its spread throughout northern Europe remains obscure. The most likely theory concerning the initial introduction into Europe was that fire blight was brought into England by contaminated fruit boxes from New Zealand or North America, since it is known that fruit boxes from these countries were used in the initial blight orchards in Kent around 1956–1957 (145). Introduction by infected propagation wood also remains a possibility. The manner in which the pathogen spread to the northern coastal regions of Europe is also conjectural. Only in the case of the outbreak at Melobadz, Poland does it appear that the disease was transmitted by importation of diseased propagation material from England (147). The secondary outbreaks in Europe are currently explained by two plausible theories.

THE WIND THEORY This suggests that prevailing winds from England either carried the pathogen directly or enabled insect vectors to reach the coastal areas

(15). One of the more intriguing possibilities is that the winds carried aerial strands which have been observed to form on hawthorn in England (35). While this hypothesis might explain the relatively short-range dissemination of fire blight across the channel to the Netherlands, it provides a less satisfactory explanation of the spread to Denmark and Poland. Aerial strands rapidly disintegrate in the presence of moisture, and it is improbable that favorable conditions would be maintained for the period of time needed for inocula to reach distant areas and a suitable infection court; also, as Bolay (15) questioned in 1972, if the wind carried inoculum from England to Denmark, why is it that fire blight is not yet established in the southeastern part of the Netherlands or in Germany, since these areas are also in line with winds from England?

THE BIRD THEORY Smith (126) noted that Waite had obtained "the strongest kind of circumstantial evidence" that fire blight was spread by birds in at least two instances in Florida and Maryland. In the case of the long-distance spread in Europe and especially to Denmark, the bird theory suggests that the pathogen was disseminated by coastal migratory birds such as starlings (15, 100) whose migration patterns closely follow the fire blight outbreak pattern. These birds, arriving from areas where the disease was prevalent, presumably spread inocula by roosting on susceptible hosts along their migration routes or by eating parts of the hosts. The flowering of hawthorns and fruit trees is four weeks earlier in southern England, where the birds overwinter, than in Denmark. By the time the birds leave Kent, the disease has appeared and there is abundant inocula to disseminate. At the same time, hosts in Denmark are reaching their most susceptible stage (15).

## MODE OF ENTRY

One of the most thoroughly studied and best understood phases of fire blight is the mode of entry of the pathogen into the host. As with other bacterial pathogens, it is generally assumed that the bacterium needs natural openings or wounds to gain entry to the intercellular spaces of the host.

### *Flowers*

The numerous natural openings occurring on flowers that provide entry points for *E. amylovora* (59, 117) include the noncutinized tissue on the tips of pistils, stigmas, and anthers; stomata on the styles (pear), fruit exterior, and sepals; hydathodes, which are sometimes present on the sepals, and the specialized stomata of the nectary, termed nectarthodes by Rosen (117). Of these, the nectaries located in the receptacle cup are the most common sites for bacterial invasion of pear (59, 115, 117). Hildebrand (59) reported that the progress of invasion and resultant infection through the nectaries was often so rapid that the entrance of the pathogen through other natural openings of the flower was obscured. Rosen (117) did not consider that the nectaries in apple were an important invasion pathway, because they are more shielded than in pears. However, this report was based on laboratory inoculations and apparently was incorrect. In studying naturally infected apple blossoms, Hildebrand (59), noted that invasion was primarily through the nectaries with invasion through stigmas being next in importance.

### *Stems*

The natural openings of young stems consist of stomata which are often replaced by lenticels as the stem matures. Rosen (114) demonstrated that invasion could occur through both of these openings although he believed that under natural conditions, wounds offered greater possibilities as infection courts than natural openings. An exception to this might be leaf axils, where numerous natural infections were observed to occur on apple and pear. These axils are lined with stomata on both the petiolar base and stem and contain glandular trichomes whose secretions might support growth of the bacteria (114). A later report (82) claiming the first demonstration of lenticels as possible portals of infection on apple has confirmed Rosen's work, as did a second paper (83).

### *Leaves*

Stomatal invasion of apple and pear leaves by *E. amylovora* was established by Miller (94), Rosen (113), and Tullis (143) in the late 1920s, after it had been suggested by the observations of Heald (54) in 1915. Heald also was of the opinion that invasion of leaves could occur through hydathodes. Tullis (143) observed structures which he called "hydathodes or larger than normal stomates" on the upper surfaces of leaves at the tips of the various serrations. It is quite probable that he included these structures in his discussion of stomatal invasion; he reported that inoculations of the upper surface of leaves were successful only when serrations, or areas containing the hydathodes, were included in the drop of water in which inoculum was placed. Recently, the glandular trichomes described by Woodcock & Tullis (160) have been postulated as an area of entry by Lewis & Goodman (82, 83). However, it is likely that most of the infections observed by these authors originated from entry through hydathodes since, although these glandular trichomes occur along the midrib and veins as well as at the serrations of the leaf blade, all the infections pictured (83) and described (82) occurred along the leaf margin. This would support the observations of Tullis (143), who used a more gentle technique.

## FIRE BLIGHT CONTROL AND FORECASTING

### *Evolution of Control*

"Driving rusty iron nails into the trees and hanging horse shoes among the branches will not control fire blight" (134).

The first attempts to control fire blight centered around orchard sanitation. Even before the causal organism was known, it was common practice to remove infected plant material from the orchard, and careful, continuous eradication and scarification (scraping bark back to live tissue) were advocated. Efforts at transfusing chemicals (e.g. calomel, formalin) into trees as a fire blight remedy (1, 16, 17) were largely unsuccessful. Burrill (23) had previously pointed out the need for a germicide to prevent infection of new wounds and to sterilize tools, and it was in 1905 that Whetzel (156) used corrosive sublimate (mercuric chloride) and copper sulfate solutions to treat scarified cankers and wounds. At the same time, Waite (155) advocated the use of corrosive sublimate on pruning tools and wounds. In summarizing control measures at that time, Whetzel & Stewart (158) stressed the importance of strict sanitation, including a systematic inspection for holdover cankers and alternate hosts during winter and early spring, followed by weekly orchard patrols once the disease appeared. All fresh wounds were covered with corrosive sublimate; Reimer (108) subsequently showed that addition of mercuric cyanide to the corrosive sublimate enhanced its effectiveness. Waite (155) considered that chemical spray protection of blossoms was impossible because "new blossoms are opening every hour of the day." However, Stewart (130) subsequently indicated that blossom spraying might decrease the amount of blight, although field evidence was lacking because it was a penal offense in New York to spray trees during the blossom period.

Other cultural practices and their effect on fire blight came under scrutiny during this period. Rapidly growing trees with succulent growth were known to be more susceptible to the disease, and recommendations were therefore made that trees should not be "over-stimulated" by excess water and fertilizer, or heavy pruning (55, 153). Since greatest losses occurred when the trunks and root systems of trees were blighted, Reimer (109) initiated investigations of rootstock and varietal susceptibility. Thomas & Parker (138) discouraged the combined planting of different varieties and species of pome fruit which would prolong the bloom period.

Scarification of cankers was time consuming, and Day (26, 27) therefore proposed that cankers be disinfected by the use of zinc chloride dissolved in water and alcohol. Although there was a narrow margin of safety between killing the pathogen and the surrounding host tissue, this method was used successfully in several areas and was faster and less hazardous for operators than previously used materials. Other materials (e.g. cadmium salts) were investigated by Parker (101).

Bloom spraying of bordeaux mixture was tried by McCown (89) in Indiana with some success. Other workers had inconsistent results, and Thomas & Ark (135) in California suggested that success may depend on spraying at or shortly before dissemination of the pathogen. However, they saw little reason to suggest the general use of blossom sprays because of the difficulty of predicting conditions for dissemination and subsequent infection and because of the phytotoxicity of copper sprays.

Control measures changed little during the late 1930s and 1940s, until spectacular advances in the use of antibiotics in human medicine set the stage for further progress. Rudolph (119) tested penicillin against fire blight without success in 1946; then Murneek (99) published results of an apple field experiment in Missouri where better than 50% blight control was obtained using a streptomycin spray at 50% bloom. In Delaware, Heuberger & Poulos (56) reported good control with five streptomycin applications during and after bloom. In California, Ark (6) achieved good control with bloom and postbloom steptomycin dusts. Goodman (44) confirmed these results, and when 100 ppm streptomycin sprays were initiated in early bloom and applied at 3–5 day intervals, control was far superior to a 1:3:100 bordeaux mixture.

Dunegan et al (33) successfully used a mixture of Terramycin® and streptomycin (Agrimycin)® several times at 2 week intervals. They reported a chlorotic leaf symptom where streptomycin was used early in the season—a response to low temperatures. Luepschen (86) showed in vitro that streptomycin was effective against *E. amylovora* over a wide temperature range; addition of adjuvants increased penetration of the antibiotic, which exhibited a protectant action in the blossoms for at least 48 hr.

During the 1960s streptomycin was widely investigated for fire blight control, and numerous reports have cited success with a diversity of adjuvants, times, and methods of application. Some of the early formulations included 1.5% Terramycin® to delay the emergence of streptomycin-resistant strains of the pathogen (38). It is significant to note that the Terramycin® component was later deleted. During the last decade, streptomycin sprays and dusts, plus copper formulations, have been the predominant bactericides employed for the blossom and twig blight phases of the disease in North America.

Although the use of streptomycin had transformed the blight control picture, erratic results still occurred on occasion, and this, coupled with its cost to the grower and concern over possible harmful spray residues at harvest time, limited its widespread use. It did, however, have a distinct advantage over copper bactericides in that it did not harm fruit finish. Copper materials used on developing fruit—even at low concentrations—damage the lenticels in certain varieties of pome fruits, thus lessening consumer acceptance of the fruit. Various studies eventually showed that streptomycin residues were short lived and that the antibiotic could be used frequently until 30 days before harvest without exceeding the tolerance of 0.25 ppm on pears (9). With the knowledge that effective blossom blight control could be achieved by frequent (3–5 day) applications of relatively low rates of streptomycin (60–240 ppm), use of streptomycin was intensified in California. Not surprisingly, a serious problem became apparent during the 1971 season; severe blight developed in pear orchards in the Sacramento Valley where growers had followed a thorough streptomycin spray program. The predominant strains of the pathogen were highly resistant to streptomycin (97). They varied in virulence about the same as streptomycin-susceptible strains (140), thus indicating that the current widespread occurrence of streptomycin-resistant strains in California was not caused by mutation of one strain. It is interesting to speculate whether this problem would have developed if the Terramycin® component had been retained in the antibiotic formulation. Considering the vast areas of the highly susceptible Bartlett pear cultivar in California and the large quantities of streptomycin applied, it was not unexpected that selection pressure favored multiplication of the resistant strains. Nevertheless, antibiotic resistance now adds a new dimension to blight control studies and underlines the need for more complete studies on the biology of the pathogen and for an integrated approach to disease control.

The role of saprophytic bacteria in reducing populations of *E. amylovora* and infections is a purely speculative subject. Rosen in 1928 (112) stated that a yellow bacterium apparently limited *E. amylovora* in the amount of host tissue it could invade, but uncharacteristically, he provided no supportive data. Since then, various reports (41, 47, 48, 124, 127) have suggested that certain yellow bacteria have a beneficial effect in reducing disease, although data to support this also are limited. Because of the common association of these bacteria with *E. amylovora* in tissues, some investigators have even conjectured that the yellow bacteria might be a phase in the life cycle of *E. amylovora* (146), or an aberrant form of *E. amylovora* (46), or one related to the *E. herbicola* group (46). Studies on so-called yellow bacteria indicated that they belong to the *E. herbicola* group (32, 78).

Parker (101) was among the earliest to consider the possibilities of biological means to control fire blight and searched for bacteria to introduce into flowers to inhibit the development of fire blight bacteria. Although he found some bacteria that were strongly antagonistic to the blight bacterium in culture, none consistently caused more than slight reduction of blossom blight in the orchard. Field studies have not yet indicated any particular relationship between the occurrence of saprophytic bacteria and populations of *E. amylovora* in pear flowers (95), but this has not been thoroughly examined in different geographical areas and in different years. Application of yellow and other bacteria to apple (110) and pear blossoms (141) resulted in only slight decrease in the incidence of fire blight infections. The biological control approach, however, could be a promising area for research, but it will require considerably more effort than has been given in the past. The key to a successful biological control will be to find those species of bacteria that will colonize flowers under the same environmental conditions as *E. amylovora* and have an inhibitory effect on the pathogen.

Other studies have centered on protecting pear or apple tissues by inoculating various bacteria into shoots either prior to or simultaneously with *E. amylovora* (46, 48, 90). These experiments are of interest to investigators of resistance phenomena, but their relevance to infection patterns in nature and disease epidemiology remains obscure.

Present-day control measures (144) are still in part based on good cultural practices; orchards should be planted on well-drained soils and caution exercised with nitrogenous fertilizers to avoid excessively vigorous tree growth. Varietal and rootstock factors are significant in some areas. Continued and vigilant sanitation is of the utmost importance, and diseased plant parts should be removed and destroyed, making cuts at least 20–30 cm below visible disease. If this pruning is accomplished during dormancy, there is no need for surface sterilization of tools and wounds. During the growing season, however, cutting tools should be dipped in a disinfestant between cuts. Although cankers have been treated with miscellaneous materials in the past, this is no longer done because of the general ineffectiveness of the treatments. Heavy reliance is still placed on spring blossom sprays with chemicals. In California (2), applications of streptomycin or copper materials begin at 5–10% bloom and are repeated between alternate rows every 5 days until at least 30 days after full bloom. If rain or hail occurs, the orchard is treated immediately. The monitoring of blossoms has shown that bactericides prevent increases in epiphytic populations of *E. amylovora,* and this presumably results in less disease (141).

The reasons for heavy reliance on chemical control of blossom blight in California are as follows: In warm valleys, mild winter temperatures encourage prolonged pear bloom, and flowers continue to appear well into the growing season. These flowers are potential infection sites whenever periods of high temperature and humidity coincide, and this time of high infection risk may extend over 12–14 weeks. Most pear producers know that neglect of diligent chemical spraying during this period can result in serious financial losses. Even so, in certain years and locations, blight infections are insignificant—even in carelessly managed orchards. This suggests that a better understanding of the factors leading to development of epidemics could enhance the possibility of more economical and ecologically desirable control measures.

### *Future Directions*

Several investigators have studied the climatic factors influencing fire blight development with a view to possible prediction of epidemics (20, 87, 96, 106, 107). However, disease prediction based on weather data alone has been of limited value and usually cannot be transposed to different geographical areas. The interactions of inoculum potential, host susceptibility, and environment that contribute to development of epidemics are complex, and epidemics occur when all variables are optimal for fire blight development. Although attempted correlations have invariably been made between weather data and disease incidence, the important parameters of inoculum potential and host susceptibility have not been studied in detail.

Brooks (20) studied the epidemiology of fire blight on apple and suggested that epidemics were favored by temperatures of 65–85°F, high relative humidity, frequent showers, vigorous growth, and heavy infestations of insects. Other investigators have examined weather conditions, not from the standpoint of forecasting, but mainly to support theories of how the pathogen was disseminated. Thus meteorological analyses showed the importance of rain (52, 93, 94, 101, 114), while other studies showed that infection occurred without rain and implicated insects as major vectors (114, 135).

Applications of chemicals during certain environmental conditions gave better control of blossom blight than at other times. This suggested that weather information possibly could be used to improve spray timing. Mills (96) appears to be the first to examine this concept. He made a statistical correlation of New York weather data and fire blight incidence over a 37 year period and found that certain climatic conditions favored disease development. Utilizing this information, plus subsequent laboratory studies, a control program was proposed whereby streptomycin was applied after bloom on the first day that (*a*) the maximum temperature exceeded 65°F and (*b*) precipitation or relative humidity of 70% or more was forecast (87). These criteria are in agreement with the earlier work of Brooks (20).

Powell (106, 107) collated data on weather and blossom blight incidence and developed the concepts that a specific temperature had to be reached before inoculum increased and also that freezing temperatures retarded inoculum buildup. He further suggested that unfavorable preblossom weather would negate the need for sprays to control blossom blight. Although this approach has limitations in certain geographical areas, and some of the assumptions need refining (e.g. those concerning bacterial growth rates at different temperatures), it has set the stage for subsequent investigations of blight forecasting.

Until monitoring techniques were developed (95), a void remained in information relating inoculum levels of the pathogen to temperature regimes and infection. Monitoring of bacterial populations revealed that *E. amylovora* was present as an epiphyte on healthy blossoms prior to disease development (95, 142, 161). This has been confirmed by Thomson et al (142) who demonstrated that the magnitude of bacterial populations in the blossoms may provide a guide to subsequent disease incidence. Epiphytic colonization of flowers accounts for the widespread availability of inocula requisite for the occurrence of a sudden epidemic.

In the spring of 1973, a number of geographical locations were monitored in California to relate weather conditions to buildup of *E. amylovora* inocula and to accumulate data for the development of a disease forecasting program. *Erwinia amylovora* was not detected on blossoms until the main bloom period was past (142, 161). However, detection of the pathogen in blossoms was prior to the occurrence of disease several days or weeks later (Figure 2). Once detected in an orchard, epiphytic populations of the pathogen consistently appeared in later blossoms even though temperatures occasionally reached 37°C.

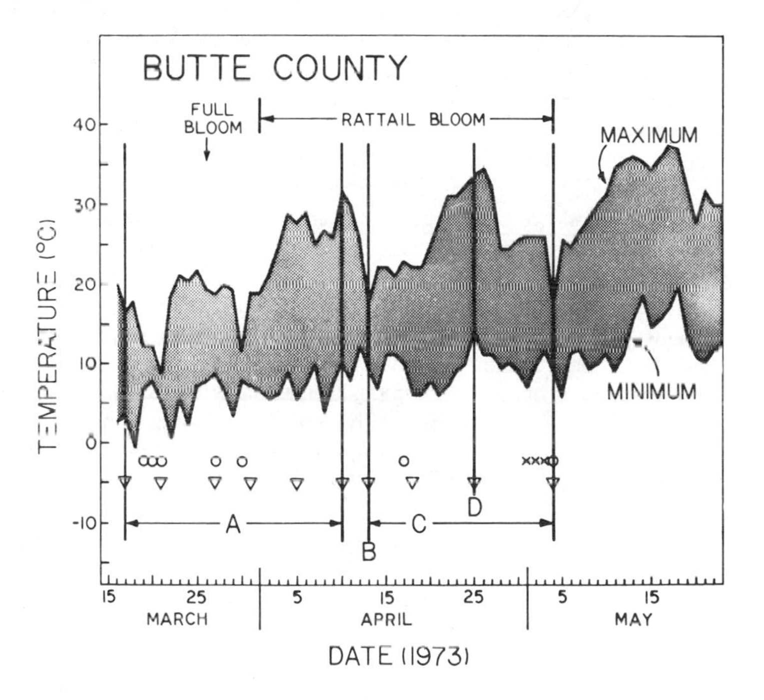


*Figure 2* Relation of weather to epiphytic populations of *Erwinia amylovora* on healthy Bartlett pear blossoms and subsequent disease occurrence. *Erwinia amylovora* bacteria were not found (*A*) in blossoms monitored from early bloom until April 13, (*B*) when populations of $3.3 \times 10^5$ bacteria/blossom were detected. Thereafter, populations (*C*) were slightly above the initial level despite high temperatures. Disease appeared (3.2 strikes/tree) April 25 (*D*), 12 days after the first detection of *E. amylovora.* Dates of precipitation = (0), irrigation = (X), and monitoring = (4). After Thomson et al (142).

Because of previous reviews (42, 63, 80, 81, 102, 137), this review excludes factors influencing the susceptibility of plants to infection. Nevertheless, it is apparent that epiphytic colonization and infection of flowers is subject to host resistance mechanisms. For example, infections sometimes do not spread past the flower. Furthermore, it is common to find small localized lesions in flowers that yield *E. amylovora* upon isolation. Rosen (116) found that 85% of the abscissed pear flowers in two counties "showed symptoms indicating infections" of *Pseudomonas syringae, E. amylovora,* or both bacteria and suggested that their importance in preventing setting of fruit was likely overlooked. There are, of course, many other causes for blossom drop.

A further indication of resistance mechanisms is the observation in California that seldom more than 100 flower strikes/tree occur during an epidemic even though every flower may be infested with *E. amylovora,* and moisture and temperature appear optimal for infection (95). Laboratory studies also showed that flowers vary in susceptibility because many may not become infected regardless of the initial population of *E. amylovora* deposited in the receptacle (142). Upon inoculating the flowers with known dosages of bacteria, the population decreased to zero in some flowers while increasing in others (142, and R. A. Lelliott, personal communication). This phenomenon varies with the year. In 1973, most primary flowers could not be infected in the laboratory by depositing in the receptacle $10^2$ to $10^4$ cells per flower, contrary to the two previous years. However, secondary flowers (rattails), which appeared later in the season, were infected readily. The disease pattern in the field was correlated with the laboratory findings; infection did not occur in orchards until secondary flowers were formed. Since pear cuttings can be brought into a laboratory several months prior to flowering and forced to bloom, it might be possible to predict the susceptibility of flowers before the season begins.

The foregoing indicates that early bloom protection procedures may be superfluous in certain years. Careful monitoring of bacterial populations in blooms plus weather data should lead to more effective blight control. Epidemics commonly are correlated with amount of rain although as numerous workers have shown, free moisture in the blossoms plus a high relative humidity are the important factors rather than rain per se (76, 87, 114). Temperatures too are important, either in predisposing the host to infection or in encouraging the dissemination of epiphytic populations of the pathogen (142, 161).

Computer simulation of epidemics has been successful with fungus pathogens, such as early blight of potato and tomato (150), which have a more complex life cycle than fire blight. With the growing body of information concerning the biotic and abiotic factors involved in fire blight development, it should soon be possible to eliminate the monitoring programs and design a computer simulation for the disease, thus enabling more accurate disease forecasting. This would have the beneficial effect of enabling growers to apply chemicals only when needed, thereby increasing the efficacy of control, reducing cost to growers, and decreasing the amount of pesticide applied to orchards.

*Literature Cited*


1. Anonymous. 1893. Remedy for pear blight. *Calif. Fruit Grower* 12:8
2. Anonymous. 1973. *Pest and disease control program for pears.* Berkeley: Univ. Calif. Div. Agr. Sci. 20 pp.
3. Anderson, H. W. 1952. Maintaining virulent cultures of *Erwinia amylovora* and suggestion of overwinter survival in mummied pear fruit. *Plant Dis. Reptr.* 36:301–2
4. Ark, P. A. 1932. The behavior of *Bacillus amylovorus* in soil. *Phytopathology* 22:657–60
5. Ark, P. A. 1937. Variability in the fireblight organism, *Erwinia amylovora. Phytopathology* 27:1–28
6. Ark, P. A. 1953. Use of streptomycin dust to control fireblight. *Plant Dis. Reptr.* 37:404–6
7. Ark, P. A., Thomas, H. E. 1936. Persistence of *Erwinia amylovora* in certain insects. *Phytopathology* 26:375–81
8. Bachmann, F. M. 1913. The migration of *Bacillus amylovorus* in the host tissues. *Phytopathology* 3:3–14
9. Bailey, J. B., Morehead, G. W. 1970. Streptomycin control of pear fireblight in California—1968 and 1969 field tests. *Calif. Agr.* 24:14–15
10. Baker, K. F. 1971. Fire blight of pome fruits: the genesis of the concept that bacteria can be pathogenic to plants. *Hilgardia* 40:603–33
11. Baldwin, C. H. Jr., Goodman, R. N. 1963. Prevalence of *Erwinia amylovora* in apple buds as detected by phage typing. *Phytopathology* 53:1299–1303
12. Bauske, R. J. 1968. Bacterial strands: a possible role in fire blight. *Iowa State J. Sci.* 43:119–24
13. Bauske, R. J. 1971. Wind dissemination of waterborne *Erwinia amylovora* from *Pyrus* to *Pyracantha* and *Cotoneaster. Phytopathology* 61:741–42
14. Billing, E., Baker, L. A. E., Crosse, J. E., Garrett, C. M. E. 1961. Characteristics of English isolates of *Erwinia amylovora* (Burrill) Winslow et al. *J. Appl. Bacteriol.* 24:195–211
15. Bolay, A. 1972. Le feu bactérien du poirier. Les étapes de sa progression dans le monde et particuliérement en Europe. *Rev. Suisse Viticult. Arboricult. Hort.* 4:137–43
16. Bolley, H. L. 1904. Tree feeding and tree medication. *Ann. Rept. North Dak. Agr. Exp. Sta.* 14:55–58
17. Bolley, H. L. 1907. Tree feeding and tree medication. *Ann. Rept. N. Dak. Agr. Exp. Sta.* 16:35
18. Borecki, Z., Basak, W., Zawadzka, B., Millikan, D. F. 1967. Fire blight in continental Europe. *Plant Dis. Reptr.* 51:3
19. Borecki, Z., Lyskanowska, K. 1968. *Erwinia amylovora* (Burr.) Winsl. in Poland. *Phytopathol. Z.* 61:157–66
20. Brooks, A. N. 1926. Studies of the epidemiology and control of fireblight of apple. *Phytopathology* 16:665–96
21. Burrill, A. C. 1915. Insect control important in checking fire blight. *Phytopathology* 5:343–47
22. Burrill, T. J. 1877. Hort. Dep. Rept., Sept. 13, 1876, to Dr. J. M. Gregory, Regent. *Dep. Board Trustees Ill. Ind. Univ.* 8:199–200
23. Burrill, T. J. 1882. Have we any new light on pear blight or yellows? *Mich. State Hort. Soc. Ann. Rept.* 11:133–39
24. Crosse, J. E., Bennett, M., Garrett, C. M. E. 1958. Fire-blight of pear in England. *Nature* 182:1530
25. Crosse, J. E., Goodman, R. N., Shaffer, W. H. Jr. 1972. Leaf damage as a predisposing factor in the infection of apple shoots by *Erwinia amylovora. Phytopathology* 62:176–82
26. Day, L. H. 1928. Pear blight control in California. *Calif. Agr. Ext. Circ.* 20: 1–50
27. Day, L. H. 1930. Zinc chloride treatment for pear blight cankers. *Calif. Agr. Ext. Circ.* 45:1–13
28. Denning, W. 1794. On the decay of apple trees. *Trans. Soc. Promot. Agr. Arts Mfg.* 2:219–22 (2nd ed., 1801. 2. 185–87)
29. DeVay, J. E., Schnathorst, W. C. 1963. Single-cell isolation and preservation of bacterial cultures. *Nature* 199:775–77
30. Dye, D. W. 1949. 23rd *Ann. Rept. Dep. Sci. Ind. Res. N.Z.* 88 pp.
31. Dye, D. W. 1968. A taxonomic study of the genus *Erwinia.* 1. The "Amylovora" group. *N.Z. J. Sci.* 11: 590–607
32. Dye, D. W. 1969. A taxonomic study of the genus *Erwinia.* 3. The "herbicola" group. *N.Z. J. Sci.* 12:223–36
33. Dunegan, J. C., Kienholz, J. R., Wilson, R. A., Morris, W. T. 1954. Control of pear blight by a streptomycin-terramycin mixture. *Plant Dis. Reptr.* 38: 666–69
34. Eden-Green, S. J. 1972. *Studies in fire blight disease of apple, pear and hawthorn [Erwinia amylovora* (Burrill) Winslow et al.]. PhD thesis. Univ. London. 202 pp.

35. Eden-Green, S. J., Billing, E. 1972. Fireblight: occurrence of bacterial strands on various hosts under glasshouse conditions. *Plant Pathol.* 21: 121–23
36. Emmett, B. J. 1971. Insect visitors to pear blossom. *Plant Pathol.* 20:36–40
37. Emmett, B. J., Baker, L. A. E. 1971. Insect transmission of fire blight. *Plant Pathol.* 20:41–45
38. English, A. R., Van Halsema, G. 1954. A note on the delay in the emergence of resistant *Xanthomonas* and *Erwinia* strains by the use of streptomycin plus Terramycin combinations. *Plant Dis. Reptr.* 38:429–31
39. Erskine, J. M. 1973. Association of virulence characteristics of *Erwinia amylovora* with toxigenicity of its phage lysates to rabbit. *Can. J. Microbiol.* 19:875–77
40. Erskine, J. M. 1973. Characteristics of *Erwinia amylovora* bacteriophage and its possible role in the epidemiology of fire blight. *Can. J. Microbiol.* 19:837–45
41. Farabee, G. J., Lockwood, J. L. 1958. Inhibition of *Erwinia amylovora* by *Bacterium* sp. isolated from fire blight cankers. *Phytopathology* 48:209–11
42. Fisher, E. G., Parker, K. G., Luepschen, N. S., Kwong, S. S. 1959. The influence of phosphorus, potassium, mulch, and soil drainage on fruit size, yield and firmness of the Bartlett pear and on development of the fire blight disease. *Proc. Am. Soc. Hort. Sci.* 73:78–90
43. Glasscock, H. H. 1971. Fireblight epidemic among Kentish apple orchards in 1969. *Ann. Appl. Biol.* 69:137–45
44. Goodman, R. N. 1954. Development of methods for use of antibiotics to control fireblight (*Erwinia amylovora*). *Mo. Agr. Exp. Sta. Res. Bull.* 540:1–16
45. Goodman, R. N. 1954. Apple fruits a source of overwintering fireblight inoculum. *Plant Dis. Reptr.* 38:414
46. Goodman, R. N. 1965. *In vitro* and *in vivo* interactions between components of mixed bacterial cultures isolated from apple buds. *Phytopathology* 55: 217–21
47. Goodman, R. N. 1966. The protection of apple tissue against infection by *Erwinia amylovora* afforded by avirulent strains of *E. amylovora,* a saprophytic bacterium and other bacterial plant pathogens. *Rev. Roum. Biol. Ser. Bot.* 11:75–79
48. Goodman, R. N. 1967. Protection of apple stem tissue against *Erwinia amylovora* infection by avirulent strains and three other bacterial species. *Phytopathology* 57:22–24
49. Goodman, R. N. 1971. Re-evaluation of the role of $NH_3$ as the cause of the hypersensitive reaction. *Phytopathology* 61:893
50. Gossard, H. A. 1916. Is the hive a center for distributing fire-blight? Is aphid honeydew a medium for spreading blight? *J. Econ. Entomol.* 9:59–64
51. Gossard, H. A., Walton, R. C. 1916. Fire-blight investigations. *Ohio Agr. Exp. Sta. Mon. Bull.* 9:274–76
52. Gossard, H. A., Walton, R. C. 1922. Dissemination of fire blight. *Ohio Agr. Exp. Sta. Bull.* 357:81–126
53. Gowda, S. S., Goodman, R. N. 1970. Movement and persistence of *Erwinia amylovora* in shoot, stem and root of apple. *Plant Dis. Reptr.* 54:576–80
54. Heald, F. D. 1915. Preliminary note on leaf invasions by *Bacillus amylovorus. Wash. Agr. Exp. Sta. Bull.* 125:1–7
55. Henderson, L. F. 1904. Fire blight—a bacterial disease of the pear and the apple. *Univ. Idaho Agr. Exp. Sta. Press Bull. New Ser.,* No. 3
56. Heuberger, J. W., Poulos, P. L. 1953. Control of fireblight and frog-eye leaf spot (black rot) diseases of apples in Delaware in 1952. *Plant Dis. Reptr.* 37:81–83
57. Hildebrand, D. C. 1971. Pectate and pectin gels for differentiation of *Pseudomonas* sp. and other bacterial plant pathogens. *Phytopathology* 61: 1430–36
58. Hildebrand, E. M. 1937. Infectivity of the fire-blight organism. *Phytopathology* 27:850–52
59. Hildebrand, E. M. 1937. The blossom-blight phase of fire blight, and methods of control. *Cornell Agr. Exp. Sta. Mem.* 207:1–40
60. Hildebrand, E. M. 1938. Growth rates of phytopathogenic bacteria *J. Bacteriol.* 35:487–92
61. Hildebrand, E. M. 1939. Studies on fire blight ooze. *Phytopathology* 29:142–56
62. Hildebrand, E. M. 1954. Relative stability of fire blight bacteria. *Phytopathology* 44:192–97
63. Hildebrand, E. M., Heinicke, A. J. 1937. Incidence of fire blight in young apple trees in relation to orchard practices. *Cornell Univ. Agr. Exp. Sta. Mem.* 203:1–36
64. Hildebrand, E. M., Phillips, E. F. 1936. The honeybee and the beehive in relation to fire blight. *J. Agr. Res.* 52: 789–810

65. Huang, P.-Y., Goodman, R. N. 1970. Morphology and ultrastructure of normal rod-shaped and filamentous forms of *Erwinia amylovora. J. Bacteriol.* 102:862–66
66. Hugh, R., Leifson, E. 1953. The taxonomic significance of fermentative versus oxidative metabolism of carbohydrates by various gram-negative bacteria. *J. Bacteriol.* 66:24–26
67. Hurst, C., Kennedy, B. W., Olson, L. 1973. Production of ammonia by tobacco and soybean inoculated with bacteria. *Phytopathology* 63:241–42
68. Ivanoff, S. S., Keitt, G. W. 1937. The occurrence of aerial bacterial strands on blossoms, fruits, and shoots blighted by *Erwinia amylovora. Phytopathology* 27:702–9
69. Ivanoff, S. S., Keitt, G. W. 1941. Relations of nectar concentration to growth of *Erwinia amylovora* and fire blight infection of apple and pear blossoms. *J. Agr. Res.* 62:733–43
70. Jones, D. H. 1909. Bacterial blight of apple, pear, and quince trees. *Ont. Dep. Agr. Bull.* 176:1–63
71. Jones, D. H. 1911. Further studies on fire blight. *Ann. Rept. Ont. Agr. College Exp. Farming* 36:163–67
72. Jones, D. H. 1911. *Scolytus regulosus* as an agent in the spread of bacterial blight in pear trees. *Phytopathology* 1: 155–58
73. Kakso, J. U., Starr, M. P. 1970. Comparative injuriousness to plants of *Erwinia* spp. and other enterobacteria from plants and animals. *J. Appl. Bacteriol.* 33:692–707
74. Keil, H. L., van der Zwet, T. 1972. Recovery of *Erwinia amylovora* from symptomless stems and shoots of Jonathan apple and Bartlett pear trees. *Phytopathology* 62:39–42
75. Keil, H. L., van der Zwet, T. 1972. Aerial strands of *Erwinia amylovora:* structure and enhanced production by pesticide oil. *Phytopathology* 62:355–61
76. Keitt, G. W., Ivanoff, S. S. 1941. Transmission of fire blight by bees and its relation to nectar concentration of apple and pear blossoms. *J. Agr. Res.* 62: 745–53
77. Klement, Z., Goodman, R. N. 1966. Hypersensitive reaction induced in apple shoots by an avirulent form of *Erwinia amylovora. Acta Phytopathol. Acad. Sci. Hung.* 1.177–84
78. Komagata, K., Tamagawa, Y., Kocur, M. 1968. Differentiation of *Erwinia amylovora, Erwinia carotovora,* and *Erwinia herbicola. J. Gen. Appl. Microbiol. Tokyo* 14:39–45
79. Lelliott, R. A. 1968. Fireblight in England. Its nature and its attempted eradication. *EPPO Pub., Ser. A.N°* 45-E:10–14
80. Lewis, L. N., Kenworthy, A. L. 1962. Nutritional balance as related to leaf composition and fire blight susceptibility in the Bartlett pear. *Proc. Am. Soc. Hort. Sci.* 81:108–15
81. Lewis, L. N., Tolbert, N. E. 1964. Nitrogen requirement and metabolism of *Erwinia amylovora. Physiol. Plant.* 17:44–48
82. Lewis, S., Goodman, R. N. 1965. Mode of penetration and movement of fire blight bacteria in apple leaf and stem tissue. *Phytopathology* 55:719–23
83. Lewis, S., Goodman, R. N. 1966. The glandular trichomes, hydathodes and lenticels of Jonathan apple and their relation to infection by *Erwinia amylovora. Phytopathol. Z.* 55:352–58
84. Lovrekovich, L., Lovrekovich, H., Goodman, R. N. 1970. Ammonia as a necrotoxin in the hypersensitive reaction caused by bacteria in tobacco leaves. *Can. J. Bot.* 48:167–71
85. Lovrekovich, L., Lovrekovich, H., Goodman, R. N. 1970. The relationship of ammonia to symptom expression in apple shoots inoculated with *Erwinia amylovora. Can. J. Bot.* 48:999–1000
86. Luepschen, N. S. 1960. Fire blight control with streptomycin, as influenced by temperature and other environmental factors and by adjuvants added to sprays. *NY Agr. Exp. Sta. Mem.* 375: 1–39
87. Luepschen, N. S., Parker, K. G., Mills, W. D. 1961. Five year study of fire blight blossom infection and its control in New York. *NY Agr. Exp. Sta. Bull.* 963:1–19
88. Martinec, T., Kocur, M. 1964. A taxonomic study of *Erwinia amylovora* (Burrill 1882) Winslow et al. 1920. *Int. Bull. Bacteriol. Nomencl. Taxon.* 14: 5–14
89. McCown, M. 1929. Bordeaux spray in the control of fireblight of apple. *Phytopathology* 19:285–93
90. McIntyre, J. L., Kuc , J., Williams, E. B. 1973. Protection of pear against fire blight by bacteria and bacterial sonicates. *Phytopathology* 63:872–77
91. Meijneke, C. A. R. 1968. Fireblight: an isolated outbreak in the Netherlands. *EPPO Publ. Ser. A. N°* 45-E:17–19
92. Merrill, J. H. 1917. Further data on the relation between aphids and fire blight

[*Bacillus amylovorus* (Bur.) Trev.]. *J. Econ. Entomol.* 10:45–46

93. Miller, P. W. 1928. A preliminary report on studies of fireblight of apple. *Science* 68:386–88
94. Miller, P. W. 1929. Studies of fire blight of apple in Wisconsin. *J. Agr. Res.* 39:579–621
95. Miller, T. D., Schroth, M. N. 1972. Monitoring the epiphytic population of *Erwinia amylovora* on pear with a selective medium. *Phytopathology* 62: 1175–82
96. Mills, W. D. 1955. Fire blight development on apple in western New York. *Plant Dis. Reptr.* 39:206–7
97. Moller, W. J., Beutel, J. A., Reil, W. O., Perry, F. J., Schroth, M. N. 1973. Streptomycin resistant fire blight control studies. *Calif. Agr.* 27:4–5
98. Morrill, G. D. 1969. *Overwintering of Erwinia amylovora inside living host tissue in Cache Valley Utah.* MS thesis, Utah State Univ., Logan
99. Murneek, A. E. 1952. Thislutin as a possible inhibitor of fire blight. *Phytopathology* 42:57
100. d'Oliveira, M. L. 1972. The threat of fireblight, *Erwinia amylovora,* for the Mediterranean region. *Actas III Congr. Un. Fitopato. Medit. Oeiras,* 22–28 October, pp. 477–81
101. Parker, K. G. 1936. Fire blight: overwintering, dissemination, and control of the pathogene. *NY Agr. Exp. Sta. Mem.* 193:1–42
102. Parker, K. G., Luepschen, N. S., Fisher, E. G. 1961. Tree nutrition and fire blight development. *Phytopathology* 51:557–60
103. Pierstorff, A. L. 1931. Studies on the fire-blight organism, *Bacillus amylovorus. Corn. Univ. Agr. Exp. Sta. Mem.* 136:1–53
104. Pierstorff, A. L., Lamb, H. 1934. The honeybee in relation to the overwintering and primary spread of the fire blight organism. *Phytopathology* 24:1347–57
105. Plurad, S. B., Goodman, R. N., Enns, W. R. 1965. Persistence of *Erwinia amylovora* in the apple aphid (*Aphis pomi* De Geer), a probable vector. *Nature* 205:206
106. Powell, D. 1963. Prebloom freezing as a factor in the occurrence of the blossom blight phase of fire blight of apples. *Trans. Ill. State Hort. Soc.* 97:144–48
107. Powell, D. 1965. Factors influencing the severity of fire blight infections on apple and pear. *Mich. State Hort. Soc. Ann. Meet.* 94:1–7
108. Reimer, F. C. 1918. A new disinfectant for pear blight. *Mon. Bull. State Comm. Hort. Calif.* 7:562–65
109. Reimer, F. C. 1925. Blight resistance in pears and characteristics of pear species and stocks. *Oreg. Agr. Exp. Sta. Bull.* 214:1–99
110. Riggle, J. H., Klos, E. J. 1972. Relationship of *Erwinia herbicola* to *E. amylovora. Can. J. Bot.* 50:1077–83
111. Riker, A. J., Banfield, W. M., Wright, W. H., Keitt, G. W., Sagen, H. E. 1930. Studies on infectious hairy root of nursery apple trees. *J. Agr. Res.* 41:507–40
112. Rosen, H. R. 1928. Variations within a bacterial species. I. Morphologic variations. *Mycologia* 20:251–75
113. Rosen, H. R. 1929. The life history of the fire blight pathogen, *Bacillus amylovorus,* as related to the means of overwintering and dissemination. *Arkansas Agr. Exp. Sta. Bull.* 244:1–96
114. Rosen, H. R. 1933. Further studies on the overwintering and dissemination of the fire-blight pathogen. *Arkansas Agr. Exp. Sta. Bull.* 283:1–102
115. Rosen, H. R. 1935. The mode of penetration of pear and apple blossoms by the fire-blight pathogen. *Science* 81:26
116. Rosen, H. R. 1936. Oversummering of fire-blight pathogen, spraying for control of fire blight, and abscission induced by *Erwinia amylovora* and *Phytomonas syringae. Arkansas Agr. Exp. Sta. Bull.* 330:1–60
117. Rosen, H. R. 1936. Mode of penetration and of progressive invasion of fire-blight bacteria into apple and pear blossoms. *Arkansas Agr. Exp. Sta. Bull.* 331:1–68
118. Rosen, H. R., Groves, A. B. 1928. Studies on fire blight: host range. *J. Agr. Res.* 37:493–505
119. Rudolph, R. A. 1946. Attempts to control bacterial blights of pear and walnut with penicillin. *Phytopathology* 36: 717–25
120. Samson, R. 1972. Heterogeneity of heat-stable somatic antigens in *Erwinia amylovora. Ann. Phytopathol.* 4: 157–163
121. Samson, R., Putier, F. 1972. Les antigenes polyosidiques d'*Erwinia amylovora. Ann. Phytopathol.* 4:197
122. Seemuller, E. A., Beer, S. V., Jones, T. M., Bateman, D. F. 1973. Nondegradation of cell wall polysaccharides by *E. amylovora. Phytopathology* 63:207 (Abstr.)
123. Shaffer, W. H. Jr., Goodman, R. N. 1962. Progression *in vivo,* rate of growth *in vitro,* and resistance to streptomycin, as indices of virulence of *Er-*

*winia amylovora. Phytopathology* 52: 1201–7

124. Shaw, L. 1929. *Studies on fireblight. I. The identity of a yellow schizomycete found associated with Bacillus amylovorus. II. The reaction of B. amylovorus to the presence of the yellow organism.* MSc thesis. Univ. Arkansas, Fayetteville
125. Smith, E. F. 1899. Dr. Alfred Fischer in the role of pathologist. *Centralbl. Bakt. II,* 5:811–17
126. Smith, E. F. 1915. A conspectus of bacterial disease of plants. *Ann. Mo. Bot. Gard.* 2:377–401
127. Smith, J. H., Powell, D. 1966. Characteristics of a yellow bacterium isolated from the tissue of Jonathan apple trees. *Trans. Illinois State Acad. Sci.* 59: 358–63
128. Southey, R. F. W., Harper, G. J. 1971. The survival of *Erwinia amylovora* in airborne particles: tests in the laboratory and in the open air. *J. Appl. Bacteriol.* 34:547–56
129. Stall, R. E., Hall, C. B., Cook, A. A. 1972. Relationship of ammonia to necrosis of pepper leaf tissue during colonization by *Xanthomonas vesicatoria. Phytopathology* 62:882–86
130. Stewart, V. B. 1913. The fire blight disease in nursery stock. *Cornell Univ. Agr. Exp. Sta. Bull.* 329:317–71
131. Stewart, V. B., Leonard, M. D. 1915. The role of sucking insects in the dissemination of fire blight bacteria. *Phytopathology* 5:117–23
132. Suzuki, Y., Uchida, K. 1965. Microbiological studies of phytopathogenic bacteria. I. On 2-ketogluconic acid fermentation by the bacteria belonging to the *Erwinia amylovora* group. *Agr. Biol. Chem.* 29:456–61
133. Suzuki, Y., Uchida, K. 1966. Microbiological studies of phytopathogenic bacteria. I. Taxonomy of genus *Erwinia* and oxidative metabolism of carbohydrate. *Ber. Ohara Inst. Landwirt. Biol. Okayama Univ.* 13:123–40
134. Talbert, T. J. 1925. Fire blight of apples and pears. *Mo. Agr. Exp. Sta. Circ.* 137:1–8
135. Thomas, H. E., Ark, P. A. 1934. Fire blight of pears and related plants. *Calif. Agr. Exp. Sta. Bull.* 586:1–43
136. Thomas, H. E., Ark, P. A. 1934. Nectar and rain in relation to fire blight. *Phytopathology* 24:682–85
137. Thomas, H. E., Ark, P. A. 1939. Some factors affecting the susceptibility of plants to fire blight. *Hilgardia* 12: 301–22
138. Thomas, H. E., Parker, K. G. 1933. Fire blight of pear and apple. *Cornell Univ. Agr. Exp. Sta. Bull.* 557:1–24
139. Thomas, H. E., Thomas, H. E. 1931. Plants affected by fire blight. *Phytopathology* 21:425–35
140. Thomson, S. V., Schroth, M. N., Moller, W. J. 1975. Streptomycin resistant strains of *Erwinia amylovora* and their occurrence in pear orchards. *Phytopathology* 65: In press
141. Thomson, S. V., Schroth, M. N., Moller, W. J., Reil, W. O. 1975. Efficacy of bactericides and antagonistic bacteria in reducing epiphytic populations of *Erwinia amylovora* and subsequent incidence of fire blight. *Phytopathology* 65: In press
142. Thomson, S. V., Schroth, M. N., Moller, W. J., Reil, W. O. 1975. Relation of weather and epiphytic populations of *Erwinia amylovora* to the occurrence of fire blight. *Phytopathology* 65: In press
143. Tullis, E. C. 1929. Studies on the overwintering and modes of infection of the fire blight organism. *Mich. Agr. Exp. Sta. Tech. Bull.* 97:1–32
144. van der Zwet, T. 1967. Review of fire blight control measures in the United States. *Trans. Ill. State Hort. Soc.* 101:63–71
145. van der Zwet, T. 1968. Recent spread and present distribution of fire blight in the world. *Plant Dis. Reptr.* 52:698–702
146. van der Zwet, T. 1969. Study of fire blight cankers and associated bacteria in pear. *Phytopathology* 59:607–13
147. van der Zwet, T. 1970. New outbreaks and current distribution of fire blight of pear and apple in northern Europe. *FAO Plant Prot. Bull.* 18:83–88
148. Veldeman, I. R. 1972. La decouverte d' "*Erwinia amylovora*" (Burrill) Winslow et al. (feu bacterien du poirier) en Belgique. *Rev. Agr. Brussels* 25: 1587–94
149. Vörös, J., Goodman, R. N. 1965. Filamentous forms of *Erwinia amylovora. Phytopathology* 55:876–79
150. Waggoner, P. E., Horsfall, J. G. 1969. EPIDEM. A simulator of plant disease written for a computer. *Conn. Agr. Exp. Sta. Bull.* 698:1–80
151. Waggoner, P. E., Horsfall, J. G., Lakens, R. J. 1972. EPIMAY, a simulator of southern corn leaf blight. *Conn. Agr. Exp. Sta. Bull.* 729:1–84
152. Waite, M. B. 1892. Results from recent investigations in pear blight. *Am. Assoc. Advan. Sci. Proc.* 1891 40:315; also in *Bot. Gaz.* 1891. 16:259 1891

153. Waite, M. B. 1896. The cause and prevention of pear blight. *US Dep. Agr. Yearb.* 1895:295–300
154. Waite, M. B. 1898. The life-history and characteristics of the pear blight germ. *Am. Assoc. Advan. Sci. Proc.* 47:427–28
155. Waite, M. B. 1906. Pear-blight-work and its control in California. *Rept. 31st Fruit Growers Conv. Calif.*, 1–20
156. Whetzel, H. H. 1905. Blight cankers of apple trees. *West. Fruit-Grower* 16: 349–53
157. Whetzel, H. H. 1906. The blight canker of apple trees. *Cornell Univ. Agr. Exp. Sta. Bull.* 236:103–38
158. Whetzel, H. H., Stewart, V. B. 1909. Fire blight of pears, apples, quinces, etc. *Cornell Univ. Agr. Exp. Sta. Bull.* 272:31–32
159. Wilde, W. H. A., Carpenter, J., Liberty, J., Tunnicliffe, J. 1971. *Psylla pyricola* (Hemiptera: Psyllidae) vector relationships with *Erwinia amylovora*. *Can. Entomol.* 103:1175–78
160. Woodcock, E. F., Tullis, E. C. 1927. Extra-floral nectar glands of *Malus malus* and *Pyrus communis*. *Pap. Mich. Acad. Sci. Arts Lett.* 8:239–43
161. Zoller, B. G., Hanke, L., Sisevich, J. 1974. *Abstracts of Reports from 48th Annual Western Cooperative Spray Project,* p. 27

# ROLE OF WILD AND CULTIVATED PLANTS IN THE EPIDEMIOLOGY OF PLANT DISEASES IN ISRAEL

❖3606

*A. Dinoor*

Department of Plant Pathology and Microbiology, Faculty of Agriculture, The Hebrew University of Jerusalem, Rehovot, Israel

## INTRODUCTION

Israel is a small country with a very intensive agriculture and a great variety of crops. Geographical and topographical conditions result in a highly variable conglomeration of relatively small but adjacent climatic regions. Add to this the application of modern technology such as irrigation and cropping under cover, coupled with very diversified crops in close proximity to large areas of uncultivated land containing a diversity of wild plants and weeds, and one ends up with a paradise of opportunities for plant diseases. An exchange of pathogens between crop and wild-plant populations is not uncommon even with specialized pathogens of a specific, narrow host range.

In many parts of Israel one crop season is well separated from the next and most if not all pathogens are faced with the problem of survival between seasons because of the discontinuity of congenial host-plant populations. For many pathogens this distinct, complete discontinuity occurs in the long, hot, dry summer, and for others, in the relatively cold and hostile winter.

Wild plants and weeds, as well as cultivated plants, may close these gaps and thus contribute to epidemics of plant diseases by bridging between seasons, between crops, and between locations. Thus, continuity in both time and space is established. In this sense, bridging hosts, including infested propagating material, may either perpetuate foci of infection for the production of additional inoculum or harbor dormant stages of the pathogen. In many diseases the sexual stage assumes an important role in terminating the dormant stage or carrying the fungus through a dormant step. Another important facet of bridging is mediating, through an alternate host, between the dormant or sexual stage and next season's crop. Bridging between seasons may be accomplished by alternative annual or perennial hosts that permit the pathogen to remain active between seasons. They may also harbor minor pathogens or newly introduced pathogens that have escaped from cultivated crops before, or without having been established in the crop. Standby, minor diseases thus constitute a constant potential danger. Bridging hosts may also supplement or genetically screen inoculum at an early stage in the epidemic.

This review is not intended to cover the whole scope of epidemiology of plant diseases in Israel, but rather to point out and discuss the role of weeds and wild plants in either supplementing or substituting for cultivated plants in establishing or promoting disease epidemics.

## HOST RANGE AND ITS IMPLICATIONS

The concept of host range is essential to understanding the role of additional hosts in the perpetuation or establishment of any plant disease. The host range of a particular parasite is defined as a list of the hosts of this parasite. The common definition of host is: "A living organism, harbouring a parasite" (1). Host ranges were first determined by examining diseased plants, determining and identifying disease-causing organisms, and listing them according to taxonomic characteristics or thc host and thc parasitc. This tcdious work, which has bccn going on cvcr sincc people became interested in plant pathogens, has resulted in the accumulation of a huge amount of information.

Numerous indices, checklists, and books have been published classifying pathogens systemically, according to hosts, without following Koch's postulates for each list as a whole. For this reason, many host range lists contain the names of plants infected by a particular parasitic species even though there is no indication whether one particular isolate of the parasite can actually infect the whole list of hosts or whether different isolates, or races, or forms of the same parasitic species are capable of parasitizing different plants in the same host range list. Such lists are incomplete also from the aspect of host uniformity. Particularly where wild plants are concerned, the hosts are variable populations rather than uniform entities, and listing a botanical species as a host does not necessarily mean that all individuals of this species will indeed be hosts (26, 30, 32).

Host range lists for epidemiological purposes should therefore be constructed with these concerns in mind. Host lists should have common strains of parasites as common denominators. Even lists based on cross-inoculations might be incomplete when there are no common denominator isolates. The most inadequate lists are those summarizing results from different laboratories, regions, countries, or periods, which will most likely have no isolate as a common denominator.

Symptomless carriers are also hosts by definition and are epidemiologically important (128), even though they are not usually included in host range lists. Indeed, since they are symptomless, they have no chance of being identified as hosts unless they are studied with their possible epidemiological significance in mind (47, 67, 129). Available information on virus diseases suggests that wild hosts often have reached some equilibrium with their pathogens and therefore disease expression and effects in them are usually more wild than when the virus is transferred to a relatively new host (127). Wild hosts also may be breeding grounds for milder strains of viruses, because aggressive strains are self eliminating (24). Such considerations have important bearings on the classical concept of host range.

Parasites may also be sustained by propagation in the rhizosphere of nonhost species (70, 107). Thus, any evaluation of the epidemiological impact of weeds and wild plants on epidemics should consider as components of the host range any plant that increases the amount of inoculum or even only maintains this amount against the natural tendency for decrease in the absence of a host.

## INVOLVEMENT OF WEEDS AND WILD PLANTS IN EPIDEMICS

As indicated earlier, the great topographic, climatic, vegetational, and plant pathological diversity within Israel has triggered many research projects aimed at understanding the sources and causes of epidemics in many crops. The challenges are still numerous, especially in carefully determining the delicate role of wild plants vs weeds or cultivated plants, as well as weeds vs cultivated plants in intensifying epidemics or amplifying their destructiveness.

## INVOLVEMENT IN EPIDEMICS OF ABOVE-GROUND PARASITES

A typical parasite of this group starts the season with a relatively small amount of inoculum. Production is either by overseasoning propagules, by infested plant residues and debris of the former season, by the sexual stage, or by a small amount of alternative hosts. The production of initial inoculum over a period of time increased the chances of hitting a congenial host during a period of suitable climatic conditions and thus assuring that "the initial spark will ignite a fire." This adaptation has been described for net blotch of barley (49), *Septoria* leaf blotch of wheat (34), powdery mildew of barley (37, 130), downy mildew of Sudan grass (51), and crown rust of oats (21), and can probably be shown for most pathogens. The initial source of inoculum operates for several weeks or even months, and in some cases it even perrenates (21, 51). The effects of such sources are overshadowed as soon as the disease starts to spread on the new crop. The initial source of inoculum is eventually exhausted at some stage, usually well into the season. In some seasons it might happen before the pathogen has had a chance to establish itself in the new crop. In the disease caused by *Sclerotinia* the exhaustion of the primary inoculum terminates the spread of the disease for that season (90, 122).

Recognition of the importance of the overseasoning sources of inoculum in initiating epidemics has stimulated interest in sanitation measures. The objective is to destroy crop residues or reduce the chances of inoculum passing from these residues to the new crop, and thus decrease the chances of an epidemic developing early enough to cause significant damage (121). A review of some of the plant diseases that have assumed epidemic proportions in Israel show that the main factor in these diseases is the cultivated host and the main effector is the farmer and his techniques and operations. This was the case with barley net blotch (49), potato and tomato early blight (102), potato and tomato late blight (125), early blight of carrots (82), bitter rot of apples (93), apple scab (94), pear scab (92), *Sclerotinia* rot of vegetables and ornamentals (90, 122), *Cerospora* of sugar beet (113), *Ascochyta* of peas (8) and chickpeas, *Septoria* leaf blotch of wheat (25, 36), downy mildew of Sudan grass (51), virus diseases (77), and others. In many examples the causal organism had been present for some time but the disease had not become epidemic until intensification of agriculture, abandoning crop rotation, doing away with sanitation, new agrotechnical measures (overhead irrigation, cropping under cover) etc contributed to the buildup of inoculum and hence to the change of a minor or moderate disease into a major problem.

What is the part played by weeds and wild plants? In certain diseases such as potato early blight (102), apple scab (94), apple bitter rot (93), and sunflower rust (3) no additional hosts are known and it cannot be claimed that wild plants have augmented the danger. In other diseases, however, wild relatives of cultivated plants grow during the same season as the cultivated hosts. The wild plants may or may not be abundant. They may or may not be infected by the same population of causal organisms. Sometimes it is difficult to determine if disease development in the wild hosts follows that in the cultivated crop or vice versa.

The impact of any form of overseasoning on epidemics is through the availability of inoculum and its quantity and quality (physiologically and genetically). Early, viable and compatible inoculum will result in a more destructive epidemic. Under marginal conditions quantity of inoculum will compensate for suboptimal conditions (103, 104); therefore, any additional development of inoculum near the crop before initiation of the epidemic or just before the season will contribute a great deal to the severity of the disease and to the damage caused.

## *Alternative Hosts*

The term *alternative host* is coined as a more general term than the alternate host of a rust fungus. Alternative host includes any host in addition to the main host, including plants of the main host out of their regular season or habitat. Alternative hosts are mainly important as epidemiological bridges for the inoculum between crops of the main hosts.

Plant pathogens capable of attacking alternative hosts may be divided into three groups: 1. pathogens cycling mainly on the agricultural crop, whose alternative hosts are occasionally found diseased during the main season; 2. pathogens whose alternative hosts play at least the same role as their main hosts in harboring and perpetuating the organisms from season to season and may play an important role in the life cycle and disease cycle of the pathogen; 3. pathogens whose alternative hosts may be found diseased during the main season of the agricultural host but whose cardinal role is between the main seasons, when they maintain and produce active inoculum.

### PATHOGENS WITH OCCASIONAL ALTERNATIVE HOSTS DURING THE SEASON

Pathogens of this group are important as causal agents of epidemics in agricultural crops, but in the wild they are reported only occasionally and do not seem to be of much importance. These pathogens might be native to this area, surviving as minor pathogens in the wild and not having the appropriate conditions to establish themselves and cause epidemics in the wild. This is probably the case in *Septoria* leaf blotch of wheat, which was isolated from wild emmer and *Aegilops* spp. (35); it is not epidemic in the wild despite its major importance in cultivated wheat (25, 36). A related species, *S. nodorum,* pathogenic on wheat, was found on wild barley (50, 52) but has not been recorded yet on wheat. Pathogeographical limitations on epidemics (99, 100) may be important in some cases, but they may be compensated in others (as in *S. tritici*) by other factors such as agrotechnology or varietal composition of the crops. Despite the fact that representative members of the genus *Pisum* are native to this region, *Ascochyta* blight does not assume epidemic proportions in the wild. *Ascochyta* is a major disease in cultivated peas. It can attack wild Leguminosae but is not important in the wild (8). It is not known whether the *Ascochyta* pathogens were introduced to this country and then spread into the wild or were originally minor pathogens in the wild.

Introduced crops of potato and tomato suffer annually from outbreaks of late blight. The pathogen was probably introduced as well, since there are solanaceous hosts native to this region and *Phytophthora* is not established in them. A continuous effort was made to unravel the life cycle and disease cycle of potato late blight in Israel (125). Infested tubers were important but not the only source of initial inoculum (73). Tomatoes are usually infected before potatoes; therefore, tubers cannot be the ultimate source of blight. Livne (75) and Sztejnberg (117) studied the host range of blight. Under controlled conditions several hosts were identified, mainly from the Solanaceae family. *Solanum* and *Withania* were also diseased under field conditions (117, 118) but by the end of the summer the infected leaves dropped and the disease could no longer be recovered. Recently it was observed[1] that canning tomatoes are sometimes diseased with blight in summer. Since the fungus thrives on tomatoes under appropriate microclimatic conditions (101) it was suggested[1] that dormant mycelium of the fungus may reside in some tomatoes during the summer and that symptoms appear when conditions are suitable. Dormant mycelium might, by the same token, be present in weed or wild Solanaceae and spread from them in autumn. In any case it appears that *Phytophthora* did not establish itself in the wild, for reasons that are probably climatic rather than genetic or evolutionary.

Some pathogens with occasional alternative hosts are in a transient group. Either they have not had a chance to spread into the wild flora and establish themselves there, or plant pathologists have not had the chance to study them thoroughly. In either case, no epidemiological significance can be attributed to such pathogens in augmenting epidemics in crops. On the contrary, the diseased crop is affecting the wild hosts. If and when such pathogens become established in the wild and cause epidemics there, they will fit into the next group.

PATHOGENS ESTABLISHED DURING THE MAIN SEASON ALSO ON WILD PLANTS Pathogens of this group cause epidemics both in cultivated crops and in

[1]J. Katan, personal communication.

wild plants. The genetical and epidemiological aspects of the mutual relationship between crops and wild populations is a new field of research that has not been explored yet.

To start with, let us assume that both cultivated crops and their diseases originated in the wild and that the pathogens carried out their life cycles at a slow pace. When hosts were cultivated, made more uniform, and crowded in large populations, the pathogens started epidemics. This aggravated situation in the cultivated hosts in turn aggravated the situation in the wild. The following descriptions show a developmental route from the source of parasites in the wild to the most complicated development of interrelationships between wild and cultivated hosts.

*Plant diseases in the wild* Wild plants constitute a reservoir of minor diseases and an important role is therefore ascribed to them as a potential source for disease outbreaks. With this view in mind, Kenneth (50, 52) set out to survey and investigate foliar diseases of wild and cultivated grasses and cereals. There is no way yet to predict when and how a minor disease of a wild plant will spread into cultivated crops and become a major problem. Checklists of pathogens in Israel include many diseases on different hosts, but detailed information on host range and potential dangers are not abundant. Kushnir (69) made a detailed study of leaf spot diseases of wild *Iris* spp. and showed differences in the susceptibility of wild species to *Septoria* and *Heterosporium.* Despite severe epidemics of these two diseases in the wild there is no indication yet of any of these diseases spreading into cultivated *Iris.* Kenneth (50, 52) pointed out the importance of *Septoria* on wild oats and the existence of *Septoria nodorum* on wild barley. None of them has been reported yet on cultivated oats or wheats.

*Wild plants: origin or rescue for a parasite?* There are several cases where both wild and cultivated plants are diseased and the actual origin is not clear at first. *Solanum nigrum* in tomato fields in Italy was found infected with tobacco mosaic virus, but it was not clear whether the disease originated from the *Solanum* seed or from tomatoes (15). In Ghana, cocoa viruses were recovered from wild host trees even far away from cocoa plantations, but no conclusion was drawn regarding the viruses' origin (72). In Canada, stone fruit viruses were found in endemic components of the flora in the orchards' region (127). The authors claim that the viruses are latent in the wild and spread out into the orchards. In the case of *Xanthomonas translucens* in wheat, wheats as well as grasses (*Aegilops, Avena,* and *Phalaris*) were found infected in Israel (123). This disease was recorded on barley a few years ago (50), but was shown then not to attack wheat. Lately it burst out as a new problem in wheat and occurs on wild barley as well. The problem was probably magnified by the introduction of a new variety in a year with conditions conducive for disease development. Ergot in wheat and grasses broke out in 1957–1958 (6, 79–81). Two possibilities were suggested: 1. The disease was introduced on wheat or rye and then escaped to grass hosts. (Twenty of the grass species tested were inoculated successfully.) The disease could have then returned to wheat under suitable conditions. 2. The disease was always a minor disease of grasses and was overlooked. Proper climatic circumstances brought it to light unexpectedly. This explanation is more plausible since the disease was found also on grasses in the sand dunes far away from any wheat fields. The role of wild plants in this case is twofold: they serve as bridges for stepwise spread from field to field (6) and as reservoirs for minor diseases until circumstances permit more abundant disease development.

Another case with some similarities has turned out to be a major phytopathological problem. Scab of apple and pear were minor diseases in Israel, found on primitive local varieties or wild plants, respectively. No disease was found for years on introduced European varieties of either apples or pears. Perlberger (96) suggested that genetic resistance is not responsible for this, but that it is a case of escape due to lack of correlation between the phenology of the host and the climatic requirements of the parasite. European varieties of both crops sprouted 4–6 weeks later than the local varieties when temperatures were higher and not optimal for infection by the pathogen. Pappo (92, 94) and Shabi et al (108) later showed that it was only a matter of physiological races. New races appeared which were capable of attacking the European varieties of both crops and of turning both diseases into major ones. The new biotypes that have appeared are not capable of attacking the old varieties and the wild pear. No genetic explanation has been put forward as to how these biotypes have evolved.

*Spread and dispersal of pathogens in propagating material of wild plants* The importance of infested propagating material in establishing disease epidemics has long been recognized, and careful precautions are taken to produce healthy propagating material. A source of inoculum introduced into the field along with the crop assures an earlier outbreak of the disease and is the cause of a more severe epidemic and losses (121). While relevant actions are being taken in cultivated crops, seed transmission of a disease is not under control in weeds or wild plants. Cucumber mosaic virus (CMV) in England was found to be transmitted in seed of *Stellaria* (119, 120) and can thus persist without a living host or vector, out of reach of control measures. CMV under Israeli conditions has numerous living hosts year round (85, 86); therefore the possibility of seed transmission was not considered.

Intriguing circumstances are described in England for tobacco rattle virus (TRV) (19). The vector is a nematode and the virus is seed transmitted in wild hosts. The virus is not only preserved in the seed but is also disseminated with the seed to new areas along with the nematodes that are disseminated with the soil by wind. In the newly invaded areas, infected hosts develop from some of the seed and the blown-in vector nematodes play their role in spreading it from the new sources. This rather fascinating example serves here only to point out potential dangers. In Israel, barley stripe mosaic virus was found to be seed transmitted in 19 out of 158 species of *Gramineae* tested (87). Through such means, the virus can be established in new regions and finally be kept in a safe reservoir from year to year (56). The potential danger of dispersal of infested seed from the wild into new areas has not yet been fully explored.

*Supplementary initial inoculum from wild plants* Pathogens with a relatively narrow host range among cultivated plants may widen their variety of hosts by attacking related plants from the wild and by these means increase the possibilities of overseasoning or extending the length of the season. Pathogens with a wide host range have many more opportunities, but even in such a case wild hosts may add means of spread and overseasoning. *Sclerotinia sclerotiorum* is such a pathogen, well known as a midwinter problem in cabbage and eggplants. With the introduction of cropping under cover and intensification of vegetable and flower growing, and with the extension of summer crops in to warmer regions in winter, the host range under field conditions has been widened to such an extent that it has become a country-wide menace. The sexual stage assumes the most important epidemiological role (122). This stage determines the spread of the disease and is probably responsible for infection on any host (84). The magnitude of the effect of the sexual stage depends on the amount of sclerotia produced, and this is a direct outcome of the extent and distribution of the disease in previous years. Within a few years there was a tremendous buildup of inoculum; it spread to adjacent and distant areas and even to isolated desert niches[2] (122). Sanitation, ploughing-in of residues (deeper than 10 cm), and chemical control will probably slow down this buildup and minimize the risk of initial infection. Even if such means are endorsed, there is still the problem of wild hosts.

Palti (90) has indicated some hosts from the wild and stressed their significance in spreading the disease into the fields. In light of the importance of the sexual stage in *Sclerotinia* and its powerful dissemination capabilities, it should not matter much if the source of inoculum is in the field or not. Wild plants as sources of inoculum will not be under agricultural control, and this is their main risk. More information is needed to predict the impact of reducing inoculum in the field on the amount of inoculum that will develop in the wild. The possibility that the amount of inoculum in the wild is merely a reflection of the amount of infestation in the cultivated crops should not yet be ruled out. If it proves to be correct, sanitation measures will contribute in the long run to the minimization of the potential danger of wild plants, in perpetuating *Sclerotinia* rot.

*Cultivated and wild plants, both involved in the epidemics* In this group of parasites both wild and cultivated plants play an important role in the life cycle of the causal organism and in the disease cycle.

Kenneth (49) has carefully analyzed the interrelationships between cultivated and wild barleys and their net blotch pathogen. He has shown that four wild species of barley are often heavily stricken by the disease. The fungus develops in the leaves and produces plenty of dormant mycelium. In late autumn and early winter, when the first rains stimulate germination of wild barleys, they germinate among last season's infected straw on which sporulation of the fungus does occur. The same happens with cultivated barley when infested straw is left in the barn or in the field, or around the field, where newly sown plants or volunteers emerge through a source of inoculum and "catch fire."

Control measures will aim at disengaging the source of inoculum from the new crop by sanitation measures or by the introduction of incompatible resistant hosts.

[2]A. Dinoor, unpublished.

Neither measure can be used with wild barleys which continue to produce inoculum and to maintain the source whatever resistant varieties are present in the field and/or however efficient the sanitation measures. With both measures the source of inoculum will thus be shifted to the fringes of the fields or to areas adjacent or farther away from the fields. The data presented by Kenneth (49) show that the spread of net blotch from infested straw within the field is quite slow because of the prevailing climatic conditions and the different requirements of host and parasite. Temperatures drop continuously during 2–3 months from the time of host germination onwards. Therefore spore production, germination, and infection are less efficient than under optimal conditions. Disease development is slowed during this period while the growth rate of the host is less affected. The slow-down of disease development is very pronounced. In such cases a "canopy effect" (49) is observed, the crop grows away from the disease, and the canopy looks clean.

Towards the end of winter, temperatures start to rise again and the rate of host growth slows since it approaches heading. Only at this stage does the disease burst out, and an epidemic develops. The epidemic explosion from the canopy effect stage depends on the amount of disease established before the onset of cold weather, and here is where the significance of the primary inoculum is manifested. Primary inoculum originating from residues within the field is the most effective inoculum since it is spread all over the field. When these residues are drastically reduced by sanitation, primary inoculum is tremendously decreased within the field. The spread of disease was shown to be slow, and since primary inoculum remains out of the field it will therefore be much less efficient in spreading the disease and in building up the subcanopy epidemic. Removal of primary inoculum by sanitation measures in wild barleys around the fields will be less efficient than by sanitation within the field but more efficient than by sanitation away from the field. It may be concluded that for the short-term epidemiological effects, wild barley plants are less important than cultivated plants, and the primary cause for that is topographical.

There is also another aspect to the problem: The pathogenic variability of the parasite. Even if resistant varieties are introduced, wild barleys, wherever they are, are still very variable and might enable the perpetuation of different biotypes of the pathogen and serve as a background for asexual as well as sexual recombination processes. Khair (58) has shown that the genetic variability of this pathogen is indeed tremendous. The relationship between genotypes in populations of this parasite on cultivated vs wild plants and the reflection of each on the other remains to be studied. Khan (59) has shown that the host range of net blotch in Australia is more restricted than in Israel and claimed that combinations of virulence are much less common away from the centers of origin of host and parasite. He claims that form species are clearly distinguished in Australia and therefore wild barleys do not contribute inoculum to cultivated barley. In Israel, however, the long-term co-evolution of host and parasite has brought together virulence capabilities towards more hosts among barley species. In the long run, and with the use of resistant varieties, wild plants are therefore more important than cultivated hosts, as more genotypes are preserved and recombined on them. This might yield future pathotypes which will eventually spread and initiate epidemics in the new varieties.

The development of powdery mildew of cereals, mainly of barley, has some similar features. Koltin (63) and Koltin & Kenneth (64) have shown that the sexual stage is the only means of overseasoning under Israeli conditions. Plenty of cleistothecia are being produced each spring and enter a dormant period at the stage when asci are produced. At this stage they are preserved until the next season. Maturation of ascospores starts in synchrony with germination of the host seedlings and continues for several weeks. No doubt such a life cycle will be self-sufficient even on cultivated hosts alone if no means of control are employed. Sanitation on the one hand and resistant varieties on the other hand will undoubtedly decrease overseasoning possibilities within the crop appreciably (chemical control here is not economical yet). The use of such measures will emphasize the relative importance of wild plants in the persistence and perpetuation of the disease. Powdery mildew cycles very efficiently on wild plants. Eshed (26) has shown a complete life cycle on 27 species of Gramineae belonging to 12 different genera. These include mildew on wild species of wheat, barley, and oats. The relative importance of overseasoning on cultivated vs wild plants has not been elucidated experimentally. The circumstances suggest (26, 31) that wild relatives of each of the cereals play an important role in the persistence and perpetuation of the disease. Here again, as in barley net blotch, the new generation of hosts germinating amid overseasoning, inoculum-producing propagules, is the most efficient mode of initial focus. The classical picture of newly germinated *Hordeum spontaneum* among the former year's straw shows a yellowish round focus of mildew (63). Going along with the comparison to net blotch, it seems that foci of initial infection of mildew in the wild have more impact on epidemics in the field than foci of net blotch. This probably has to do mainly with the dependence of spore production in net blotch on high humidity which results in a different rate of spore production, and it is also due to differences in spore weight, distance of dissemination of each pathogen, and limiting factors for germination and infection. The amount of mildew-infested tissue in the field is usually smaller than the amount of net blotch, and the relative production of overseasoning structures (cleistothecia of mildew vs dormant mycelium of net blotch) is also smaller, so that the relative importance of perpetuation within vs outside the field is greatly shifted toward the latter as far as mildew is concerned.

Apart from short-term effects of infected wild hosts on the annual initiation of the epidemic there is also the long-term effect on the pathogenic characters of the inoculum. The host range of mildew was found to be relatively wide under experimental conditions (26, 32), with several hosts common to wheat, barley, and oat mildews. The specialization into forms may be bridged genetically. Under field conditions, however, even though under somewhat suboptimal conditions, the host range was narrower. Many isolates from several hosts were tested on grasses and cereals using both conidial and ascosporic inoculum. Both inocula from *Aegilops, Bromus, Dactylis, Phalaris,* and *Scleropoa* were not virulent on the cereals. Each cereal was susceptible to both types of inoculum only from species of its own genus. With the ascosporic inoculum, the frequency of virulent isolates was higher than with conidial inoculum. No genetic analysis of the potential of virulence has thus far been undertaken. The opportunities for different pathotypes to exist and recombine are better on a variable population of wild hosts than on uniform cultivated hosts, and this is again where wild plants assume more importance.

The role of the sexual stage in overseasoning differs with climatic conditions. The sexual stage provides an oversummering bridge, and as such is more important in semi-arid zones (64, 89) than in northern countries,[3] where the perpetuation from season to season does not depend on it exclusively. The life cycle of mildew in northern countries, where overwintering is only a prelude to the main season, is a different version of that in semi-arid conditions, where it is part of the main season. Main season epidemics in the north depend on the amount of overwintering asexual inoculum and on the type of the overwintering bridge, and therefore control measures (chemical and biological) may depend on the handling of winter crops.[3] Under our semi-arid conditions the main crop is exposed directly to the variable product of the sexual stage and later to a variable propagation from it maintained on the variable wild populations. Any control measures here would therefore have to rely on considerations different from those in the north.

The possibility that a pathogen can persist on wild plants was also shown for carrot early blight (82, 83) and *Cercospora* of sugarbeet (113, 114). *Diplodia* fruit rot of citrus was shown to spread from *Acacia* hedges, ornamentals, and other plants including *Ricinus* (78). The relative epidemiological importance of persistence on wild as compared to cultivated plants has not yet been shown in these cases.

Very complex life and disease cycles are found among the rusts. The overseasoning bridge of net blotch and mildew supplies inoculum directly to the same host in the next season. In many rusts, however, an alternate host is involved between the overseasoning bridge and the main host of the next season. This complicated life cycle seems to be more reliable under semi-arid conditions than the type of bridge in mildew. Alternate hosts are mainly perennials, and their existence at the beginning of the season is much more reliable and less ephemeral than the grass hosts, for example. Irregularities in rainfall at the beginning of the season may endanger much more the bridging host of mildew than that of rust. Alternate hosts of rusts in our region are all wild, and therefore annual epidemics are, at least theoretically, dependent exclusively on wild plants.

Several investigators (3, 4, 9, 21, 23, 95, 132, and personal communication with Y. Anikster and Z. K. Gerechter-Amitai) point out that a complete life cycle is not necessary for a rust in Israel even to develop even into an epidemic. The key to the explanation of these phenomena lies in the fact that under our climatic conditions the absence of compatible hosts is the limiting factor, rather than unfavorable climatic conditions for the parasite. Another key factor is the need for a continuous source of viable inoculum. Whenever discontinuity of compatible host disrupts the development of the parasite, the asexual cycle is stopped and cannot be resumed until compatible host and viable inoculum are again available together. If the asexual cycle is interrupted and then the host is available again, it will be free from disease until inoculum is being supplied by the sexual stage or by a foreign asexual source. When the activity of the sexual stage coincides with the appearance of a compatible main host, synchronization is perfect and the disease cycle is linked with

[3]M. S. Wolfe, personal communication.

the life cycle. This is the case with leaf rust of barley and *Ornithogalum* (Y. Anikster, personal communication), with *Uromyces* rusts of wild barleys (5), and with rust of stone fruits and *Anemone* in the mountains (132).

In several rusts, no sexual stage has yet been found in Israel: wheat stem rust (39), leaf rust[4] and yellow rust;[4] sunflower rust (3, 4); and oat stem rust.[4] Outbreaks of these diseases are nevertheless quite common, or even annually occurring. In an epidemiological sense, other rusts with an active sexual stage also seem to bypass the sexual stage: stone fruit rust (9, 95, 132) and crown rust of oats (21). The explanations for bypassing the sexual stage or for existence without a sexual stage are different for each rust. Rust on stone fruits remains viable on leaves that have not been shed by late autumn or midwinter. Early sprouting almonds, especially in the coastal plain, are then infected from this late viable asexual inoculum (9, 95, 132). The situation for yellow rust of wheat and especially for wheat leaf rust, which is very common almost annually, has not yet been clarified. With both rusts, it seems that an external source is responsible and the link here is the wheat crop itself with no mediation.

The frequent occurrence of both rusts on grasses is probably an effect of the crop on the wild hosts rather than the reverse. The situation described in detail for yellow rust in Oregon (109) also suggests that yellow rust in the wild is a reflection rather than the cause of rust in the fields. On the other hand it is reported from the USSR (68) that grasses might be an additional source of inoculum for wheat leaf rust.

Stem rust of wheat had 93 grass species in its host range (39) while oat stem rust had 75 (124). All these plants are winter hosts; they are probably infected with inoculum from the cereals in most cases. Small, oversummering foci of both rusts were found in many parts of the country throughout summer, and occasionally a much larger focus was found (39, 124). The key to survival of stem rust from season to season is the possibility of establishing an infection in autumn before the onset of cold weather. Inoculum for such infection could come from small oversummering foci (39) and by long-distance dissemination to wild plants, volunteers, and early crops in warmer regions. Long distance dissemination was thought to be the cause for the appearance of new races (38). Volunteers of both wheat and barley were first thought to be the most important autumn host (39) responsible for augmenting stem rust near cereal fields, but emphasis was later shifted to wild plants.[4] It was also concluded that small oversummering foci show the ability of rust to exist in summer but cannot be responsible for next year's epidemics.[4] It is our opinion (A. Dinoor, unpublished) that volunteers (39) and early crops rather than wild plants in the warmer regions are the most important targets for wheat stem rust inoculum arriving from an outside source in autumn.

Crown rust of oats usually bypasses the sexual stage in establishing annual epidemics (21). The sexual stage on the alternate host provides a bridge between the dormant oversummering stage and the compatible main host during winter (42, 98); but this bridge does not explain the disease cycle under local conditions (21).

[4]Z. K. Gerechter-Amitai, personal communication.

Compatible hosts are available from October on while the sexual stage operates from mid-December till March. The gap from October to December is not an idle period. Crown rust epidemics starts in the warm autumn and establish themselves in wild and cultivated oats. No epidemiological role can therefore be assigned to bridging alternate hosts; this is quite different from the pattern in other countries (115, J. C. Santiago, personal communication). In this respect crown rust is not different from rusts lacking a sexual stage.

In a detailed study on the life cycle and disease cycle of crown rust, Dinoor (21) tried to point out the links between the two. It was concluded that crown rust can develop on compatible hosts throughout summer and even endures short gaps (two weeks) in the presence of compatible hosts. Grasses were found to harbor the disease during summer under different climatic conditions. Oversummering foci could not be responsible for fall-established epidemics. The spread of rust from small local foci was restricted even within one location. Thus, a large external source of inoculum was postulated that overrides the effects of small local ones: it is probably located in East Africa. Rust exists there almost all year round with a peak in October (E. J. Guthrie, personal communication) [see also Green et al (40) for stem rust]. Samples from prevalent crown rust in East Africa in October were identified here as belonging to the same physiologic races that are prevalent in Israel. Proper synoptic situations during autumn were claimed to have contributed to long distance dissemination to the Middle East (23). Wild plants (mainly oats) as well as cultivated oats are the target for the arriving inoculum.

On the basis of these conclusions an intercontinental cycle for the exchange of microorganisms between East Africa and the Middle East was suggested. It was also suggested that this cycle may be linked to another route of spread from the Middle East to Northern Europe through Southeastern Europe (21).[5] The importance of wild plants in this respect is that they serve as a huge trap for landing inoculum, and the larger the trap (wild oats are much more widespread here than cultivated oats) the better are the chances for a landing inoculum to hit a compatible host and establish itself in it.

The role of wild plants in the life cycle of crown rust is not only epidemiological as described above. The alternate host *Rhamnus* serves as a breeding ground for the rust and variable rust populations emerge therefrom (21). It also serves, at least under experimental conditions, as a host for crosses between different forms of crown rust which result in the exposure of new virulence genes in the population (30). The genetic impact of wild hosts on crown rust development is therefore double: production of biotypes on the alternate host and maintenance of variable populations on the main wild host.

The cases of downy mildews of pearl millet (*Sclerospora graminicola*) and sorghum (*S. sorghi*) (53) are used to illustrate a potential role for weeds and wild grasses which has fortunately not materialized. These two diseases are new to Israel and were first fully described in 1966 (51). They were probably introduced with the seed, *S. graminicola* spreading on pearl millet, while *S. sorghi* became established in

[5]J. C. Zadoks, personal communication.

Sudan grass and later on in sweet corn (57). *S. sorghi* did not establish itself in sorghum thanks to climatic conditions and agrotechnology used with this crop (53, 54). *S. graminicola* is known from the literature to include several summer Gramineae in its host range, including weeds and wild grasses very common in Israel. Despite the fact that *S. graminicola* was encountered in different parts of the country, it did not spread into wild hosts and did not become established there (51). *S. sorghi* has only one wild host—*Sorghum halepense* (Johnson grass). Slow spread of the disease by means of infested straw and seed probably would have occurred if the disease could have established itself in the wild. Kenneth (51) first reported establishment of the disease in Johnson grass, with the pathogen invading the perennial rhizome of the host. Later reports (54, 55) based on many more observations indicated that the great majority of Johnson grass clones in this country are not infected and are probably resistant to the disease.

Perennial hosts have assumed an important role in the establishment of virus diseases. Maize dwarf mosaic virus (MDMV) has been established in Johnson grass (41, 62) and was recently reported to have established also in Bermuda grass (*Cynodon dactylon*).[6] These two perennial hosts therefore serve as an annual source for MDMV. It was suggested[6] that vectors should be chemically controlled on these alternative hosts towards the beginning of the season.

PATHOGENS HARBORED ON ALTERNATIVE HOSTS BETWEEN SEASONS In the former group of pathogens the alternative hosts, when infected during the season, mostly reflected the disease situation in the main host. When the alternative hosts were infected between seasons they were occasional main season hosts that managed to develop between seasons, and their epidemiological role in most cases has been uncertain. Alternative hosts of the group that harbors pathogens between seasons serve as bridges in their main growing season. This is an entirely different adaptation of a parasite to a range of hosts; plants that are distinctly unrelated serve as hosts to a common pathogen. Many virus diseases belong to this group, and in this case a distinction should be made between diseases where both virus and vector are adapted to both hosts and diseases where only the vector or only the virus is adapted to both hosts. Pathogens of this group serve as clear-cut proofs against the contention that hosts of a particular parasite are phyllogenetically rather than only evolutionarily related and support the contrary contention that co-existence without any phyllogenetic relations may result in compatibility of different hosts with the same pathogen. Only a few pathogens of this group will be dealt with in detail to illustrate the case.

Powdery mildew of Cucurbitaceae is important annually in Israel. This disease, caused by *Sphaerotheca fuliginea* (27), does not produce the sexual stage and an alternative host was not known. Even at times when seasons of cucurbit growth were more spaced, mildew constituted a problem, let alone lately when cropping under cover or in the desert during the winter closes the gap between the seasons. To unravel this enigma was quite an undertaking. All mildews look alike and many of

[6]M. Klein, personal communication.

them do not even produce the sexual stage. Closing the gap between checklists and carefully determined host range means isolating mildew from each host and inoculating it to all hosts—an endless job. Thanks to the method developed by Hirata (45) for the identification of mildews at the asexual stage, a short cut could be employed. Mildews from different hosts were examined and grouped according to type of spore germination (27). Cross inoculations and host range determination were then performed only within each group (27, 28). It was thus found that the mildew of cucurbits can develop on several different hosts, and inoculum from these hosts does infect cucurbits (27–29). Some of these hosts, like *Xanthium, Senecio,* and *Papaver* grow throughout winter and may well serve as alternative hosts for the mildew. Epidemiological verification of these findings is still lacking. Studies in England (116) have shown *Sonchus* to be a likely alternative host of cucumber mildew. *Sonchus* is found infected here as well, but it is parasitized by *Erysiphe cichoracearum* which does not show up on cucurbits in Israel.

*Oidiopsis* mildew is a severe problem in peppers, eggplants, tomatoes, and artichokes. Many hosts are infected, many of them wild (91, 131). Results of cross-inoculation tests for the accurate determination of host range are few (91, 106, 131). Overseasoning is attributed to perennial hosts like eggplants and artichokes and to the continuous cultivation of susceptible crops throughout the year, rather than to wild plants (91). The following evidence suggests that wild hosts from other families, some of them perennials, are probably responsible for overseasoning: reports from other countries (see Ref. 91), the severity of the disease in isolated places in the desert, and the frequent occurrence of the disease also at times when there was no winter bridging with crops under cover.

Mosaic viruses of cucurbits have a wide host range, extending well into the wild flora (16, 17, 85, 86, 88). The widest host range so far was determined for cucumber mosaic virus (16, 88)—it extends to 15 crop species and 9 wild species from different families. Winter hosts were reported for mosaic of cucumber (CMV), melons (MMV), and squash (SMV).

The importance of wild winter hosts in the epidemiology of the diseases can be demonstrated by comparing CMV to MMV. CMV is much more widespread and more important in the spring, since it has many winter hosts. MMV on the other hand builds up slowly in spring and early summer and becomes important only then, since it has few winter hosts (16, 88). Nitzany (86) has reviewed the problem of wild plants as sources for virus diseases, stressing the following points: virus diseases have a wide host range; many potential hosts in the wild are not recognized because most of them are symptomless carriers or have no known symptoms; a constant supply of inoculum is contributed from the huge area of fringes around small field plots; exchange of viruses with the wild by vectors might be erratic and inconsistent, but this enables the virus to become established in the wild and invade the crop continuously.

The role of weeds in virus disease incidence was reviewed and analyzed by Duffus (24), who presented the importance of weeds (wild plants would be similar) as reservoirs of vectors, viruses, or both. Wild plants are an undelimited source of viral diseases (14) since many wild hosts are either symptomless (2, 127) or show different symptoms from those on cultivated plants (20). Bacterial (10) or viral diseases (127) may be preserved in the annual vegetation of orchards or vineyards (20), in perennial hosts like grazing fields (74), in related ornamental species (76), or in weeds like Johnson grass (41, 62). The discovery in Sweden that large areas of leys are infected with barley yellow dwarf (74) illustrates the unknown potential in alternative hosts. More relevant to our conditions is the description from British Columbia (127) of virus diseases, unnoticed in the wild, that spread into orchards. The new crop is intolerant of the virus, it reacts severely, and a disease, unnoticed in the tolerant wild hosts, manifests itself. Any new introduction, especially to a region rich with native flora, is therefore potentially dangered when no investigation of disease in the native flora has been made.

Special attention should be drawn to maize rough dwarf virus (MRDV). This disease is encountered in some countries of the Mediterranean basin and southern Europe, but is not recorded in the New World. Available information is presented in a book by Harpaz (43). MRDV causes disease in a planthopper, its transmission from insect to insect occurring via a plant host. Several cereals and winter grasses are compatible but almost symptomless hosts for the virus (60, 61). Some summer grasses are also almost symptomless carriers (43, 60, 61). In colder regions, overwintering is achieved in the diapaused planthopper (18). Maize is a newcomer to this region but is highly susceptible to MRDV. It is a temporary and occasional host for the planthopper, which in spring moves from winter to summer grass hosts. If only maize is available, the planthopper feeds on it even though it cannot propagate on it. The virus provokes a severe reaction in maize and a disease shows up. Research on MRDV has changed some fundamental ideas about host-vector-parasite relationships. The host range, for example, is widened for a diseased insect. Bermuda grass, a nonhost, will be compatible with an inoculative vector; more than that, it may infect a healthy vector and thus enable the vector to feed on the host. With sorghum the relationships are even more particular: the virus cannot develop in the host, nor reinfect a vector through sorghum; the vector cannot propagate on sorghum, but an inoculative vector can propagate on sorghum. Under our climatic conditions the disease is perpetuated from year to year through vectors and through almost symptomless winter and summer hosts. Winter and summer grasses are therefore alternative hosts to each other, but neither is damaged enough to render the other the ultimate role of an alternative host. Somewhere in the cycle (April-May), a short-term bridging host is encountered: the maize plant. Being intolerant of the virus, it reacts abnormally, and an epidemic of plant disease is created.

## INVOLVEMENT IN EPIDEMICS OF SOIL-BORNE PARASITES

Contrary to the above-ground parasites, in which the inoculum builds up to epidemic proportions each season, buildup of inoculum in soil-borne parasites is a lengthy process extending over several years. Another difference is that soil-borne parasites are less conspicuous in the wild, since they are more localized and in most cases are difficult or impossible to identify since most hosts are symptomless carriers.

The key point in understanding the problem of soil-borne parasites is the amount of inoculum available, propagated, or maintained in the soil. Diseases caused by soil-borne parasites are not regarded as serious until their inoculum concentration in soil reaches a certain threshold. Inoculum buildup below this threshold is not recognized as significant. The potential danger of a particular parasite in a particular field can be predicted only by monitoring its inoculum concentration in that field (Y. Henis and J. Katan, personal communication). Concentrations in the field will depend on propagation of the parasite in cultivated plants and/or in the wild flora.

From information gathered in Israel (46–48, 65–67, 112) and from abroad (11–13, 33, 44, 70, 97, 110, 129) it appears that cultivated hosts and cultivation methods are the main factors determining the severity of soil-borne diseases. Wild plants, on the other hand, serve only as reservoirs, bridges, and rescue hosts for the pathogens, supplementing their growth and propagation only to a limited extent. Wild plants, therefore, mainly insure the persistence of the parasite in the soil, but have not been shown yet to contribute to a dangerous inoculum buildup.

The contribution of various hosts to propagation of inoculum in the field differs with different hosts. A distinction should first be made between susceptible and resistant cultivars of the same host (46, 70, 107, 126). Lacy (70) has shown that the propagation of *Verticillium* in the rhizospheres of susceptible and resistant cultivars of mint was similar, while in immune cultivars, even though inoculum increase was significant, it was less than the increase on susceptible or resistant cultivars. Number of invasion sites on resistant and susceptible cultivars was also similar, the only difference being that colonization of the susceptible plants was greater than that of the resistant ones. Katan (46) has shown that colonization of *Fusarium* resistant tomatoes is less than that of susceptible cultivars. Later, quantitative determinations of propagules in the host tissue have shown that the number of propagules developed in the roots of a resistant tomato cultivar was 0.3% of that developed by a susceptible cultivar (H. Alon, J. Katan, and N. Kedar, unpublished), and in the lower stem it was $10^6$ times more in the susceptible compared to the resistant cultivar. This difference has an additional, very important epidemiological effect. Hosts that produce large amounts of dormant propagules above the ground spread the pathogen in space (47). For a soil-borne parasite with a restricted potential for spread through the soil, this is an advantage that should be avoided agrotechnically by sanitation or restrictions on residue dissemination (11, 44). It is also avoided biologically by the use of resistant varieties that greatly decrease the above-ground production of propagules. This is also one reason a soil-borne pathogen does not reach epidemic proportions in the wild but does become a problem after the introduction of a susceptible agricultural crop.

In *Sclerotium rolfsii* it was shown that there is no correlation between host resistance and sclerotia production (7). In *Verticillium* (33) some of the immune hosts, even when colonized, are not congenial for the production of microsclerotia, while others are congenial.

Schroth & Hildebrand (107) have explained the resistance of hosts and the decreased propagation of parasites by differences in root exudates, some of them promoting and some suppressing propagation of parasites in the rhizosphere. In another account (126) it was shown that penetration into roots is different between peas resistant and susceptible to *Fusarium.* Since germination in the rhizosphere was not affected in this case, such resistant hosts might depress the population of their parasites. This will occur if abortive germination reduces the number of propagules more than the limited propagation in the resistant host sustains it. Resistant hosts may sustain or renew the propagules of the parasites if abortive germination results in the production of new dormant propagules. Rovira (105) also ties specificity to root exudates and explains that host resistance is overcome by adaptation of the parasite to formerly unfavorable exudates. Adaptation to new microhabitats due to nutritional or chemical factors might be a very successful means of becoming an inhabitant of a new alien or hostile surrounding. The great variety of wild plants exposes multitudes of chemically different microhabitats in which a newcomer could become established even without an appropriate host. Thus the question of whether a parasite is endemic or an introduction is more complicated with a soil-borne parasite than for above-ground parasites.

Apart from resistant hosts, there are many symptomless carriers in which parasites develop and propagate. Symptomless carriers may include disease-free plants of a susceptible host (46) in which disease has not shown up because they are young and/or not sufficiently colonized for a disease to be induced. For several parasites, symptomless carriers mean a whole array of hosts from other species and genera. The difficulty with these hosts is that they cannot be visually detected in the field. Several investigators have shown a wide host range of *Verticillium* among weeds and wild plants (12, 33, 44, 65–67, 97, 110, 111, 129). The role of wild plants in the establishment of *Verticillium* was described in at least two cases. Evans (33) maintained that weed seed, as well as fleshy fruits, spread the disease into a new area in Australia where it propagated on wild plants and then attacked cotton. Krikun (65–67 and unpublished) claimed that *Verticillium* was most likely introduced with potato tubers into agricultural areas in the Negev desert of Israel, established there in wild plants, and then magnified tremendously by rotation of cotton and potatoes to become the most damaging pest in the area. *Fusarium* wilt of tomatoes also has other symptomless carriers among the wild flora (47). The amount of parasite propagules developed within symptomless carriers of *Fusarium* is in the range of 1–4% of propagules developed in a susceptible host (46, 47)—much more than the amount developed even in the roots (the most infected organ) of a resistant tomato cultivar. This means that the contribution of a symptomless carrier to the buildup of inoculum is more important than that of a resistant cultivar. Powelson (97), on the other hand, claimed that at least with *Verticillium,* the contribution of resistant cultivars is much the same as that of wild hosts.

## CONCLUDING REMARKS

Plant diseases found in the wild are sometimes indigenous but are often common to cultivated crops and even spread from cultivation into the wild. The many examples of diseases found and studied in Israel in the wild indicate at least one important point: there is a huge and variable reservoir of parasites deeply rooted in the wild which can occasionally or constantly serve as a source of initial inoculum for many crop diseases. The next question is whether these initiation sources have a cardinal role in initiating or causing epidemics. For parasites harbored on wild plants between seasons, wild plants might well have an important role. The importance of wild hosts of this group may sometimes be masked by modern agrotechnology. In the past, bridging between seasons could be accomplished only on wild hosts. Cropping out of season sometimes makes bridging possible without wild hosts. For parasites established in the wild during the main cropping season, it is difficult to distinguish the epidemiological contribution of the wild plants from that of the cultivated plants. The use of sound cultural control measures against a disease in a crop will eventually expose most of the net effects of wild plants. So far there has been no attempt to expose wild plants as a single epidemiological factor under an experimental design to evaluate its net role in epidemics. There is much information on the contribution of wild hosts to the propagation of parasites and to their disease cycle, but there is no information on the importance of this contribution relative to that from cultivated plants.

Evaluation of the importance of symptomless carriers in epidemics is even more complex. Hosts without symptoms cannot be visually identified, and even if a species is known already to be a host, the frequency of colonized hosts in an area is very difficult to determine. The amount of virus inoculum in a population of symptomless carriers seems impossible to estimate. With soil-borne parasites the amount of inoculum produced by symptomless carriers can be estimated either indirectly, by determining the amount of inoculum in the soil where they grew, or directly, by sampling the particular host and estimating its extent of colonization (46). The dynamics of inoculum buildup by soil-borne parasites has not been studied adequately (71). The dynamics of inoculum propagation within the season and from season to season is assumed to aid appreciably in identifying and understanding the contribution of different factors to the buildup of inoculum in soil. The study of dynamics will also indicate how rapidly inoculum accumulates to a level above the critical threshold, and how different hosts and successions of hosts contribute to the increase or decrease of inoculum concentration. This general claim is no less important for symptomless carriers.

The whole subject of host range needs evaluation for its significance in inoculum production and the development of epidemics. The epidemiological importance of hosts may be measured by the amount and timing of available, compatible inoculum they supply.

Studies of host range of a pathogen among wild plants and its involvement in the disease cycle are only a prologue to the fascinating subject of relationships between populations of wild and cultivated plants. Crop cultivars exposed adjacent to variable native vegetation constitute a proper background for studying the effects of the genetic architecture of the host population on development of epidemics. Mapping the distribution of resistance in the host and virulence in the parasite (22) in this situation will add a new dimension to classical epidemiological studies. After all, the role of wild plants in epidemics is not limited to effects on the amount of inoculum; it also embraces the quality of inoculum supplied. Only a combined epidemiological and genetic approach will shed the proper light on the interactions of wild and cultivated plants.

*Literature Cited*


1. Ainsworth, G. C. 1961. *Dictionary of the Fungi.* Kew, Surrey, England: Commonw. Mycolog. Inst. 547 pp.
2. Aleshin, E. P., Zemlina, A. G., Galochkina, L. A., Kanevcheva, I. S., Toropstev, V. V. 1968. Reservoirs of wheat streak mosaic virus. *Zashch. Rast. Mosk.* 13(8):13–14
3. Anikster, Y. 1966. *Studies on the biology of sunflower rust Puccinia helianthi Schw. in Israel and its chemical control.* MSc thesis. Hebrew Univ., Jerusalem, Israel. 67 pp. (In Hebrew)
4. Anikster, Y., Dinoor, A. 1967. The biology and control of sunflower rust in Israel. *Proc. 1st Israel Congr. Plant Pathol.* pp. 48–49
5. Anikster, Y., Wahl, I. 1966. *Uromyces* rusts on barley in Israel. *Israel J. Bot.* 15:91–105
6. Avizohar-Hershenson, Z. 1957. The ergot disease of wheat and rye in Israel. *Hassadeh* 37:879–80 (In Hebrew)
7. Avizohar-Hershenson, Z. 1962. The influence of some summer crops on the sclerotia population of *Sclerotium rolfsii* in the soil. *Nat. Univ. Inst. Agr. Rep.* No. 384
8. Barash, I. 1960. *The causal organism of Ascochyta diseases on peas in Israel.* MSc thesis. Hebrew Univ., Jerusalem, Israel. 65 pp. (In Hebrew)
9. Ben-Arie, R. 1961. *Phenology of the rust on stone fruits in Israel.* MSc Thesis. Hebrew Univ., Jerusalem, Israel. 60 pp. (In Hebrew)
10. Berg, L. A. 1971. Weed hosts of the SFR strain of *Pseudomonas solanacearum,* causal organism of bacterial wilt of bananas. *Phytopathology* 61:1314–15
11. Brooks, D. H., Dawson, M. G. 1968. Influence of direct drilling of winter wheat on incidence of take-all and eyespot. *Ann. Appl. Biol.* 61:57–64
12. Brown, F. H., Wiles, A. B. 1970. Reaction of certain cultivars and weeds to a pathogenic isolate of *Verticillium alboatrum* from cotton. *Plant. Dis. Reptr.* 54(6):508–12
13. Canova, A. 1964. Researches on the virus diseases of grasses. I. Wheat soilborne mosaic. *Phytopathol. Mediter.* 3:94
14. Chessin, M., Lesemann, D. 1972. Distribution of Cactus viruses in wild plants. *Phytopathology* 62:97–99
15. Cirulli, M. 1972. Pathogenicity of tobacco mosaic virus isolates collected from four species of Solanaceae. *Proc. 3rd Congr. Mediter. Phytopathol. Union,* 1–9
16. Cohen, S., Nitzany, F. A. 1962. Survey of the distribution of Cucurbitaceae viruses in Israel. *Hassadeh* 42:1140–42 (In Hebrew)
17. Cohen, S., Nitzany, F. A. 1962. Two additional viruses on Cucurbitaceae in Israel. *Hassadeh* 42:842–43 (In Hebrew)
18. Conti, M. 1972. Investigations on the epidemiology of maize rough dwarf virus. I. Overwintering of virus in its planthopper vector. *Proc. 3rd Congr. Mediter. Phytopathol. Union,* 11–17
19. Cooper, J. I., Harrison, B. D. 1973. The role of weed hosts and the distribution and activity of vector nematodes in the ecology of tobacco rattle virus. *Ann. Appl. Biol.* 73:53–66
20. Dias, H. F. 1963. Host range and properties of grapevine fanleaf and grapevine yellow mosaic viruses. *Ann. Appl. Biol.* 51:85–95
21. Dinoor, A. 1967. *The role of cultivated and wild plants in the life cycle of Puccinia coronata Cda. var. avenae F&L and in the disease cycle of oat crown rust in Israel.* PhD thesis. Hebrew Univ., Jerusalem, Israel. 373 pp. (In Hebrew with English summary)
22. Dinoor, A. 1974. Evaluation of sources for disease resistance. In *Genetic Resource in Plants* (FAO-IBP), ed. O. Frankel, J. G. Hawkes. Cambridge, Engl.: Cambridge Univ. Press.
23. Dinoor, A., Levi, M. 1967. The relation of oat crown rust outbreaks in autumn in Israel to long distance dissemination. *Proc. 1st Israel Congr. Plant Pathol.,* 13–14
24. Duffus, J. E. 1971. Role of weeds in the incidence of virus diseases. *Ann. Rev. Phytopathol.* 9:319–340
25. Ellal, G., Dinoor, A. 1973. Chemical control of *Septoria* leaf blotch in wheat. *Proc. 7th British Insecticides and Fungicides Conf.*
26. Eshed, N. 1966. *Host range of powdery mildews of barley, wheat and oats, and the interrelationship of varieties of Erysiphe graminis DC. The role of wild grasses in epidemiology of powdery mildew on winter cereals.* MSc thesis. Hebrew Univ., Jerusalem, Israel. 146 pp. (In Hebrew)
27. Eshed, N. 1972. The host range of several powdery mildews in Israel. *Hassadeh* 52:911–15

28. Eshed, N. 1972. Determination and host range of some powdery mildews in Israel. *Proc. 3rd Congr. Mediter. Phytopathol. Union,* 165–70
29. Eshed, N. 1972. Sources of inoculum for some powdery mildew diseases. *Proc. 3rd Israel Congr. Plant Pathol.,* 31–32
30. Eshed, N., Dinoor, A. 1973. Genetic studies on the specialization of crown rust into pathogenic varieties (formae speciales). *Proc. 2nd Int. Congr. Plant Pathol.,* 0709
31. Eshed, N., Koltin, Y., Kenneth, R., Wahl, I. 1967. The manner of overseasoning of the barley powdery mildew fungus and the role of wild Gramineae in the epidemiology of the disease. *Proc. 1st Israel Congr. Plant Pathol.,* 19–20
32. Eshed, N., Wahl, I. 1970. Host range and interrelations of *Erysiphe graminis hordei, E. graminis tritici* and *E. graminis avenae. Phytopathology* 60:628–34
33. Evans, G. 1971. Influence of weed hosts on the ecology of *Verticillium dahliae* in newly cultivated areas of the Namoi Valley, N.S.W. *Ann. Appl. Biol.* 67: 169–76
34. Eyal, Z. 1971. The kinetics of pycnospore liberation in *Septoria tritici. Can. J. Bot.* 49:1095–99
35. Eyal, Z. 1972. *Septoria* leaf blotch of wheat in Israel. *Proc. 3rd Israel Congr. Plant Pathol.,* 54–56
36. Eyal, Z. 1972. Effect of *Septoria* leaf blotch on yield of spring wheat in Israel. *Plant Dis. Reptr.* 56:583–86
37. Eyal, Z., Silberstein, O., Moseman, J. G., Wahl, I. 1972. Mode of oversummering and parasitic specialization of the barley powdery mildew fungus *Erysiphe graminis hordei* in Israel. *Proc. 3rd Congr. Mediter. Phytopathol. Union,* 145–46
38. Gerechter-Amitai, Z. K. 1966. Physiologic races pf *Puccinia graminis tritici* in Israel in 1965. *Robigo* 18:9–10
39. Gerechter-Amitai, Z. K., Wahl, I. 1964. Wheat stem rust on wild grasses in Israel. Role of wild grasses in the development of the parasite and in breeding for resistance. *Proc. 3rd Cereal Rusts Conf., Cambridge,* 207–17
40. Green, G. J., Martens, J. W., Ribeiro, O. 1970. Epidemiology and specialization of wheat and oat stem rusts in Kenya in 1968. *Phytopathology* 60: 309–14
41. Greenberger, A. 1969. *Maize dwarf mosaic virus on maize and sorghum in Israel.* MSc thesis. Hebrew Univ., Jerusalem, Israel. 36 pp. (In Hebrew)
42. Halperin, J. 1957. *The role of Rhamnus palaestina Boiss. in the production and spread of physiologic races of Puccinia coronata.* MSc thesis. Hebrew Univ., Jerusalem, Israel. 27 pp. (In Hebrew)
43. Harpaz, I. 1972. *Maize Rough Dwarf.* Jerusalem: Israel Univ. Press. 251 pp.
44. Heale, J. B., Isaac, I. 1963. Wilt of lucern caused by species of *Verticillium.* IV. Pathogenicity of *V. albo-atrum* and *V. dahliae* in soil and in weeds; effect upon lucerne production. *Ann. Appl. Biol.* 52:439–51
45. Hirata, K. 1942. On the shape of the germ-tubes of *Erysiphaceae. Bull. Coll. Hort.* 5:34–49 (In Japanese with English summary)
46. Katan, J. 1968. *Studies on the life cycle and multiplication in soil of the fungus Fusarium oxysporum f. lycopersici (Sacc.) Snyder & Hansen, the incitant of the wilt disease in tomatoes.* PhD thesis. Hebrew Univ., Jerusalem, Israel. 109 pp. (In Hebrew with English summary)
47. Katan, J. 1971. Symptomless carriers of the tomato *Fusarium* wilt pathogen. *Phytopathology* 61:1213–17
48. Katan, J., Sofer, S. 1972. Inoculum build-up of soil-borne pathogens. *Proc. 3rd Israel Congr. Plant Pathol.,* 3
49. Kenneth, R. 1960. *Aspects of the taxonomy, biology and epidemiology of Pyrenophora teres Drechsl. (Drechslera teres (Sacc.) Shoemaker), the causal organism of net blotch disease of barley.* PhD thesis. Hebrew Univ., Jerusalem, Israel. 226 pp. (In Hebrew with English summary)
50. Kenneth, R. 1965. *Survey of leaf diseases (excluding rusts) on wild and cultivated cereals and grasses in Israel.* Final Report of Research to Ford Foundation (Project A-13). 43 pp.
51. Kenneth, R. 1966. Studies on downy mildew diseases causec by *Sclerospora graminicola* (Sacc.) Shroet and *S. sorghi* Weston & Uppal. *Scripta Hierosolymitana* 18:143–70
52. Kenneth, R. 1967. New diseases of wild and cultivated Gramineae in Israel, and their present and potential importance. *Proc. 1st Israel Congr. Plant Pathol.,* 41–42
53. Kenneth, R. 1974. Diseases of summer Gramineae. In *The Encyclopedia of Agriculture.* Tel-Aviv: Encycl. Agr. Publ. In press (In Hebrew)
54. Kenneth, R., Klein, Z. 1969. Sorghum downy mildew in Israel. *Hassadeh* 49:17–20 (In Hebrew)
55. Kenneth, R., Klein, Z. 1970. Epidemiological studies of sorghum downy mil-

dew (*Sclerospora sorghi* Weston & Uppal) on sorghums and corn in Israel. *Israel J. Agr. Res.* 20:183
56. Kenneth, R., Nitzany, F. A., Pinthus, M. 1959. Barley stripe mosaic virus. *Hassadeh* 39:755–56 (In Hebrew)
57. Kenneth, R., Shahor, G. 1971. Sorghum downy mildew on sorghums and maize in Israel. *Hassadeh* 51:853–56 (In Hebrew)
58. Khair, J. 1965. *Investigations on physiologic specialization in Drechslera teres (Sacc.) Shoemaker (Helminthosporium teres Sacc.) the causal agent of barley net-bloch.* MSc thesis. Hebrew Univ., Jerusalem, Israel. 77 pp. (In Hebrew)
59. Khan, T. N. 1973. Host specialization by western Australian isolates causing net blotch symptoms on *Hordeum. Trans. Brit. Mycol. Soc.* 61:215–20
60. Klein, M. 1962. *Studies on maize rough dwarf virus.* MSc thesis. Hebrew Univ., Jerusalem, Israel. 82 pp. (In Hebrew)
61. Klein, M. 1967. *Studies on the rough dwarf virus disease of maize.* PhD thesis. Hebrew Univ., Jerusalem, Israel. 138 pp. (In Hebrew with English summary)
62. Klein, M., Harpaz, I., Greenberger, A., Sela, I. 1973. A mosaic virus disease of maize and sorghum in Israel. *Plant Dis. Reptr.* 57:125–28
63. Koltin, Y. 1963. *The role of the sexual stage of Erysiphe graminis hordei in oversummering of the disease in Israel.* MSc thesis. Hebrew Univ., Jerusalem, Israel. 84 pp. (In Hebrew)
64. Koltin, Y., Kenneth, R. 1970. The role of the sexual stage in the oversummering of *Erysiphe graminis* D.C. f. sp. *hordei* Marchal, under semi-arid conditions. *Ann. Appl. Biol.* 65:263–68
65. Krikun, J. 1967. *Verticillium* wilt in the Negev region of Israel. *Proc. 1st Israel Congr. Plant Pathol.*, 34
66. Krikun, J., Chorin, M. 1966. Verticillium—A new soil pathogen in the Negev region of Israel. *Israel J. Agr. Res.* 16:177–78
67. Krikun, J., Chorin, M. 1969. *Verticillium* build-up and survival in the soil, and its control. *Proc. 2nd Israel Congr. Plant Pathol.*, 15
68. Kulikova, G. N., Borisenko, A. N. 1969. Races of the causal agent of brown rust on wheat in 1967 and on wild grasses in 1966–7. *Vestn. Selskokhoz. Nauki Alma Ata* 12:88–92 (*Rev. Appl. Mycol.* 49:1329)
69. Kushnir, T. 1949. Leaf spot diseases of *Iris* in Palestine. *Palestine J. Bot. Jerusalem Ser.* 4:230–33
70. Lacy, M. L., Horner, C. E. 1966. Behaviour of *Verticillium dahliae* in the rhizosphere and on roots of hosts and non-hosts. *Phytopathology* 56:427–30
71. Last, F. T. 1971. The role of the host in the epidemiology of some non-foliar pathogens. *Ann. Rev. Phytopathol.* 9: 341–62
72. Legg, J. T. Bonney, J. K. 1967. The host range and vector species of viruses from *Cola chlamydontha* K. Schum., *Adansonia digitata* L. and *Theobroma cacao* L. *Ann. Appl. Biol.* 60:399–403
73. Lev, I. 1962. *Experimentation on the control of Phytophthora late blight by the use of disease-free tubers.* MSc thesis. Hebrew Univ., Jerusalem, Israel. 54 pp. (In Hebrew)
74. Lindsten, K., Gerhardson, B. 1969. Investigations on barley yellow dwarf virus (BYDV) in leys in Sweden. *Meddn. St. Växtsk Anst.* 14:261–80
75. Livne, A. 1960. *Modes of transmission of potato late blight from one season to the next, under Israeli conditions.* MSc thesis. Hebrew Univ., Jerusalem, Israel. 43 pp. (In Hebrew)
76. Loebenstein, G. 1959. *The degeneration disease of sweet potatoes.* PhD thesis. Hebrew Univ., Jerusalem, Israel. 65 pp. (In Hebrew with English summary)
77. Loebenstein, G. 1961. Let us reduce virus damage. *Hassadeh* 41:274–75
78. Minz, G., Ben-Meir, Y. 1944. Pathogenicity of *Diplodia* from various hosts to Citrus fruits. *Palestine J. Bot. Rehovot Ser.* 4:157–61
79. Minz, G., Gerechter, Z. 1958. The ergot disease of wheat. *Hassadeh* 38:95–97 (In Hebrew)
80. Minz, G., Gerechter, Z., Avizohar-Hershenson, Z. 1959. Ergot disease of wheat: Experiments on artificial inoculation of wheat and wild grasses. *Ktavim* 9:269–71
81. Minz, G., Gerechter, Z., Avizohar-Hershenson, Z. 1960. Inoculation experiments with ergot on wheat and wild grasses. *Ktavim* 10:29–32
82. Netzer, D. 1970. *Studies on the biology of the fungus Alternaria dauci (Kühn) Groves and Skolko, the causal agent of leaf blight of carrots.* PhD thesis. Hebrew Univ., Jerusalem, Israel. 101 pp. (In Hebrew with English summary)
83. Netzer, D., Kenneth, R. 1969. Persistence and transmission of *Alternaria dauci* (Kühn) Groves & Skolko, in the semi-arid conditions of Israel. *Ann. Appl. Biol.* 63:289–94
84. Newton, H. C., Sequeira, L. 1972. Ascospores as the primary infective propa-

gules of *Sclerotinia sclerotionum* in Wisconsin. *Plant Dis. Reptr.* 56: 798–802

85. Nitzany, F. A. 1958. Some virus diseases of vegetables in Israel. *Hassadeh* 38:872–74 (In Hebrew)
86. Nitzany, F. A. 1964. Wild plants as sources for the infestation of cultivated plants with virus diseases. *Hassadeh* 44:1322–24
87. Nitzany, F. A., Gerechter, Z. K. 1963. Barley stripe mosaic virus host range and seed transmission tests among Gramineae in Israel. *Phytopathol. Mediter.* 2:11–19
88. Nitzany, F. A., Wilkinson, R. E. 1960. The identification of cucumber mosaic virus from different hosts in Israel. *Photopathol. Mediter.* 1:71–76
89. Padalino, O., Antonicelli, M., Grasso, V. 1970. Biology of *Erysiphe graminis* DC f. sp. *tritici* Marchal, in the conditions of southern Italy. *Phytopathol. Mediter.* 9:122–35
90. Palti, J. 1963. *Sclerotinia sclerotiorum* in Israel. *Phytopathol. Mediter.* 2:60–64
91. Palti, J. 1971. Biological characteristics, distribution and control of *Leveillula taurica* (Lév.) Atn. *Phytopathol. Mediter.* 10:139–53
92. Pappo, S. 1965. Scab disease of pears. *Hassadeh* 46:669–74 (In Hebrew)
93. Pappo, S. 1965. *Apple bitter rot, caused by Gloesporium fructigenum* (Berk.). PhD thesis. Hebrew Univ., Jerusalem, Israel. 134 pp. (In Hebrew with English summary)
94. Pappo, S. 1966. A new strain of *Fusicladium dendriticum* (Wal.) Fel. on apples in Israel. *Phytopathol. Mediter.* 5:149
95. Perlberger, J. 1943. The rust disease of stone fruit trees in Palestine. *Jewish Agency for Palestine, Agr. Res. St. Bull.* 34
96. Perlberger, J. 1944. The occurrence of apple and pear scab in Palestine in relation to weather conditions. *Palestine J. Bot. Rehovot Ser.* 4:157–61
97. Powelson, R. L. 1970. Significance of population level of *Verticillium* in soil. In *Root Diseases and Soil-borne Pathogens,* ed. T. A. Toussoun, R. V. Bega, P. E. Nelson, 31–33. Berkeley: Univ. Calif. Press. 252 pp.
98. Rayss, T., Habelska, H. 1942. *Rhamnus palaestina* Boiss.—A new host of crown rust. *Palestine J. Bot. Jerusalem Ser.* 2:250
99. Reichert, I. 1950. A biogeographical approach to phytopathology. *Trans. NY Acad. Sci.* 20:333–39
100. Reichert, I., Palti, J. 1966. On the pathogeography of plant diseases in the Mediterranean region. *Phytopathol. Mediter.* 5:129
101. Rotem, J. 1955. *Studies on Phytophthora infestans.* MSc thesis. Hebrew Univ., Jerusalem, Israel. 44 pp. (In Hebrew)
102. Rotem, J. 1961. *Studies on the early blight disease of tomato and potato caused by Alternaria porri (Ell.) Neerg. f. sp. solani (E & M).* PhD thesis. Hebrew Univ., Jerusalem, Israel. 191 pp. (In Hebrew with English summary)
103. Rotem, J., Cohen, Y., Putter, J. 1971. Relativity of limiting and optimum inoculum loads, wetting durations and temperatures for infection by *Phytophthora infestans. Phytopathology* 61: 275–78
104. Rotem, J., Palti, J. 1969. Factors affecting development of plant disease epidemics in Israel. *Proc. 2nd Israel Congr. Plant Pathol.,* 33
105. Rovira, A. D. 1965. Plant root exudates and their influence upon soil microorganisms. In *Ecology of Soil-Borne Pathogens,* ed. K. F. Baker, W. C. Snyder, 170–94. Berkeley: Univ. Calif. Press. 571 pp.
106. Saad, A. T. Abul-Hayja, Z., Sonmez, M. M. 1972. Investigations on *Leveillula* species in Lebanon. *Proc. 3rd Congr. Mediter. Phytopathol. Union,* 147–54
107. Schroth, M. N., Hildebrand, D. C. 1964. Influence of plant exudates on root infecting fungi. *Ann. Rev. Phytopathol.* 2:101–32
108. Shabi, E., Rotem, J., Loebenstein, G. 1973. Physiological races of *Venturia pirina* on pear. *Phytopathology* 63: 41–43
109. Shaner, G., Powelson, R. L. 1973. The oversummering and dispersal of inoculum of *Puccinia striiformis* in Oregon. *Phytopathology* 63:13–17
110. Skadow, K. 1969. Investigations on the wilting agents *V. albo-atrum* and *V. dahliae.* I. Weeds as host plants. II. Weeds as reservoirs of infection. *Zentralbl. Bakteriol. Parasitenk. Infektionskr.* Abt. 2, 123:715–65
111. Shoham, H., Krikun, J. 1970. Verticillium disease and the means to prevent its spread. *Hassadeh* 50:1392–95
112. Sofer, S. 1972. *The effects of ploughing-in of plant residues on the build-up of root diseases.* MSc thesis. Hebrew Univ., Jerusalem, Israel. 54 pp. (In Hebrew)

113. Solel, Z. 1967. *Physiological specialization of the fungus Cercospora beticola Sacc., the causal agent of sugarbeet leaf spot, and mechanisms influencing plant infection.* PhD thesis. Hebrew Univ., Jerusalem, Israel. 90 pp. (In Hebrew with English summary)
114. Solel, Z. 1970. Survival of *Cercospora beticola,* the causal agent of sugarbeet leaf spot in Israel. *Trans. Brit. Mycol. Soc.* 54:504–06
115. Stakman, E. C., Harrar, J. G. 1957. *Principles of Plant Pathology.* New York: Ronald Press. 581 pp.
116. Stone, O. M. 1962. Alternate hosts of cucumber powdery mildew. *Ann. Appl. Biol.* 50:203–10
117. Sztejnberg, A. 1963. *Modes of persistence of Phytophthora infestans in wild Solanaceae between crop seasons.* MSc thesis. Hebrew Univ., Jerusalem, Israel 53 pp. (In Hebrew)
118. Sztejnberg, A., Wahl, I. 1966. Studies on the host range of the potato late blight fungus and its oversummering on wild and ornamental Solanaceae plants. *Proc. 1st Congr. Mediter. Phytopathol. Union,* 452–60
119. Tomlinson, J. A., Carter, A. L. 1970. Studies on the seed transmission of cucumber mosaic virus in chickweed (*Stellaria media*) in relation to the ecology of the virus. *Ann. Appl. Biol.* 66:381–86
120. Tomlinson, J. A. Carter, A. L., Dale, W. T., Simpson, C. J. 1971. Weed plants as sources of cucumber mosaic virus. *Ann. Appl. Biol.* 66:11–16
121. Van der Plank, J. E. 1963. *Plant Diseases: Epidemics and Control.* New York/ London: Academic. 349 pp.
122. Vigosky, H. 1969. Methods for controlling *Sclerotinia* rot of Gerbera, and studies on the disease cycle in Israel. *Israel J. Agr. Res.* 19:185–93
123. Volcani, Z., Kenneth, R. 1972. A new bacterial disease of barley and wheat. *Hassadeh* 52:795–98 (In Hebrew)
124. Wahl, I., Dinoor, A., Gerechter-Amitai, Z. K., Sztejnberg, A. 1964. Physiologic specialization, host range and development of oat stem rust in Israel. *Proc. 3rd Cereal Rust Conf., Cambridge,* 242–53
125. Wahl, I., Lev. I., Sztejnberg, A. 1966. Effect of sowing healthy tubers on potato late blight incidence. *Scr. Hierosolymitana* 18:171–208
126. Whalley, W. M., Taylor, G. S. 1973. Influence of pea-root exudates on germination of conidia and chlamydospores of physiologic races of *Fusarium oxysporum* f. *pisi. Ann. Appl. Biol.* 73:269–76
127. Welsh, M. F., Lott, T. B., Wilks, J. M., Keane, F. W. L. 1963. Stone fruit viruses as latent infections in indigenous plants in British Columbia. *Phytopathol. Mediter.* 3:199–202
128. Wilhelm, S. 1959. Parasitism and pathogenesis of root-disease fungi. In *Plant Pathology, Problems and Progress,* ed. C. S. Holton, et al, 356–66. Madison: Univ. Wis. Press. 588 pp.
129. Woolliams, G. E. 1966. Host range and symptomatology of *Verticillium dahliae* in economic, weed and native plants in interior British Columbia. *Can. J. Plant Sci.* 46:661–69
130. Zilberstein, O., Eyal, Z. 1972. The dynamics of cleistothecial maturation in *Erysiphe graminis* f. sp. *hordei. Proc. 3rd Israel Congr. Plant Pathol.,* 65–66
131. Zwirn-Hirsch, H. E. 1943. Studies on *Leveillula taurica* (Lév.) Arn. *Palestine J. Bot. Jerusalem Ser.* 3:52–53
132. Zwirn-Hirsch, H. E. 1945. Infection experiments with aeciospores of *Tranzschelia pruni-spinosae* (Pers.) Diet. in Palestine. *Palestine J. Bot. Jerusalem Ser.* 3:178–79

# SOIL FACTORS INFLUENCING FORMATION OF MYCORRHIZAE

❖3607

*V. Slankis*

c/o Ministry of Natural Resources, Forest Research Branch, Maple, Ontario, Canada

## INTRODUCTION

Research on mycorrhizal (M) symbiosis, particularly during the last three decades, has provided convincing evidence that most higher plants are mycotrophs. While the majority of forest trees have ectomycorrhiza (ECM) (*148*),[1] most nonwoody species, including a large number of agricultural plants, form endomycorrhizae (EDM) usually of the vesicular-arbuscular type (*29, 53, 109a*).[2]

ECM benefit forest trees by (*a*) aiding in absorption of inorganic and organic nutrients (2, 44, *89*), (*b*) supplying trees with growth-regulating substances (*139*), (*c*) deterring root pathogens (*81, 166*), (*d*) decreasing soil toxicity (168), and (*e*) increasing host plant resistance to drought (*7*, 71, 146) and extreme soil temperatures (82). Inoculating soil with proper M fungi has aided in establishment of forests in previously treeless parts of the world (*100*). Recently, EDM have been shown to benefit their hosts in a manner generally similar to ECM (*29, 109a*).

These findings have transformed M symbiosis from a subject of only academic interest to a biological factor of major importance in forestry and agriculture. If the recent rate of progress in M research is sustained, the threshold of a new era will soon be reached when, based on sound knowledge, this symbiosis will be employed extensively in forestry and agriculture to facilitate crop production and to establish economically significant plants in areas with adverse soil and climatic conditions. This chapter has been written in part to stimulate and to encourage the development of that expertise.

In this review, only some of the soil factors influencing M formation can be considered, and only selected references from the voluminous literature are cited. Additional references can be found in the bibliographies of earlier reviews (5, 15,

[1]Italicized numbers in parentheses refer to reviews and other publications in which the pertinent data from primary research articles are cited.

[2]The terms ectomycorrhiza, ectendomycorrhiza, and endomycorrhiza used in this review have been proposed by Peyronel, Fassi, Fontana & Trappe (118a) to replace the formerly used terms ectotrophic, ectendotrophic, and endotrophic mycorrhiza.

29, *42*, 52, 53, *71*, 80, 88, 94, 131, 166). This chapter deals mainly with soil factors influencing ECM. Comprehensive reviews on EDM have been published by Gerdemann (29) and Mosse (109a).

## EFFECT OF SOIL NUTRIENTS

After Stahl (141) postulated in 1900 that forest trees have abundant M only in nutritionally poor soils, the effect of inorganic and organic fertilizers on M formation was studied intensively (*7, 42,* 43, 44, *71, 89, 131*). Very large studies were carried out in conjunction with efforts to establish forest shelterbelts and forest plantations on the steppes of the Soviet Union (*71, 131*).

### *Inorganic Nutrients*[3]

Early laboratory experiments with axenic cultures led to the view that M infection is stimulated by an imbalance or limited availability of certain elements, particularly N and P (2, 44). High concentrations of N arrest M infection (2, 44) and cause well-established M roots to become nonmycorrhizal (137). Hatch (44) concluded that formation of ECM is determined by the internal nutrient status of short roots, which in turn depends on nutrient availability. Björkman (2) postulated that M formation requires a surplus of soluble sugars in the roots—a condition determined by the availability of N and P and by light intensity.

In natural soils, the effect of inorganic nutrients on M formation is more complex than previously thought. Neither relatively high concentrations of N and P nor severe deficiencies of these elements necessarily arrest M infection. Meyer (92) found that fertilization of already fertile soils had little or no effect on M formation in seedlings of beech and spruce. In another experiment (94), half the root system of individual beech seedlings was allowed to grow in mull from a nutrient-rich soil and the other half in mor from a nutrient-poor podsol. Although the more fertile soil had about 24 times more P and a 3.2 times greater nitrification rate, 86% of short roots were converted into M in the mull section as compared with 60% in the podsol. Koberg (55) found similar results for Scots pine potted in forest soil: the treatment that provided the maximum amount of N and P produced both the greatest frequency of M and the largest seedlings.

Similarly, in field experiments, M infection was not arrested by excessive fertilization (3, 31, 102, *130, 131*). Pushkinskaya & Valikova (*130*) demonstrated that fertilization of nutritionally poor forest soil with P, P + K, N + P + K, N + 3P + K, and N + 3P + 2K stimulated better growth and more abundant M on one-year-old oak seedlings. The fertilizer amounts introduced were: 20 kg/ha N as $NH_4NO_3$, 45 kg/ha $P_2O_5$ as superphosphate, 30 kg/ha K as KCl. Fertilization with P + K and N + 3P + 2K induced the most abundant M, whereas application of N + 3P + K produced the tallest seedlings with the greatest dry weight but less abundant M. Similar results were found with *Pinus cembra* in two high altitude

[3]Recently, this subject has been comprehensively reviewed by Bowen (7). Therefore, only selected aspects of inorganic nutrients in relation to mycorrhiza are discussed in this chapter.

nurseries in Austria (31). The highest incidence of M was found in plots treated annually during a 4 yr experiment with 40 kg/ha N as ammonium sulfate, 50 kg/ha P as superphosphate, and 70 kg/ha K as patent potassium. Dosages of N, P, and K that were increased by 53, 60, and 77.4% or by 107, 60, and 77.4%, respectively, did not arrest M formation but only lowered M frequency by 20 and 39% below that of the control.

However, by combining the first of the higher concentrations with an inoculation of *Suillus plorans* spawn, M frequency was raised 16% above control and maximum weight seedlings were obtained. Göbl & Platzer concluded that an increase in the population of M fungi increases the host plant's ability to utilize nutrients (31).

The above data demonstrate that the effects of high concentrations of inorganic nutrients on M development in natural soils are not the same as in axenic cultures (2, 44). In natural soils, presumably due to leaching and strong competition between soil microorganisms for nutrients, introduced N seldom reaches concentrations totally inhibitory to M infection unless fresh nutrient is periodically flushed through the medium. Under the latter conditions, mycorrhizal seedlings of *Pinus strobus*, transplanted from forest nursery into washed fine granitic gravel, did not form new M, and the symbiotic relationship in previously established M was terminated (137).

M infection also occurs when nutrient deficiencies are severe. This has been demonstrated with *P. strobus* seedlings transplanted from a nursery into washed granitic gravel in which the N supply was decreased to 2.3 mg/l (137), and with *P. elliottii* and *P. clausa* seedlings grown in sand in which one or more of the elements N, P, and S were deficient (16). In both cases the stunted seedlings with yellowish needles had abundant M. Similar phenomena may occur in forest nurseries (15). The above data should caution against the generalization that prevalence of M assures normal development of host plants.

Harley (42) was correct in warning that M development should not be viewed as a cure for all ills since this symbiosis is not the only factor that determines successful growth. Morrison (105) demonstrated that in some cases, M provide no obvious advantage to the host plant. Although M seedlings of *Pinus radiata* showed greater uptake of $^{32}P$ than nonmycorrhizal seedlings, no such difference was found for *Nothofagus menziesii.*

The relationship between ECM and host-plant development is further complicated by the existence of peritrophic mycorrhizae (PM)—a type of symbiosis in which the hyphae of the fungal symbiont only envelop the rootlets but do not invade their tissues (49). Wilde (160) claims that abundant ECM is an incidental phenomenon largely confined to seedlings grown under high light intensity such as in greenhouses, nurseries, and young plantations. He maintains that PM dominate in older forests in Wisconsin. In Poland abundant PM have been reported on roots of healthy 60- to 200-year-old oak trees (115). Roots of beech at the Baltic coast also were found to be surrounded but not invaded by hyphae of ECM-forming *Cenococcum graniforme* (13). Without establishing a symbiotic relationship, growth of endotrophic tree species was stimulated by mycelium known to form ECM (64). Similarly, growth of Scots pine was stimulated by a nonsymbiotic strain of ECM-forming *Xerocomus (Boletus) subtomentosus.* This stimulatory effect exceeded that of N fertilization alone (4). The distribution and importance of PM in other parts of the world deserve further attention. Since the fungus is restricted to root surfaces, PM may be of greater benefit to the host than to the fungus. If this is true, then in older forests the host plant dominates the fungal associate. The mechanism of formation and significance of PM needs to be studied.

### *Organic Nutrients*

Contrary to Frank's (27) opinion that ECM of forest trees is formed only in soils containing humus, it has been proven that ECM also develop in exclusively inorganic media. In nature, however, most ECM occur in the upper humus layer (*42, 71, 88, 131*).

Experiments with organic fertilizers show that humus stimulates both M formation and host-plant growth. This has been demonstrated with composts (*71,* 120, *131*), plowed-in heather and lichens (17), decomposed cow and horse manure (*131*) and peat (31).

Humus from established forest stands inoculated in steppe soils enhanced M formation of oak seedlings (*71, 131*). At first, this was attributed to M fungi introduced with the humus, but steppe soils, especially in the Ukraine, already harbor M-forming fungi (95, 169). Mikhowitsch (95) showed that addition of forest humus stimulates M infection of oak seedlings in humus-poor soil but not in humus-rich chernozem. Mirchink (*131*) found that pine seedlings did not form M in soils containing less than 2% humus, whereas 60 and 83% of the seedlings had M when the humus content was 2.2, 2.7 and 4%, respectively. In Czechoslovakia, field surveys revealed that the type and amount of humus was decisive for formation of beech M (86). Differential effects of various types of humus have been shown with pine seedlings grown in blocks of woodland and hyphnum peat. Both media were fertilized with the same combination of N, P, and K, but woodland peat produced greater seedling dry weight and at least three times more M (131).

Since most ECM are formed in the humus layer (*42, 88, 131*), questions have been raised about the ability of M fungi to utilize complex organic matter. In pure culture, most ECM fungi apparently cannot utilize cellulose and lignin (*36*). But isolates of *Xerocomus* (*Boletus*) *subtomentosus* and *Lactarius delicious* produced as much extracellular polyphenol oxidase as wood-destroying fungi (68). Also, *Tricholoma fumosum* produced enough cellulase to use cellulose as a carbon source, and two other *Tricholoma* spp. were also able to decompose cellulose slightly if a small amount of "start" glucose was provided (114). More recently, the number of species of ECM fungi known to produce extracellular cellulase in pure culture has increased (73, 75, 125). However, except for a *X. (Boletus) subtomentosus* strain (73) their enzymatic activity was found to be far below that of litter-decomposing fungi (75).

Similarly, pure cultures of M fungi apparently are deficient in enzymes essential for decomposition of humus-bound N (*42, 71, 88, 131*). Among 26 species of ECM and litter-decomposing fungi only one strain of M-forming *Xerocomus (Boletus) subtomentosus,* known to be a facultative symbiont, utilized $^{15}N$ bound in raw humus. The remaining M formers and also about half of the litter decomposers apparently were poorly adapted for utilizing this N source.

Several investigators (17, 18, 20, 21, 158, 159, 165) maintain that M fungi decompose litter in their natural habitat. Indeed, there is a vast difference between an artificial medium in vitro and nutrients provided in natural habitats by the soil, the soil-inhabiting microorganisms, and host plant root exudates. A similarly vast difference may exist between isolated M fungi and M. In M the symbiotic partners may cooperate to form adaptive enzymes and other metabolites not produced by either associate alone. For example, the pathogenic fungus *Pythium debaryanum* produced pectolytic enzymes only in the presence of plant extracts (35). In soil where more than one carbon source is available, lignin and cellulose may be decomposed more rapidly than when provided to pure cultures as sole carbon source (11). Since "start" glucose enabled *Tricholoma vaccinum* to form a small amount of cellulase in vitro, Norkrans (111) postulated that cellulolytic enzymes are adaptive and that litter-decomposing and M-forming species differ quantitatively, rather than qualitatively, in their ability to produce cellulase.

Lundeberg (74) showed that M-forming *Suillus (Boletus)* spp. and *Rhizopogon roseolus* caused mineralization of humus-bound $^{15}N$ when grown in association with Scots pine seedlings. The most efficient in this respect were *S. granulatus* and *S. variegatus.* In the absence of a host plant, however, these two species assimilated little or no N from the humus. Lundeberg also suggested that intracellular enzymes leached from dead and autolyzed hyphae may have participated in the humus mineralization. However, good mycelial growth was obtained in pure culture with the above two *Suillus* species and also with *S. bovinus* on media consisting of 1.5% glucose and either an extract of shed pine needles or sterilized forest litter as the sole N and P source (18). Similar results were reported in a medium of peat and needles of pine or fir with a 2% glucose supplement (*18*). Differences in nutrient supplied and strains used may have caused some of the apparently contradictory results discussed above. For instance, recently a strain of M-forming *X. (Boletus) subtomentosus* was found which was unable to form M (73) but did utilize raw humus efficiently (4, 73).

More research is needed to explain the stimulatory influence of humus on M formation and host-plant growth. The view that stimulation is related to mineral nutrients, particularly phosphoric acid (12) seems to be an oversimplification, since Melin and others (*44, 88*) have demonstrated a growth-promoting effect of humus on mycelia of ECM fungi that is not related to mineral nutrients. Due to microbiological processes, humus may contain (*a*) various organic breakdown intermediates, (*b*) growth-stimulating and -regulating substances released by soil microflora, and (*c*) a myriad of other organic compounds leached from dead microbial cells. Since many of these organic compounds can be assimilated directly by M fungi, they probably affect both symbiotic associates. Introduction of humus may also improve the soil reaction, temperature, aeration, and its moisture content.

## EFFECT OF SOIL pH

ECM fungi are generally assumed to be acidophilic. Most species tested in pure culture have shown maximum mycelial growth at pH 4–6 (104). But some *Suillus* species prefer pH of 3 (47), some *Paxillus involutus* strains pH 3.1 or 6.4, although some of these strains showed good growth even at pH 2.7 (63). M fungi that thrive over a wider pH range include the following: *Cenococcum graniforme,* 2.4–7; *Corticium bicolor,* 2.9–5.7 (96, 97); and *Armillaria robusta,* 3–7 (62).

The pH of soil adjacent to sporophores of ECM fungi in Sweden corresponds closely to their preferred pH in pure culture (104, 111). The optimum pH for M formation in vitro frequently corresponds with that for growth of mycelium in pure culture (85, 132). But the pH and growth relationships for ECM fungi in culture do not necessarily reflect the tolerance of these fungi to soil reaction in natural habitats. In the Soviet Union pine M were found in soils ranging from acid mor with pH 3.8 to alkaline chestnut-brown and humus-chalk type soils with pH above 8. In the latter soil even rootlets in direct contact with the chalky marl showed M. Oaks growing in alkaline chestnut-brown soil also had abundant M (*71*). In Sweden, pine, spruce, and birch have M in mor-type soils with the pH often above 7 (1), and in Poland profuse M development occurs in soils with pH 7.5 (14). Forest nurseries in Nebraska (USA) with soil of pH 7.8 produced M seedlings of ponderosa pine (33).

Melin (88) suggested that fungi forming ECM in slightly acidic and neutral soils differ from those in acid soil. Support of this view is provided by the fact that white-colored ECM and brown-colored ectendomycorrhizae were formed with *Pinus radiata* seedlings in podsolic sand at pH 4.5 or 6.2, but only the latter type was formed when the soil pH was adjusted to 8 (145). But the pH requirements of the ECM fungi can be influenced by nutrients. *Suillus (Boletus) granulatus* and *S. variegatus* isolates produced maximum mycelial growth at pH 5–6 in medium composed of inorganic salts, glucose, asparagine, peptone, and thiamine (104); but in a medium containing inorganic salts, glucose, asparagine, thiamine, and a specially prepared potato extract, maximum growth was obtained at pH 3 (47).

Variation in pH optimum among different strains of the same species suggests that M fungi may adapt to soil pH. Nine strains of *Paxillus involutus* grown in identical nutrients had optimum pH values from 3.1–6.4 (63). Also, *Suillus granulatus,* known to be acidophilic in pure culture (47, 87, 104) and commonly associated with pine in acid forest soils of the northern hemisphere, associates with *Pinus halepensis* on lime rocks in the Mediterranean area and in alkaline soils of La Pampa Province, Argentina (99). In some cases, however, even a slight change in soil pH can effect M infection. Tserling found that a shift from pH 7.4 to 7 increased the percentage of M larch seedlings from 0.8 to 6.5 (*131*).

Recently, Richards & Wilson (124) have questioned the view that sparse development of ECM in neutral or alkaline soils is due to the acidophilic nature of M fungi. They suggest that indirect effects of soil pH on the availability of soil N may be more important to M formation than the direct effects on the fungi per se. *Pinus carribaea* seedlings, potted in topsoil taken from a young pine plantation, produced more M at pH 7.5 with less soil N than at pH 5.8 with more N. The latter treatment increased the N content of the roots. They ascribe the inhibitory effect of alkaline soils on M formation to increased availability of N (cf. 2, 44) resulting from enhanced nitrification, rather than from high pH per se. Also, Richards (123) concluded that normal M development with *Pinus taeda* is possible at pH 7.2 and 7.5, provided that Fe deficiency is corrected and soil nitrate concentration is kept low. He also proved that the inhibitory effect of $NaNO_3$ on M formation is due to the $NO_3$ ion, since addition of $Na_2CO_3$ did not reduce M infection although the soil pH increased. Since only forked M were counted in the above experiments, the statistical data may not fully reflect M development. A recent report on basswood M (117) shows that in ten sites with soil pH from 7.1–8.4, no correlation was found between either the soil pH and the $NO_3$-N content or the prevalence of M and the $NH_4$-N content of the soils. Theodorou & Bowen (145) have attempted to clarify the effect of soil pH and N availability on colonization and infection of pine roots by M fungi. They infer that in acid soils with abundant $NO_3$, M formation is hampered mainly by the inhibitory effect of $NO_3$ on infection rather than on the external colonization of root surfaces, whereas regardless of N concentration in alkaline soils, inhibition of M development is caused mainly by the inhibitory effect of soil pH on surface colonization. These conclusions must be questioned because of the shortness of their experiment (4 weeks) using pine seedlings at radicle stage (cf. *42*) and the high mortality in certain treatments (up to 75%).

## EFFECT OF TEMPERATURE

Optimal temperatures for colonization and infection of roots by M fungi may differ considerably from those for mycelial growth in culture. Optimum temperatures for mycelial growth in culture vary with the medium used (38) and have limited value in predicting the behavior of the fungus in the rhizosphere (*9*). A decrease in temperature from 25 to 16°C reduced the mycelial growth of a *Rhizopogon luteolus* strain in pure culture by 50%, while almost no hyphal growth was observed in the rhizophere of *Pinus radiata* seedlings (147). On the other hand, a strain of *Suillus granulatus* showed markedly greater colonization of root surfaces at 16°C than expected from its mycelial growth in pure culture at this temperature. Although mycelial growth of *Pisolithus tinctorius* in pure culture was maximal at 28°C, M frequency with *Pinus taeda* increased at temperatures above 14°C with maximum infection at 34°C (84).

In natural habitats, M formation often occurs at temperatures well below those required for maximum growth of mycelium in vitro. At 10–15°C, mycelial growth of *Cenococcum graniforme* in culture is poor, yet this fungus is common in spruce forests of northern Finland where the ground remains frozen until midsummer and soil temperatures rarely reach 20°C (96). Formation of ECM may begin in the spring when the temperature in the upper soil layer reaches 10–11°C (*71*), and in some instances root growth and M infection occur at soil temperatures as low as 3–9°C (112).

Although M infection may occur at a relatively low temperature, higher temperatures generally accelerate this process. At 16–18°C, both M frequency and number of M oak seedlings was nearly twice that observed at 12–14°C (*131*). Under laboratory conditions, M frequency of *Pinus taeda* inoculated with *Pisolithus tinctorius* increased above 14°C and was maximum at 34°C (84). *Rhizopogon luteolus* formed M with *P. radiata* more rapidly at 20 and 25°C than at 15°C.

It is not known whether this was due solely to enhanced colonization or to stimulation of infection (147). A different pattern was shown by *Telephora terrestris;* this fungus converted about 45% of the *P. taeda* short roots into M at 14–20°C and only 30% at 29°C. No M were formed at 34°C (84). In coal waste areas of Pennsylvania (USA) ECM formed by *Pisolithus tinctorius* and other M fungi tolerate soil temperatures that may exceed 60°C (129). In the laboratory, hyphae of *P. tinctorius* were killed in 48 hr at 45°C. Temperatures of 28–38°C for 48 hr killed hyphae of *Suillus granulatus* and *Rhizopogon roseolus* (*9*).

Mikola (96) suggested that M fungi adapt to temperatures. *Cenococcum graniforme* occurs over a wide range of climatic zones (149). Mycelia of strains of the same M fungus grow at different rates on same medium at the same temperature (84, 96, 107, 111, 147). Strains of M-forming *Suillus variegatus* and *Paxillus involutus* obtained from sporophores growing in valleys grew at 2–8°C while those obtained from mountains grew at –2 to 4°C (107). Similar differences between strains in root colonization efficiency have been reported (147).

In natural habitats, soil temperature may be selective for the fungal associate in M (30, 107). This suggests that in attempts to establish forests in new areas the soil should be inoculated with M fungi having the proper temperature tolerances. This approach has been very promising in the Austrian Alps where protective shelterbelts are needed to prevent avalanches (106).

Since formation of M involves complex metabolic interactions between fungus and host, any change in root physiology induced by temperature could affect both colonization and infection of roots. Soil temperature affects the amount and composition of root exudates (*127*) as well as the elongation and maturation of roots and thus the length of the M infection zone (79). A departure of 2°C in soil temperature from the ca 18.5°C optimum for growth of *Betula (lutea) alleghaniensis* seedlings resulted in a considerable decrease in M frequency (122). Because this change increased rootlet mortality (60% compared with the normal 6%), M infection in this case was presumably affected more by disturbances in root physiology than directly by temperature.

## EFFECT OF SOIL MOISTURE AND AERATION

Soil moisture is very important for M formation. Although soils in the steppes of the Ukraine contain M fungi (155, 169), M are rare on planted tree seedlings because of insufficient soil moisture (71, *131*). At 30–60% of field capacity, oak seedlings potted in these soils develop M (169). Similarly, deep soaking of the ground twice during the growth season induced abundant M formation on planted oak seedlings (77). Frequency of M in steppe soils increases during rainy years (103, 128, 155), but only 5% of oak seedlings showed M when the upper soil layers were markedly dessicated (34).

ECM generally are more abundant in moist locations and lowland areas than in drier sites (140, 155). With daily watering of potted *Pinus virginiana* seedlings, about 83% of short roots became M compared to 2.4% with watering every fourth day (162). Soil moisture also affects EDM formation (60, 133).

Excessive soil moisture affects M formation severely. In nutrient solutions *Suillus* spp. formed deep intracellular infections and no Hartig net in attached (72) and excised (Melin & Slankis, unpublished) roots of Scots pine.

In natural habitats, however, some M relationships endure prolonged exposure to excess moisture. Stagnant water only partly inhibited M formation in spruce; new M were still formed although without mantles (161). After six months of flooding, *Liquidambar* had 26% EDM roots with viable fungi compared with 31% in non-flooded areas (22). By contrast, *Pinus radiata* M were significantly affected after 4–16 weeks of flooding (28).

In soils with excess water, oxygen deficiency limits development of both the symbiotic fungi and tree roots. M fungi are strongly aerobic (42, *88*). Respiration rates in M are higher than in noninfected roots (*88,* 98). Dogwood roots are severely injured after one week of flooding (118). Roots of yellow poplar and *Ligustrum* may be killed within three weeks (56) but roots of loblolly, shortleaf, and pond pine are highly resistant to flooding (48). Prolonged anaerobiosis profoundly changes root physiology, increasing phosphate leakage (*42*), decreasing permeability to water and nutrients, arresting growth, and eventually killing roots (57).

The tolerance of M for excess moisture depends on the physiological state of the roots and their ability to supply oxygen to the fungus. Read (121) found that M fungi colonized only those root regions of conifer roots that evolved significant amounts of $O_2$. After two weeks of water logging, the succinic dehydrogenase activity in M of Douglas fir decreased and they absorbed less $^{32}P$ than control M roots (28).

In Finland's water-logged peat bogs, tree M occur only near the soil surface and are often dark and slender (45). After drainage, however, their vertical distribution increased and their appearance improved (114). Improving aeration in forest soil by adding sand increased the number of M larch plants from 0.8 to 24% (*131*). M fungi differ in tolerance of soil moisture. *Cenococcum graniforme* grew on laboratory media at much lower humidity than *Rhizopogon luteolus* (*9*). Colonization of *Pinus radiata* roots by *R. luteolus* declined both above field capacity and below 50% field capacity (9).

But moisture deficits that decreased the abundance of M formed by other symbiotic fungi did not decrease the frequency of M formed by *C. graniforme* (93). This may explain why imposed moisture deficiency can alter the fungal species found in M (50, 162). Successive M associations were found rather frequently with *P. radiata* (78). This phenomenon needs to be studied further.

## EFFECT OF FUNGAL GROWTH HORMONES AND PLANT ROOT EXUDATES

Björkman's (2) view that the concentration of soluble sugars in roots is the sole factor controlling ECM formation has been challenged recently (*43, 139*). From experiments with *Fugus sylvatica* seedlings, Meyer (92, 94) concluded that increased content of reducing sugars in M is not the cause but the result of M infection and is due to the hydrolyzing action of the fungal auxins on carbohydrates. Also, no correlation was found between the abundance of M and the amount of reducing sugars in roots of *Pinus taeda* and *P. carribaea* (124).

In the above experiments, soluble sugars were not analyzed individually; but their total amount was assumed to correspond to the amount of reducing substances in roots. Thus, the data obtained may not reflect the actual sugar economy in the roots (cf *43*).

In a recent experiment (70) in which $^{14}C$-labeled sugars were detected chromatographically, free soluble sugars were present in roots of *Pinus strobus* seedlings grown in Björkman's (2) nutrients with either very low (2.5 mg/l) or very high (265 mg/l) N concentrations. The radioautograms showed that soluble sugars were present in the root systems of all test seedlings. Excepting sucrose, which showed a tendency to decrease as the concentrations of N and P increased, the amounts of radioactive glucose, fructose, and raffinose were greatest in roots of seedlings that received either a moderate (173 mg/l) or very high (692 mg/l) P concentration in combination with the highest N concentration (265 mg/l), i.e. concentrations which in Björkman's experiments (2) arrested M infection. Also in short roots, whether from M low-N seedlings (2.5 mg/l) or from uninfected high-N seedlings (265 mg/l), the amounts of sucrose, glucose, and fructose were very similar (138).

With the discovery that ECM fungi produce extracellular growth hormones (*139*) the role of hormones in ECM formation mechanism has become increasingly apparent: (*a*) ECM-like root structures can be induced in pine roots by either exudates from ECM fungi (135) or by supraoptimal concentrations of synthetic auxins (116, 136); (*b*) stopping a periodic supply of exudates or auxins results in renewed elongation of swollen M-like short roots followed by formation of root hairs (135, 136); and (*c*) ECM undergo similar reverse morphogenesis when mycotrophic pine seedlings are subjected to N concentrations that arrest new M formation and, concurrently the symbiotic relationship also terminates (137).

Indoleacetic acid and other indole compounds occur abundantly in M roots of pine seedlings (76, 143), but only traces are present in nonmycorrhizal roots (143). This indicates that the excess auxin in M results from the presence of the fungus. Palmer (116) found that an external supply of synthetic auxins at supraoptimal concentrations enhanced M formation with *Pinus virginiana* seedlings. The above findings support my view that the specific structures of ECM (*79*) reflect a specific physiological and metabolic state that is induced and maintained by the symbiotic fungus auxins and that this state is a prerequisite for the establishment and functioning of the symbiotic relationship (137). The above data also suggest that only suboptimal N concentrations enable the fungus to impose the symbiotic relationship on the host plant, because such conditions presumably stimulate the fungus to produce extracellular auxin in sufficient amounts.

In the creation of hyperauxiny in M roots, the symbiotic fungus may be assisted by the synergistic action of factors other than N: (*a*) ECM fungi produce extracellular inhibitors that inactivate auxin oxydases exuded by pine and birch roots (126, 147a); (*b*) Fortin (25) reported that excised pine roots rapidly absorb and accumulate IAA from nutrient solutions. Since high auxin concentrations inhibit mycelial growth of M fungi (24a) he suggests that this removal of auxin assures continuous auxin production by the fungus without causing autoinhibition.

The mode of action of low and high N concentrations on the fungus-auxin production in the rhizosphere and in M roots of the host plant is still unknown. In pure cultures auxin production evidently decreases with increased supply of both inorganic N (108, 147a) and organic N (108). Isolates of *Suillus bovinus* produced IAA abundantly at $NH_4NO_3$ concentrations of 100 mg and 350 mg/l but 1000 mg/l totally inhibited auxin production despite the luxurious growth of the mycelium (147a). Also an increased supply of certain amino acids suppressed utilization of tryptophan as precursor for auxin synthesis (108).

In the rhizosphere, the organic N supply for the fungus, before infection, is derived mainly from the host plant in the form of root exudates. Almost 50 years ago Melin (87) suggested that the virulence of ECM fungi is affected by root exudates. Variations in their composition may stimulate or inhibit fungus virulence and auxin production and thus control M formation. Evidently, roots of forest trees exude many organic compounds, such as sugars, amino acids, vitamins, and a large number of organic acids, cytokinins, and a stimulant designated as M factor[4] (*9, 139*). Some amino acids and certain combinations of them stimulate as well as strongly inhibit mycelial growth (*88*) and auxin synthesis (108) by ECM fungi. In culture the amount of tryptophan (108, 152) and the concentration of certain N sources (108, 147a) determine whether or not extracellular auxins are produced. However, M fungi are able to synthesize auxins from compounds other than tryptophan (46, *142*). Also increased cytokinin concentrations can inhibit while low concentrations stimulate auxin synthesis (32).

M development in pine roots was more closely correlated with the ratio of soluble sugars to total N in the roots than to soluble carbohydrates alone (123, 124). An increase in exogenous inorganic N concentration increases the number of different amino acids in root systems (70) and especially the amount of certain amino acids in short roots (138) of *Pinus strobus* seedlings. Any change in nutritional conditions affects both the rate of root exudation and the composition of exuded metabolites; both in turn exert different effects on the fungus. For instance, $PO_4$ deficiency increased exudation of amino acids and amides from roots of *Pinus radiata* seedlings while $NO_3$ deficiency greatly reduced it (6). Termination of the symbiotic relationship in established ECM with an increase in N concentration (137) indicates that also intracellular root metabolites affect the fungal associate from the moment its hyphae invade the root tissues. Presumably, the rejection is due to the inhibitory effect of high N on auxin synthesis by hyphae in M roots as well as by hyphae in the soil. Also, the fact that synthetic auxin induced more pronounced ECM-like swellings in roots of *Pinus strobus* seedlings at higher N concentration (138) suggests that in natural habitats high N more likely inhibits the fungus-auxin synthesis rather

[4]Nilsson (*91*) obtained evidence that the stimulatory effect of the M factor can be replaced by diphosphopyridine nucleotide. Gogala (*32*), however, maintains that the M factor is related to cytokinins.

than inactivates the auxin produced. Also, the increased cytokinin concentrations in the host plant inhibit auxin production by the fungus (32). Whether increased N concentrations concurrently stimulate production of fungistatic substances (*79,* 91a) by the host roots needs to be elucidated.

Apparently the specific physiological state in ECM is induced through an interaction between the fungus-auxin and root metabolites (32, *139*). The composition root metabolites, on the other hand, is controlled by chemical, physical, and microbiological factors in the soil as well as by light intensity. For instance, roots of Scots pine explants formed M when organic nutrient was supplied through the still attached remnant of the hypocotyl (24). Short roots of excised and attached root systems of Scots and white pine, when subjected to supraoptimal auxin concentrations, showed great variations in the formation of M-like swelling and forking even within the same root system (*139*). At low light intensity (5% of daylight), supraoptimal auxin concentrations did not induce M-like swellings even with an adequate supply of sugars in the nutrient solution, but such swellings were readily formed at 25% light intensity without a sugar supply (*138*). This finding and others (40) contradict Björkman's (2) view that inhibition of M formation at low light intensity is caused only by sugar deficiency in the roots due to reduced photosynthesis.

Recent discoveries that some ECM fungi in culture produce cytokinins (32, 101) lead to speculation that these hormones known to be linked with many metabolic processes (*142*) also may regulate M formation and greatly influence the growth and development of host plants. An excess of these hormones in infected roots should profoundly influence their physiology and metabolism thus complicating further our understanding of M formation mechanisms.

## EFFECT OF RHIZOSPHERE

Harley (41) suggested that nonmycorrhizal soil organisms can stimulate plant growth. Extensive rhizosphere studies have shown not only that this is true (*8*) but also that they influence M formation. A comprehensive review on this subject was published recently by Rambelli (119).

Forest soil enhanced M formation and seedling growth more than soil inoculated with pure cultures of ECM fungi (*71, 119, 131*). Inoculation with *Trichoderma* and *Azotobacter* promoted M formation with oak and pine seedlings (77, 151, 154, 156). *Trichoderma lignorum* was more stimulatory than *Azotobacter* on pine. Greatest development of oak M was achieved with the above organisms combined with fluorescent bacteria (131). *Azotobacter chroococcum, T. lignorum,* and *Pseudomonas fluorescens* are closely associated with oak roots throughout the year (77). Similarly, axenic cultures of clover, inoculated with disinfected spores of EDM-producing *Endogone,* were only slightly infected unless a third organism, a species of *Pseudomonas,* was present in the rhizosphere (109). Failure to establish a symbiotic relationship unless a third organism is present suggests that some fungi found to be nonsymbiotic in vitro might establish a symbiotic relationship in natural habitat.

The stimulatory effect of soil microorganisms on M infection is attributed to the extracellular substances they release. Of 153 cultures of different soil microbes tested, 73 cultures produced thiamin and 65 produced auxins. The most efficient producers were *Pseudomonas* and *Azotobacter.* Among freshly isolated actinomycetes, 80–85% produced Vitamin $B_{12}$ (58). *Azotobacter* also produces gibberillic acid (153). ECM-forming Basidiomycetes are thiamin heterotrophs (*88*); thus a supply of this vitamin from soil microorganisms could greatly stimulate fungal growth and M formation.

Some soil microorganisms inhibit M formation. Rayner (*120*) demonstrated that pine and some other conifer seedlings planted in heathland soil failed to establish M even though the soil harbored the ECM fungus *Suillus (Boletus) bovinus.* This inhibition has been attributed to glyotoxin produced by certain *Penicillium* species (10) and to the antagonistic action of the *Calluna* endophyte (39). M formation also may be inhibited by rhizosphere-inhabiting pathogenic bacteria (134), actinomycetes (110), and other fungi (66). Theodorou (144) succeeded in establishing symbiosis between *Pinus radiata* and *Rhizopogon luteoulus* in nursery soil only after eliminating antagonists by partial soil sterilization.

The microbial flora in the mycorrhizosphere (zone in the soil surrounding M roots) often differs from that in the rhizosphere of nonmycorrhizal roots (*42, 71, 119, 131*). For instance, pine mycorrhizosphere contained 9–10 times larger populations of fungi (151), and that of yellow birch contained about 10 times larger bacterial populations (51) than the rhizosphere of non-M roots.

Difference also exists in microbial populations in the mycorrhizosphere and that on M surfaces. Thus, only 12,000–60,000 denitrifying bacteria per gram were found in the mycorrhizosphere of oak M while the number reached 2.5 million on M surfaces (157). The outer layers of *Pinus radiata* M mantle had bacterial population 16 times larger than that present in the outermost region of the mycorrhizosphere (26).

Foster & Marks (26) found that different types of microbial populations associated with different types of M on the same host plant. They suggest that mycorrhizal roots exert selective action on the rhizosphere-inhabiting microorganisms. Such selectivity is also reported by others (23, 26, 51, 113). Apparently, the specific composition of microflora in the mycorrhizosphere and on the surfaces of M develops under the influence of two main factors: the root exudates and extracellular metabolites of M fungi. Roots exude numerous and diverse substances (*9, 119, 127*); their influence on soil microflora is profound (*127*). For instance, they can induce morphological and physiological changes in *Azotobacter* (58) and can inhibit *Rhizobium* nodulation (19). ECM fungi excrete growth-promoting substances (32, 101, 108, *131, 139,* 152) and antibiotics (*61, 81, 166*), some of which are volatile (59). Such antibiotics have been detected in both M and in the mycorrhizosphere (*59, 81*).

Khudyakov (54) has proposed that soil bacteria not only break down organic matter but also transport the breakdown products via cells to the roots. He has designated this mode of symbiosis as bacteriotrophy and presumes that a sufficient density of bacterial cells exists in the soil so that bacteriotrophy operates together with mycotrophy.

## CONCLUDING REMARKS

The information reviewed in this chapter and in the recently published comprehensive review on EDM (109a) shows that during the last few decades our knowledge about M symbiosis has advanced considerably. Due to this progress ECM is now successfully employed in forest nurseries to obtain healthy, large size M seedlings and to establish new plantations in previously treeless areas or in areas with adverse ecological conditions (100). Similarly promising but less advanced is the practical application of ECM in agriculture (109a).

The reviewed data on ECM also provide deeper insight into the symbiotic relationship. Although this symbiosis generally is viewed as mutual it does not necessarily mean that the benefits are in equilibrium. It seems that under certain nutritional conditions one of the symbiotic partners may gain the upper hand. Melin (91) emphasized that M fungi are parasites and that their aggression is curbed to mutual symbiosis by the host plant's protective measures. The termination of the symbiotic relationship when N supply reaches the optimum for the host (137) confirms the view of earlier investigators (2, 44) that the fungus gains entrance only at suboptimal nutritional conditions, namely, when its aid to the host in obtaining soil nutrients is essential. It follows, however, that with gradually improving nutritional conditions the host plant begins to dominate while the virulence of the fungal associate and presumably also its beneficial gains decrease. Finally this dominance can reach the stage when the arrested M roots regain elongation and the fungal associate is totally rejected from the invaded roots. A reverse situation may occur when moderate nutritional conditions for the M host change to severe deficiency. Then, apparently, the fungal associate begins to dominate; new M with well-developed mantles continue to form in abundance despite the suppressed growth and severe nutritional deficiency symptoms of seedlings (15, 16, 137). Dominik (15) has reported a case in forest nurseries where, under adverse growth conditions, the dominance of M-forming *Suillus (Boletus) luteus* became fatal for M pine seedlings.

The above findings suggest that in natural habitats the beneficial gains of ECM symbiosis are not necessarily always in equilibrium for both associates and that this imbalance cannot be determined from the prevalence of M only. Also it has become apparent that intelligent regulation of nutritional conditions is one of the ways by which the maximum benefits from this symbiosis may be assured to the host. To achieve full control of this symbiosis, our knowledge of several aspects of M needs to be augmented. Of particular importance is further information about the physiological and metabolic factors that control fungal colonization of root surfaces, their invasion into roots, and development of the Hartig net. Evidently, the ECM-formation mechanism is far more complex than earlier anticipated since metabolites of both symbiotic partners interact in this process. Therefore, more information is needed about host plant metabolites that stimulate or inhibit hormone production by the fungus and determine whether or not these hormones induce and maintain the specific physiological state in infected roots. Since the composition of endogenous and extracellular metabolites produced by hosts and M fungi is greatly influenced by nutrients, this important relationship requires further investigation to obtain more information about the metabolites produced. In natural habitats, this relationship is very complex due to the presence of metabolites released by the rhizosphere-inhabiting microflora. Although these metabolites significantly influence M formation, their mode of action on both symbionts is little understood.

Recent findings indicate that, in addition to soil factors and light intensity, the formation and also the prevalence of M are influenced by (*a* the age (92, 102, *131*), provenance (69, 73, 163), and parentage of the host (83), and (*b*) the efficiency of M fungi that may differ between species (37, 65, 106, 146, 164), and their strains (63, 67, 74, 146, 164). To avoid inconsistency in experimental results affected by so many variables and to facilitate progress in M research, standardization of at least some experimental methods is needed urgently. The recently suggested principles for classification of M fungi and maintenance of their isolates in stock cultures (150, 167) deserve serious consideration. More attention should be given to the genetic origin of seeds used for experiments. Alternately, since refrigerated forest tree seeds stay viable for years, large quantities of seeds from one source could be collected and stored for future experiments and to be available for other researchers. To avoid differences in nutrient supply the amount of potted soil and the volume of nutrient solution per seedling should be standardized. Also the nutritional value of the potted soils and the composition of soil-inhabiting microflora should be known.

The many methods used to express M prevalence (*131*) also produce confusing inconsistency in experimental results. At present, the most acceptable method is to express M frequency as the number of mycorrhizal roots per unit of root length (2), provided that statistics about the total number and length of long roots of the root system are given. The occurrence of dichotomy still creates a most serious problem in this respect. Should a forked or a repeatedly forked M root be considered as a simple M or should each of the multiple tips be counted? From experiments with synthetic auxins it is apparent that the frequency of root dichotomy in pine is largely determined by the hereditary properties of the seedlings and that considerable variation among short roots may occur even within the same system whether excised or attached (136, *139*). J.A. Menge suggests that repeatedly forked M are formed more readily by certain M fungi under certain nutritional conditions (personal communication). For these reasons it would be advisable in the future to provide separate statistics about the simple and branched M as well as about the number of tips formed by the latter.

Satisfactory solutions to the above problems will contribute greatly to the advancement of research on mycotrophy.

ACKNOWLEDGMENTS

I am very indebted to Dr. W. R. Henson, Director of the Research Policy Branch, and Mr. D. H. Burton, Director of the Forest Research Branch, Ministry of Natural Resources, who kindly provided me with office space after my retirement, access to library facilities and who were generous with their assistance whenever necessary. This help greatly facilitated the preparation of this chapter.

*Literature Cited*


1. Björkman, E. 1941. Die Ausbildung und Frequenz der Mykorrhiza in mit Asche gedüngten und ungedüngten Teilen von entwässertem Moor. *Medd. Skogsförsöksanst* 32:255–96 (In Swedish, English summary)
2. Björkman, E. 1942. Über die Bedingungen der Mykorrhizabildung bei Kiefer und Fichte. *Symb. Bot. Upsal.* 6(2):1–191
3. Björkman, E. 1956. Über die Natur der Mykorrhizabildung unter besonderer Berüchsichtigung der Waldbaüme und die Anwendung in der forstlichen Praxis. *Forstwiss. Zentralbl.* 75:265–86
4. Björkman, E. 1970. Mycorrhiza and tree nutrition in poor forest soils. *Stud. Forest. Suec.* 83:1–24
5. Boullard, B. 1968. *Les Mycorrhizes.* Paris: Masson. 135 pp.
6. Bowen, G. D. 1969. Nutrient status effects on loss of amides and amino acids from pine roots. *Plant Soil.* 30:139–42
7. Bowen, G. D. 1972. Mineral nutrition of ectomycorrhizae. In *Ectomycorrhizae. Their Ecology and Physiology,* ed. G. C. Marks, T. T. Kozlowski, 151–97. New York: Academic. 444 pp.
8. Bowen, G. D., Rovira, A. D. 1969. The influence of micro-organisms on growth and metabolism of plant roots. *Root Growth.* In *Proc. 15th Easter Sch. Agr. Sci. Univ. Nottingham,* ed. W. J. Whittington, 170–201. London: Butterworth
9. Bowen, G. D., Theodorou, C. 1972. Growth of ectomycorrhizal fungi around seeds and roots. See Ref. 7, pp. 107–45
10. Brian, P. W., Hemming, H. G., McGowan, J. C. 1945. Origin of a toxicity to mycorrhiza in Wareham heath soil. *Nature* 155:637–38
11. Cochrane, V. W. 1958. *Physiology of Fungi.* New York:Wiley. 524 pp.
12. Crowther, E. M. 1951. Nutrition problems in forest nurseries (Summary report on 1949 experiments). *Rept. Forest Res., London.* 1949–1950:97–105
13. Dominik, T. 1957. Investigations on the mycotrophy of beech associations on the Baltic coast. *Ecol. Pol. Ser. A* 5:213–56 (In Polish, Transl. Sci. Publ. Foreign Coop. Center, Cent. Inst. Sci. Technol. Econ. Inform., Warsaw, 1964)
14. Dominik, T. 1961. Investigations on the mycotrophy of plant associations in the Pieniny National Park and on Mt. Skalka above Lysa Polana (Clearing) in the Tatra, with special reference to the mycotrophy of relict pines. *Prace Inst. Badawcz. Leśn.* 208:31–58
15. Dominik, T. 1961. Studies on Mycorrhizae. *Folia Forest Pol. Ser. A,* No. 5:1–160 (In Polish, Transl. Sci. Publ. Foreign Coop. Center, Cent. Inst. Sci., Technol. Econ. Inform., Warsaw, 1966)
16. Dumbroff, E. B. 1968. Some observations on the effects of nutrient supply on mycorrhizal development in pine. *Plant Soil* 28:463–66
17. Eglite, A. K. 1955. Experimental mycorrhization of pine. In *Mycotrophy in Plants. Trudy Konf. Mikotrofii Rastenii.,* ed. A. A. Imshenetskii, 200–10. Moscow:Acad. Nauk SSSR. 312 pp. (Transl. from Russian, Israel Program Sci. Transl., Jerusalem, 1967)
18. Eglite, A. K. 1958. Slightly soluble minerals and organic substances as nutrients of mycorrhizal fungi. *Tr. Inst. Mikrobiol. Akad. Nauk Latv. SSR.* 7:67–75 (Transl. from Russ., Israel Program Sci. Transl. Cat. No. 668, Jerusalem, 1963)
19. Elkan, G. 1961. A nodulation-inhibiting root excretion from a non-nodulating soybean strain. *Can. J. Microbiol.* 7:851–56
20. Falck, R., Falck, M. 1954. *Die Bedeutung der Fadenpilze als Symbionten der Pflanzen für die Waldkultur.* Frankfurt: Sauerländer. 92 pp.
21. Fassi, B. 1963. Die Verteilung der ektotrophen Mykorrhizen in der Streu und in der oberen Bodenschicht der Gilbertiodendron-Dewevrei-(Caesalpiniaceae) -Wälder im Kongo. In *Mycorrhiza. Proc. Int. Mycorrhiza Symp. Weimar,* ed. W. Rawald, H. Lyr, 297–302. Jena: VEB Fischer. 482 pp.
22. Filer, T. H. Jr., Broadfoot, W. M. 1968. Sweetgum mycorrhizae and soil microflora survive in shallow-water impoundment. *Phytopathology* 58(8):1050 (Abstr.)
23. Fontana, A., Luppi, A. M. 1966. Funghi saprofiti isolati da ectomicorrize. *Allionia* 12:39–46
24. Fortin, J. A. 1966. Synthesis of mycorrhiza on explants of the root hypocotyl of *Pinus sylvestris* L. *Can. J. Bot.* 44:1087–92

24a. Fortin, J. A. 1967. Action inhibitrice de l'acide 3-indolyl-acètique sur la croissance de quelques Basidiomycètes mycorrhizateurs. *Physiol. Plant.* 20: 528–32

25. Fortin, J. A. 1970. Interaction entre Basidiomycètes mycorrhizateurs et ra-

cines de Pin en présence d'acide indol-3 yl-acétique. *Physiol. Plant.* 23:365–71

26. Foster, R. C., Marks, G. C. 1967. Observations on the mycorrhizas of forest trees. II. The rhizosphere of *Pinus radiata* D. Don. *Aust. J. Biol. Sci.* 20:915–26
27. Frank, A. B. 1885. Neue Mitteilungen über die Mycorhiza der Bäume und der *Monotropa hypopitys. Ber. Deut. Bot. Ges.* 3:27–33
28. Gadgil, P. D. 1972. Effect of waterlogging on mycorrhizas of radiata pine and Douglas fir. *N.Z.J. Forest Sci.* 2:222–26
29. Gerdemann, J. W. 1968. Vesicular-arbuscular mycorrhiza and plant growth. *Ann. Rev. Phytopathol.* 6:397–418
30. Göbl, F. 1965. Die Zirbenmykorrhiza im subalpinen Aufforstungsgebiet. *Zentralbl. Ges. Forstw.* 82:89–100
31. Göbl, F., Platzer, H. 1967. Düngung und Mykorrhizabildung bei Zirbenjungpflanzen. *Mitt. Forstl. Bundes-Versuchsanst.* Vienna 74:1–63
32. Gogala, N. 1971. Growth substances in mycorrhiza of the fungus *Boletus pinicola* Vitt. and the pine-tree *Pinus sylvestris* L. Slovenska Akad. Class IV: Nat. Sci. Med. Diss. XIV 5:1–82
33. Goss, R. W. 1960. Mycorrhizae of ponderosa pine in Nebraska grassland soils. *Univ. Nebr. Coll. Agr. Exp. Sta. Res. Bull.* 192:1–47
34. Grudzinskaya, I. A. 1955. Data on the mycotrophy of tree species under steppe conditions. See Ref. 17, pp. 354–57
35. Gupta, S. C. 1956. Studies in the physiology of parasitism. XXII. The production of pectolitic enzymes by *Pythium de Baryanum* Hess. *Ann. Bot. London* 20:179–90
36. Hacskaylo, E. 1972. Carbohydrate physiology of ectomycorrhizae. See Ref. 7, pp. 207–27
37. Hacskaylo, E., Bruchet, G. 1972. Hebelomas as mycorrhizal fungi. *Bull. Torrey Bot. Club* 99:17–20
38. Hacskaylo, E., Palmer, J. G., Vozzo, J. A. 1965. Effect of temperature on growth and respiration of ectotrophic mycorrhizal fungi. *Mycologia* 57: 748–56
39. Handley, W. R. C. 1963. Mycorrhizal associations and *Calluna* heathland afforestation. *Bull. Forest. Comm. London* 36:1–70
40. Handley, W. R. C., Sanders, C. J. 1962. The concentration of easily soluble reducing substances in roots and the formation of ectotrophic mycorrhizal associations. A re-examination of Björkman's hypothesis. *Plant Soil.* 16:42–61
41. Harley, J. L. 1948. Mycorrhiza and soil ecology. *Biol. Rev. Cambridge Phil. Soc.* 23:127–58
42. Harley, J. L. 1969. *The Biology of Mycorrhiza,* 2nd ed. London: Hill. 334 pp.
43. Harley, J. L., Lewis, D. H. 1969. The physiology of ectotrophic mycorrhizas. *Advan. Microb. Physiol.* 3:53–81
44. Hatch, A. B. 1937. The physical basis of mycotrophy in *Pinus. Black Rock Forest. Bull.* 6:1–168
45. Heikurainen, L. 1955. Der Wurzelaufbau der Kiefernbestände auf Reisemoorböden und seine Beeinflussung durch die Entwässerung. *Acta Forest. Fenn.* 65(3):17–45
46. Horak, E. 1963. Untersuchungen zur Wuchsstoffsynthese der Mykorrhizapilze. See Ref. 21, pp. 147–63
47. Hübsch, P. 1963. Die Beeinflussung der Myzelwachstums von Reinkulturen von Boletazeen durch Kartoffelextrakte. See Ref. 21, pp. 101–21
48. Hunt, F. M. 1951. Effect of flooded soil on growth of pine seedlings. *Plant Physiol.* 26:363–68
49. Jahn, E. 1934. Die peritrophe Mykorrhiza. *Ber. Deut. Bot. Ges.* 52:463–74
50. Katenin, A. E. 1965. Ectotrophic mycorrhiza of tree spp. in the Eastern-European forest tundra. *Bot. Zh. SSSR* 50:434–40 (In Russian)
51. Katznelson, H., Rouatt, J. W., Peterson, E. A. 1963. Der Rhizosphäreneffekt mykorrhizer und nichtmykorrhizer Wurzeln von Sämlingen der Gelbbirke. See Ref. 21, pp. 273–78
52. Kelley, A. P. 1950. *Mycotrophy in Plants.* Waltham, Mass: Chronica Botanica. 206 pp.
53. Kelley, A. P. 1972. Mycotrophic Nutrition. Mt. Jackson, Va.: Kelley 68 pp.
54. Khudyakov, Ya. P. 1955. The nature of mycotrophy. See Ref. 17, pp. 80–86
55. Koberg, H. 1966. Düngung und Mykorrhiza. Ein Gefässversuch mit Kiefern. *Forstwiss. Zentralbl.* 85:371–79
56. Kramer, P. J. 1951. Causes of injury to plants resulting from flooding of the soil. *Plant Physiol.* 26:722–36
57. Kramer, P. J., Kozlowski, T. T. 1960. *Physiology of Trees.* New York: McGraw. 642 pp.
58. Krasilnikov, N. A. 1955. The relationships between soil microorganisms and higher plants. See Ref. 17, pp. 1–16
59. Krupa, S., Fries, N. 1971. Studies on ectomycorrhizae of pine. I. Production of volatile organic compounds. *Can. J. Bot.* 49:1425–31

60. Krushcheva, E. P. 1955. Mycorrhizas of agricultural plants. See Ref. 17, 142–51
61. Krywolap, G. N. 1971. Production of antibiotics by certain mycorrhizal fungi. In *Mycorrhizae. Proc. 1st N. Am. Conf.* 1969, ed. E. Hacskaylo, 219–21. Washington DC: USDA–Forest Service. 255 pp.
62. Kuraishi, H. 1953. Studies in mycorrhizal fungi I. Occurrence, environment and development of *Armillaria robusta. Ecol. Rev.* 13:159–67
63. Laiho, O. 1970. *Paxillus involutus* as a mycorrhizal symbiont of forest trees. *Acta Forest. Fenn.* 106:1–72
64. Levisohn, I. 1953. Growth response of tree seedlings to mycorrhizal mycelia in the absence of mycorrhizal association. *Nature* 172:316–17
65. Levisohn, I. 1957. Differential effects of root-infecting mycelia on young trees in different environments. *Emp. Forest. Rev.* 36:281–86
66. Levisohn, I. 1957. Antagonistic effects of *Alternaria tenuis* on certain root-fungi of forest trees. *Nature* 179: 1143–44
67. Levisohn, I. 1959. Strain differentiation in a root-infecting fungus. *Nature* 183:1065–66
68. Lindeberg, G. 1948. On the occurrence of polyphenol oxidases in soil-inhabiting Basidiomycetes. *Physiol. Plant.* 1: 196–205
69. Linnemann, G. 1960. Rassenunterschiede bei *Pseudotsuga taxifolia* hinsichtlich der Mykorrhiza. *Allg. Forst. Jagdztg.* 131:41–47
70. Lister, G. R., Slankis, V., Krotkov, G., Nelson, C. D. 1968. The growth and physiology of *Pinus strobus* L. seedlings as affected by various nutritional levels of nitrogen and phosphorus. *Ann. Bot.* 32:33–43
71. Lobanov, N. W. 1960. *Mykotrophie der Holzpflanzen.* VEB Berlin: Deut. Verlag Wiss. 352 pp.
72. Lundeberg, G. 1960. The relationship between pine seedlings (*Pinus sylvestris* L.) and soil fungi. *Sv. Bot. Tidskr.* 54:346–59
73. Lundeberg, G. 1968. The formation of mycorrhizae in different provenances of pine (*Pinus sylvestris* L.). *Sv. Bot. Tidskr.* 62:249–55
74. Lundeberg, G. 1970. Utilization of various nitrogen sources, in particular bound soil nitrogen, by mycorrhizal fungi. *Stud. Forest. Suec.* 79:1–95
75. Lyr, H. 1963. Zur Frage des Streuabbaues durch ektotrophe Mykorrhizapilze. See Ref. 21, pp. 147–63
76. MacDougal, D. T., Dufrenoy, J. 1944. Mycorrhizal symbiosis in *Aplectrum, Corallorhiza,* and *Pinus. Plant Physiol.* 19:440–65
77. Malyshkin, P. E. 1955. Stimulation of tree growth by microorganisms. See Ref. 17, pp. 211–20
78. Marks, G. C., Foster, R. C. 1967. Succession of mycorrhizal associations on individual roots of Radiata Pine. *Aust. Forest.* 31:193–201
79. Marks, G. C., Foster, R. C. 1972. Structure, Morphogenesis, and ultrastructure of ectomycorrhiza. See Ref. 7, pp. 2–35
80. Marks, G. C., Kozlowski, T. T., Eds. 1972. See Ref. 7
81. Marx, D. H. 1972. Mycorrhizae and feeder root diseases. See Ref. 7, pp. 351–77
82. Marx, D. H., Bryan, W. C. 1971. Influence of ectomycorrhizae on survival and growth of aseptic seedlings of loblolly pine at high temperature. *Forest Sci.* 17:37–41
83. Marx, D. H., Bryan, W. C. 1971. Formation of ectomycorrhizae on half-sib progenies of slash pine in aseptic culture. *Forest Sci.* 17:488–92
84. Marx, D. H., Bryan, W. C., Davey, C. B. 1970. Influence of temperature on aseptic synthesis of ectomycorrhizae by *Telephora terrestris* and *Pisolithus tinctorius* on loblolly pine. *Forest Sci.* 16:424–31
85. Marx, H. D., Zak, B. 1965. Effect of pH on mycorrhizal formation of slash pine in aseptic culture. *Forest Sci.* 11:66–75
86. Mejstrik, V., Dominik, T. 1969. The ecological distribution of mycorrhiza of beech. *New Phytol.* 68:689–700
87. Melin, E. 1925. *Untersuchungen über die Bedeutung der Baummykorrhiza. Eine ökologisch-physiologische Studie.* Jena: Fischer. 152 pp.
88. Melin, E. 1953. Physiology of mycorrhizal relations in plants. *Ann. Rev. Plant Physiol.* 4:325–46
89. Melin, E. 1958. Die Bedeutung der Mycorrhiza für die Versorgung der Pflanze mit Mineralstoffen. In *Handbuch der Pflanzenphysiologie,* ed. W. Ruhland, 4:281–88. Berlin: Springer. 1210 pp.
90. Melin, E. 1959. Mykorrhiza. See Ref. 89, 11:605–38. 1033 pp.
91. Melin, E. 1963. Some effect of forest tree roots on mycorrhizal Basidiomycetes. In *Symbiotic Associations,* ed. P. S. Nutman, B. Mosse, 125–45. London: Cambridge Univ. Press. 356 pp.

91a. Melin, E., Krupa, S. 1971. Studies on ectomycorrhizae of pine: II. Growth inhibition of mycorrhizal fungi by volatile organic constituents of *Pinus sylvestris* (Scots pine) roots. *Physiol. Plant.* 25:337–40
92. Meyer, F. H. 1962. Die Buchen-und Fichtenmykorrhiza in verschiedenen Bodentypen, ihre Beeinflussung durch Mineraldünger sowie fur die Mykorrhizabildung wichtige Faktoren. *Mitt. Bundesforschungsanst. Forst Holzwirtsch., Reinbek,* 54:1–73
93. Meyer, F. H. 1964. The role of the fungus *Cenococcum graniforme* (Sow.) Ferd. et Winge, in the formation of mor. In *Soil Micromorphology,* ed. A. Jongerius, 23–31. Amsterdam: Elsevier. 540 pp.
94. Meyer, F. H. 1966. Mycorrhiza and other plant symbiosis. In *Symbiosis,* ed. S. M. Henry, 1:171–255. New York: Academic. 478 pp.
95. Mikhowitsch, A. I. 1963. Untersuchungen zur Infizierung der Eiche mit Mykorrhizapilzen in der Waldsteppe und in der Halbwüste. See Ref. 21, pp. 441–59
96. Mikola, P. 1948. On the physiology and ecology of *Cenococcum graniforme* especially as a mycorrhizal fungus of birch. *Inst. Forest. Fenn. Commun.* 36(3):1–104
97. Mikola, P. 1962. The bright yellow mycorrhiza of raw humus. *Proc. 13th Congr. Int. Union Forest. Res. Organ., Vienna.* Part 2, Sect. 24–4. 12 pp.
98. Mikola, P. 1967. The effect of mycorrhizal inoculation on the growth and root respiration of Scotch Pine seedlings. *Proc. 14th Congr. Int. Union For. Res. Organ., Munich.* Part V, Sect. 24: 100–11
99. Mikola, P. 1969. Afforestation of treeless areas: importance and technique of mycorrhizal inoculation. *Unasylva* 23: 35–48
100. Mikola, P. 1972. Application of mycorrhizal symbiosis in forestry practice. See Ref. 7, pp. 383–406
101. Miller, C. O. 1971. Cytokinin production by mycorrhizal fungi. See Ref. 61, 168–74
102. Mishustin, E. N. 1951. Soil microflora and mycorrhiza formation with oak. *Agrobiologiya* 2:27–39 (In Russian)
103. Mishustin, E. N. 1955. Mycotrophy in trees and its value in silviculture. See Ref. 17, pp. 17–31
104. Modess, O. 1941. Zur Kenntniss der Mykorrhizenbildner von Kiefer und Fichte. *Symb. Bot. Upsal.* 5(1):1–146
105. Morrison, T. M. 1956. Mycorrhiza of silver beech. *N. Z. J. Forest.* 7:47–60
106. Moser, M. 1956. Die Bedeutung der Mykorrhizen für Aufforstungen in Hochlagen. *Forstwiss. Centralbl.* 75: 8–18
107. Moser, M. 1958. Der Einfluss tiefer Temperaturen auf das Wachstum und die Lebenstätigkeit höherer Pilze mit spezieller Berücksichtigung von Mycorrhizapilzen. *Sydowia* 12:386–99
108. Moser, M. 1959. Beiträge zur Kenntnis der Wuchsstoffbeziehungen im Bereich ectotropher Mycorrhizen. *Arch. Mikrobiol.* 34:251–69
109. Mosse, B. 1962. The establishment of vesicular-arbuscular mycorrhiza under aseptic conditions. *J. Gen. Microbiol.* 27:509–20
109a. Mosse, B. 1973. Advances in the study of vesicular-arbuscular mycorrhiza. *Ann. Rev. Phytopathol.* 11: 170–96
110. Müller, I.. 1963. Untersuchungen über die Wirkung von Antibiotika und antibiotisch aktiven Aktinomyzeten auf Basidiomyzeten. See Ref. 21, pp. 85–99
111. Norkrans, B. 1950. Studies in growth and cellulolytic enzymes of *Tricholoma. Symb. Bot. Upsal.* 11(1):1–126
112. Orlov, A. Y. 1957. Observations on absorbing roots of spruce (*Picea excelsa* Link) in natural conditions. *Bot. Zh. SSSR.* 42:1172–81 (Transl. from Russian, Israel Program Sci. Transl., Jerusalem)
113. Oswald, E. T., Ferchau, H. A. 1968. Bacterial associations of coniferous mycorrhizae. *Plant Soil* 28:187–92
114. Paavilainen, E. 1966. On the effect of drainage on root systems of Scots pine on peat soils. *Commun. Inst. Forest. Fenn.* 66 (1):1–100
115. Pachlewski, R., Gagalska, J. 1953. Investigations on the mycotrophy of oaks in Poland, bioecological conditions being considered. *Acta Soc. Bot. Pol.* 22:561–76 (In Polish, Transl. Center Inst. Sci. Technol. Econ. Inform., Warsaw, 1964)
116. Palmer, J. G. 1954. *Mycorrhizal development in Pinus virginiana as influenced by growth regulators.* PhD thesis. George Washington Univ., Washington DC. 114 pp.
117. Park, J. Y. 1971. Some ecological factors affecting the formation of *Cenococcum graniforme* mycorrhizae on basswood in southern Ontario. *Can. J. Bot.* 49:95–97
118. Parker, J. 1950. The effect of flooding on the transpiration and survival of

some north-eastern forest tree species. *Plant Physiol.* 25:453–60

118a. Peyronel, B., Fassi, B., Fontana, A., Trappe, J. H. Terminology of mycorrhiza. *Mycologia* 61:410–11

119. Rambelli, A. The rhizosphere of mycorrhizae. See Ref. 7, pp. 299–343

120. Rayner, M. C., Neilson-Jones, W. 1944. *Problems in Tree Nutrition.* London: Faber. 184 pp.

121. Read, D. J., Armstrong, W. 1972. A relationship between oxygen transport and the formation of the ectotrophic mycorrhizal sheath in conifer seedlings. *New Phytol.* 71:49–53

122. Redmond, D. R. 1955. Studies in forest pathology. XV. Rootlets, mycorrhiza and soil temperatures in relation to birch dieback. *Can. J. Bot.* 33:595–627

123. Richards, B. N. 1965. Mycorrhiza development of loblolly pine seedlings in relation to soil reaction and the supply of nitrate. *Plant Soil* 22:187–99

124. Richards, B. N., Wilson, G. L. 1963. Nutrient supply and mycorrhiza development in Caribbean pine. *Forest Sci.* 9:405–12

125. Ritter, G. 1964. Vergleichende Utersuchungen über die Bildung von Ectoenzymen durch Mycorrhizapilze. *Z. Allg. Mikrobiol.* 4:295–312

126. Ritter, G. 1968. Auxin relations between mycorrhizal fungi and their partner trees. *Acta Mycol. Warszawa* 4: 421–31

127. Rovira, A. D. 1969. Plant root exudates. *Bot. Rev.* 35:35–57

128. Samtsevich, S. A. 1955. The importance of ectoendotrophic mycorrhizas in tree nutrition. See Ref. 17, pp. 152–73

129. Schramm, J. R. 1966. Plant colonization studies on black wastes from anthracite mining in Pennsylvania. *Trans. Am. Phil. Soc.* 56(1):1–194

130. Shcherbakov, A. P., Mishustin, E. N. 1950. Nutritional conditions as means of enhancing growth of oak seedlings and development of mycorrhiza with their roots. *Agrobiologiya* 5:121–27 (In Russian)

131. Shemakhanova, N. M. 1962. *Mycotrophy of Woody Plants.* Moscow: Acad. Nauk. SSSR. 329 pp. (Transl. from Russian, Israel Program Sci. Transl., Jerusalem, 1967)

132. Shemakhanova, N. M. 1963. Bedingungen der Mykorrhizabildung bei *Pinus silvestris* mit *Boletus luteus* (Linn.) Fr. in Reinkultur. See Ref. 21, 191–202

133. Shemakhanova, N. M., Mazur, O. P. 1968. Mycorrhizae of *Juglans regia* and the conditions for their formation. (In Russian, English Summary) *Izv. Akad. Nauk. SSSR Ser. Biol.* 4:517–29

134. Sideri, D. I., Zolotun, V. P. 1952. Improvement of conditions for oak development in the eroded soils of Zaporozh'e. *Les i Step* No. 8 (In Russian)

135. Slankis, V. 1948. Einfluss von Exudaten von *Boletus variegatus* auf die dichotomische Verzweigung isolierter Kiefernwurzeln. *Physiol. Plant.* 1:390–400

136. Slankis, V. 1951. Uber den Einfluss von $\beta$-Indolylessigsaure und anderen Wuchstoffen auf das Wachstum von Kiefernwurzeln. I. *Symb. Bot. Upsal.* 11(3):1–63

137. Slankis, V. 1967. Renewed growth of ectotrophic mycorrhizae as an indication of an unstable symbiotic relationship. *Proc. 14th Congr. Int. Union forest. Res. Organ., Munich* Pt. V, Sect. 24:84–99

138. Slankis, V. 1971. Formation of ectomycorrhizae of forest trees in relation to light, carbohydrates, and auxins. See Ref. 61, pp. 151–67

139. Slankis, V. 1972. Hormonal relationships in mycorrhizal development. See Ref. 7, pp. 232–91

140. Sobotka, Z. 1965. A contribution to the problem of mycorrhizae on maples in some forest types. *Commun. Inst. Forest. Czech.* 4:205–17

141. Stahl, E. 1900. Der Sinn der Mycorhizenbildung. *Jahrb. Wiss. Bot.* 34: 539–668

142. Steward, F. C., Ed. 1972. *Physiology of Development: The Hormones. Plant Physiol. A Treatise.* New York London: Academic. 365 pp.

143. Subba-Rao, N. S., Slankis, V. 1959. Indole compounds in pine mycorrhiza. *Proc. 9th Int. Bot. Congr., Montreal.* 2:386

144. Theodorou, C. 1967. Inoculation with pure cultures of mycorrhizal fungi of radiata pine growing in partially sterilized soil. *Aust. Forest.* 31:303–9

145. Theodorou, C., Bowen, G. D. 1969. The influence of pH and nitrate on mycorrhizal associations of *Pinus radiata* D. Don. *Aust. J. Bot.* 17:59–67

146. Theodorou, C., Bowen, G. D. 1970. Mycorrhizal responses of radiata pine in experiments with different fungi. *Aust. Forest.* 34:183–91

147. Theodorou, C., Bowen, G. D. 1971. Influence of temperature on the mycorrhizal associations of *Pinus radiata* D. Don. *Aust. J. Bot.* 19:13–20

147a. Tomaszewski, M., Wojciechowska, B. 1974. The role of growth regulators released by fungi in pine mycorrhizae.

*Proc. 8th Int. Conf. Plant Growth Subst. Tokyo.* In press

148. Trappe, J. M. 1962. Fungus associates of ectotrophic mycorrhiza. *Bot. Rev.* 28:538–606
149. Trappe, J. M. 1964. Mycorrhizal hosts of *Cenococcum graniforme. Lloydia* 27:100–6
150. Trappe, J. M. 1967. Principles of classifying ectotrophic mycorrhizae for identification of fungal symbionts. See Ref. 98, pp. 46–59
151. Tribunskaya, A. Ya. 1955. Investigation of the rhizosphere microflora of pine seedlings. *Mikrobiologiya* 24: 188–92 (In Russian)
152. Ulrich, J. M. 1960. Auxin production by mycorrhizal fungi. *Physiol. Plant.* 13:429–43
153. Vančura, V. 1961. Detection of gibberillic acid in *Azotobacter* cultures. *Nature* 192:88–89
154. Vedenyapina, N. S. 1955. Effect of *Azotobacter* on the growth of oak seedlings. See Ref. 17, pp. 253–59
155. Vlasov, A. A. 1955. Importance of mycorrhiza for forest trees and procedures for its stimulation. See Ref. 17, pp. 101–17
156. Voznyakovskaya, Yu., M. 1954. Mycorrhiza and its practical value. *Mikrobiologiya* 23:204–20 (In Russian)
157. Voznyakovskaya, Yu. M., Ryzhkova, A. S. 1955. Microflora-accompanying mycorrhizas. See Ref. 17, pp. 320–23
158. Went, F. W. 1971. Mycorrhizae in a montane pine forest. See Ref. 61, pp. 230–32
159. Went, F. W., Stark, N. 1968. Mycorrhiza. *BioScience.* 18:1035–39
160. Wilde, S. A. 1968. Mycorrhizae: their role in tree nutrition and timber production. *Res. Bull. Coll. Agr. Life Sci., Univ. Wis.* 272:1–30
161. Wojciechowska, H. 1960. Studies on the mycotrophy of the spruce (*Picea excelsa* Lk) in its northern range with particular regard to the mycotrophy of plant communities in the forest district Sadlowo, subdistrict Lipowo near Biskupiec Reszelski. *Folia For. Pol. Ser. A.,* 2:123–66 (In Polish, Transl. Sci. Publ. Foreign Coop. Center, Cent. Inst. Sci., Technol. Econ. Inform., Warsaw, 1968)
162. Worley, J. F., Hacskaylo, E. 1959. The effect of available soil moisture on the mycorrhizal association of Virginia pine. *Forest Sci.* 5:267–68
163. Wright, E., Ching, K. K. 1962. Effect of seed source on mycorrhizal formation on Douglas-fir seedlings. *Northwest Sci.* 36:1–6
164. Young, H. E. 1940. Mycorrhizae and the growth of *Pinus* and *Araucaria.* The influence of different species of mycorrhiza-forming fungi on seedling growth. *J. Aust. Inst. Agr. Sci.* 6:21–25
165. Young, H. E. 1947. Carbohydrate absorption by roots of *Pinus taeda. Queensl. J. Agr. Sci.* 4:1–7
166. Zak, B. 1964. Role of mycorrhizae in root disease. *Ann. Rev. Phytopathol.* 2:377–92
167. Zak, B. 1971. Characterization and identification of Douglas-fir mycorrhizae. See Ref. 61, pp. 38–53
168. Zak, B. 1971. Detoxication of autoclaved soil by a mycorrhizal fungus. *US Forest Serv. Pac. Northwest Forest Range Exp. Sta. Res. Note* No. 159:1–4
169. Zerova, M. Ya. 1955. Mycorrhiza formation with forest trees in the Ukrainian SSR. See Ref. 17, pp. 39–60

## SOME RELATED ARTICLES IN OTHER ANNUAL REVIEWS

# REPRINTS

The conspicuous number aligned in the margin with the title of each article in this volume is a key for use in ordering reprints. Available reprints are priced at the uniform rate of $1 each postpaid. Effective January 1, 1975 the minimum acceptable reprint order is 10 reprints and/or $10.00, prepaid. A quantity discount is available.

The sale of reprints of articles published in the Reviews has been expanded in the belief that reprints as individual copies, as sets covering stated topics, and in quantity for classroom use will have a special appeal to students and teachers.

# AUTHOR INDEX

D

E



F

M

R

# SUBJECT INDEX

G

# CUMULATIVE INDEXES

## CONTRIBUTING AUTHORS VOLUMES 8-12

## CHAPTER TITLES VOLUMES 8-12